Property Valuation

Property Valuation

Third Edition

Peter Wyatt

WILEY Blackwell

Registered Office
John Wiley & Sons, Inc., 111 River Street, Hoboken, NJ 07030, USA
John Wiley & Sons Ltd, The Atrium, Southern Gate, Chichester, West Sussex, PO19 8SQ, UK

Editorial Office
111 River Street, Hoboken, NJ 07030, USA

For details of our global editorial offices, customer services, and more information about Wiley products visit us at www.wiley.com.

Wiley also publishes its books in a variety of electronic formats and by print-on-demand. Some content that appears in standard print versions of this book may not be available in other formats.

Limit of Liability/Disclaimer of Warranty
In view of ongoing research, equipment modifications, changes in governmental regulations, and the constant flow of information relating to the use of experimental reagents, equipment, and devices, the reader is urged to review and evaluate the information provided in the package insert or instructions for each chemical, piece of equipment, reagent, or device for, among other things, any changes in the instructions or indication of usage and for added warnings and precautions. While the publisher and authors have used their best efforts in preparing this work, they make no representations or warranties with respect to the accuracy or completeness of the contents of this work and specifically disclaim all warranties, including without limitation any implied warranties of merchantability or fitness for a particular purpose. No warranty may be created or extended by sales representatives, written sales materials or promotional statements for this work. The fact that an organization, website, or product is referred to in this work as a citation and/or potential source of further information does not mean that the publisher and authors endorse the information or services the organization, website, or product may provide or recommendations it may make. This work is sold with the understanding that the publisher is not engaged in rendering professional services. The advice and strategies contained herein may not be suitable for your situation. You should consult with a specialist where appropriate. Further, readers should be aware that websites listed in this work may have changed or disappeared between when this work was written and when it is read. Neither the publisher nor authors shall be liable for any loss of profit or any other commercial damages, including but not limited to special, incidental, consequential, or other damages.

Library of Congress Cataloging-in-Publication Data
Names: Wyatt, Peter, 1968- author.
Title: Property valuation / Peter Wyatt.
Description: Third Edition. | Hoboken, NJ : Wiley, 2023. | Revised edition of the author's Property valuation, 2013. | Includes bibliographical references and index.
Identifiers: LCCN 2022030096 (print) | LCCN 2022030097 (ebook) | ISBN 9781119767411 (paperback) | ISBN 9781119767428 (adobe pdf) | ISBN 9781119767435 (epub)
Subjects: LCSH: Commercial real estate–Valuation–Great Britain. | Real estate investment–Great Britain.
Classification: LCC HD1393.58.G7 W93 2023 (print) | LCC HD1393.58.G7 (ebook) | DDC 333.33/8720941–dc23/eng/20220804
LC record available at https://lccn.loc.gov/2022030096
LC ebook record available at https://lccn.loc.gov/2022030097

Cover Design: Wiley
Cover Image: © VideoFlow/Shutterstock

Set in 10/12pt Sabon by Straive, Pondicherry, India
Printed and bound by CPI Group (UK) Ltd, Croydon, CR0 4YY

C9781119767411_031022

Contents

Preface

The legal ownership of land and buildings, also known as 'real property' and collectively referred to as *property* throughout this book, confers rights that enable it to be developed, occupied or leased. The physical occupation of property is essential for social and economic activities including shelter, manufacture, commerce, recreation and movement. Typically, physical property ownership is not desired as an end in itself, although a prestigious site or landmark building can confer non-financial value. Rather, demand for property is a derived demand; occupiers require property to help deliver social and economic activities, and investors require property as an investment asset. This concept of derived demand has a direct bearing on its valuation. The interaction between the supply of and demand for property generates exchange prices and valuations are estimates of those market prices. Valuers interpret the way in which market participants measure and quantify value. Increasingly, there is recognition of value beyond 'market' value; referred to as non-market value. The valuer's skill set has conventionally focused on the estimation of market value but in some cases social and environmental assessments can complement market valuations to ensure that owners, occupiers and other affected parties are aware of and, if appropriate, fairly compensated for any interference with their property rights.

Section A covers valuation principles. Chapter 1 begins with a discussion on property rights and the way in which those rights can confer value to their holders. The chapter ends by explaining how and why valuation has evolved as a discipline to help estimate the value of those property rights. Chapter 2 outlines the economics or property value relevant to property markets and estimates of exchange price. It introduces economic terms and concepts associated with the supply of and demand for property, the concept of rent as a payment for their use and some land-use theory. It explains how property values arise using economic principles and theories that have been developed and expounded over the past century and a half. Building on the theories relating to the agricultural land market, the causes and spatial distribution of urban land and property uses and rents are described. The chapter examines the economics of supply and demand and the establishment of equilibrium exchange price in the property market and its constituent sectors. The chapter explains the causes of price differentials between land uses. Chapter 3 describes the main property market sectors – occupation, investment and development – and the way that they interact with one another. Chapter 4 explains the mathematics that underpin the valuation methods described in Section B.

Section B covers valuation approaches. It begins with Chapter 5, which describes the valuation process, from confirmation of the instruction to value a property through to the publication of a valuation report. Chapters 6, 7 and 8 explain the valuation methods and techniques that fall under the three internationally recognised approaches to property valuation – market, income and cost. Whichever method is employed, it should reflect the behaviour of market participants. The figure below shows that in active markets, where there is a large quantity of transactions involving properties with similar characteristics, the role of the valuer is essentially to interpret market signals and apply them to the subject property – a comparison approach. With limited availability of market information, market-valuation methods are increasingly cost-based. Valuers are required to make more assumptions, and this increases valuation uncertainty.

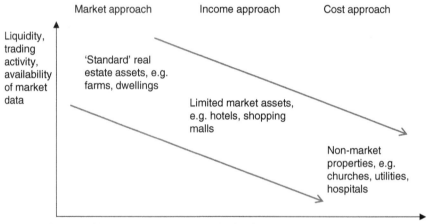

Using the market approach (Chapter 6), the valuer examines recent market transactions involving comparable properties and uses this market intelligence to help estimate value. The income approach (Chapter 7) considers the net income that a property might generate, typically in the form of rent, and this income can be capitalised to estimate a capital value. Both the rent and the capitalisation rate will be estimated using comparable evidence. For properties where comparable evidence is lacking, the cost approach (Chapter 8) estimates the cost of replacement, perhaps incorporating a depreciation rate for older or less-functional properties. Building costs, depreciation rates and land values are all estimated by referring to comparable evidence. The cost approach also covers the valuation of development land. The choice of valuation approach depends on the type of property that is to be valued and the purpose of the valuation.

Section C considers the application of these valuation approaches in practice. Chapter 9 examines the valuation of investment properties, where the ownership and use of a property is split so that the occupier pays a rent to the owner, representing the return on the owner's investment. Valuations are required to determine the level of rent that might be charged and the price that an investor might pay when buying or selling a property investment. Chapter 10 covers the valuation of development property, an important and challenging area of valuation work. The valuer is required to estimate the cost of construction and the market value of a development that is hypothetical, since it has not been built yet. This is done so that either the value of development land or development profit can be calculated. Chapter 11 explains how valuations are used in financial statements of companies and public bodies to report the market value of their property assets. The chapter also explains how valuations are relied upon by lenders to assess the value of property that is used as loan security. Most countries have some form of tax that is based on property values so Chapter 12 considers how valuations are undertaken for tax purposes. Similarly, most countries have legislation that allows them to compulsorily acquire property for government purposes. Chapter 13 explains how valuations are central to the assessment of compensation to owners, occupiers and other affected parties. Finally, Chapter 14 reviews the issues of risk, flexibility and uncertainty and the way that they affect valuations.

About the Companion Website

Property Valuation is accompanied by a companion website:

www.wiley.com/go/wyatt/propertyvaluation3e

The website includes:

- Excel templates and PowerPoint slides

Scan this QR code to visit the companion website.

Section A
Valuation Principles

Chapter 1
Property Rights and Property Value

1.1 Property rights

Real Property refers to the physical land (including minerals and other natural resources, soil, trees, plants and growing fruits) and improvements to land (including infrastructure, buildings and other facilities). *Tenure* refers to the relationship, either formal or informal, between members of a society and real property (UN FAO 2002). Tenure determines who can hold and use resources, for how long and under what conditions. These 'holding and using' arrangements are referred to as *property rights*, and the way in which they are formed, recognised and transferred between members of society depends on whether tenure systems are based on written policies and laws or unwritten customs and practices (UN FAO 2012).

For example, an individual may hold or 'own' a piece of land in perpetuity, providing the holder with an extensive set of property rights, free of disturbance. This *freeholder* may decide to lease or rent the land to another person for a period of time. The *leaseholder* would benefit from use rights (rather than ownership rights), which would end when the lease expires. There can be any number of arrangements, depending on the tenure systems in place. For example, *shared equity* refers to a combination of leasehold for a term of years but with the ability (usually subject to conditions) to acquire the freehold. This kind of arrangement has been used to enable occupiers of residential dwellings to rent initially but then buy their homes later.

The ways in which real property can be held have been the focus of a great deal of philosophical thought. A key debate is whether property rights should be held communally (or in common) or whether they can be legitimately held individually (or privately). Private property rights are advocated on libertarian and utilitarian grounds. From a libertarian perspective, there are those who advance a natural right to property. Aristotle argued that, although privately owned, individuals will learn to exercise generosity and moderation and share their property. In the thirteenth century, Thomas Aquinas argued that while property should be private, the *use* of things should remain common. 500 years later in the eighteenth century, Edmund Burke believed that private property provided the foundation for a just social order and a spur to personal industry and national prosperity. He argued in favour of widespread access to acquiring property, which he regarded as a check on encroachments by the state. Similarly, Georg Hegel and his followers probed relations between the existence of private property and things such as self-assertion, mutual recognition, stability of will, and the establishment of a sense of prudence and responsibility. They argued that not only is private property morally legitimate, but it also contributes to the ethical development of the individual.

Advancing a natural right to property, in the seventeenth century, John Locke maintained that an individual has a right to property on account of what he has done to it, so long as his appropriation does not violate the property rights of others. Thus, improving land in some way, perhaps by draining it or constructing buildings on it, imparts a degree of private control that others should refrain from interfering with. Utilitarian arguments for private property purport to show that the happiness or welfare of a society will be improved if resources (and, in particular, the means of production) are owned and controlled privately rather than by the state or a community. Developing a utilitarian theory of real property, Jeremy Bentham argued that the state should protect and promote private property ownership because landowners are more likely to invest in their land and to manage it better than public owners. Similarly, Demsetz (1967) argues that, by excluding others, private ownership internalises 'many of the external costs associated with communal ownership'. This concentration of benefits and costs on owners incentivises more effective use of resources.

Perhaps unsurprisingly, as Linklater (2015, p. 12) observed, living in a private property society encourages self-interest, in contrast to clan values, family values and communal structures found in other societies. Locke observed that with markets and prosperity come inequality and insecurity. To overcome these insecurities, people form governments to protect their private property, in return for which landowners should pay tax. Not every individual will or must establish such a relationship, so natural rights are not universal like a natural right to life and liberty (Waldron, 1990). This contrasts with the more contemporary view set out in Article 17 of the Universal Declaration on Human Rights, which states that 'everyone has the right to own property alone as well as in the association of others'. And of course, it contrasts with the case for communal property rights and the abolition of private property, advanced by Marx and Engels in the nineteenth century. To summarise, the way that property rights are held shapes the way society is organised.

1.1.1   Tenure

This 'holding' of property rights or *tenure* can be communal or individual. An example of communal tenure would be an apartment block that is jointly owned by all of the occupants, while each apartment is leased to individual occupants. Also, tenure can be customary, perhaps vested in a village community and administered by elders. Customs and traditions may allocate rights of private residence, communal grazing and so on.

Tenure is a social construct that regulates how property rights are allocated among members of society. Property rights are myriad, but the main ones are: *use*, *take* (fishing rights for example), *extract* (minerals or water for instance), *enjoy* (or usufruct), *access* (such as a right of way), *transfer* (buy, sell, lease, inherit, assign), impose *charges* (seize, foreclose, and so on, perhaps in cases of default), and *options* (such as a right to buy). There may also be rights granted to the state on behalf of the people, which can take precedence over other property rights. Examples of these *overriding interests* include land and property taxes; occupiers' rights such as *adverse possession*, *easements* and *wayleaves*; land recovery in cases of fraud and forgery; and acquisition by the state. Also, so that they can plan the physical environment of their jurisdictions, many states require landowners to ask permission before they develop or change the use of their land.

Property rights may be held by the state, by a community or privately. The definition of real property interests in the latest edition of the International Valuation Standards now includes communal, community and collective rights held in an informal, traditional, undocumented and unregistered manner (IVSC 2021). This includes informal tenure rights over tribal land and in informal settlements, which can take the form of possession, occupation and use rights. In the case of state and communal property, *use rights* are important, in terms of who is entitled to them and what they entitle holders to do. Many African countries have customary tenure systems based around communal ownership where the rights are held jointly *in common*. Decisions about allocation of rights among members of the community, how land might be used and so on are made by agreement. It is important to ensure that rights are allocated in a way that minimises external costs on neighbours and on future generations.

Under private ownership, it is argued that the economic use of land can be optimised and costs that are incurred by the owner (internal costs) will be minimised. But each owner has no direct incentive to minimise costs that affect other landowners (external costs). Each owner may negotiate with other landowners to minimise these external costs, and this is likely to be cheaper than negotiations between communal landowners since there are fewer stakeholders involved. To conduct these kinds of negotiations, societies often establish centralised bodies such as planning authorities to represent affected parties, which will include affected landowners as well as other members of society.

Barry (2015) notes that this economic perspective on property rights and their ownership ignores social, cultural, and political dimensions of land tenure, particularly in the case of private ownership. Land may form part of a person's identity, and communal property rights may foster connections between people and between people and their physical environment.

In westernised economies, private property rights prevail but the consequence of this approach to land tenure has come under scrutiny (see Picketty 2014 for example). It appears insensitive to the distribution of property rights and associated welfare or wealth in society. A situation can arise where net wealth is maximised but enjoyed by a few land-owning elites is superior to a situation in which there is slightly less wealth but enjoyed by everyone equally. As globalisation continues, market integration, technological advances, population growth and mass migration lead to rapid urbanisation and growing pressure on land and growing pressure on institutions[1] to protect property rights. This pressure is acutely observed in developing countries where private property rights tend to lack legal formality and are under threat from those who fail to recognise the rights of others and seek to deprive citizens of their customary, ancestral and religious rights in terms of both peaceful enjoyment and economic benefit. The threat is not new; state colonisation is replaced by corporate land grabbing, but the impact remains the same.

It is important to recognise all legitimate tenure right holders, whether formal or informal. This is the goal of the Voluntary Guidelines on the Responsible Governance of Tenure Rights (UN FAO 2012), which aim to:

- Safeguard tenure rights against threats, infringements and arbitrary loss, including forced evictions,
- Provide justice to deal with infringements, including compensation where rights are taken for public purposes, and
- Promote and facilitate enjoyment of legitimate tenure rights, and prevent disputes, violent conflicts and corruption.

In much of the world, tenure, and thus property rights, is not recorded in documentary form. Instead, tenure, such as it might be known, may take the form of verbal agreements, social relationships, customs and traditions. There is much debate among land professionals as to whether these unregistered property rights should be registered in a way that is recognisable to western-led developed nations, or whether they should be recognised for what they are and societies' institutions that deal with them (such as lenders, businesses, developers, and investors) adapt their practices to handle this diversity of property rights. There are arguments for and against both approaches. Some argue that registration of property rights releases hidden wealth or capital that is latent within land (see, for example, de Soto 2001). Others see registration as an aid to big business in exploiting the more minor interests of smallholders and communal landowners. In this respect, land is no different from other resources; consolidation of ownership can reduce competition and increase returns. It is a matter for each society and nation to decide how it wants to organise its land resource and the ownership of property rights.

A single parcel of real property may have many rights relating to it, and these rights may be held by many different members of society. For example, the state may hold an overriding right to expropriation, a freeholder may hold transfer rights, a lender may hold repossession rights, a leaseholder may hold use rights and society as a whole may have a right of access. Relationships between rights and those who hold those rights can be complex and difficult to identify, particularly if they are unregistered.

1.1.2   Property rights in England

English common law, as it relates to property, is derived from the system of feudal land tenure by which the monarch and his or her lords ruled the land. In the United Kingdom, only the Crown can own land and historically lords merely 'held' their land under a system of tenure. The lords, in turn, granted lesser rights to hold property to others in return for loyalty, services or *rent*. The monarch or superior *land*lords could withdraw their patronage and reclaim their land at any time. This holding of land was categorised according to its duration and, because of its derivation in the doctrine of legal estates, it is more accurate to speak of someone holding an *estate* rather than owning physical land (Card et al. 2003). The two most important estates are freehold and leasehold. A freeholder holds land in perpetuity from the Crown and is at liberty to use it for any purpose subject to statutory regulation and the legal protection afforded to third parties. The freeholder (landlord or *lessor*) may be an occupier or may be an investor deriving a rental income from a lease granted to an occupier. A leaseholder (tenant or *lessee*) holds a property for a term of years, the duration of which is usually specified in or implied by the terms of the *lease* granted by the landlord. Under certain conditions tenants can obtain legal rights that protect their occupational interest and investment that they may have made to improve the premises.

In England, there are four principal types of lease set out here in order of decreasing length:

- Long *ground leases* are typically for a term of more than 100 years where the landlord grants a lease of, say, a vacant site to a tenant who in turn may construct a building on it and enjoy the economic benefits of doing so during the term of the lease. Historically these ground leases required a rent to be paid that typically remained the same during the entire term and, as time passed, the real value of this rent diminished. Nowadays, it is common to find rent review mechanisms inserted into ground leases that enable the landlord to participate in rental value growth.
- Long leases of say 99- or 125-years' duration are granted in respect of residential apartments and dwellings. At the end of these residential leases, tenants may have a right to renew for a further term.
- Shorter leases of say 5–25 years' duration are granted in respect of commercial premises. These leases usually include periodic reviews of the rent and, at the end of the lease, business tenants may have a legal right to renew their leases.
- Assured shorthold tenancies, which are the predominant form of lease for residential property. They can be for a fixed term – usually 12 months – or they can be periodic, i.e. they continue from period to period (usually monthly) until determined at the end of any period by a 'notice to quit' issued by either party.

Depending on the conditions set out in these leases, tenants may be able to subdivide and sublet but only for durations of less than the length of any head-lease. Thus, a single unit of property may comprise several legal estates, each with a market value, provided they are capable of being exchanged. Figure 1.1 provides an example of the way in which these estates might be structured, but there is no theoretical limit to the number of leasehold and subleasehold estates that may be created in this way.

Other important ownership and financial interests include *trusts*, where the interest of a beneficiary under a trust is an equitable interest as opposed to the legal interest of a trustee, and financial interests that are created by a legal *charge* if the property is used as collateral to secure finance (the owner's equity position is considered a separate financial interest). There are other more minor legal

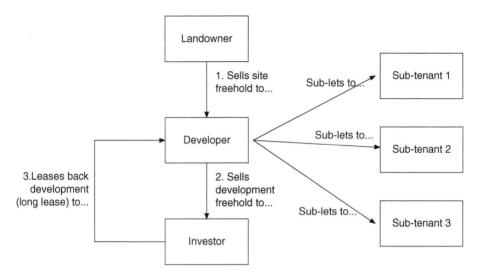

Figure 1.1   Legal estates in a property.

interests in land such as *covenants*, which run with the land and may affect the use, development and transfer of ownership, or *easements*, which convey use but not ownership of property, such as a right of way.

1.2   Property value

Property value derives from the benefit of holding property rights. Value may be derived from the economic benefits that property rights confer, but value may also be derived from social, cultural, religious, spiritual or environmental benefits too. If property rights provide an ability to use land for subsistence farming, for manufacturing, as sacred ground or for amenity, then they have a value-in-use. If those rights are transferable, then they may have a *value-in-exchange* too. Exchanges of property rights usually take place in *markets*, so value-in-exchange is often referred to as *market value*. *Market prices* are revealed when property rights are exchanged.

Market value is mainly influenced by the economic benefits that property rights are capable of generating. For example, an ancient monument embodies non-economic social and cultural values and may be prevented from being used for other purposes or from being sold. Consequently, its value-in-exchange (market value) would be nil, but its value-in-use (in the form of social and cultural values) may be high.

Because these non-market values are not revealed in markets, they are difficult to quantify, but that does not mean they should be ignored. Usually, it is when property rights are being acquired by the state compulsorily that non-market value needs to be estimated as a basis for compensation purposes. Non-market value is also a consideration when it comes to the strategic management of state-controlled land and property such as national forests and coastal areas. For example, if an investor wishes to acquire mineral extraction rights, as well as the loss of non-market value to a local community, there may be a wider cost to society as a whole. A forest provides timber and other forest products but perhaps offers social (amenity) value to the region and global environmental value. Governments may decide to set policies or introduce regulations to deal with these costs to society and the environment.

Since a parcel of real property can have many property rights associated with it and because they can be held by more than one person or group, it can have more than one value at the same time. For example, value might refer to the rent that a tenant pays to a landowner (a rental value) or it may be an estimate of the price paid for the land if it was sold (a capital value). The state might assess the value of the land for taxation purposes, which may be based on legally defined factors such as soil quality, agricultural productivity, or the floor area of buildings. In this way, value is dependent upon its context and definition.

How much a property might be worth, property value in other words, will be determined by the extent and security of property rights, and by the physical and geographical characteristics of the underlying real property.

1.2.1   Extent of property rights

The extent of property rights determines the degree to which holder(s) can legitimately benefit from land. Rights that are more comprehensive, exclusive, longer, and more secure are the most valuable. Unencumbered ownership of perpetual property rights will be more valuable than a short terminable right to occupy the same property. There are three principal *types* of property rights:

[1]   The right to *use*; reside, use for commerce and manufacturing, withdraw (to farm natural products for example), extract (to mine natural resources for example)

[2]   The right to *transfer*; alienate (perhaps via a lease), bequeath, donate, sell, assign, mortgage

[3]   The right to *control*; access, manage, change use, improve or develop (physically alter, subdivide or assemble into larger units), include or exclude others

Generally speaking, the more rights that are held, the more valuable the interest. There may be circumstances where specific rights are not held. For example, informal tenure holders may not have the right to sell or lease their rights. This might be because the state retains ownership or it might be to prevent an imbalance of negotiating power with external actors (elites and large corporations for example) in an attempt to reduce the risk of the holders being misrepresented, manipulated, or coerced into relinquishing their rights. The prevention of transfer of real property rights may also help maintain a community's collective identity, particularly in cases where territories of indigenous peoples are based on a communal tenure system. Some jurisdictions may grant alienation rights but with conditions. For example, in Papua New Guinea the law prevents customary landowners from leasing land directly to outsiders, so they have to first lease it to the State, which can then sublease.

As well as the quantity of rights, the level of those rights affects value. There are three principal *levels* of rights:

[1] *Superior* interests provide absolute possession and control in perpetuity, subject to any subordinate interests and statutory constraints

[2] *Subordinate* interests provide exclusive possession and control for a defined period (a lease for example, with the rent reflecting the value that can be derived from the land)

[3] *Rights to use* but without exclusive possession or control (a right of way for example)

Superior (perpetual) real property rights are often more valuable than subordinate rights, and long-term subordinate interests are often more valuable than short ones.

When considering the holding of property rights, two important factors are gender and exclusivity. Regarding gender, the UN provides a checklist for gender-equitable valuation (UN FAO 2013). Regarding exclusivity, property rights may be held jointly, where parties share the whole interest or, severally, where each party holds a defined portion of the whole interest. If property rights are shared, the value may be apportioned between tenure rights holders. Identifying the holders of these various property rights can be challenging, particularly if the property is vacant or the occupiers are uncooperative. Additional resources may be required to collect and verify information in such circumstances. As an aid, tenure registration systems should record not only high-level rights, owners of perpetual interests for example, but also ownership of subsidiary interests such as leases, licences and charges over land and property.

With community-owned rights, the community may be able to exclude others and collectively determine how land is used. Each occupier may enjoy valuable use rights but cannot sell or otherwise transfer their rights. A disadvantage of this approach is that occupiers may not be able to mortgage the property, and this can deter investment.

In countries where the state retains ownership, individuals and organisations may be allocated land-use rights, whereby the state owns the land and citizens own improvements to the land. These land-use rights may be 'sold' by the state on an indefinite basis for a fee or leased for fixed periods of time at an annual rent. Split-tenure rights, such as these, have two important value consequences.

First, valuers may be required to separate the value of the land from the value of the improvements even though they may be functionally linked and difficult to distinguish. For example, a farmer may have invested labour and capital to irrigate land and construct sluices and pumping stations. If the land is to be expropriated, compensation may be payable for the sluices and pumps but not for the irrigation channels.

Second, valuers should consider the market value of the original location when estimating compensation to holders of land-use rights. This can be difficult in a planned economy where free market trading of land is constrained; there would be little evidence of the market value of the original land.

In rural areas, property rights may be held by the state, particularly in national parks and other protected areas. Forestland may also be state-owned, although it can be beneficial to jointly manage such natural resources with local communities.

A particular element of value, known as *reversionary value*, emanates from land and property that is subject to concurrent superior and subordinate property rights. Reversionary value is realisable once a lease ends, the tenant vacates, and the superior interest holder takes possession of the property. Sometimes, this value can be very high, particularly if tenants have made improvements to the land, increasing its value. There can be a strong incentive, therefore, for owners to terminate leases at the earliest opportunity. Valuers should check whether the law allows tenants to continue in occupation beyond the end of the original lease, as this will have a significant impact on the value of their tenure rights, and leases should be clear about what happens to the value of any improvements at the end of a lease.

When property rights are alienated, by leasing occupation rights for example, rent payments are usually reviewed to market levels periodically. Sometimes, a low rent may be agreed alongside the payment of a lump sum or premium at the start of a lease. Similarly, a rent-free period at the start of a lease might be offered as an incentive to the tenant. The rent could be based (at least partly) on turnover, thus allowing the owner an element of profit share. Valuers should check for payments (monetary and in-kind) that are in lieu of price or rent.

1.2.2   Security of property rights

There is a spectrum of security ranging from legally registered titles of demarcated land parcels to customary tenure and informal occupation consisting of little more than de facto recognition of occupation via political patronage, receipt of utility bills or payment of property taxes. The degree to which property rights are formalised has a significant impact on value because occupiers, investors, and lenders regard formal, registered title as less risky.

In many countries, rural land is held under customary rights. Often involving social transitions of entitlement in unregulated markets, land held under this form of tenure is passed from generation to generation via family lineages and inheritance overseen by traditional leaders. Typically, formal land markets do not exist, and ownership details and transaction information can be difficult to obtain.

Nevertheless, customary and informal property rights are often the most valuable assets possessed by rural communities, forming the base for subsistence and cash crop agriculture. These rights may be created through informal or oral agreements or arise from the customs and practices of local communities and relate to individuals, households and extended families. It is these household or extended family level rights and agreements that often drive economic use of land for both subsistence and cash crop agricultural production.

Valuing customary and informal rights is challenging, identifying the specific nature of the rights (use, transfer, control), the owners of those rights, and the wider political context. The price that someone pays for informal property rights might, for example, reflect the prospect of formalising them sometime in the future. Unregistered land may be transacted informally and undocumented, and prices are often low due to the risk of conflicting claims on possession and occupation. A valuer may be confronted with multiple oral customary interests that create rights ranging from perpetual interests to time-limited interests and other forms of agreement that may be determined on the occurrence of certain events.

It can be difficult to identify the true extent of interests to be valued. It is important to ensure that legal constructs of customary rights accord with observed rights. For example, constitutional provisions may restrict customary land grants to leasehold rather than transfers in perpetuity. Yet, if informal or oral grants of crop and farm share agreements strongly resemble perpetual interests, should the interest be valued as encountered or as constrained by the constitutional provision? In other words, based on a hypothetical leasehold interest, which implies setting, arbitrarily, commencement and termination dates, rent and other terms?

Valuations can be particularly challenging where ownership and occupation are split between many holders. Property rights need to be defined precisely so that buyers and sellers know what they can and cannot do with land and property. For example, a household or extended family that holds customary rights over forestland may enter into a farming operation agreement with a non-community member who clears the forest and cultivates a cash crop. Depending on custom, the third party may gain perpetual or terminable rights. A valuer would need to identify which interests are perpetual, which are terminable, and disentangle them to identify and value each one. Another difficulty stems from the absence of market data. Transactions, apart from being undocumented, tend to be complex, involving crop or land-sharing agreements.

Notwithstanding the valuation challenges posed by customary and informal property rights, independent and objective valuations of these rights have the potential to improve the welfare of rural inhabitants by:

- Helping to eradicate fraudulent and corrupt rent-seeking practices;
- Revealing prices of customary tenure rights, thus triggering exchanges that could optimise investment of capital and labour, moving families from subsistence to cash crop agriculture and other welfare-enhancing uses of land;
- Providing fair compensation for expropriation of customary tenure rights; and
- Supporting government land administration practices, including land taxation and state-land leasing or sales.

Therefore, valuers need to understand customary and informal markets. Important questions include:

- Can the property rights be categorised using the types and levels described above? For example, how many layers of interests exist between occupier and landowner? Are there any use rights, such as migratory pastoralists, periodically using land?
- Is there any evidence, perhaps written or observed verbal agreements, to corroborate identified property rights?
- Where does finance come from? Purchasers of informal rights may be unable to borrow money to help finance their acquisitions, so sellers sometimes accept instalment payments. Such arrangements are risky for both borrower and lender and therefore the cost of the loan (the interest rate) can be high.
- Is there an accessible legal system to resolve disputes? Is there a functioning insurance system? These may not be recognisable in the formal style of developed markets but often exist informally.
- Can informal rights be compared with similar rights held on a formal basis and establish benchmarks for returns and benefits?

1.2.3   Physical and geographical characteristics

The value of land will vary depending on where it is, and the value of improvements to land will vary depending on their nature and extent.

For land, accessibility is a key influence on value, the importance of which is dependent upon the use to which the land is put. It is important to consider the location of land-use activities in relation to travel infrastructure and neighbouring uses. The various needs for access result in a process of competitive bidding between different land uses. A price pattern emerges that is correlated with the pattern of accessibility. Farmland or forestland that is more accessible to markets for produce and livestock is likely to be more valuable than farmland further away, all else equal.

Other important location considerations are the benefits that can accrue when similar land uses cluster together. Once land in an area has been assigned to a particular use, this will largely determine the best use for adjacent land due to advantages of clustering. The extent of the benefit depends on the need for contacts. Shops tend to group together. Offices cluster near shopping facilities and residential neighbourhoods. Industry benefits from grouping production sequences. Smaller firms tend to group together but larger firms are less dependent on clustering because they are able to internalise their production processes. Some land uses, heavy industry and residential dwellings, for example, prefer to locate apart to minimise social costs.

The ability to improve land can enhance its value. For example, soil can be improved or drained; buildings and infrastructure can be constructed. The nature of, and the extent to which, improvements influence value will vary depending on the use to which the land is put. Each land use is often valued separately. A high-level classification of uses might be agriculture and fisheries, forest, minerals, recreation and leisure, transport, utilities and infrastructure, residential, community services, retail, industry and business, defence, vacant, derelict and unused land. Below this, there are likely to be many categories and subcategories, the detail of which will be largely dependent upon the value differentials between uses. A land information system can be a valuable tool for recording current land use.

Value will also be influenced by the degree to which any improvements can be adapted to other uses, in other words, their flexibility for change of use. It is important to consider potential alternative uses, as these may be more valuable than the current use.

1.3   Property valuation

By examining the conduct of societies, we can see that *value* is a widely adopted basis for making decisions about how property rights are allocated among individuals and communities. Globally, and particularly in developing countries, there is a pressing need to understand how the value of property rights arises and can be estimated, particularly in situations where there are vulnerable populations and where there is unequal access to information, knowledge, and power.

Property valuers are asked for advice on the capital and rental value of property rights and the service is often closely associated with agency work where a client seeks advice on the appropriate asking price (in the case of a vendor) or the accuracy of an asking price (in the case of a prospective purchaser) and the terms of the transaction are negotiated. This close association allows valuers to have a strong link to current market activity and helps spot price signals.

Valuation practice, then, focuses on the estimation of market value or value-in-exchange, based on price discovery and the analysis of market evidence. Market transparency is key: the market learns from itself (Fisher et al. 2004) and valuers learn from the market. Active and transparent markets are able to facilitate more reliable valuations whereas valuations in emerging markets, where information is difficult to obtain, are more uncertain.

The term *investment valuation* refers to a wider consideration of client-specific issues as well as market signals. It broadens the viewpoint from *market* value-in-exchange towards *investment* value-in-use, whether that use is for occupation or investment purposes. More recently, valuation practice has been broadened further to encompass non-market value. Not all of the advantages of holding property rights are priced in markets: social connections, an identity, subsistence farming in times of need, a connection with the environment and the stewardship that it engenders. Markets struggle to price these qualities, so the concept of non-market value is a first step towards recognising, identifying and accounting for them. Estimating the non-market value of property rights is a challenging, but nevertheless, vital undertaking since it attempts to quantify the non-economic benefits that they offer.

Valuation standards support quality, integrity and consistency of valuation process. They promote good practice and provide defence against claims of negligence. They mean valuations are reported in a consistent way.

The purpose for which the valuation is required and the type of property that is to be valued will determine the nature of the valuation task, including the techniques employed and the BASIS on which value is to be estimated. Independent estimates of value that meet recognised standards may be required for many purposes relating to the development and subsequent occupation and ownership of property, including:

- Market transactions: buy/sell decisions, letting/reletting decisions, and property management;
- Investment decisions, including development appraisal and performance measurement;
- Compensation of expropriated property rights;
- Land and property taxation; and
- Reporting the value of property assets held by companies, estimating the value of property rights being used as loan security and assessing insurance risk.

1.3.1   Market transactions

Property rights may be traded in order to relocate or expand a residence or place of business. Land and property may be bequeathed to children. Owners of property rights may wish to lease land and property to others. Valuations provide the necessary information to allow parties to negotiate an agreeable price or rent.

A property owner who wishes to sell would need to advertise an asking price that will attract potential purchasers and the level is clearly dependent on market conditions. If the owner wishes to lease the property, then advice will be sought regarding the level of rent that could be obtained, the lease terms that should be sought and the type of tenant that can be expected. Rent reviews ensure that the rent paid by the tenant is periodically reviewed to market value, and it is necessary (usually as a condition of the rent review clause in the lease) to employ a valuer to estimate the revised rent. If the property is already leased and the tenant wishes to dispose of the lease, then the lease must be assigned to a new tenant and a *premium* or reverse premium might be paid.

In some countries the state owns a lot of land and, in order to put it to good use, the state may decide to sell or lease it to farmers, forest enterprises and so on. Valuers can advise on the best means of disposal, how much the land should be sold for and the terms of any lease that may be agreed.

There is a growing movement, particularly in Africa, for large-scale investors to acquire substantial tracts of land for agricultural use or commercial development. Valuations help ensure owners and occupiers receive fair market value for their property rights. This can be very important in cases where the rights are *customary* or *informal*. Once legitimate property rights are recognised, markets can develop to enable the buying, selling and leasing of these rights.

1.3.2   Investment decisions

Holders of property rights may identify opportunities to develop or redevelop their land or use it for another purpose. Valuations help determine what land is worth in its current use as a farm or forest for example and whether it might have a higher value if it can be redeveloped for, say, housing or commercial use. Developers need to know how much they should bid for a piece of development land or a building that is in need of redevelopment.

When an investor purchases a property and leases it to a tenant, the expectation is that it will generate sufficient income in the form of rent payments and capital appreciation to provide an adequate rate of return in comparison to other investment opportunities such as equities and bonds. After a period of time, the investor may sell the property to another investor at a value that has risen over the holding period. Properties held as investments are valued on a regular basis as a means of monitoring investment performance. Indeed, many property investors are legally required to revalue their property investment assets regularly and annual, often monthly, valuations of properties in the portfolios of these investors are undertaken. Many of these investment valuations may be recorded in databases, and this enables investors to benchmark the performance of their property investment portfolios.

1.3.3   Compensation

A key motivation for developing an objective and impartial valuation capacity is to ensure that when property rights are expropriated, fair compensation is awarded to affected parties. Consider, for example, a country that is in the process of upgrading its public transport infrastructure and wishes to acquire land for a new railway line. How much should the acquiring authority pay for the land required to construct the railway? How do landowners know whether their land is being acquired at a fair price?

Valuers are employed to assess the amount of compensation that should be paid to landowners whose land is compulsorily acquired to make way for public sector and utility network projects. These include major transport infrastructure projects, urban public transport networks and airport construction, for regeneration projects where sites in fragmented ownership need to be assembled, and for minor works such as the realignment of a road junction to improve sight lines. Compensation may also be paid to landowners where none of their land has been acquired but there has been a reduction in the value because of nearby public works, such as noise from a new road.

1.3.4   Land and property taxation

Many countries regard land and property as a legitimate source of tax revenue, particularly for local government expenditure such as medical, police and fire services, and maintenance of infrastructure and amenities. Land and property values are widely used as a means of allocating tax liability fairly among owners and occupiers of rural and urban real property rights. Regular revaluations can be undertaken to ensure fairness is maintained as values change over time, and public access to these tax valuations means that taxpayers can appeal against their payment liability if they wish to.

The fair valuation of legitimate real property rights is now widely recognised as a global concern. For example, the New Urban Agenda (United Nations Conference on Housing and Sustainable Urban Development 2016) encourages countries to capture and distribute any uplift in land value that results from public investment, and also offers support for the development of land information records (including records of legitimate customary rights as well as sales and lease records) to assist valuation.

1.3.5   Accounting, lending and insurance

Historically, companies reported the original cost of property assets in their balance sheets. This led to considerable under-valuation of company assets. Entrepreneurs could buy these businesses for a price that reflected their historic asset value and then release real value by disposing of valuable assets, including property, at current prices (a process known as asset stripping). Companies may now elect to report the current value of their property assets in their annual accounts and valuers are required to perform these valuations for corporate disclosure purposes. As businesses are acquired or merged, valuers are often asked to value the property assets of the companies concerned.

If property rights can be used as security for loans, the value of those rights is a key factor in deciding whether and how much to lend. For example, a farmer plans to invest capital to irrigate, level, and fence some land and wishes to borrow money from a bank using the land as collateral for the loan. The bank needs to know whether the land is worth more than the amount of the loan and the farmer wants to know whether the money invested will add more value than the cost of the loan. If a borrower defaults, then the lender may wish to take possession of the property and sell it in order to realise its value and thus recover the debt. A lender who is lending money for property development will clearly wish to be suitably reassured (with adequate allowance for the risk taken) as to the expected value of the completed development.

Finally, land and property are valuable assets, but they are real, physical entities too and therefore vulnerable to all sorts of risks such as flooding, fire, earthquakes, subsidence, contamination or invasion. It is important to mitigate these risks and insure against them. Insurers undertake risk assessments and valuations to check that their premiums and level of cover are appropriate. Strictly speaking, this is less of an estimate of market value in the sense of an exchange price and more an assessment of the cost of replacing buildings and other improvements to the land.

Key points

- Societies value those things that are important to them: property rights are fundamental to society and are highly valued.
- The value of a property reflects its capacity to fulfil a function: key determinants include the extent and security of property rights and the physical characteristics of the underlying real property. Valuation is the process of identifying and quantifying the effect these determinants have on the value of property rights.
- There is no 'one-size-fits-all' concept of value: depending on economic conditions over time, on individual or business perceptions of value and on assumptions made by valuers, property rights can have more than one value.
- There are two key concepts of value, value-in-use and value-in-exchange.
- Property valuation has a key role to play in determining the value of real property rights in a wide variety of contexts.
- The diversity of real property rights and land-use arrangements means that valuations can be complicated. This underlines the need for knowledgeable, skilled and experienced valuers and explains why standards are essential.
- States should recognise the importance of valuations of all property rights – formal, customary and informal – and valuers should recognise that the degree to which property rights are formalised could have a significant impact on value.
- Estimating the market and non-market value of property rights is difficult when they are not clearly defined, when markets are not well developed and when valuation professionals and standards are absent.
- Valuers should be able to apportion value among holders who enjoy different levels of rights in the same property: if property rights are shared, then a valuer should be able to apportion value to each holder.

Note

1. Depending on the social context, these institutions might be state or community level.

References

Barry, M. (2015). *Property Theory, Metaphors and the Continuum of Land Rights*. United Nations Human Settlement Program (UN-Habitat) Global Land Tools Network.

Card, R., Murdoch, J., and Murdoch, S. (2003). *Law for Estate Management Students*, 6e. Oxford: Oxford University Press.

Demsetz, H. (1967). Toward a theory of property rights. *Am. Econ. Rev.* 57 (2): 347–359.

Fisher, J., Gatzlaff, D., Geltner, D., and Haurin, D. (2004). An analysis of the determinants of transaction frequency of institutional commercial real estate investment property. *Real Estate Econ.* 32: 239–264.

IVSC (2021). *International Valuation Standards (IVS)*, IVS 400: Real Property Interests, Effective 31 January 2022, International Valuation Standards Council.

Linklater, A. (2015). *Owning the Earth: The Transforming History of Land Ownership*. London, UK: Bloomsbury.

Picketty, T. (2014). *Capital in the Twenty-First Century*. Cambridge, Massachusetts, US: Harvard University Press.

de Soto, H. (2001). *The Mystery of Capital: Why Capitalism Triumphs in the West and Fails Everywhere Else*. Black Swan.

UN FAO (2002). *Land Tenure and Rural Development*, FAO Land Tenure Studies, vol. 3. Rome: United Nations Food and Agriculture Organisation.

UN FAO (2012). *Voluntary Guidelines on the Responsible Governance of Tenure of Land, Fisheries and Forests in the Context of National Food Security*. Rome: United Nations Food and Agriculture Organisation.

UN FAO (2013). *Governing Land for Women and Men: A Technical Guide to Support the Achievement of Responsible Gender-Equitable Governance of Land Tenure*, Governance of Tenure Technical Guide, vol. 1. Rome: United Nations Food and Agriculture Organisation.

United Nations Conference on Housing and Sustainable Urban Development (2016). *Habitat III: New Urban Agenda*, Draft outcome document for adoption in Quito.

Waldron, J. (1990). *The Law*, 1e. Abingdon, Oxon, UK: Routledge.

Chapter 2
The Economics of Property Value

2.1 Introduction

Valuation practice tends to focus on the estimation of market value or value-in-exchange, but increasingly, there is a need to consider a broader concept of value. The value of natural resources and ecosystem services (often referred to as natural capital), the value of land and property rights that do not trade in markets (as defined in western economies), and the value of environmental, social, cultural, religious, and spiritual assets more generally are of growing interest to many stakeholders around the world.

This chapter explains the economic principles that underpin property value and how these can be revealed in markets. It describes how property is different from other tradeable commodities due to its fixed location and intensity of use. The chapter explains the economic motives of market participants, such as land-owners, developers, occupiers and investors. It then considers concepts of value outside of the market.

2.2 Land as a resource

Land (and the property rights that determine its occupation and use) is a resource that can be combined with other resources to produce goods and services that consumers desire. Economists refer to these resources as *factors of production* to emphasise that various factors need to be combined to produce goods or services. The factors of production are usually classified into three groups; land, capital, and labour, and sometimes entrepreneurs are specifically identified as a fourth category. To construct a building, labour is required to develop a plot of land, and plant and equipment, which may be hired or bought,

Property Valuation, Third Edition. Peter Wyatt.
© 2023 John Wiley & Sons Ltd. Published 2023 by John Wiley & Sons Ltd.
Companion website: www.wiley.com/go/wyatt/propertyvaluation3e

are required to facilitate the process. These manufactured resources are called capital or, more precisely, *physical capital*. Each factor of production receives a specific kind of payment. Landowners, who provide the use of land over time, receive rent; owners of physical capital receive interest; workers receive wages and the entrepreneur gains profit. It is interesting that Marxists challenge the logic of this model as they understand land to be a gift of nature, a non-produced resource, which exists regardless of payment. From a pure Marxist perspective, therefore, land has no value, and all property is regarded as theft. Indeed, in many countries the state or some collective arrangement owns and allocates land.

The world's land, labour and capital resources are used to create economic goods to satisfy human desires and needs and *economics* is concerned with the allocation of these finite resources to humanity's infinite wants: a problem formally referred to as *scarcity*. To reconcile this problem, economists argue that people must make choices about what is made, how it is made and for whom; or in terms of property, choices about what land should be developed, how it should be used and whether it should be available for purchase or rent. Economics, therefore, is the 'science of choice'. Because resources are scarce, their use involves an *opportunity cost*; resources allocated to one use cannot be used simultaneously elsewhere, so the opportunity cost of using resources in a particular way is the value of alternative uses foregone. In other words, in a world of scarcity, for every want that is satisfied, some other want remains unsatisfied. Choosing one thing inevitably requires giving up something else; an opportunity has been forgone. This fundamental economic concept helps explain how economic decisions are made, for example how property developers decide which projects to proceed with and how investors select the range of assets to include in their portfolios. To avoid understanding opportunity cost in a purely mechanistic way – where one good is simply chosen instead of another, we need to clarify how decisions between competing alternatives are made.

Before proceeding, four economic characteristics should be noted regarding property rights.

- First, property rights are demanded not for their own sake but as a means to an end. In other words, it is a *derived demand*. A manufacturer might need a production facility, a retailer a trading location, a household requires living accommodation and an investor desires an income in the form of rent. So, property rights can be demanded by different types of user for different purposes.
- Second, property rights are rarely required to satisfy a single type of utility; they usually fulfil a variety of needs. For example, a commercial building provides a range of services for the tenant, including office space for employees, a certain image and a specific location relative to transport, supplies and customers.
- Third, land is immoveable. This means that one parcel of land is not perfectly substitutable for another, and scarcity value can arise.
- Land is also permanent, so it can be reused or redeveloped repeatedly. This means that the economics of property development has an important role.

2.3   Supply and demand, markets and equilibrium price determination

An assumption must be made at this stage: consumers of resources seek to maximise their welfare. A budget constraint limits the choices of consumers when choosing between resources in a market; in effect, desire, measured by opportunity cost, is limited by a budget. The existence of a budget constraint reflects the distribution of resource-buying capacity throughout an economy. In some economies this distribution might be state controlled; in others it is left to competitive forces. In a market economy, the allocation of scarce land resources and the property rights over them is facilitated by means of a *market*. In economic terms, a market has particular characteristics: there are lots of participants and they behave competitively, and any advantage some might have in terms of access to privileged information for example does not continue beyond the short run. Each consumer will have preferences or requirements and a budget, and these will influence the price that can be offered for property rights and the quantity obtained. Landowners, or suppliers of property rights, interact with consumers in a marketplace where property rights are exchanged, usually indirectly by means of money.

Applying marginal productivity theory, a welfare-maximising consumer will buy property rights up to a point at which additional revenue from using another 'unit' is exactly offset by its additional cost. The additional revenue obtainable from a unit of property rights is referred to as the *marginal revenue product* (MRP) and it is calculated by multiplying the *marginal revenue*[1] (MR) obtained from selling another unit of output by the *marginal product*[2] (MP) of the property rights. If other labour and capital are fixed, as more and more property rights are used, MP decreases due to the onset of *diminishing returns*. So, if MR is constant and MP declines, the MRP of land will decline as additional units of land are used ceteris paribus. The declining MRP can represent a firm's demand schedule (curve) for property rights as shown in Figure 2.1.[3] If the price of property rights falls relative to labour and capital, demand will increase; that is why the demand curve in Figure 2.1 is downward sloping.

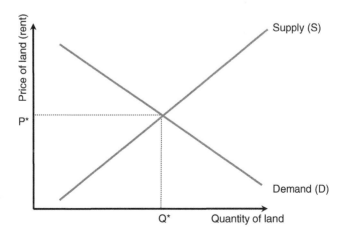

Figure 2.1   Short-run supply of and demand for property rights.

The short-run[4] demand schedule illustrated in Figure 2.1 represents consumer behaviour and is a downward-sloping curve to show that possible buyers and renters of property demand a greater quantity at low prices than at high prices (assuming population, income, future prices, consumer preferences, etc., all remain constant). The short-run supply curve maps out the quantity of property interests available for sale or lease at various prices (assuming factors of production remain constant).[5] The higher the price that can be obtained, the greater the quantity of property that will be supplied. Equilibrium price P* is where demand for property equals supply at quantity Q*. Price varies directly with supply and indirectly with demand. Valuation is the process of estimating equilibrium price.

The result of an efficiently functioning property market in the long run should be economic efficiency, achieved when resources have been allocated optimally; welfare is maximised and property rights could not be reallocated without making at least one consumer worse off, a concept known as *Pareto optimality*.

The supply of land is not infinite and its *scarcity*, along with its *utility* to occupiers and investors described above, means that it has economic value. Consumers do not have unlimited means to consume land; they have a *budget constraint*, and this means they must choose how to allocate their budgets among land and other resources that they need. This choice involves weighing up the opportunity costs of the various resources. The way these factors interact to create value is reflected in the basic economic principle of supply and demand, and valuation is the process of estimating the equilibrium price at which supply and demand might take place under normal market conditions. Capitalist market economies have developed systems of private property rights and market trading of those rights between owners and occupiers as a means of competitive allocation. Economists try to understand the nature of payments that correspond with the trading of these property rights and this is, from an economic perspective at least, the essence of market valuation.

2.4   The property market and price determination

2.4.1   *The property market*

Real property has certain economic characteristics that distinguish it from other factors of production. In physical terms, it actually has two components: the land itself and (usually) improvements that have been made to the land in the form of buildings and other man-made additions. This has several implications, not least the existence of a separate market in land for development, discussed later. Each unit of property is heterogeneous, if only because each land parcel occupies a unique geographical position. This means that it will vary in quality; for urban land, this is largely due to accessibility differences but will also differ in terms of physical attributes, legal restrictions (different lease terms for example) and external influences such as government intervention in the form of planning. Property tends to be available for purchase in large, indivisible and expensive units so financing plays a significant role in market activity. Also, because of its durability, there is a big market for existing property and a much smaller market for development land on which to build new property.

In addition to physical characteristics, the nature of property rights makes land very different from other resources. A key distinction can be made between ownership of property rights in perpetuity and for a fixed duration. For example, in the United Kingdom, around half of the total stock of commercial property is owned by investors who receive rent paid by occupiers in return for the use of property. The other half own the property that they occupy. For residential property, the split is approximately 2/3 owner-occupied and 1/3 rented. Assuming land provides the same utility and welfare to both owners and renters, price and rent should be related, and this relationship is explained in Chapter 5.

2.4.2   Price determination in the land market

Early classical economists regarded rent as a payment to a landlord by a tenant for the use of land in its 'unimproved' state (land with no buildings on it), typically for farming. David Ricardo (1821) set out a basic theory of agricultural land rent. The theory implied that land rent was entirely demand-determined because the supply of land as a whole was fixed and had a single use (to grow corn). The most fertile or productive land is used first and less productive land is used as the demand for the agricultural product increases. Rent on most of the productive land is based on its advantage over the least productive, and competition between farmers ensures the value of the 'difference in productivity of land' is paid as rent (Alonso 1964). Rent is therefore dependent on the demand (and hence the price paid) for the output from the land – a derived demand.

Since the supply of all land is fixed (perfectly inelastic) and cannot be increased in response to higher demand, the only response is higher price. Figure 2.2 illustrates this with a vertical supply curve (S), meaning that no matter what price consumers are prepared to pay for more land, the quantity supplied cannot be altered. Price is solely demand-determined; the equilibrium price (P) of land can only change due to shifts in the demand curve, D. For example, if the productivity of land or the price of the commodity produced increases, demand for land would rise, illustrated in Figure 2.2 by shifting the demand curve upwards and to the

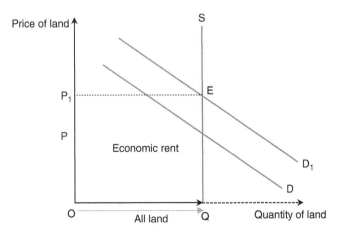

Figure 2.2   Equilibrium price determination with a fixed supply of land.

right from D to D_1. With fixed supply, S, the equilibrium price of land increases from P to P_1. Whatever the level of demand, because supply remains fixed, the opportunity cost of using land is zero and all earnings from the land (represented in Figure 2.2 by the area OP_1EQ) result from its scarcity and are referred to as *economic rent*.

This theory assumes that land has only one use, so the theory is really only relevant when land use is strictly controlled. In reality land can be used for various uses, and each potential user will bid a price that is determined by their estimated net utility/welfare/profit. When there is more than one possible use for land, the supply of land for a particular use will not be fixed because, in response to an increase in demand for one use, additional supply could be bid from and surrendered by other uses if the proposed change of use has a value in excess of its existing use value. The payment to the landowner for the use of land is still made in the form of rent but, since land can be used for alternative uses, supply is no longer perfectly inelastic and has an opportunity cost. Land rent, rather than comprising economic rent only, can now be considered to consist of two elements: *transfer earnings*; a minimum sum or opportunity cost to retain land in its current use, which must be at least equal to the amount of rent that could be obtained from the most profitable alternative use, and *economic rent*, a payment in excess of transfer earnings that reflects the scarcity value of the land.

Diagrammatically, the supply curve is no longer vertical; instead, it is upward sloping. Figure 2.3 illustrates the demand for and supply of land for a particular use. Assuming competition between users of land, interaction of supply and demand will lead to a supply of Q* land for this particular use, all of which will be demanded and for which the market equilibrium rent will be P*. Because supply is not perfectly elastic, some of this rent is transfer earnings and the rest is economic rent. If the rent falls below the transfer earnings, the landowner will transfer from this land use or at least decide to supply less of it. Q* is the marginal land and is only just supplied at price P* and all of the rent is transfer earnings. Assuming a homogeneous supply, the interaction of supply and demand leads to an equilibrium market rent for this type of land use and competition between uses ensures that this rent goes to the optimum use (Harvey 1981).

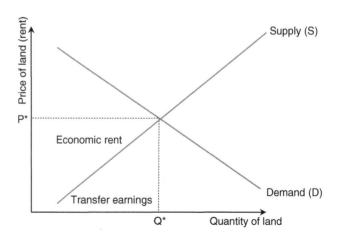

Figure 2.3   Elastic supply and elastic demand.

The amount of price shift in response to a change in supply will depend on the elasticity of supply – the more inelastic the supply, the greater the change in price. Using this neoclassical land-use rent theory, it is possible to look at the interaction between supply and demand more closely in order to understand the nature of the rent payments for different land uses. Figure 2.4 shows that the rent for land in the centre of an urban area is almost entirely economic rent because of its scarcity.

The more elastic supply of land on the edge of an urban area means that the lower rent is for largely transfer earnings (Figure 2.5). The proportion of transfer

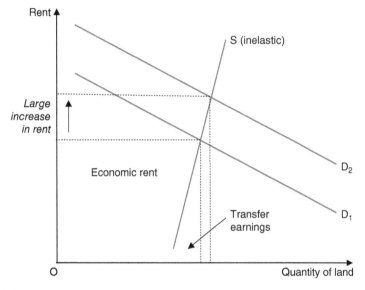

Figure 2.4   Rents for land in the central area under conditions of inelastic land supply.

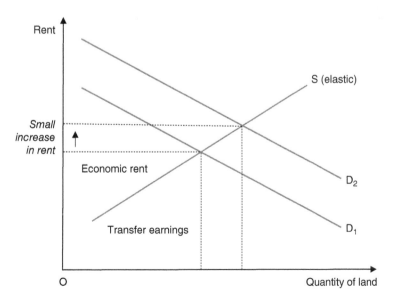

Figure 2.5   Industrial land rents on the edge of an urban area under conditions of elastic land supply.

earnings and economic rent depends on the elasticity of supply of land: the more inelastic the supply, the higher the economic rent while the more elastic the supply, the higher the transfer earnings. Because urban land is fairly fixed in supply (inelastic) and is increasingly so near the centre, economic rent forms an increasing proportion of total rent as the centre of an urban area nears. So, any increase in demand (or reduction in supply) for central sites is reflected in large rises in rent, but on the outskirts an increase in demand (or decrease in supply) for land for a specific purpose produces a small change in economic rent (and thus total rent as a whole) because land is more plentiful.

Before moving on, consider the effect of time on the elasticity of supply of and demand for land. Using conventional equilibrium analysis, in the short-run, supply for a particular use – say housing – will be inelastic[6] (S in Figure 2.6) and demand represented by D will be elastic, producing an equilibrium rent, r^*. If demand for housing land increases to D_1, rent will rise to r_1. In the long run, supply adjusts in response to this increase in demand (and rent) by switching from other uses. The assumption of inelasticity can therefore be relaxed, and the supply of land will increase to say S_1, settling rents back to r_2, assuming no further change in demand. It should be noted that this is a very simple model of a complex market that is seldom in a state of equilibrium (Fraser 1993).

2.4.3   Price determination in the property (land and buildings) market

It is now time to consider the use of land *and* buildings (property) as a combined factor of production. Regarding rent, rather than a payment for the use of land,

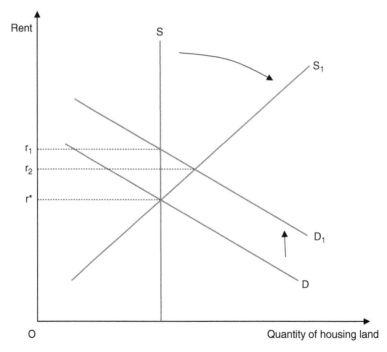

Figure 2.6   Equilibrium analysis of rent for office space. Source: Modified from Fraser, W. (1993).

property rent is a payment for 'improved' land, typically land that has been developed in some way so that it now includes buildings too. If the property is leased to a tenant, then the rent would include not only a payment for the use of the land but also some payment for the interest and capital in respect of the improvements that have been made to the land. But it is not easy to distinguish the rent attributable to buildings from that attributable to land. Land is permanent and although buildings ultimately depreciate, they do last a long time. It can be assumed therefore that land and buildings are a fixed factor of production in any time frame except the very long run, which the user can combine with variable amounts of other factors of production.

In absolute terms, the physical supply of *all land* is *completely* inelastic and the supply of land for *all uses* is *very* inelastic. The supply of land and buildings (or property) for *specific uses* is *relatively* inelastic in the short run due to the requirement for planning permission to change use and the time it takes to develop new property, but less so in the long run. Also, because property is durable, it accumulates over time and new developments add only a small amount to the existing stock. Consequently, new supply has negligible influence on price, and demand is the main determinant of property rent.

The amount of land that a user demands depends not only on its price but also on the price of the final (derived) product. In response to increased demand (and rent), the productivity of land can be increased by using it more intensively through the addition of capital. In other words, it is possible to add units of other factors of production, such as labour but particularly capital, to the fixed amount of land to increase the amount of building area or floorspace in the same way that a farmer might add fertiliser to farmland. By providing more accommodation on a site, building area is acting as a substitute for land area. The relative price of land and building will determine the extent of this substitution. If land is cheap, it will not take much extra building before it becomes more cost effective to acquire more land to provide additional space. If land is expensive, a large amount of building may take place before building costs increase to a level at which it pays to acquire more land to provide extra accommodation. It must be borne in mind that the process of adding more capital to a fixed amount of land is subject to the principle of diminishing returns.

So, for each unit of land, a land-use rent theory must simultaneously allocate the optimum (profit maximising) use *and* intensity of that use. The *allocation of land use* has already been considered, so now the focus is on the *intensity of land use*. Assume that the optimum land use of a particular site has already been determined. This means that land is a factor of production that has a fixed cost. What is unknown is the optimum amount of capital (in the form of building floorspace) to add to the land. Assuming a competitive capital market keeps the cost per unit of capital the same regardless of the quantity required, as more capital (floorspace) is added to the fixed amount of land, initially the MRP of the land might increase because of economies of scale but the law of diminishing returns means that eventually it will fall. Profit is maximised where the MRP of a unit of capital equals the MC of a unit of capital. In Figure 2.7, this is when OX units of capital are employed and is the optimum amount of capital to combine with the land. The total revenue earned is represented by the area QYXO. Total cost (including profit) is area PYXO and surplus revenue is therefore QYP.

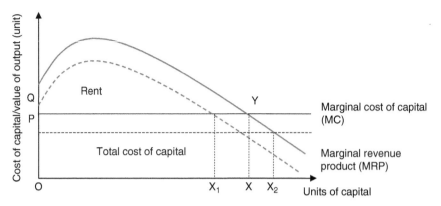

Figure 2.7 Optimum combination of land and capital. Source: Adapted from Fraser (1993).

If the current land use is the most profitable, then land rent is QYP, i.e. the surplus remaining after deducting costs of optimally employed factors of production from expected revenue (Fraser 1993).

The amount of land that a user demands depends on its price relative to other factors of production, the price of the good or service produced on or provided from the land and the productivity of the land. If the price obtained for goods and services produced from the land falls, the MRP curve will drop from the solid line to the dashed line. Alternatively, if production cost (the cost of each unit of capital) falls, perhaps due to an improvement in construction technology or a fall in the cost of borrowing capital, this would shift the MC of capital line downwards. Either case will, ceteris paribus, affect the margin at which it is profitable to use the land, the property rent that can be charged and the intensity of use of the land. Similarly, a more profitable use would have a higher MRP curve and could therefore afford to bid a higher rent. Competition between different land uses ensures that the land is allocated to its most profitable use and the land rent surplus QYP is maximised.

In terms of land-use intensity, Figure 2.7 and the underlying land-use rent theory show that, to maximise revenue from a site, capital must be added to the point where MRP equals MC. This also has the effect of maximising the surplus revenue that is available to pay as rent: the highest bidder or rent payer is also the most intensive user of the land. This assumes that competition between land uses ensures that the use of each site will be intensified up to a point at which it is no longer profitable to add any more capital. In a market where supply is inelastic, as demand for floorspace in a locality increases, its price rises. At the same time, the higher price of land means that it makes sense to intensify its use up to the point where the production costs (excluding rent) are so high that it is more cost-effective to purchase additional land than use the existing site more intensively. So, a land user in a central location may find that, on account of the high rent for the site, the revenue generated will not cover production costs and may decide to relocate and sell the site to a higher bidding user. Harvey and Jowsey (2004) illustrate this point by comparing two sites of the same size: (i) one in the city centre and (ii) one in a suburb. Figure 2.8 shows that it is the strength of demand (represented by the MRP curve) that determines land rent and intensity of land use.

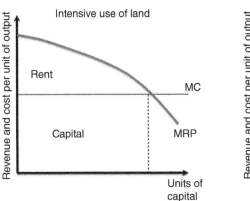
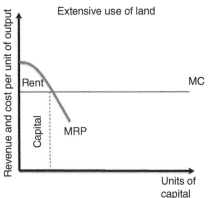

Figure 2.8   Demand and its effect on rent and intensity of land use: intensive use of land and extensive use of land.

For reasons that will become clear in the next section, it is the city centre site from which a user is able to extract more revenue per unit of output. From the owner's perspective, where demand (reflected in the property rent obtainable) is high (high MRP curve), a more intensive use of land is possible, and rents are high.

This is a very simple model, which will be developed later in the context of property development. Specifically, it will be assumed that MC is not constant, as increasing amounts of capital are added to a fixed piece of land it becomes progressively more expensive to do so, as is the case when building a high-rise office building. The MC curve therefore rises.

2.5   Location and land use

The discussion so far suggests that different users of land might bid different rents for a land parcel because it offers different levels of welfare or profit. A key driver of this differential utility is the location of the land parcel. Land is immoveable so the location of each site influences the way in which it is used and its welfare-enhancing or profit-making potential.

As well as formulating a theory of agricultural land rent based on fertility, Ricardo also recognised that the produce from land located near a market incurs lower transport costs and so generates more revenue. Surplus revenue (over and above costs and normal profit) would be paid as land rent. In 1826, German landowner von Thünen applied Ricardo's rent theory in a spatial context and demonstrated the relationship between agricultural rent and distance from the market. The theory assumes that farmland exists in a boundless, featureless plain over which natural resources and climate are uniformly distributed and that produce is traded at a central market that is connected to its catchment area by a uniformly distributed transport network. It was also assumed that although different agricultural produce can be produced that differ in production costs and bulk so that cost of transportation varies, revenue from each product per unit area of land is the same; in other words, von Thünen's theory was a cost-based

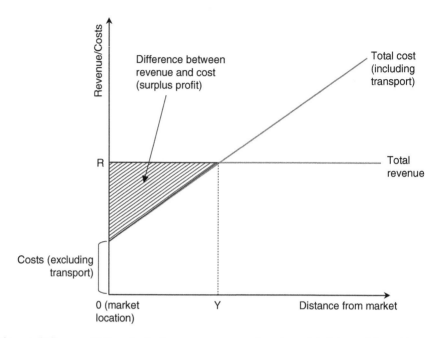

Figure 2.9 von Thünen's single-use revenue and cost model. Adapted from Harvey and Jowsey, 2004

model, which ignored intensity of land use and revenue differentials. Fixing all other costs, Figure 2.9 shows that, for a single land use, transport costs increase as distance from the central market increases. Assuming competition between uses, any surplus profit over and above costs (which include normal profit to the farmer) is paid as rent to the landowner. As the theory assumes total revenue remains constant, the rent (surplus profit[7] in Figure 2.9) decreases as the distance to the market increases. Beyond distance Y, this use is no longer profitable as costs exceed revenue.

This theory can be applied to other land uses. Land that is close to a market or labour supply (a 'prime' site) might yield the same revenue as land further away (a 'secondary' site) but would incur lower labour and capital costs due to accessibility advantages. Therefore, the surplus revenue of the prime site is higher and is transferred via competitive bidding from user to owner in the form of rent.

Figure 2.10 introduces a second land use (A) for which fixed production costs are lower, OA, but the final product is bulkier than the original land use (B) and therefore incurs more steeply rising transport costs as distance to the market increases. Assuming revenue is the same from both products, close to the market, land use A has the greatest surplus (revenue less costs) available to bid as rent (AR as opposed to BR). Land use A outbids land use B up to distance X from the market, after which, because B's total production costs do not rise so steeply, it is able to outbid A.

As more land uses are added with different levels of fixed costs and different rates of rising transport costs, an agricultural land-use rent theory is obtained by rotating Figure 2.10 and considering the rent-earning capacity (revenue less cost) of each land use on the y axis. In Figure 2.11, which is adapted from

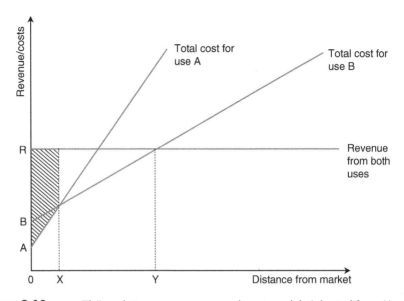

Figure 2.10   von Thünen's two-use revenue and cost model. Adapted from Harvey and Jowsey, 2004

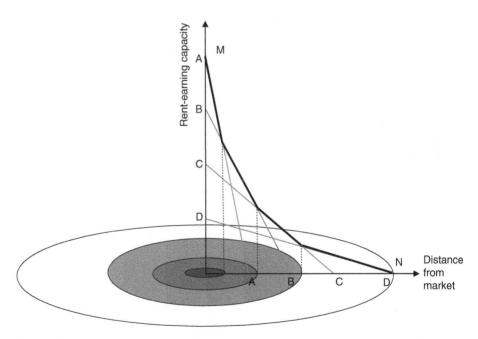

Figure 2.11   Land-use bid-rent theory. Adapted from Harvey and Jowsey, 2004

Harvey and Jowsey (2004), the shaded areas represent rent-earning capacity and the sizes of these are maintained for each land use. The revenue line is dropped as it is constant for all land uses. Rent curve MN is derived, showing land rent declining as distance from the market increases. Given a central market and a homogeneous agricultural plain, a series of concentric zones of land use results, and the relationship between location, land use and rent is evident. Of course, reality confounds all of the simplifying assumptions made by von Thünen's and we do not see concentric rings in the real world. Instead, natural features, the layout of the transport network and other irregularities, such as government trade policy, break up this simple pattern.

Building on Ricardo and von Thünen's work, Mill (1909) argued that, in a country where land remains to be cultivated, the worst land in actual cultivation pays no rent and it is this marginal land that sets the baseline for estimating the rent yielded by all other land (beyond D in Figure 2.11). So, whatever revenue agricultural capital produces, beyond what is produced by the same amount of capital on the worst land, or under the most expensive mode of cultivation, that revenue will be paid as rent to the owner of the land on which it is employed. In other words

> 'Rent, in short, merely equalises the profits of different farming capitals, by enabling the landlord to appropriate all extra gains occasioned by superiority of natural advantages' (Mill 1909).

Urban land uses, like agricultural land uses, desire accessibility: locations that minimise transport costs associated with marshalling factors of production (particularly labour but capital too) but that maximise access to the market and to complementary land uses.[8] With a radial transport network around a central market and the other simplifying assumptions, von Thünen's model can be applied to urban land uses. In explaining the cause of different land values within an urban area, Hurd (1903) suggested that

> 'since value depends on economic rent, and rent on location and location on convenience, and convenience on nearness, we may eliminate the intermediate steps and say that value depends on nearness.'

Theoretically, as Kivell (1993) points out, in a monocentric urban area the centre is where transport facilities maximise labour availability, customer flow and proximate linkages and therefore attracts the highest capital and rental values. Haig (1926) suggested that

> 'rent appears as the charge which the owner of a relatively accessible site can impose because of the saving in transport costs which the use of the site makes possible.'

His theory emphasised the correlation between rent and transport costs, the latter being the payment to overcome the 'friction of space': the better the transport network, the lower the friction. The theoretically perfect site for an activity is that which offers the desired degree of accessibility at the lowest costs of friction. Haig's hypothesis was therefore

> '. . .the layout of a metropolis . . . tends to be determined by a principle which may be termed the minimising of the costs of friction' (Haig 1926).

Like von Thünen, Haig's hypothesis concentrated on the cost side of profit maximisation, but some land uses such as retail are able to derive a revenue-generating advantage from certain sites, particularly those most accessible to customers.

Therefore, the revenue-generating potential of a site must be weighed against the costs of friction for these land uses. Marshall (1920) noted that demand for the highest value land comes from retail and wholesale traders rather than manufacturers because they can fit into smaller sites (i.e. develop land more intensively) in places where there are plenty of customers. Therefore

'In a free economy, the correct location of the individual enterprise lies where the net profit is greatest' (Losch 1954).

In attempting to quantify spatial variation in rent and land use, Alonso (1964) adapted von Thünen's agricultural land-use model to urban land use. Alonso suggested that activities can trade off falling revenue and higher costs (including transport) against lower site rents as distance from the centre increases. This can be illustrated by defining 'bid-rent' curves (similar in nature to indifference curves), which indicate the maximum rent that can be paid at different locations and still enable a business to earn normal profit, as shown in Figure 2.12. In other words, the lines join equilibrium locations where access and rent are traded off against each other. In a monocentric city market, the rent curve derived in Figure 2.11 can be superimposed. Land users will endeavour to locate on the bid-rent curve nearest the origin, and the equilibrium location is at X as this is the most profitable location at the market rent.

Some urban land uses place greater emphasis on accessibility than others and these will have steeper bid-rent curves since a considerable drop in rent will be necessary to compensate for the falling revenue as distance from the central business district (CBD) increases. Rent gradients emerge, as illustrated in Figure 2.13, for each land use where the steepest gradient prevails. Retailers outbid office occupiers because they are particularly dependent on a central location where access to the market is maximised and transport costs are minimised. The availability of

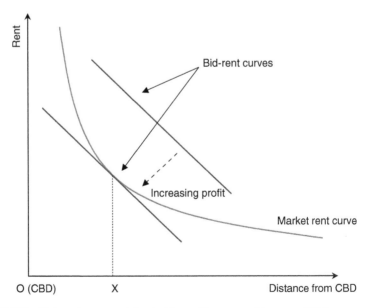

Figure 2.12   Bid-rent curves. Adapted from Harvey and Jowsey, 2004

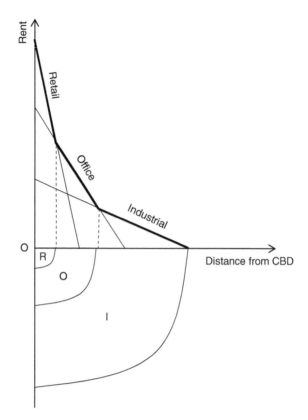

Figure 2.13   Alonso's bid-rent concept.

such sites is limited and therefore supply is inelastic. Office occupiers, in turn, out-bid industrial occupiers. Consequently, rents decline as distance from the central area increases. To summarise, greater accessibility leads to higher demand, which, in turn, causes rents to rise and land-use intensity to increase. Competitive bidding between land users allocates sites to their optimum use.

Alonso's theory rests on simplifying assumptions: a central market in an urban area and a perfect market for urban land. Agglomerating forces, spatial interdependence, special site characteristics, and topographical irregularities are ignored. If the main determinant of differences in urban rent in a city was accessibility and if transportation were possible in all directions and the transport cost–distance functions linear, there would be a smooth land value gradient declining from the centre. In reality, the gradient falls steeply near the centre and levels off further out (Richardson 1971). Other distortions result from trip destinations to places other than the centre such as out-of-town office, retail and leisure, and a non-uniform network of transport infrastructure.

Despite the simplifying assumptions, this *bid-rent* theory is still regarded as a good explanation of spatial variation in the demand for property. As Ball et al. (1998) argue, the rent or price paid for an owner-occupied property reflects its utility to the user. This utility is a function of land and building characteristics and location. Rents and capital values thus vary spatially, and occupiers will choose a location based on an analysis of profit they can make at different locations. Competitive pricing should ensure that, in equilibrium, land is allocated to its most profitable use, but inertia and planning controls influence this. In reality, competitive bidding between users of land often results in mixed use on sites, retail on the ground floor and offices above (Harvey and Jowsey 2004).

As Richardson (1971) notes, the central feature of the market is that land rent is an inverse function (typically a negative exponential function) of distance from the centre. This function is primarily a reflection of external and other agglomeration economies and transport costs.

> 'The significance of transport costs is obvious. People and activities are drawn into cities because of the need for mutual accessibility, especially between homes and workplaces. Even within cities, the distances between interrelated activities have to be minimised, and the existence of transport costs tends *ceteris paribus* to draw activities together' (Richardson 1971).

The role of external economies and agglomeration economies is generally less obvious but probably more significant. Agglomeration economies include scale economies at the firm or industry level. External economies include access to a common labour market, benefits from personal contacts and environmental factors.

So, according to Geltner et al. (2007), equilibrium in a well-functioning land market is attained when aggregate transport costs are minimised, and aggregate land value is maximised. Bid-rents represent the maximum land rent that a user would be willing to pay for a location and a bid-rent curve shows how the bid-rent from a user falls as distance from some central point increases. This central point is the point at which costs are minimised and value is maximised for a given use, each of which has its own bid-rent curve (and central point).

The classical economic theories of urban rent and land use have been criticised primarily due to their simplifying assumptions and the increasing influence of modern working practices and living habits on the way urban land use is organised. These criticisms are summarised as follows:

- Land use changes infrequently because of the long life of buildings, lease contracts, neighbourhood effects, expectations and uncertainty. Consequently, adjustments in supply and demand towards an equilibrium are slow.
- The process of allocating a use to a site is constrained by inertia (preventing a high proportion of land that is in suboptimal use from coming on to the market) and high mobility costs (preventing users from relocating) (Richardson 1971).
- A change in the distribution or level of income or a change in the spatial pattern of consumer demand will cause a change in urban land values and the pattern of uses.
- A change in transport costs will have a greater effect on those uses that depend more heavily on transport.
- The theories have no regard for land-use interdependence or complementarity between neighbouring land uses. The spatial differentiation of land use becomes more marked and complex as the degree of specialisation increases in significance and complementarity linkages become more commonplace.

- There is no uniform plane; geographical and economic factors, the rank and size of urban areas, proximity to other centres, history, favoured areas, cultural dispositions, existence of publicly owned land and ethnic mix all distort the perfect market assumption.
- The theories unrealistically assume a free market with no intervention and perfectly informed market players. In reality, a major restriction on the competitive allocation of land uses to sites is planning control. This may restrict supply for some uses (leading to artificially higher rents) and over-supply other uses (leading to artificially lower rents).
- Owners of property have monopoly power due to heterogeneity of property.
- The theories ignore spill-over effects such as the filtering of land uses and property types and diseconomies such as traffic congestion.

The emergence of greater spatial flexibility as a result of increased car use, lower transport costs, and better information and communications technology meant that, in the 1960s, the classical economic approach to explaining land-use allocation, growth and pricing was challenged (see Meier 1962 for example). Indeed, ubiquitous car ownership led to the growth of out-of-town leisure, retailing and office activity, causing rents to rise in outer areas. Developments in information and communication technology, which facilitate home-working and internet shopping, may have similarly dramatic impacts on land-use patterns in the future.

2.6   Economics of property development

The development or supply of new property resulting from activity in the development sector adds only a small fraction to the existing stock of commercial property each year. The supply of new property has little impact on overall supply in the short run: property is a durable good. This helps explain why property prices are largely explained by demand-side factors. Price signals from buying and selling investments and occupational interests in the *existing* stock influence the supply of and demand for *new* stock.

2.6.1   Type and density of development

As demand for property increases, it becomes worthwhile to pay more for land (land rent increases). This stimulates supply in the form of new construction. Sub-marginal land might become marginal (break-even) or even super-marginal (profit-making) if demand increases sufficiently. This process is subject to the principle of diminishing returns, which can be delayed by more intensive use of land. If the fixed unit of land is expensive or less marginally productive in comparison with the variable units of capital, then a developer will employ more capital on the fixed unit of land. In other words, use it more intensively, perhaps by building at a higher plot density – a high-rise building for example. This is why land in the city centre is more intensively developed than land in more peripheral urban locations (Fraser 1993).

Marshall (1920) was the first to consider how the principle of diminishing returns may be applied to the intensity of development on a parcel of land. If land is plentiful, the amount of capital employed per unit area, which would yield the maximum return, varies with the use to which the site is put. So, the use that yields the maximum return for a given amount of capital per unit area will tend to be the use to which the site is put, all other things being equal. But when land is scarce, it may be worthwhile to go on applying capital beyond this maximum rather than pay for more land to extend the site. In places of high levels of scarcity (and therefore high land value), this intensified use of land will be greater than on sites used for similar purposes but where land is less scarce (and therefore of lower value). Marshall used the phrase 'margin of building' for that floorspace, which it is only just worth adding to a site and which would not be added if the land were less scarce. The example he used was the top floor of a building: by adding this floor instead of building on extra land, a saving equivalent to the cost of that land is obtained, which just compensates for the expense of constructing the extra floor. Two processes are occurring simultaneously: competition between land uses allocates land to its optimum use and intensity of use (maximising return for a given amount of capital per unit area) up to the margin of building, at which point it is no longer profitable to apply more capital to the same site.

Fraser (1993) illustrated Marshall's ideas in a diagram similar to that shown in Figure 2.14. The MC curve shows the additional cost for each extra unit of floorspace added to a site of fixed size. At low density levels, there are economies of scale to be reaped by adding more floorspace. Therefore, the cost per unit of floorspace initially falls. At a certain point, it becomes more expensive to add more floorspace to the fixed amount of land. For example, a high-rise building will need bigger foundations, faster lifts and so on, and the time taken to build it will be longer so finance costs will be higher. Moreover, uncertainty over what the market will be like at the time of completion will be greater and this will mean that the risk and hence profit required by the developer will be higher. All of this means that the cost of adding each extra unit of floorspace starts to increase. The MR curve is the addition to revenue that is obtained from each additional unit of floorspace. It slopes downwards because diminishing returns mean that

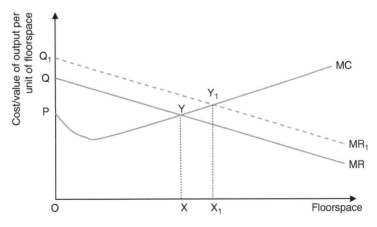

Figure 2.14 Optimum development density. Source: Modified from Fraser, W. (1993).

users will obtain less utility for each additional unit of floorspace. The optimum amount of floorspace is OX units of accommodation and the area bounded by PQY represents the price the developer would pay for the site, i.e. the capital value of the site for this particular development.

Harvey and Jowsey (2004) also reiterate Marshall's ideas and note that by building higher, the developer is effectively saving on land cost. Consequently, a developer will only build more intensively so long as it is cheaper than acquiring extra land. So, there is a margin of building in terms of the intensity of use of each piece of land (or density of development) and the extent to which additional land is used. Under free market conditions, competition for land between different developers ensures that, in the long run, development everywhere will be pushed to the point where MR is equal to MC of capital.

Fraser (1993) extends his analysis of development density by demonstrating that site values and development density are affected by changes in costs and revenue. For example, an increase in property values will cause the MR to increase to MR_1, raising the optimum density to OX_1 and increasing site value to Q_1Y_1P. Fraser also argues that the diagram can be used to explain differences in site value and building density that are observed in different locations. Quite simply, if more revenue can be obtained from a particular site, perhaps because of its accessibility advantages in a city centre for example, then its marginal revenue will be higher at say MR_1. The value and development density of such a site will be high. A less accessible site on the edge of town would yield less marginal revenue at say MR and its value and density of development will be lower.

The type of development that is allowed to take place on a site and the intensity to which that site is developed are not determined solely by free market economics; they are regulated by planning policy and development control. Evans (1985) demonstrated how government controls intervene to determine land use independent of the market. Landowners may also dictate the type, density and timing of development.

2.6.2   Timing of development

According to Fraser (1993), there are two conditions necessary for property development to be economically viable. First, the value of the completed development must exceed the costs of the development, including the price of the land and the developer's profit. Second, development site value (the value of the site cleared and ready for development to its optimum use) must exceed the existing use value, otherwise the developer would be unable to purchase the site at a price that would allow sufficient profit to be made and the owner would be unlikely to sell. Assuming competition among developers to acquire a site, this development site value will be the highest bid from the most efficient developer (Fraser 1993), subject to planning constraints and inertia of ownership and occupation.

The value of a site that has just been developed to its optimum use will be the highest value that could be obtained for that site. Over time, the value of that developed site (or, more specifically, the buildings and other improvements on the site) will decline as depreciation takes hold, maintenance costs increase relative to rental value and a better standard of accommodation is expected. At some point,

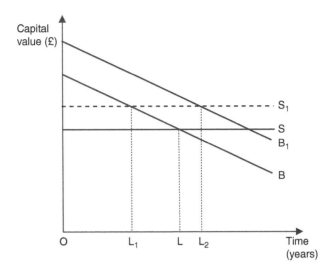

Figure 2.15   The economic life of a building. Source: Modified from Lean, W. and Goodall, B. (1966).

this existing use value will fall below the value of the site cleared and ready for a new development: the trigger for redevelopment. In Figure 2.15, line S shows the capital value of a cleared site assuming development value and cost remains constant over time. Line B shows how the capital value of a building on the site falls over time. It is not economically viable to redevelop the site until time L. In reality, redevelopment is likely to occur sometime after L, perhaps when the lease ends, and the decision is subject to planning constraints and sunk investment in the existing use (Lean and Goodall 1966).

The economic life of the building depends primarily on its earning power and only secondarily on its structural durability. S may increase to S_1, perhaps due to infrastructure improvements, and this will reduce the economic life of the building to L_1. Similarly, B may increase to B_1 due to refurbishment or conversion to a more valuable use and this will increase the economic life of the building to L_2. The model can also be used to explain urban structure. In the central area, buildings fall into disrepair as owners anticipate redevelopment (B_1 to B) while at the same time site values may increase (S to S_1). Further out from the centre the built environment is characterised by lots of conversions and refurbishments, increasing building values (B to B_1) but the infrastructure usually worsens (S_1 to S). In the suburbs, buildings tend to be well maintained (B to B_1), but development forces are strong (S to S_1).

In reality, according to Fraser (1993), development site value will have to exceed existing or alternative use value sufficiently to overcome landowners' inertia. Evans (1985) expands on this theme: expectations of landowners as to what might be the 'right' price for land may lead to a refusal of a bid that is different from expectations either now or in the future. This is known as speculation if the price expectation is higher and inertia if it is lower. Evans (1985) notes two landownership issues that may delay development. The first issue is tenure: whereas landlords may readily displace tenants[9] and sell, owner-occupiers must displace themselves.

The second issue is fragmentation of ownership: trying to assemble a large development site from several smaller sites that are separately owned can be costly. Sometimes, developers will work with local authorities – which have powers of compulsory purchase – to ensure that these types of development can proceed.

2.7   Non-market concepts of value

Property rights are capable of bestowing non-market value on individuals, communities and society as a whole. This value can take the form of social and environmental benefits. Social benefits may be cultural, religious, spiritual, recreational, aesthetic, inspirational, educational, communal or symbolic. Environmental benefits include regulation of climate, flood and disease mitigation, detoxification, carbon sequestration, soil and water quality as well as supporting biodiversity, nutrient cycling and primary production. These benefits may be 'consumed' by current incumbents or retained as an option for use by future generations. Their 'value' to the individuals and communities using them goes beyond monetary value and takes the form of 'importance'.

It is argued that market values and non-market values cannot be reduced to a single value (Barton et al. 2014). Instead, value pluralism captures the needs and wants of society and individuals, including physiological and subsistence, safety and protection, affection, sense of belonging, esteem, identity and other components of the quality of life. Value pluralism, a core concept of ecological economics, groups values into categories, although the boundaries between them are often blurred:

- Monetary: including use values (direct and indirect use, option values) and non-use values (existence values, altruist values, bequest values)
- Sociocultural values
- Ecological value

Certain land and property assets provide economic benefits that may not be traded in a market and so they do not have market prices. Consequently, their economic value may be different from their market value (Myrick Freeman et al. 2014). For example, forestland provides economic value in terms of its timber, and this also has a market value, but the forest also has economic value as a wildlife habitat, as a carbon store and as a public amenity space. These values are not traded and so are not reflected in any market value. Nevertheless, the total economic value of the forest includes both marketed and non-marketed benefits.

Some might regard the monetisation of non-market values as unethical but a decision *not* to value them may imply zero value. That is, if a transaction is going to take place, either through negotiation or expropriation, and no replacement can be provided other than in monetary terms, it might be preferable to assess monetary value as accurately as possible. Nevertheless, placing a value on non-market assets is a challenging and often highly subjective task. The principal difficulty is not knowing whether a valuation is 'right' – there is no comparable evidence (or revealed preferences) to draw upon. The act of valuing non-market land and property may establish benchmarks, but these are likely to be localised and heterogeneous.

Key Points

- The concepts of scarcity, choice, opportunity cost and rent form the basis of property economics. A definition may describe it as: a social science that studies how individuals choose to allocate scarce resources to satisfy the competing needs of society for various goods and services.
- Land has *economic* value because it is *useful* and is *scarce*. Demand for land, which is a derived demand in the main, is limited by *budget constraint*, which means users must weigh up the *opportunity cost* of alternative spending choices.
- *Scarcity* and *utility* of land give rise to its value: scarcity of all land in terms of its limited supply relative to other factors of production and the unique spatial characteristics of each site, and utility of all property in terms of durability and the specific physical and legal attributes of each site.
- Economic rent is the income from a site net of production costs (labour and capital). It arises once the use of land creates a surplus over what the landowner needs to subsist. Who receives it depends on who holds the relevant property rights.
- Supply of land is fixed in a global sense, but not in a land-use sense, in that land can be transferred from one use to another and can be used more intensively. The process of transferring between land uses comes with transfer costs (it takes time and money and probably requires some form of consent from a regulatory authority).
- In a capitalist economy, the *market* is the distribution mechanism, allocating scarce resources to satisfy competing needs. Competition ensures that land is allocated to the optimum use and intensity, subject to regulation.
- The market for property rights is a decentralised market with fewer transactions than markets for many other goods and services. Each transaction usually involves a lot of money, so finance is important.
- Each parcel of land is geographically unique and immobile and so will vary in terms of spatial characteristics such as accessibility, physical attributes and institutional restrictions such as planning.
- The immobility of land and the durability of improvements to land (i.e. buildings) mean that the market is primarily concerned with the exchange of existing rather than new developments.
- Also, because land is immobile, the economic rent that it can earn is determined not only by supply and demand but also by its location. In a purist sense, each parcel of land occupies a unique location but in reality, there will be many parcels that are substitutable to a greater or lesser extent. When considering urban land, sites in the centre are less substitutable than those on the outskirts simply because there are less of them. Consequently, the supply of these sites is more inelastic. But these sites are the ones in greatest demand because they are the most accessible to labour and capital and to the market (consumers) so their rents are higher, and they tend to be the most intensively developed. This inelastic supply means that economic rent is high in a central area and may even represent 100% of the total rent.
- Property comprises land and improvements to land (e.g. buildings), so there can be a separate market in land for development. Development of a site is economically viable when its value cleared and ready for development is greater than its existing use value.

Notes

1. In a competitive product market, price is constant, so MR is also constant and equal to price.
2. MP of a factor is the addition to total product (output) obtained from using another unit of that factor.
3. Technically, the MRP schedule is equal to the demand schedule only if the firm uses a single factor, but it can be proven that when more than one factor is used the demand schedule for each slopes downwards.
4. In economics the short-run is the decision-making time frame of a firm in which at least one factor of production remains fixed while in the long-run all factors of production may be varied, and firms can respond to price changes.
5. Supply and demand schedules are referred to as curves, but, for illustration purposes, these curves are normally depicted as straight lines because they are simple representations of the general form of the schedule rather than an empirically based one.
6. Even if supply was not fixed/perfectly inelastic in the short run, the longevity of property means that new stock is a very small proportion of total stock and therefore stock availability/supply depends much more on the availability of existing stock, either via vacant premises or the ability of uses to change easily (Ball et al. 1998). Consequently, new supply has negligible influence on price. This explains why r1 rarely falls back to r2.
7. The rent paid in respect of any particular use of the land is therefore a geared residual payment (unless there is monopoly ownership of land), but its volatility is reduced as the land can be transferred to the next most profitable and thus restrict drops in rent. Also, land rent is based on expectations of profitability rather than actual year-to-year profit revenue and this tends to reduce the volatility of land rent in the short-term (Fraser 1993).
8. Complementary land uses include things such as comparison shopping and symbiotic business activities.
9. Tenants may have rights that legally secure occupation beyond the end of the current lease.

References

Alonso, W. (1964). *Location and Land Use: Toward a General Theory of Land Rent.* Cambridge, US: Harvard University Press.

Ball, M., Lizieri, C., and MacGregor, B. (1998). *The Economics of Commercial Property Markets.* London, UK: Routledge.

Barton, D., Braat, L., Kelemen, E. et al. (2014) State-of-the-art report on integrated valuation of ecosystem services, July 2014, Openness Project, European Union Seventh Framework Programme.

Evans, A. (1985). *Urban Economics: An Introduction.* UK: Blackwell.

Fraser, W. (1993). *Principles of Property Investment and Pricing*, 2e. UK: Macmillan.

Geltner, D., Miller, N., Clayton, J., and Eicholtz, P. (2007). *Commercial Real Estate Analysis and Investments*, 2e. US: Thomson South Western.

Haig, R.M. (1926). Toward an understanding of the metropolis. *Q. Econ. J.* 40: 421–423.

Harvey, J. (1981). *The Economics of Real Property.* UK: Macmillan.

Harvey, J. and Jowsey, E. (2004). *Urban Land Economics*, 6e. Basingstoke, UK: Palgrave Macmillan.

Hurd, R. (1903). *Principles of City Land Values.* US: Cornell University Library.

Kivell, P. (1993). *Land and the City.* London: Routledge.

Lean, W. and Goodall, B. (1966). *Aspects of Land Economics*. London, UK: Estates Gazette Ltd.

Losch, A. (1954). *The Economics of Location*. New Haven, US: Yale University Press.

Marshall, A. (1920). *Principles of Economics*, 8e. London UK: Macmillan.

Meier, R.L. (1962). *A Communications Theory of Urban Growth*. Cambridge, US: MIT Press.

Mill, J.S. (1909). *Principles of Political Economy with Some of their Applications to Social Philosophy*, 7e (ed. W.J. Ashley). London: Longmans, Green and Co.

Myrick Freeman, A., Herriges, J., and Kling, C. (2014). *The Measurement of Environmental and Resource Values: Theory and Methods*, 3e. Routledge, New York: RFF Press.

Ricardo, D. (1821). *On the Principles of Political Economy and Taxation, (First Published in 1817)*, 3e. London: John Murray.

Richardson, H. (1971). *Urban Economics*. Middlesex, UK: Penguin.

Chapter 3
Property Markets

3.1 Introduction

When property rights are transferred from one party to another in a market where the medium of exchange is monetary, the amount is referred to as a *price*. A vendor might advertise an asking price, a bidder might suggest an offer price; then, usually after some period of negotiation, there might be an agreed exchange price. Thus, under normal market conditions, the economic outcome of interaction between the supply of and demand for property rights is an *exchange price*.

The concept of *value* can be difficult to pin down. Adam Smith[1] first noted the ambiguity surrounding the word 'value', which can mean usefulness in one sense and purchasing power in another, referring to them as value-in-use and value-in-exchange respectively. Value-in-exchange is an estimate of exchange price, typically an estimate of the most likely price to be concluded at a specific point in time by buyers and sellers of property rights that are assumed to be available for purchase. Consequently, exchange prices are useful indicators of value-in-exchange. Individual properties will have different values-in-use depending on the user but, insofar as these benefits are reflected in market prices, competing users should converge on a consensus exchange price.

If information on user benefits and exchange prices is readily available, it will reveal user preferences and the values users ascribed to them. Property rights, however, are traded less frequently than many other resources and commodities because they are expensive and are held for relatively long periods of time. Also, they relate to heterogeneous and durable land and property assets. This means that price information is scarce. *Valuers* fill these price information gaps by analysing market information and making estimates of exchange (or market) prices, a process referred to as *valuation*. A *market valuation* attempts to quantify the aspirations of buyers and sellers of a property in an 'open market' situation at a

Property Valuation, Third Edition. Peter Wyatt.
© 2023 John Wiley & Sons Ltd. Published 2023 by John Wiley & Sons Ltd.
Companion website: www.wiley.com/go/wyatt/propertyvaluation3e

particular moment in time. It has a formal basis and a methodology, which is grounded in the analysis of market transactions, whether they are rental or capital transactions. The concept of a market is, therefore, key to valuation.

3.2   Property markets

In economies where property rights can be traded, markets emerge to allocate them to various needs and desires: economic production, social welfare (housing, leisure, well-being, amenity) and investment (capital accumulation, wealth creation). A market is an environment in which property rights are exchanged between buyers and sellers through a price mechanism, usually without undue restriction, known as an 'open market'. Buyers and sellers interact and respond to supply and demand stimuli as well as to their own constraints (such as budget and desired risk exposure), knowledge and understanding of the relative utility of the property for its intended purpose. The level of efficiency of a market is determined to some extent by the standardisation of the product and the degree of efficiency with which it functions. The stock market in a typical developed economy, for example, provides instant information worldwide about the prices and quantities of shares being bought and sold during the current trading period. By contrast, property markets are more informal, less structured and more diverse; in many ways, each property market transaction can be regarded as unique. Consequently, property markets are more complex; buyers and sellers rarely come together and simply strike a bargain as they usually need to appoint agents with local knowledge to act on their behalf and commission independent valuations to verify asking prices. As the Appraisal Institute (2001) points out: the property market has never been considered as strongly efficient due to decentralised trading, the heterogeneity and high cost of each unit of product, the high cost and lengthy transaction process that is common when buying and selling property (referred to as illiquidity), the relatively few buyers and sellers at a single point in time in one price range and location, the paucity of market information at the individual property level, and the opportunity to exercise monopoly power.

For a property market to be effective, the property rights that are being traded should be legitimate and recognised as such. A register of property rights can be helpful in providing security of tenure and enabling effective transfer of rights. Regulation and enforcement of property rights are important. This can be achieved using laws relating to the ownership and use of land, such as land-use planning. As well as facilitating optimum land use and density (reflected in value), such regulations help control development, protect common rights and customary rights and prevent unregulated land use. They also help record and protect historical sites and buildings, enable compulsory purchase of private land for public purposes and help protect the environment. According to Bell et al. (2005), the high fixed cost of institutional infrastructure needed to establish and maintain property rights favours public provision, or at least regulation. Without central provision, individuals and businesses would need to spend resources defending themselves against claims to their property and this is socially wasteful and disadvantages the poor.

Land can be used in different ways, so this attracts different users. In a competitive 'use' market, the highest bidder will secure the land. But as property rights govern access to that land, the user may bid for those rights in an 'occupier' market or an 'ownership' market. Bids in the former may take the form of periodic rent and, in the latter, capitalised price. The occupier market will thus have a companion 'investment' market, where investors bid for ownership rights and rent is the return on the investment. The capital prices that investors pay are determined by expectations of occupier activity and its forecast effect on rents.

In general terms, it is possible to define three interlinked property markets:

[1]   The occupier market, which can be broken down into sub-markets, such as agriculture, housing, commerce, industry and leisure. Furthermore, because land is immoveable, each sub-market may operate differently depending on location. For example, the office market in central London is regarded as distinct from the office market in Edinburgh. More specialised sub-markets may also be identified, such as city centre residential apartments blocks, distribution warehousing or out-of-town retail.

[2]   The investment market, where a property is regarded as a financial asset. Rational investors seek to maximise returns on a range of assets and compare the risks of holding property against other investment opportunities. Investors aim to maximise return for a minimum acceptable level of risk. This market is driven by the opportunity cost of investment capital.

[3]   The development market, where land is developed to create new stock or redeveloped to replace existing stock. A developer conceives a scheme, acquires the land, negotiates finance and organises construction. Completed development schemes may be retained for occupation or let to occupiers and sold or retained as an investment. Prices in the investment market and rents in the occupier market provide signals to developers regarding the supply of new space.

For succinctness, from now on, property rights will be referred to simply as property and thus markets in property rights will be referred to as property markets.

3.2.1   Occupier market

The price of property in the occupier market is determined by the interaction of demand and supply. Users bid against one another as a means of allocating land to a particular use, subject to any regulations that may be in place. Once land is allocated to a use, users will bid against each other to allocate each parcel of land (or unit of floorspace or property) among themselves. Take the housing market as an example. There will be a total stock of dwellings at any one time, and a proportion of them will be supplied for either rent or purchase. They are likely to be a mixture of existing dwellings and new dwellings created by developers. Depending on the nature of property rights, the ability to trade them and the role of the state, buyers and sellers will gather information about these dwellings and use it to price differences in location, size, quality and other characteristics.

The quantity that users demand and the price that they are willing to pay will be limited by their budgets and by the opportunity cost of other expenditure items. Budgets will, in turn, be affected by the ability to borrow as a means of purchase. This, in turn, will be affected by the availability of and access to finance. Budget plays a key role in determining whether an occupier rents or purchases. It is important to note that, for property that is owned and occupied, it provides a utility in terms of use and wealth. Expectations of how these determinants might change in the future is also an important factor affecting demand.

Occupier sub-markets are usually described by reference to their use. The sections below summarise the main categories of land use and the factors that influence value. Appendix A at the end of the book suggests appropriate approaches and methods for valuing them.

3.2.1.1 Agriculture and fisheries

Agriculture includes horticulture, fruit growing, seed growing, dairy farming, breeding and keeping of livestock, grazing land, meadow land, osier land, market gardens and nursery grounds, woodlands where the use is ancillary to the farming of land for other agricultural purposes, all ancillary land including uncultivated patches, banks, footpaths, ditches, headlands and shoulders, all associated buildings and hard surface areas on farm holdings. Fisheries include hatcheries and farms in inshore and freshwater areas and other fishing activities such as inshore or estuarial fishing using nets and pots (where these are the primary use of land and associated waters and can be clearly delineated).

Value is influenced by the productivity capacity or potential of the land. Key determinants are (RICS 2011a):

- Climate, topography (slope, aspect) and water availability (drainage and irrigation);
- Boundaries, size, and layout of land (field sizes, ease of cultivation);
- Buildings (use, age, construction, layout, adaptability, provision of power and other services, potential for redevelopment/conversion);
- Soil type, fertility and quality;
- Flooding and erosion, pollution, infestations and contamination;
- Obstructions (pylons, rocks, ditches), monuments, trees and woodland;
- Occupation by other parties;
- Amenity, sentiment and family connections;
- Provision of services (water supply, sewerage, electricity, telecommunications);
- Third-party rights, easements and wayleaves, which may be legal or informal (e.g. rights of way, sporting rights, riparian rights, mineral rights);
- Land-use regulations, development opportunities and potential for expropriation by the state;
- Location within the region, in particular access to markets, population centres and amenities;
- Yield, frequency of harvest and unit value of the product, feeding/carrying capacity for livestock;
- Estimate of stabilised income based on crop patterns and cycles prevailing in the market area and
- Subsidies, grants, support payments, quotas and contracts.

3.2.1.2 Forests, woodland and trees

Forests and woodland may be managed for timber production, seasonal grazing and foraging by livestock, recreation and amenity, conservation and environmental uses. In terms of location and size, for large, mainly commercial woods, access and proximity to timber markets (including processing plants or shipping facilities) are important. For small, mainly amenity woods, proximity to urban areas can increase value due to recreational potential (or other forms of non-timber income) but can reduce value due to trespass and encroachment. Smaller woods tend to yield a higher value per hectare.

Important characteristics include:

- Physical site characteristics: Soil type and drainage, climate, elevation, exposure and threat of wind damage (can affect thinning and rotation length). All these affect the species that can be grown and their growing potential. Also, site terrain, topography, external access (including proximity to adequate public roads), internal access, quality of internal road surface, existence and diversity of flora and fauna (indicates soil type and habitat value), non-woodland features such as dwellings and other buildings, streams and ponds, ancient hedges, banks and archaeological features (add to amenity and habitat value)
- Crop details: Timber volumes, species, tree size, age, health, timber quality, length of vegetation period, and long-term yield
- Management: Thinning and appearance and logging conditions (drainage, swamps, rocks)
- Environmental designations: Voluntary or compulsory, restrictive or supportive
- Forestry policy, felling regulations, fiscal incentives, grants and tax concessions
- Market factors: Supply of and demand for timber and other forest products, average yield value, interest rates, availability and cost of finance, economy, performance of other investments, planning and fiscal policies, and woodland economics (woodlands acquired for investment, altruistic, operational reasons)
- Legal and regulatory: Title, access, fencing covenants, mineral and sporting rights, environmental regulations and felling regulations
- Forestry policies and fiscal incentives

Forest investments are mostly long term, with little if any financial return until cropping, which may be decades in the future. The valuation calculation, therefore, requires detailed forecasting of future wood flows, costs and cash flows, and the accuracy of the valuation will be extremely sensitive to the discount rates adopted. For amenity and related non-timber uses, valuers may need to reflect the presence and impact of trees in valuations and may be asked to value trees as separate assets (RICS 2010a and FAO 1999).

Establishing the facts involves the following tasks (RICS 2010b):

- Determine the site boundary, prepare a schedule of trees, determine where trees straddle boundaries or when more than one interest exists over the land. Adjoining owners of a tree that is bisected by a boundary may be tenants in common of the tree, with responsibilities to each other under the ancient doctrine of waste.

- Check for any protected status, including for proposed developments.
- Note species, location, age, condition, life expectancy and service life, soil conditions, current and future 'contribution' of trees to a site, evidence of work undertaken, safety inspections and management activity.
- Threats, stress to trees, die-back, recent activities nearby (e.g. excavation), disease, etc., extent to which tree(s) may be a liability, for example to nearby structures, underground services, overhead wires and roads.

3.2.1.3   *Natural resource extraction – water, minerals and other materials*

3.2.1.3.1   Water extraction

Water provides environmental benefits (biodiversity, landscape, amenity) and utility (domestic, human consumption, agricultural, industrial). Valuers may be asked to advise on the utility value of water-extraction rights. There is a statutory and regulatory environment to consider, in terms of ownership, use and 'consumption' of water.

An inspection will identify the type of water resource (natural or statutory provider), current and potential use, extent of resource (area, volume), relevant consents and licencing provisions, location within context of river basin or catchment and demand situation within context of surroundings, other users in catchment, infrastructure, occupation and use by other parties, surrounding landscape, designations, drainage, boundaries, dwellings, irrigation, access, amenity and leisure, contamination, third-party users, covenants, obligations, security, health and safety, and energy production/potential.

The economic value of water extraction rights is likely to be in the form of income and therefore the income capitalisation method would be appropriate. It may also be appropriate to make certain assumptions relating to title, licensing, infrastructure, quality, quantity, surety of course and supply, other users in catchment and environmental impacts. Special assumptions may also be appropriate too, such as: future legislation and regulation, climate change, anticipation of physical change, security of future supply, future licensing, charges for regulated water supplies, brand, reputation, goodwill, establishment of high-volume traded water rights, future value of energy production, future marriage value and future designation (RICS 2011b).

3.2.1.3.2   Mineral extraction

A mineral property is any contractual or permanent right to explore, mine or otherwise extract minerals from the earth, and any interest in such a right, and any land ownership that includes or inherently provides that right. This includes surface mineral workings and quarries including waste-disposal areas together with buildings and installations for surface and underground mineral extraction and handling. Value will be influenced by the volume and quality of the resource, accessibility and the regulatory environment.

When valuing mineral-extraction rights, the following are material considerations: surface, natural resource and/or operation, the ownership of other minerals and rights to disturb, rights to work and withdraw support, and the type of natural resource to be extracted. Inspection should focus on geology and hydro-geology of the natural resource, and planning, permitting and licensing of the mining operation. Value will be dependent upon the annual quantity and quality of materials being or proposed to be extracted, production yields achieved/achievable after processing and saleable output. The financial analysis will therefore examine selling prices, operating costs and profits margins. Valuations should include a market feasibility study for saleable products, a statement on rehabilitation/restoration requirements, residual income or end-use alternative use value, subsidence or withdrawal of support liabilities and/or discharge liabilities. Residual value (or negative value if a liability is anticipated) of a site/working once it reaches the end of its economically useful life will depend on the extent of workings such as mine shafts, open pit and quarry voids, tailings and dams. Much depends on nature and extent of the restoration/rehabilitation programme and this will, in turn, be dependent upon planning permission and regulations. This residual value should be separately identified (RICS 2011c; International Mineral Valuation Committee 2018).

3.2.1.4   Recreation and leisure

This covers a plethora of land uses including:

- Outdoor amenity and open spaces, including gardens, parks, zoos, aquaria, agricultural show grounds, picnic areas and play areas, civic spaces, heritage sites and monuments.
- Places for amusement and entertainment such as visitor centres, cinemas, theatres, opera houses, concert halls and arenas, conference venues, broadcast studios, dance halls, nightclubs, gaming and gambling clubs and premises, amusement arcades, fun fairs and circuses. Size and capacity of venue, number of seats or screens, ancillary facilities, admission prices, parking availability, location (accessibility), competition, etc., are all important influences on value. There may be some attractions that are regulated such as casinos and gambling clubs. Value will depend on the existence and continuation of relevant certificates and licenses as well as number and type of gaming tables, machines and so on.
- Buildings, places or institutions devoted to the acquisition, conservation, study, exhibition and educational interpretation of objects having scientific, historical or artistic value, for example museums, libraries, art galleries, and public and exhibition halls.
- Sports facilities and grounds for land and water sports.
- Holiday parks, activity centres, sites for tents, touring caravans and camper vans.
- Allotments and city farms.

These activities usually take place in specialised properties that have been designed and constructed for the specific activity.

3.2.1.5   Utilities and infrastructure

3.2.1.5.1   Energy production and distribution

This includes power stations, cableways and transformer stations for the distribution of electricity, gas manufacture and storage facilities, and pipelines and pumping stations for oil and gas. Renewable energy installations include biomass, biogas from anaerobic digestion, combined heat and power (CHP), geothermal, hydroelectric, landfill gas, sewage gas, solar photovoltaic (PV), tidal power, wave power and wind power. Valuation considerations include proximity to grid connection, site size and constraints, surrounding land uses, ability to use heat or electricity at point of generation and access to energy resource (including wind speed, water flow, insolation, etc.).

3.2.1.5.2 Water-treatment and -purification facilities

These include extraction from springs, rivers, or aquifers; water storage and distribution places, for example reservoirs, water towers and pumping stations; and sewage disposal and treatment works, including drains, pumping stations and sewage farms. Key influences on the value of water supplies include:

- Type, extent and location of the resource;
- Legislation and regulatory framework surrounding ownership, use and 'consumption' of water;
- Containment of resource, drainage and irrigation;
- Boundaries, infrastructure and buildings (including dwellings);
- Occupation and use of the resource within the catchment (including energy production/potential and amenity/leisure) and
- Access, security, health and safety provisions and contamination.

3.2.1.5.3 Waste-management facilities

There are four main types of waste-management facilities, although some sites will be a combination: waste collection (e.g. transfer stations, amenity sites), waste treatment (e.g. recycling, green waste composting, liquid waste treatment), energy from waste plants (e.g. thermal, mechanical, biological), and waste disposal (landfill/raise sites and underground storage) – this is a wasting asset. The first three types will generate a residual waste that can only go to landfill so their life may be linked to that of waste-disposal sites.

Valuation considerations include the amount of approved landfill void, type(s) of waste licenced to be accepted at facility, annual quantities of waste being taken/to be taken, compaction ratios, geology, hydrogeology of the void, engineering requirements, licencing, environmental consents, financials, taxes, levies, market appraisal, discharge requirements, other ongoing costs and residual value. Legislation and regulations have a big influence. The valuer should consider actual or perceived (stigma) hazards that may result in negative residual value, and the cost of rehabilitation, restoration and aftercare management (RICS 2011c).

3.2.1.5.4 Places for storage and disposal of human remains

These include mortuaries, chapels of rest, crematoria, cemeteries and churchyards.

3.2.1.5.5 Postal and telecommunication facilities

Postal service places, including depots and sorting and delivery offices. Telecommunication installations are facilities for transmitting and receiving messages by telephone, radio, radar, cable, television, microwave and satellite including radio and TV transmitting/receiving stations and masts (including microwave masts) and cable networks. Telecommunication installations include the masts and associated plant. They are usually valued using the income capitalisation method in respect of the rent for the site, plus the replacement cost method for facilities not reflected in the land rent, namely buildings, plant and machinery. Specific factors that affect the value of telecommunication installations include surrounding telecommunication sites, height and line of sight, site-sharing and mast-sharing rights, extendibility, power requirement, visual amenity and local community issues, decommissioning responsibilities, equipment rights, wayleave rights and property tax liability.

3.2.1.5.6   Transport infrastructure

Transport infrastructure includes passenger routes such as tracks and ways, roads, railway lines, cycle tracks, footpaths and bridleways as well as terminals and transport interchanges for people, including airports, ship passenger terminals, railway, bus and coach stations, and petrol-filling and service stations. Other passenger transport infrastructure includes car parks and 'Park and Ride' terminals, storage places for vehicles such as lorry parks, bus and coach depots, railway sidings and aircraft hangars. Transport property also includes terminals and transhipment places for goods such as airfreight terminals, rail freight terminals, container depots, docks, railway yards and depots, and customs depots, and the mechanised handling of goods and raw materials, such as aerial ropeways, conveyors and lifts. Waterway transport includes canals and navigable rivers, moorings, marinas, boat yards and anchorage for watercraft.

3.2.1.6   Residential

This category includes:

- Houses and flats for individuals and families living as a single household, including adjoining garages, gardens, non-thoroughfare service, and distribution roads and pathways;
- Caravan sites and mobile homes used as permanent dwellings;
- Sheltered residential accommodation with separate front entrances;
- Hotels, boarding and guest houses;
- Residential accommodation for care provision, including old peoples' homes, children's homes and other non-medical homes;
- Residential schools and colleges and training centres, including university and hospital residences, and
- Communal residences such as barracks, monasteries and convents. Influences on value include location, size, layout, construction type and quality. Important information to collect includes number of rooms and quality of accommodation, occupancy rates and extent of facilities (parking, dining, leisure, etc.)

Generally speaking, locational characteristics that influence the value of residential land use are access to employment opportunities, transport, education and health facilities, leisure and amenities. In terms of the dwelling itself, the quality of the immediate surroundings is important, together with physical attributes such as size, type and age, aspect and outlook, condition of structure and services, facilities and energy efficiency.

3.2.1.6.1 New-build dwellings

New dwellings may attract a price premium compared to second-hand dwellings to reflect certain attributes, such as innovative forms of construction and specification, lower maintenance costs and better environmental performance. They may be sold using incentives such as (RICS 2012):

- Payment of legal fees, survey fees and stamp duty;
- Reimbursement of deposit;
- Shared ownership, where the purchaser buys part of the property with an option to purchase the remainder later;
- Purchase of buyer's existing property (part exchange);
- Payment of mortgage for a period and
- Cashback, fixtures, fittings, furnishings or other material gifts such as electrical goods or a car.

They also come with guarantees, warranties and an insurance scheme. The premium would only relate to the first owner and is likely to vary with supply and demand.

3.2.1.6.2 Affordable housing

Affordable housing comprises several tenures:

- 'Social rent': Housing provided by a landlord where access is based on housing need, and rents are no higher than target rents set by government for housing associations and local authority rents.
- 'Affordable rent': The 'Affordable Rent' tenure is let by Local Authorities or Registered Providers to households who are eligible for social rented housing. Affordable rent is subject to rent controls that require a rent of no more than 80% of the local housing market (including service charges where applicable).
- 'Shared ownership': Low-cost home ownership housing provided by registered social landlords in which the occupier owns a percentage of the property (normally 30–50% but no less than 25%) and the remainder is owned by the landlord (and a rent is normally charged to the occupier at say 2.75% of the unsold equity).
- 'Equity share': The occupier owns a percentage of the property (typically around 70%) and the remainder is owned by a third party (typically the developer, landowner, employer or their agent). No rent is charged on the outstanding equity, but the purchaser may be expected to buy at the market value at a specified date in the future.
- 'Low-cost sale': The property is sold outright either at a discount on the market value of the property or at a lower price than other properties available in the area.
- 'Intermediate rent': Property that is available for rent at a cost, which is at or below 80% of the market rent. Occupancy may be restricted to certain income or occupational groups such as keyworkers.

3.2.1.6.3 Residential property purpose built for renting

Most properties will be occupied by tenants on periodic tenancies or contractual tenancies (e.g. company lets). The rent and term are agreed by parties and repair obligations tend to rest with the landlord, but the tenant might be responsible for internal repairs and furniture.

Valuation considerations include: security and growth potential of income, likelihood of tenant change, speed of letting, strength of occupier market, void rates, likely expenditure necessary to maintain rent, assessment of other value factors such as planning, legal issues and assessment of break-up potential assuming it is permitted. Look at accounts where available, and whether management and lettings are undertaken in-house or by an agent (cost difference).

Sources of potential ancillary income include parking, storage, services (such as cleaning) and utilities (such as broadband), and furniture hire. Value would only be attributed if it is likely to be reflected in market bids. It is also important to consider the potential for lost income via bad debts, payment delays, non-recoverable arrears, voids and tenancy 'churn'.

Other considerations include the proportion of affordable housing, commercial uses, clawback provisions and break-up value. For the latter, a valuer might need to consider absorption rates, the ability to break up the asset, the cost of sales, buyers desired margin, potential price shifts and the cost of holding vacant units (RICS 2018).

3.2.1.6.4   Student accommodation

There are various forms of student accommodation. Direct let units are where the developer or investor takes all risk but has flexibility for rent and lease terms. Then there are university halls and houses in multiple occupation. Sources of income include student lets, holiday lets, sale of insurance, vending machines and laundry facilities. Rent may or may not include heating and lighting (if not, then these should be added as an expenditure) and non-returnable deposits. Expenditure includes services, maintenance, sinking fund, direct costs and wages.

Valuation considerations include:

- Room types (e.g. ensuite, studio bedrooms) and sizes, the number of bathrooms and toilet facilities per bed for non-ensuite rooms, kitchen and dining facilities.
- Accommodation type (traditional corridors or modern cluster flats, fitted furniture, internet access, TV point, telephone point, security). Luxury facilities include swimming pool, gym, sauna, solarium, Sky TV, allocated parking and superior rooms.
- Ratio of ensuite to non-ensuite rooms and of standard to luxury accommodation.
- Typical length of tenancy and availability of longer lets.
- Occupancy data, expenditure data, and revenue data (including rent, vending, holiday lets, etc.).
- Health and safety.

3.2.1.7   Community services

These include:

- Medical and health care services, such as medical diagnosis and treatment centres, auxiliary medical centres, clinics and day centres hospitals and convalescent homes. Other medical and health services include dentists, doctors, chiropodists and opticians.

- Places of worship including churches, mosques and synagogues.
- Educational establishments such as schools, colleges, higher and further education centres, universities and other learning places.
- Community protection and justice administration services such as law courts, records and archive facilities, police, fire and ambulance stations, prisons and detention centres, coastguard and lifeboat stations.
- Community meeting places such as public halls and other community centres.
- Public sanitation facilities.
- Animal welfare facilities.

3.2.1.8   Commerce and industry

Location is driven by linkages to people and other uses, measured in terms of accessibility to markets and factors of production. Accessibility refers to the ease with which contacts can be made considering the number, frequency and urgency of those contacts. If there is greater reliance on access to customers, there is a greater desire to locate at the position of maximum accessibility. The layout of transport routes and the cost of traversing them influence the pattern of accessibility.

Other important location considerations are agglomeration economies and complementarity, collectively known as neighbourhood effects. These are the benefits that can accrue when properties of a similar nature cluster together. The amount of benefit depends on the need for contacts. Once sites in an area have been developed for a particular use, this will largely determine the best use for remaining sites due to advantages of concentration. Offices cluster near shopping facilities and desirable residential neighbourhoods. Industry benefits from clustering the production sequence, which in turn lowers costs due to external economies of scale. This explains the success of industrial estates. Smaller firms locate near the centre, but larger firms have less dependency on agglomeration economies and complementarity because they are able to internalise their production processes. Incompatibility is the inverse of complementarity where properties locate apart to prevent higher costs or loss of revenue, for example an obnoxious industry and food production.

3.2.1.8.1   Offices
Office space includes:

- Offices for research and development and testing of products or processes,
- Offices hosting scientific facilities and laboratories,
- Business meeting places and centres,
- Art studios, music recording, film and television studios and
- Computer centres.

Influences on value include size, floor layout, flexibility, building services, specification (including air conditioning), service charge level, transport facilities, car parking, energy efficiency and environmental sustainability.

Headquarters and large branches of international firms regard accessibility and a prestigious address as very important. Professional institutions require similar attributes but often fail to outbid the first category and therefore locate near

Table 3.1   Desirable attributes of office space.

Quality of accommodation	Fitness for purpose	Temperature and ventilation, lift(s) where appropriate, daylight
	Design	Longevity, durability, life-cycle cost Aesthetic, corporate identity, energy efficiency
Flexibility	Space	Uninterrupted floor layout, raised floors, suspended ceilings
	Future contractual (lease) liability	Lease length that fits with business plan, option(s) to break lease, wide user clause, standard assignment provision
Location	Access: By staff and suppliers To clients and complementary businesses	Private transport: parking, motorway network, congestion, cycle routes Public transport: air, rail, buses
	Quality of surroundings	Low incidence of crime, attractive appearance, prestige address Access to: open space, retail, leisure, amenities (post office, doctors, schools, opticians, dentists, pharmacists, etc.)

Section A

parks, squares or buildings of interest. Small professional firms and branch offices require access to a resident population and usually locate in a high street, suburb or near a public transport node. Local government and civil service offices used to be centrally located but now tend to occupy cheaper sites on the edge of the central area. Office uses that are not reliant on people linkages – 'back office' functions as they are sometimes referred to – might locate in out-of-town business parks. Property attributes that office occupiers seek are listed in Table 3.1. Occupiers will select offices according to their preferences, and these will be reflected in the weight they assign to each attribute.

3.2.1.8.2   Industrial premises
Industrial space includes:

- Heavy manufacturing (refineries for processing of coal, petroleum, metals and other raw materials, foundries, brick and cement works, shipbuilding and marine engineering, chemical processing);
- Light industrial (brewing, milling, printing) and high tech (computing, data warehousing);
- Mechanical, instrument and electrical engineering;
- Places for packing agricultural and food products (separate from farm holding) and
- Storage and distribution facilities including warehouses, repositories and open storage land.

Heavy industry requires access to raw material and heavy freight, while light industries are often located in, or on the periphery of, an urban area. If the firm's market is outside the urban area, then intra-urban location is irrelevant with regard to sales but will differ on costs due to land value variation, access to the labour market and the transport network. Other location considerations include access to materials, parts and components, skilled labour, ancillary activities, owner's preferences, utilities and services. Business and science parks require motorway access and close proximity to academia. For warehousing and distribution, important influences on value accessibility to major transport links, site access and loading facilities, building layout offering clear space, eaves height, floor loading, power supply, office content and site coverage.

In terms of the specification of the premises, occupiers of industrial property favour an uninterrupted ground floor area with good load-bearing capacity, generous eaves height, easy loading and access. Generous car parking, good ventilation and canteen facilities might also be desirable. For manufacturing, the handling of materials, products and maintenance arrangements are important. High-tech industrial units might require a campus style accommodation with good communications, generous car parking, and close proximity to skilled labour and amenities, and high specification space with the flexibility to cope with changes in information, communication, production and distribution technologies. Generally, the more space extensive the industry, the less demand for central sites. Compared to other land uses, industrial relocation is uncommon due to inertia and sunk costs.

3.2.1.8.3 Retail property

Retail premises, patronised by the public, come in many shapes and sizes: kiosks, market stalls, standard retail units on the high street or in shopping centres, department stores, retail warehouses, food stores and showrooms. There are also more service-led operators such as betting shops, banks, building societies, hairdressers, travel agents, insurance brokers, estate agents, funeral directors, dry cleaners, etc. Finally, there are trade-related properties such as restaurants and cafes, public houses and bars, hotels, petrol-filling stations, etc.

Retail property is highly dependent on market accessibility, and it is a key objective to locate a shop where it has vehicular or pedestrian access to the greatest number of customers. Differences can be observed at the individual property level and are caused by the type of district, street, position in the street, and whether there are 'anchor' stores, car parks or public transport nodes nearby. Large multiple retailers and chain stores tend to cluster to provide comparison shopping and complementary shops cluster to offer a wider range of goods and services. Certain types of office premises such as building societies, employment agencies and estate agents also require particularly accessible locations in order to attract customers. They try to locate at ground level in those locations where they are not outbid by retailers.

Retail property value can be influenced by what would appear to be minor physical considerations such as aspect, lighting, internal configuration (including frontage length, depth, ground floor area, capacity for display, sale and storage space) and delivery facilities. Non-retail uses in a shopping area, such as a civic building or a church, can have a detrimental impact on value due to different opening hours and a lack of display frontage. In shopping malls, the quality of centre management may also be relevant. Out-of-town shop values are influenced by location within the retail park (particularly visibility), access to the road network, car parking and public transport, mix of adjacent retailers, building layout, size, height, loading and access.

3.2.1.9   Land and buildings with (re)development potential

Subject to relevant regulatory requirements and consents, land is capable of switching from one use to another, and this has implications for value because development or redevelopment usually enhances value. Key influences on development value are:

- Demand for the proposed land use;
- Supply of nearby land allocated to the same use;
- Developable area, permitted use and density;
- Accessibility and nature of nearby land uses and
- Ground conditions, building and other costs.

3.2.2   Investment market

Economists refer to investment as anything that adds to productive capacity, in other words, activities that make use of resources today in order to secure greater production in the future. In financial terms, investment is the sacrifice of present capital for future gain, typically in the form of income and/or capital.

There are certain attributes that are desirable regardless of the type of investment: the level or amount of return on capital invested (this return may take the form of income or growth in capital value or a combination of the two), the security of capital and income (typically regarded as the risk inherent in an investment), the accessibility of the invested funds (often referred to as the liquidity of an investment) and tax efficiency. Some investments will produce little or no income but will provide a return to the investor by way of capital growth, such as gold, works of art and precious gems. Other investments produce a high income but little or no capital growth. Inflation is a major factor affecting security of capital and income. High inflation quickly erodes capital and will also affect income if it is not regularly revised to ensure parity with real income levels. The fiscal implications of any investment need to be considered, especially when comparing the returns across different investment assets as their tax treatment may differ; for example, the importance placed on capital or income growth can depend on the tax position of the investor. Convenience refers to the amount of management that an investment asset requires; can an investor leave the investment to look after itself or does the investor, perhaps with recourse to expert advice, need to keep an eye on performance?

Investors rely on a combination of income and capital growth to generate required return and property benefits from real growth in rent and capital value; each operates in a separate sub-market and is affected by different forces, so it is possible that rental growth may be strong because of high demand by tenants while at the same time capital growth may be limited because of sluggish demand from prospective investors. Property can be invested in directly through ownership as an investor, developer or occupier. As with equities and bonds, property investments can be traded second-hand and indeed this market, rather than the market for new property, is where the vast majority investment trading activity takes place. It is also possible to invest in property indirectly by purchasing shares in property companies or companies that deal with property, property unit trusts and other securitised investment vehicles. The advantage of indirect property investment is that many of the problems associated with direct property investment such as illiquidity, high transaction costs and lengthy sale time disappear, but the portfolio diversification benefits are reduced. On balance, indirect property investment is a good way of allowing small investors to pool their funds so that property can be acquired that could not be done so by these investors individually.

On the supply side, property investments take the form of properties that are already in existence and occupied by one or more tenants paying rent. These 'standing investments' form the majority of assets in the property investment sector, but new ones come along all the time in the form of newly developed properties and transfers from owner occupation (perhaps as sale and leasebacks, but other financial instruments are used too).

As well as standing investments and new developments, property investments can be classified in terms of their perceived investment qualities. *Prime* space includes investment-grade buildings, generally the most desirable in their markets, offering an excellent location and first-rate design, building systems, amenities and management (at least at the time they were built). These buildings command the market's highest rent and attract creditworthy tenants. While some older buildings can be renovated and repositioned as prime, prime space is usually limited to new, high specification buildings. Secondary space includes buildings in good locations, sound management and construction, and little functional obsolescence or deterioration. Such space is found generally in well-located buildings of an earlier generation that have been maintained to a high standard. Tertiary buildings are often substantially older than prime and secondary buildings and have not been modernised. They may be functionally obsolete and contain asbestos or other environmentally hazardous materials. Their low values make many tertiary buildings potential candidates for demolition or conversion to other uses.

While data for prime and secondary space are available in most markets, tertiary space is seldom tracked with any accuracy. Indeed, definitions of prime, secondary and tertiary, even within a single market, are not standard; they are difficult concepts to pin down. It involves breaking prime down into legal (lease), physical and locational attributes and considering them from the point of view of the owner (investor landlord) and occupier (tenant). When does prime space cease to be prime?

Property investments can also be classified by their ownership characteristics. Freeholds offer a pure equity interest to the owner-occupier and an equity/bond mix to an investor because of the stepped income growth pattern obtained from properties let at rents that are reviewed periodically. Leasehold investments come in two main types. The first type of leasehold investment is long leases on ground rents where the reversion is a long way off – like long-dated or undated gilts but without the same level of liquidity and with higher management and transaction costs, causing yields to be slightly higher. The second type of leasehold investment is shorter leases, but these are not very popular.

On the demand side, property competes against other forms of investment, primarily bonds and equities. Perhaps as a consequence of the unique investment characteristics of commercial property, investment is dominated by large financial institutions such as pension funds, insurance companies, investment and unit trusts.

These organisations traditionally invested in property as a hedge against inflation, but nowadays it is the relatively favourable return that provides the incentive. Pension funds (which have long-term inflation-linked liabilities) and life assurance companies (which have long-term fixed-interest liabilities) seek to match their liability profiles with suitable investment assets (Sayce et al. 2006). In addition to institutional investors, other investors include public and private property companies, overseas investors, investment and high street banks and building societies, private individuals and charitable organisations.

The majority of commercial property investments can be placed in one of three principal sectors: retail (shopping centres, retail warehouses, standard shops, supermarkets, department stores), offices (standard offices, business parks) and industrial (standard industrial estates, distribution warehousing). Investment market subsectors are often defined using a combination of this sector classification and their location, 'City of London offices' or 'south west high street retail' for example. There are also smaller, more specialised, sectors of the property market that attract investment interest such as leisure facilities, hotels, student accommodation and serviced offices. Property investments might be let to a single tenant or to several tenants, perhaps in an office building, a residential apartment block or a shopping centre, and the cost of maintenance of the building as a whole is recovered via a service charge to each tenant in addition to their rent.

Property investments are capable of maintaining their value in real terms (keeping pace with inflation) and hopefully growing in real terms. This is achieved through growth in capital value and income. With regard to capital growth, as standing property investments trade on a second-hand market, capital values rise and fall depending on economic activity, just as they do for equities. But, unlike equities, the capital value of a property will not fall below its inherent land value regardless of the rent-earning capacity of the occupier. Regarding income growth, this is realised whenever the rent is revised. In contrast to bonds and equities (paper investments), property represents a tangible investment asset that needs to be managed and maintained in order to secure a steady income stream.

For property investments, capital risk is low because property is a tangible asset where proof of ownership is usually registered by law and its usefulness ensures a high opportunity cost (transfer earnings). Income risk is reduced by rent reviews helping to keep rent in line with inflation. Security of income is affected by factors such as the quality of the tenant and the nature of the lease terms; for example, how likely is the tenant to default on the rent and thus undermine the investment value or is there a break clause in the lease that may lead to a void or gap in rental income? Rent is a prior charge above dividend payments should a tenant go into receivership.

As well as providing a real return and offering a relatively secure investment opportunity, property can provide corporate identity, there may be tax advantages and it is a useful portfolio diversifier. This means that levels of property risk can be hedged against non-correlated investment asset classes such as equities and bonds. An influence that causes a change in gilt yields may lead to an opposite change in property yields. For example, a rise in inflation can lead to a higher gilt yields and therefore higher property yields (as the risk-free component of the latter is often based on the former). But the higher rate of inflation may also lead to a higher rental growth expectation and thus reduce property yields as property investment becomes more attractive and investors bid up prices. This reduction might cancel the increase and might explain why property yields are relatively stable when compared to yields from gilts and only follow significant trends (Fraser 1993).

Many of the larger, institutional portfolios contain a mix of investment types as a means of hedging against adverse market conditions in any single sector or location and portfolio managers rebalance their assets from time to time as a response to market conditions. It is important to note that whereas all shares in a company are the same, property investments are heterogeneous and vary by size, location, use, age, construction and tenant (Sayce et al. 2006). So, investing in property as part of a mixed portfolio of investments can help reduce the amount of risk that the portfolio as a whole is exposed to. Even within a property portfolio it is useful to hold a mix of property types because the returns may not always move in same way. A final but important feature of property as an investment vehicle is the ability to borrow money to help purchase property investments. This allows investors to combine their equity with debt finance and thus invest in either bigger properties or in a larger number of properties than they would otherwise be able to do. This debt financing represents an advantage over the equity and bond investment markets.

As an investment, property has a number of disadvantages too. First, it comes in large indivisible heterogeneous units that suffer from deterioration and obsolescence. Its lumpiness makes it difficult for smaller investors to acquire big, prime investments. It also means that only larger investors can afford to assemble balanced and sufficiently diversified portfolios (Sayce et al. 2006). Ways round this are to syndicate investment acquisitions or use debt finance. Second, property is an illiquid investment asset. This means transactions take time and money to complete. A sale of an investment property usually takes weeks or months rather than days – the norm in equity and bond markets. The purchase of a property investment sometimes involves the acquisition of complex legal interests and the arrangement of complicated finance structures. Transfer costs and taxes are usually higher for property investments than they are for other investment assets, surveyors are employed to survey the property and negotiate price and general lease terms and legal advisors are required to draft the lease and oversee the conveyance. However, lot size and holding period are often higher and longer respectively than for bonds and equities so the annual equivalent of costs is lower but probably still higher than for equities and bonds (Fraser 1993). Third, there are high management costs to cover rent collection, ensure compliance with lease terms, negotiate rent reviews and lease renewals, revaluations, performance analysis and so on, and this means that net income might be significantly below gross income. But, on the plus side, pro-active management, which might include refurbishment and renewal, can enhance income and capital value (Sayce et al. 2006). An investor will seek to minimise these costs and transfer liability, wherever possible, to the tenant. For example, a typical UK lease requires the tenant to be responsible for internal and external repairs and insurance of the premises.

To summarise, property investments are traded in a decentralised and cyclical market in which a high degree of market knowledge is required. It is susceptible to external influences and government intervention in the form of planning, environmental controls, building regulations and tenant protection. All aspects of property dealings, whether occupation, investment or development take time to respond to changes in economic activity and this leads to periods of over- and under-supply and hence greater volatility and risk (Ball et al. 1998). Consequently, property is typically a long-term investment because a long holding period reduces the problems associated with illiquidity, and the emphasis is on security of income and capital, especially in real terms.

3.2.3   Development market

Property development may be defined as a process by which land is improved (usually by constructing buildings) either for occupational or for investment reasons. From an economic perspective, development timing is driven by the lifecycle of land and property: when the net benefit of development outweighs the net benefit of existing use, then development is viable. In a market economy, the optimum scheme will be the one that yields the highest margin over development costs, which include the acquisition and preparation of the site, construction materials and labour costs, finance and a suitable profit element.

In common with other economic activities, the process of development requires the integration of land, capital, labour and enterprise, and the process takes time, sometimes several years. A site needs to be acquired and prepared, legal and regulatory consents must be obtained, financial arrangements may need to be arranged and construction must be undertaken.

It is important to identify the optimum use that can be envisaged for a site and choices about design, planning, funding, construction, renting or selling need to be made. Development success depends to a large extent on piecing together the various factors; the right site with relevant permission to develop, access to sufficient finance, labour and construction materials, a market for the completed development. Only if these ingredients lead to a successful development can a developer extract a profit as payment for the enterprise. The profit is a residual sum and is therefore very risky. As a way of reducing competition risk, secrecy surrounds the assembly of development sites, the securing of planning permission and finance. A characteristic of the property development market, therefore, is the paucity of transaction information. Not only do development land transactions occur infrequently (especially when compared to the number of transactions that takes place in other financial markets or, indeed, the occupation and investment sectors of the property market) but also the market participants are not inclined to share this information.

3.2.3.1   Property development process

Various actors and institutions play their roles in the development process, including:

- Developers,
- Owners and occupiers,
- Lenders,
- Contractors and sub-contractors,
- Architects and technologists,

- Government (decision-makers [councillors], planners, highways, building control),
- Consultants (design, development, environmental, etc.),
- Engineers (structural, mechanical, electrical, civil, etc.) and
- Surveyors (land, quantity, building, valuation, planning and development).

Each has a specific role to play and a time to play it. Typically, a property developer coordinates these roles. Developers can be categorised as trader developers, who develop and sell completed developments or investor developers, who develop and retain property as an investment asset. There is also a wider economic context to this process: long-term social trends, economic conditions, and government policies set the scene for property development.

The property development process comprises a series of events, some of which occur sequentially, others simultaneously. The events may include:

- Feasibility (financial viability, environmental impact, regulatory approval, etc.);
- Acquisition of property rights in land;
- Designing, costing, financing and procuring construction work;
- Construction and
- Marketing and disposal.

A completed development might undergo a number of changes until obsolescence sets in and the cycle begins all over again. The cycle can last for decades, even centuries but can be much quicker. Retail, leisure and industrial property might be redeveloped within a decade, a house might stand for a hundred years or more before it is redeveloped.

3.2.3.2   Feasibility and land acquisition

There are many events that can trigger development. A development plan may identify land for new housing or commercial use, or a private sector organisation may decide to construct new premises as an investment asset or for occupation themselves as part of a relocation or expansion decision. These changes are expressed in a variety of statements about land requirements and allocations (in plans from both public and private sectors) and in applications for planning permission. To the developer, this is the beginning of one of many concurrent tests of the feasibility of their aspirations. The intention of these tests is to inform the developer of the likely success (viability) of their proposition.

A developer may own the land they intend to develop, or they may own an 'option' or some other right that would allow them to acquire the site at some point in the future. Alternatively, a developer may work with a landowner to develop a site, acting as a development or project manager. Whatever the arrangement, this stage will be a time of careful calculation and planning by the developer, in consultation at some points with the public sector. A number of tasks might need to be undertaken, which may take place concurrently:

- Assemble the development site. A large development may need the purchase of several adjoining sites or developing in partnership with adjacent owners.
- Formulate a design brief to set main criteria against which the development project can be evaluated in terms of time, cost, quality and performance.

- Seek legal and regulatory permissions and consents in order that the development may be allowed to proceed.
- Undertake site investigations to determine suitability of the sub-soil for foundations, possible contaminants, boundary and access issues, legal covenants and easements. The unexpected discovery of ancient archaeological remains while excavating foundations is a sure way to slow progress.
- Commission an environmental impact assessment to ensure that the likely effects of a new development on the environment are fully understood before the development proceeds. The assessment describes the nature of such effects on the environment and proposes mitigation measures to reduce the impact.

3.2.3.3   Design, costing and financing

The design will have been developed in outline, setting the overall scale and layout of the buildings and perhaps some of the materials. As time progresses, decisions will be made that firm up elements of the design. Other specialist consultants may become involved in the design of elements, such as the structure of the building, electrical and mechanical services, including heating, cooling, water supply, lifts, cladding systems and internal components, such as suspended ceilings and raised access floors. Some of the consultants used will be responsible for the design and construction. The extent to which a main contractor (builder) or sub-contractor, for example cladding manufacturer, might be responsible for the design will depend upon the procurement route used and the nature of the construction.

As the team grows, then so the coordination between them grows and the contractual relationships become more complex. Drawings and specifications will be produced to communicate these specialist designs, and these must all interrelate with each other. Traditionally, management and coordination have been undertaken by the architect, but a project manager may be appointed to perform this complex task. The estimated cost of the project needs to be monitored as the design progresses. The relative cost of separate elements of the construction becomes significant and an elemental cost plan is often prepared by the quantity surveyor. Estimated costs can be calculated by reference to other projects while making adjustments for the difference in quality or date when the project was undertaken. The RICS Building Cost Information Service (BCIS) provides this information. Documents will be prepared to enable the final design to be tendered for competitively by a number of contractors.

Development finance is an essential element of any project, and its availability can enable or restrict the opportunity for development. Short-term finance is often used to pay for the land, construction of buildings, professional fees and marketing: it is the developer's working capital. These loans of three to five years' duration are paid back upon completion of the development from the sale proceeds. So, money markets that deal in short-term loans are important. Loans with variable interest rates mean that if the rate increases, construction becomes expensive and reduces the number of viable projects. Also, the longer the development period, the more uncertain developers are about future costs and the riskier it is to predict them.

3.2.3.4   Construction

Often, the construction stage will follow 'traditional' procurement route (essentially the completion of a design by an architect before then submitting documents to contractors for competitive tendering). Alternative procurement routes bring in the technical and management expertise of main contractors and sub-contractors early in the design phase. Packages are increasingly used, such as: design and build, design, build and manage, and management contracting. Generally, it is accepted that quick, low cost and high-quality developments are not possible. Unlike other manufactured products, buildings:

- Take a long time to put together,
- Are not (normally) factory assembled,
- Involve many individuals and skills,
- Require substantial financing,
- Are often unique and
- Enjoy a relatively long life although they require periodic maintenance and repair or renewal to stave off premature obsolescence.

The construction stage can be a complex management task that requires considerable planning and control to coordinate the combined activities of a large workforce.

3.2.3.5   Occupation

Completion of construction sees the end of the project, but it is the start as far as the occupier is concerned. Ownership of the project may also change at this point. This stage may also involve the taking of a profit by the developer either through the sale of the building or as an income stream from occupier(s). Little development would ever happen if this could not be achieved.

Long-term finance may be obtained on completion and sale of a development to the occupier or to a financial institution. Long-term finance pays off the short-term development loan and provides a profit for the developer. The financial institutions of insurance companies and pension funds are the traditional providers of long-term finance and influence the disposal of development by purchasing or investing in the completed product. These institutions create the market for the product and, in so doing, provide developers with the long-term finance that is necessary for their operations. This may be through a direct purchase of a completed development scheme, which the institutions then go on to let and collect the proceeds over a long period, or through the lending of money to prospective occupiers to enable them to purchase schemes. These agreements may well have been made at the outset of the development process.

The longest period in the process now begins. The normal expectation would be for a development to last for perhaps 80 years, although this may be much shorter. During this period, essential maintenance, repairs and replacements will be necessary to avoid premature failure or obsolescence. The predicted life of the building, and the expenditure needed for maintenance, will depend greatly on the decisions that were made at the feasibility and design stages. Less-durable materials are often specified to reduce the initial (capital) costs of the scheme.

The requirements of the occupier will change over the life of a development resulting in the alteration, extension or conversion of the building. A decision to sell and find or procure new premises may be an alternative. The property may be bought and sold several times over its life and pass into a number of different uses.

3.2.3.6 External influences on development

Even though total land supply is fixed, it is possible to alter the supply of land for different uses, perhaps converting agricultural land to residential for example. It is possible, and many would argue desirable to re-use derelict, vacant and under-utilised urban land. Also, developing land at higher density increases the supply of usable space per unit of land.

Social trends lead to changing requirements for such items as living space, personal mobility and possibly a changing employment structure. All of these create demand for land for houses, infrastructure and new forms of industry and commerce. Development plans, produced by local planning authorities, generally try to reflect these trends. For example, plans need to allocate land for housing, both to cope with population growth and to satisfy changing aspirations in household formation. It is at the development plan preparation stage that land (use) allocations are made, based upon perceived needs, which in turn are linked to population forecasts and calculations about the growth (or otherwise) of industry and commerce in an area. The allocations themselves will be discussed through formal procedures of consultation. This is a crucial stage because it will determine the amount of land (and the sites themselves) available for the respective uses over the next 10 or 20 years. Future development control decisions will be based upon the allocations contained in the plans.

Economic trends also have a more immediate impact on the development process in the shorter term through government monetary and fiscal policy, and the financial institutions. This aspect embraces the process as a whole in setting the macroeconomic environment in which it operates. The state, via its central bank, has a major role to play, for example by raising interest rates to bring inflation down or lowering interest rates to encourage growth. In times of low economic growth, with perhaps high unemployment, low business confidence and falling house prices, few individuals or firms will be contemplating major investments, such as house moves, factory construction and office renting. In times of high economic growth, with low unemployment, rising house prices and a general feeling of hope for the future, many more investors, individuals and businesses will be contemplating major new schemes and there will be a healthy market for property developments. More schemes create more work for the construction industry, and the 'boom' feeds on itself.

It is commonly believed that the property development industry's fortunes lag behind those of the economy in general. This is because development projects take a long time to complete. Thus, for example, when the economy goes into a slump, sufficient development projects are in existence from the boom times to shield the property development industry from the general economic conditions. The slump does not hit the development industry until these projects are completed (and no replacements come forward). Consequently, when the economy begins to improve, there will be a time lag before investors have the confidence again to commission major development projects. Thus, the recovery comes later to the development industry. Governments are generally very interested in the fortunes of the construction industry. When it begins to recover, they can conclude that general economic recovery is well under way.

Other factors setting the context for the development process are those relating to the construction industry. The main characteristics of the construction industry are:

- Preponderance of small firms plus some very large international companies,
- Labour-intensive and under-mechanised: A high proportion of casual labour that may be hired (and released) on a project-by-project basis and
- Very sensitive to changes in economic conditions.

These characteristics may be seen as products of the conditions of supply and demand. On the supply side, construction is an assembly industry in which operations are largely consecutive. Any break in the chain of events is usually detrimental to the eventual output. For example, the walls of a building cannot be built before the foundations are complete, the windows cannot be installed before the walls are erected. A hold-up in the supply of bricks, or cement, may mean that the entire development project grinds to a halt. During periods where construction activity is high, shortages do occur and lead times grow (lead time is the time between ordering materials and their arrival on site). Construction involves a great deal of outdoor work, so the industry is vulnerable to weather conditions. There are certain conditions in which it cannot operate or at least not efficiently. Also, labour costs are a high proportion (1/3 to 1/2) of total costs. The unfortunate side of this, of course, is that a construction firm wishing to cut costs will look at its labour costs first and foremost. The industry is very susceptible to 'hiring and firing'.

On the demand side, the development is very dependent upon borrowing. Such is the value of its output that the money that underpins and finances the industry will normally have been borrowed from a financial institution. Thus, the industry is very sensitive to conditions affecting the cost and availability of that finance. A rise or a fall in interest rates, for example, will normally have a major effect on the construction industry. Also, demand is subject to seasonality and cyclical fluctuations. Over the long term, investment prospects and thus demand for the output of the construction industry will rise and fall with broad economic cycles.

The property development industry has a very important role to play in providing new and refurbished buildings that are sustainable. This can be achieved in three ways:

[1] Design and construction: the environmental impact of large and small building projects can be reduced by producing buildings that are constructed using sustainable techniques.
[2] Design and operational use: the long-term use of the building itself can be improved environmentally by:
   - Specifying sustainable materials,
   - Engaging with energy efficient designs, for example in heating systems, and

- Adopting sustainable operational policies, for example not using air conditioning unless absolutely necessary.
[3] Location: Ensuring that buildings are located on brownfield land and in places with sustainable transport links.

3.3   Property markets interaction

Decisions on the development and occupation of, and investment in, property require an assessment of current and future macroeconomic conditions and an understanding of the related markets. For example, if interest rates rise sharply, consumer spending tends to decline and the demand for retail and manufacturing property reduces and, in some instances, may even become surplus to requirements. Property market activity responds to short- and long-term macroeconomic stimuli: the former is largely a function of availability of debt finance and the latter a function of changes in employment, population, income and shifts in consumer preferences. Consequently, property markets do not operate in isolation; they are influenced by, but tend to lag, movements in the economy as a whole and in the financial markets in particular.

It is important therefore that valuers monitor key macroeconomic indicators and understand how their movements may influence the supply of and demand for different types of property in different locations. Knowing this will facilitate more informed judgements about rental and capital values, rental growth, investment and occupier demand, and development activity. Key macroeconomic indicators include gross domestic product, trade deficit, tax-to-GDP ratio, inflation, employment and unemployment figures, oil prices, house prices, household debt and debt as a percentage of income. A key money market indicator is the price of money or interest rate, which is influenced by supply and demand and set by national banks. The interest rate is very important to the property market as most investment and development activity is a combination of debt and equity finance, typically a large amount of the former and a small amount of the latter. The cost and availability of equity and debt finance influence demand for and supply of property. The interest rate is also a component of yields and discount rates used in valuation and so directly affects property values (Appraisal Institute 2001). It is also important to monitor government policy not just in relation to planning and development control but also legislation and other statutory controls regarding the environment, workplace, landlord and tenant relationship, licencing and so on.

The property market is really a set of interrelated sub-markets, and, like most markets, they are rarely in equilibrium themselves, let alone with one another. Because of longevity and fixed location of property, its high unit price and the terms of lease contracts, markets take time to adjust (Ball et al. 1998), so market prices tend to lag changes in buying and selling pressure, a feature of an imperfect market. This means that the property markets, like the economy, are prone to cyclical fluctuations, displaying successive periods of expansion, decline, recession and recovery. Figure 3.1 illustrates how this cycle operates.

The position of property in its cycle is determined by supply and demand in the occupier market (measured by stock availability, rental value, rental growth) and

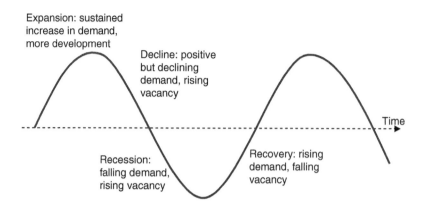

Figure 3.1   Property market cycle. Source: Modified from Appraisal Institute (2001).

supply and demand in the investor market (measured by yields and capital values). So, according to the Appraisal Institute (2001), trends in the property markets can be observed by measuring vacancy rates, rental growth rates, yields and changes in supply but remembering that the property market is slow to react to new information. For example, the vacancy rate may begin to rise and rental growth to stagnate, but new buildings will still be constructed in the short to medium term and landlords tend to be very reluctant to reduce rents unless they must.

Property market movements can be identified by monitoring changes in key indicators of property market activity, such as investment returns. This gives an indication of the way in which the sub-markets interact during these fluctuations – the leads and lags. In the short term of three years or so, the supply of property is relatively inelastic, so disequilibrium can characterise the market when demand increases or decreases. For there to be equilibrium in the overall property market, all sub-markets must be in equilibrium simultaneously, but markets are continually trying to adjust to new supply and demand conditions and with inherent lags they are unlikely to be in equilibrium at any one time.

This continuing movement between property markets is illustrated by DiPasquale and Wheaton's (1992) four-quadrant model, shown in Figure 3.2. The figure depicts two interlinked markets: a space (user) market and an asset (investment) market. Starting with quadrant A in the top right corner, a reduction in demand for space reduces the rent paid for any given quantity. This transfers a reduction in the value of that space in the asset market (quadrant B, top left). In turn, this translates to a reduction in the price of new space in the asset market (quadrant C, bottom left) and a consequent reduction in the amount of new space that is constructed (quadrant D, bottom right). In this way, shifts in the market for existing space signal changes in the market for new space (the development market).

Fuerst and Grandy (2012) found that, for the central London office market, developer decisions are explained largely by current and historic local market conditions and suggested that this is due to the long lead-in period associated with new development. The lag in construction activity can lead to over-supply and raise vacancy rates in times of reduced market activity. This, in turn, causes a drop

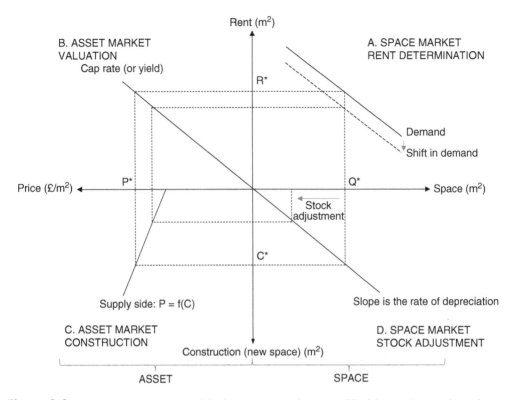

Figure 3.2 The four-quadrant model of property markets. Modified from DiPasquale and Wheaton, 1992.

in rents and an increase in yields until such time as demand increases to remove any surplus. However, development activity introduces only a small amount of new property each year in comparison to the size of the total stock and so tends not to significantly influence the property market as a whole. At the start of a market upturn supply lags the increase in demand, which causes the vacancy rate to drop and rents to rise and yields to fall. In the medium term, developers increase supply in response to rising demand. Building costs tend to follow general price levels over the long term but may vary in the short term and geographically. High building costs lead to increased demand for existing buildings and more refurbishment of existing buildings.

It is important for valuers to understand the position of the economy in its cycle because different types of valuation work might predominate at certain stages. For example, valuations in relation to foreclosures, bankruptcies and tax appeals might be more prevalent during a declining market or recession, valuations for lending purposes and in connection with investment and occupation market transactions would tend to dominate during a recovery or expansion phase, and at the peak of the market valuers may lead with consulting on investments as investors want to know when to buy and sell or redevelop their assets.

According to the Appraisal Institute (2001), although the general economic cycle influences the property cycle, it is typically not synchronised with it. The property cycle is the compounded result of cyclical influences from the wider economy, which are coupled with cyclical tendencies inherent to property markets. The critical linkage between property and economic cycles can be, in the main, captured in simple models, which are intuitively plausible and statistically sound (RICS 1999) and more sophisticated commercial rent-determination models are also possible (see McCartney 2012 for an example).

Key points

- The exchange of information between buyers and sellers about factors such as price, quality and quantity takes place in a market. Property is made up of a diverse range of market sectors and, relative to all other markets, they have distinguishing characteristics: the market is decentralised and restricted to fewer transactions than consumer goods or services, the product is heterogeneous, physically immobile, durable and of finite supply.
- Property exists to serve the needs of its users. It is a derived demand that can be classified by property type. A lot of property is not actually owned by occupiers themselves but by investors instead.
- Property investments tend to be of interest to a wide range of institutional investors seeking real income and capital growth. There is a broad range of opportunities to choose from, each comprising a different set of attributes.
- Property, as an investment medium, exhibits some of the characteristics of equities and bonds. The risks and returns associated with property and other investment assets continually shift in absolute and relative terms as economic conditions change, driven by the level of the interest rate and the opportunity cost of capital invested elsewhere (Ball et al. 1998).
- Developers play a key role in assembling sites and procuring the services of a professional team to bring forward property for investment and occupation.
- As in the general economic cycle, the property cycle consists of recurrent upswings and downswings, which vary in length, scale, and composition.

Note

1. (1723–1790) In 1776, Smith published 'Inquiry into the Nature and Causes of the Wealth of Nations', which helped create the academic discipline of economics.

References

Appraisal Institute (2001). *The Appraisal of Real Estate*, 12e. Chicago, USA: The Appraisal Institute.

Ball, M., Lizieri, C., and MacGregor, B. (1998). *The Economics of Commercial Property Markets*. London, UK: Routledge.

Bell, M., Bowman, J. and Clark, L. (2005). Valuing land for tax purposes in traditional tribal areas of South Africa where there is no land market, Lincoln Institute of Land Policy Working Paper.

DiPasquale, D. and Wheaton, W.C. (1992). The market for real estate assets and space: a conceptual framework. *Real Estate Econ.* 20 (2): 181–197.

FAO (1999). *FAO guide on forest valuation*. United Nations Food and Agriculture Organisation http://www.fao.org/docrep/008/v7395e/v7395e00.htm#Contents.

Fraser, W. (1993). *Principles of Property Investment and Pricing*, 2e. Basingstoke, Hampshire, UK: Palgrave Macmillan.

Fuerst, F. and Grandy, A. (2012). Rational expectations? Developer behaviour and development cycles in the Central London office market. *J. Prop. Invest. Finance* 30 (2): 159–174.

International Mineral Valuation Committee (2018). *International Mineral Property Valuation Standards Template*, 3e. International Mineral Valuation Committee.

McCartney, J. (2012). Short and long-run rent adjustment in the Dublin office market. *J. Prop. Res.* 29 (3): 201–226.

RICS (1999). *The UK Property Cycle - a History from 1921 to 1997*. Royal Institution of Chartered Surveyors and the Investment Property Databank.

RICS (2010a). *Valuation of Trees for Amenity and Related Non-Timber Uses. RICS Guidance Note*, 1e. Royal Institution of Chartered Surveyors.

RICS (2010b). *Valuation of Woodlands. RICS Guidance Note*, 1e. Royal Institution of Chartered Surveyors.

RICS (2011a). *Valuation of Rural Property. Guidance Notes*. Royal Institution of Chartered Surveyors.

RICS (2011b). *Valuation of Water as a Separate Resource. RICS Information Paper*, 1e. Royal Institution of Chartered Surveyors.

RICS (2011c). *Mineral-Bearing Land and Waste Management Sites, RICS Guidance Note*, 1e, GN84/2011. Royal Institution of Chartered Surveyors.

RICS (2012). *Valuation of Individual New Build Homes, RICS Professional Guidance UK*, 2e, GN52/2012. Royal Institution of Chartered Surveyors.

RICS (2018). *Valuing Residential Property Purpose Built for Renting. Guidance Note*, 1e. Royal Institution of Chartered Surveyors.

Sayce, S., Smith, J., Cooper, R., and Venmore-Rowland, P. (2006). *Real Estate Appraisal: From Value to Worth*. Oxford, UK: Blackwell Publishers.

Section A

Chapter 4
Valuation Mathematics

4.1   Introduction

Property is usually demanded not as an end in itself but as a means to an end – as a factor of production or as an investment asset – it is a derived demand and the opportunity cost of capital invested in property must be measured against other factors of production for occupiers and other investment asset types for investors. Valuers rely on this feature of property demand when attempting to quantify financially the opportunity cost of owning or leasing property. Economists (and valuers) use financial mathematics when measuring the opportunity cost of capital spent on property, and this is necessary because property usually requires large amounts of money to be invested over periods lasting several years, so the 'time value of money' should be factored into calculations. This time value of money is an expression used to refer to the fact that, although in nominal terms £1000 tucked under the mattress today will be £1000 in say 10 years' time, in real terms it will be worth less because inflation will have partially eroded its real (purchasing) value. This means that the further into the future an amount of money (rent for example) is received, the less it is worth in today's terms.

This chapter begins by introducing formulae for calculating investment value that take into account the time value of money. It then describes simple ratios of the price paid to the financial return expected from a property acquisition. The focus is on acquisition as an investment, but the theory is applicable to acquisitions for occupation or for development, where the investor's target rate of return is replaced by the opportunity cost of capital and the developer's target rate of return, respectively.

Property Valuation, Third Edition. Peter Wyatt.
© 2023 John Wiley & Sons Ltd. Published 2023 by John Wiley & Sons Ltd.
Companion website: www.wiley.com/go/wyatt/propertyvaluation3e

4.2   The time value of money

In order to be able to estimate the economic value of a property, it is necessary to understand how future economic benefits, usually in the form of a cash flow, can be expressed in terms of present value (PV). After an initial expenditure on acquisition, a property cash flow typically takes the form of rental income.[1] This would be a real rent to an investor and an imputed rent to an owner-occupier.

The *time value* of money reflects the fact that a sum of money to be received in the future may not be worth the same as it is now but a sum less than that, the actual amount depending upon the rate at which money is deemed to decline in value over time. This rate is known as the *discount rate*.

Mathematical formulae measure the time value of regular income cash flows such as rent. These formulae are founded on the premise that rational purchasers of property, whether for ownership, investment or development, would prefer to have money now rather than later because money has an opportunity cost. In other words, money could be spent on alternatives that may deliver a better return. The real value of money may be eroded by the general rise in the cost of all goods and services (inflation) over time. Also, this time value is a function of property investment characteristics described in Chapter 1, namely loss of liquidity and costs associated with the management of the investment, inflation and risk.

The principles of compounding and discounting measure the value of money over time (forwards and backwards, respectively). By compounding it is possible to calculate the *future value* (FV) of income or expenditure and by discounting it is possible to calculate the *PV* of future income or expenditure. Compounding and discounting measure the value of money over time in opposite directions, shown in Figure 4.1.

When compounding, the size of the future sum is determined by a combination of the compound interest rate (the rate at which the original sum earns interest) and the deferment period (the delay in payment), illustrated in Figure 4.2.

Discounting is the opposite of compounding and is a basic concept underpinning valuation, shown in Figure 4.3.

Some notation will help present the formulae in a succinct and consistent form, shown in Table 4.1. The formulae assume, unless otherwise stated, that investment deposits are made at the start of each period and interest is accrued at the end of each period (in arrears).

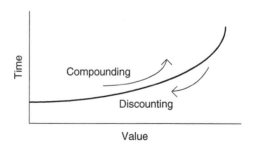

Figure 4.1   Compounding and discounting.

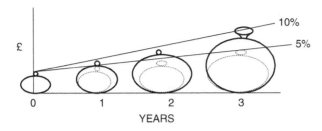

Figure 4.2 Compounding.

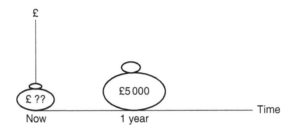

Figure 4.3 Discounting.

Table 4.1 Frequently used notation.

Variable	Description
PMT	Constant periodic payment
n	Number of periods over which the cash flow is estimated
r	Rate of return or discount rate per period
y	Yield

4.3   Single-sum investments

The *FV of a single sum* is the amount an investment will accumulate to at r rate of return after n periods. For example, if £1 is invested at the beginning of year one at r rate of return, the capital accrued at the end of the year will be $1 + r$. If £1 is invested for two years, the *FV* will be $(1 + r) + r(1 + r)$ or $(1 + r)^2$ and if it is to be invested for n periods:

$$(1+r)^n \tag{4.1}$$

If PV is the sum originally invested, rather than £1, the formula to calculate the FV is:

$$FV = PV(1+r)^n \tag{4.2}$$

For example, the roof of a factory will need replacing in four years' time as part of a rolling programme of maintenance. The current cost of the work is estimated to be £25 000. Building costs are forecast to increase at an average annual rate of 3.5% per annum over this period of time. The cost of the repair in four years' time will be:

$$FV = PV(1+r)^n = £25\,000(1+0.035)^4 = £28\,688$$

The *PV of a single sum* is the amount that needs to be invested now to accumulate to a specified *FV* by the end of *n* periods at *r* rate of return. If an amount of money is invested for *n* periods, earning *r* rate of return per period so that at the end of the investment period the investor receives £1 (which includes the original amount *PV* plus the accrued return), we can solve for *FV* using Eq. (4.2) as follows:

$$1 = PV(1+r)^n \tag{4.3}$$

So,

$$PV = \frac{1}{(1+r)^n} = (1+r)^{-n} \tag{4.4}$$

For any *FV* other than £1, the formula would be:

$$PV = FV(1+r)^{-n} \tag{4.5}$$

For example, if money can be invested in a secure investment and receive an annual return of 4% per annum, how much capital should be invested now to meet the estimated future expenditure calculated in the roof repair example above?

$$PV = FV(1+r)^{-n} = £28,688(1+0.04)^{-4} = £24,523$$

4.4   Multi-period investments

Property investments typically provide a regular or multi-period return or annuity.

4.4.1   Level annuities

The *FV of a level annuity* is the amount to which a series of identical payments invested at the end of each period will accumulate at *r* rate of interest after *n* periods. It is based on multiple deposits rather than a single deposit. The formula is derived by adding the single-sum *FVs* for each successive period, remembering that, when interest accrues in arrears, the last payment accrues no interest. Take a regular series of *n* £1 payments:

$$FV = (1+r)^{n-1} + (1+r)^{n-2} + \cdots + (1+r)^2 + (1+r) + 1 \tag{4.6}$$

This is an example of a geometric progression and we can use some of the recurring terms to simplify matters when calculating its sum. This is achieved by applying the general form of a geometric progression, $a, ar, ar^2, ar^3, ar^4, \ldots, ar^{n-1}$ where there are n terms, a is the first term and scale factor and r ($\neq 0$) is the common ratio. The sum of a geometric progression in its general form looks like this:

$$\sum_{i=0}^{n} ar^i = a + ar + ar^2 + ar^3 + ar^4 + \cdots + ar^{n-1} \tag{4.7}$$

If both sides of the above equation are multiplied by r:

$$r\sum_{i=0}^{n} ar^i = ar + ar^2 + ar^3 + ar^4 + ar^5 + \cdots + ar^n \tag{4.8}$$

and Eq. (4.8) is deducted from Eq. (4.7), we are left with the following since all the other terms cancel:

$$\sum_{i=0}^{n} ar^i - r\sum_{i=0}^{n} ar^i = a - ar^n \tag{4.9}$$

Rearranging Eq. (4.9), we get the following formula for the sum of a geometric progression:

$$\sum_{i=0}^{n} ar^i (1-r) = a(1-r^n), \tag{4.10}$$

which simplifies to:

$$\sum_{i=0}^{n} ar^i = \frac{a(r^n - 1)}{r - 1} \tag{4.11}$$

This equation for calculating the sum of a geometric progression can now be used to construct a formula for the FV of a level £1 annuity by inserting 1 as the first term and $(1 + r)$ as the common ratio:

$$FV = \frac{1(1+r)^n - 1}{(1+r) - 1} = \frac{(1+r)^n - 1}{r} \tag{4.12}$$

For any series of payments (PMT) other than £1, the FV for n periods is:

$$FV = PMT \left[\frac{(1+r)^n - 1}{r} \right] \tag{4.13}$$

For example, there are major repair works planned in eight years' time for the entire industrial estate that you hold in your investment portfolio. If you can invest money at a rate of return of 6.5% per annum, how much will accrue if you invest £50 000 at the end of each year for the next eight years?

$$FV = PMT \left[\frac{(1+r)^n - 1}{r} \right] = £50\,000 \left[\frac{(1+0.065)^8 - 1}{0.065} \right]$$
$$= £50\,000 \times 10.0769 = £503\,845$$

If we know what the future amount is going to be, the *FV of a level annuity* formula can be rearranged to calculate the *PMT*. When rearranged like this, the series of payments is known as a *sinking fund* (SF), being the *PMT* which must be invested at the end of each period, accumulating at r rate of return, to provide a known amount after n periods. So, if *PMT* must accumulate to £1, Eq. (4.13) is rearranged, substituting £1 as the amount to which the annuity must accrue:

$$1 = PMT\left[\frac{(1+r)^n - 1}{r}\right] \tag{4.14}$$

Rearranging this equation to isolate *PMT*:

$$PMT = \frac{r}{(1+r)^n - 1} \tag{4.15}$$

The formula for an *SF* is the reciprocal of the *FV of a level annuity* formula.

For example, rather than set aside a single capital amount now for the roof repair as we did in the *PV £1* example above, you decide to set aside equal annual instalments. What should these instalments be, assuming the repair will still cost £28 688 in four years' time and you can invest money at a rate of return of 4% per annum?

$$PMT = FV\left[\frac{r}{(1+r)^n - 1}\right] = £28\ 688\left[\frac{0.04}{(1+0.04)^4 - 1}\right]$$
$$= £28\ 688 \times 0.2355 = £6\ 756$$

In other words, £6756 should be invested at the start of each of the next four years to accrue £28 688 assuming an interest rate of 4% per annum paid annually in arrears. This can be checked using the *FV of £1 pa* formula to calculate the FV of £6756 invested in each of the next four years at 4% per annum. The answer should be £28 688.

The *PV of a level annuity* is the present value of the right to receive a series of payments at the end of each period for n periods at r rate of return. It is the addition of the single-sum *PVs* over n periods. So, the *PV of £1 per annum* is:

$$PV = (1+r)^{-1} + (1+r)^{-2} + (1+r)^{-3} + \cdots + (1+r)^{-n} \tag{4.16}$$

This is another geometric progression where the first term is $(1+r)^{-1}$ and the common ratio is also $(1+r)^{-1}$. So, substituting these terms into Eq. (4.11) we get:

$$PV = \frac{(1+r)^{-1}\left[1 - \left((1+r)^{-1}\right)^n\right]}{1 - (1+r)^{-1}} = \frac{(1+r)^{-1} - \left((1+r)^{-1}\right)^{n+1}}{1 - (1+r)^{-1}} = \frac{(1+r)^{-1} - \dfrac{1}{(1+r)^{n+1}}}{1 - (1+r)^{-1}}$$

$$PV = \frac{(1+r)^{-1} - \dfrac{1}{(1+r)^{n+1}}}{1 - (1+r)^{-1}} \tag{4.17}$$

If we multiply both sides of this equation by $(1 + r)$, it simplifies to[2]:

$$PV = \frac{1-(1+r)^{-n}}{r} \qquad (4.18)$$

For any series of payments other than £1[3]:

$$PV = PMT\left[\frac{1-(1+r)^{-n}}{r}\right] \qquad (4.19)$$

For example, how much would you pay for the right to receive £50 000 per annum over the next 15 years assuming average investment returns of 8% per annum?

$$PV = £50\,000\left[\frac{1-(1+0.08)^{-15}}{0.08}\right] = £50\,000 \times 8.5595 = £427\,975$$

The *PV of a level annuity* formula is used to calculate the present capital value of regular cash flows, which, of course, includes rent payments. If we replace the word 'calculate' with 'value' in the preceding sentence, the mathematical essence of valuation should now be apparent. The valuation of a finite (terminable) cash flow involves capitalising the net income at a suitable rate of return (or discount rate) r for the duration n that the income is received. In other words, the formula is used to convert a series of regular rent payments into a capital value.

Conventionally the *PV* of a level annuity formula is referred to as the *years purchase* by valuers, being the multiplier applied to the annual rent to calculate the capital value of a property. It is called the 'years purchase,' or YP for short, because the multiplier represents the number of years that will pass before the income equals the capital value, like a payback period but taking account of the time value of money, so a discounted payback period. In the example above, it will take approximately 8.56 years of receiving £50 000 per annum to recoup the original outlay of £427 975 at the prevailing interest rate of 8% per annum.

In property valuation, the discount rate is often referred to as the *cap rate* (short for capitalisation rate). This is because it is the rate that capitalises the rental income to arrive at a capital value.

Now consider an investment that provides a constant annual rent of £1 in arrears *in perpetuity*. If we assume a discount rate of 10% per annum, as the investment period goes beyond about 60 years, the value of this investment levels out to a fraction under £10. Mathematically, as n gets bigger the $(1 + r)^{-n}$ term in Eq. (4.19) gets smaller and smaller until eventually the equation simplifies to:

$$PV = \frac{PMT}{r} \qquad (4.20)$$

So, in terms of the mathematical accuracy typically required for property valuation, any stream of income receivable for 60 years or more may be regarded as receivable in perpetuity because the PV of income received after this time is negligible. This means that freehold (perpetual) and long leasehold (i.e. leases with 60 or more years to run) property interests can be valued to an acceptable degree of accuracy by dividing the income by the rate of return r. For example, a freehold shop investment that currently produces an annual rent of £80 000 per annum is for sale. If investors generally require a 5% return on investments of this sort, what is the capital value of this investment?

$$PV = \frac{80\,000}{0.05} = £1\,600\,000$$

When examining property investment transactions that have occurred in the marketplace, Eq. (4.20) can be rearranged to identify the market rate of return. Because the rate of return is now an output from the model rather than an input, it is given a different name. It is referred to as a yield, y (more of which is in Section 4.6). So, starting with Eq. (4.20) but changing the notation:

$$P = \frac{PMT}{y} \tag{4.21}$$

Then rearranging it so that the unknown y is on the left side and the two knowns, rent (PMT) and price paid (P), are on the right side:

$$y = \frac{PMT}{P} \tag{4.22}$$

These market yields, derived from market transactions, can be used to estimate a cap rate with which to value a property. For example, when valuing (i.e. calculating the present value or PV of) freehold properties where the $PMTs$ are a market-level rent assumed to be received in perpetuity, the market rent (MR) is divided by the yield y, as in Eq. (4.22), but substituting P for V (value).

$$V = \frac{MR}{y} \tag{4.23}$$

4.4.2   From a level annuity to a growth annuity

The rent from a property, like the income from many other types of investment, might be expected to grow over time and the PV of a level annuity formula can be adapted to incorporate this growth:

$$PV = \frac{PMT}{(1+r)} + \frac{PMT(1+g)}{(1+r)^2} + \frac{PMT(1+g)^2}{(1+r)^3} + \cdots + \frac{PMT(1+g)^{n-1}}{(1+r)^n} \tag{4.24}$$

$$= PMT\left[\frac{1-\left(\frac{(1+g)}{(1+r)}\right)^n}{r-g}\right]$$

where g is an annual growth rate. If the payments are receivable in perpetuity, Eq. (4.24) simplifies to:

$$PV = \frac{PMT}{r - g} \tag{4.25}$$

4.5 Timing of receipts

The formulae presented so far assume that the rent is received annually in arrears. If payments from a level annuity are receivable in advance (at the start of each period) for n periods, the first payment is received immediately so there is one less time period over which a payment is discounted, and the last payment is received after $n - 1$ periods. Therefore, the series of PVs that comprise the *PV of a level annuity* with payments received at the beginning of each period becomes:

$$PV_{adv} = PMT + \frac{PMT}{(1+r)} + \frac{PMT}{(1+r)^2} + \frac{PMT}{(1+r)^3} + \cdots + \frac{PMT}{(1+r)^{n-1}}$$

$$= PMT \left[\frac{1 - \dfrac{1}{(1+r)^{n-1}}}{r} + 1 \right] = PMT \left[\frac{1 - (1+r)^{-n}}{r} \right] (1+r)$$

$$\tag{4.26}$$

This is the same as Eq. (4.19) (the *PV of a level annuity*) but multiplied by $(1 + r)$. If payments are receivable in perpetuity, the formula can be simplified to:

$$PV_{adv} = \frac{PMT(1+r)}{r} = \frac{PMT}{r} + PMT \tag{4.27}$$

Most leases on commercial property in the United Kingdom require the tenant to pay rent in quarterly instalments at the beginning of each quarter, usually on 'quarter days' at the end of December, March, June, and September. Because the income is received sooner than if it was paid annually in arrears, these arrangements have a small but beneficial impact on the value of the investment. So, although rents are quoted as annual figures and used in valuations in this way, the actual return that an investor receives is enhanced by this payment method but not quite to the same extent as having all of the annual rent at the start of each year. To illustrate this, compare the PV of two investments that both yield a 6% annual return on an income of £10 000 for the next five years, but one pays this income annually in advance and the other annually in arrears. Using Eq. (4.19), with the income receivable annually in arrears, the PV is:

$$PV = PMT \left[\frac{1 - (1+r)^{-n}}{r} \right] = £10\,000 \left[\frac{1 - (1+0.06)^{-5}}{0.06} \right] = £42\,124$$

With the income receivable annually in advance, using Eq. (4.26), the PV is:

$$PV = PMT\left[\frac{1-(1+r)^{-n}}{r}\right](1+r) = £10\,000\left[\frac{1-(1+0.06)^{-5}}{0.06}\right](1+0.06) = £44\,651$$

If the income is paid in instalments of £2500 at the beginning of each quarter, the denominator r in Eq. (4.19) is altered to reflect these four receipts in each year as follows:

$$PV = PMT\left[\frac{1-(1+r)^{-n}}{4\left(1-(1+r)^{-0.25}\right)}\right] \tag{4.28}$$

Substituting:

$$PV = £10\,000\left[\frac{1-(1+0.06)^{-5}}{4\left(1-(1+0.06)^{-0.25}\right)}\right] = £43\,692$$

In the United Kingdom, residential leases usually require rent to be paid monthly in advance. Continuing the example above, if it is assumed that £833 is paid at the beginning of each month, the PV would be:

$$PV = £10\,000\left[\frac{1-(1+0.06)^{-5}}{12\left(1-(1+0.06)^{-0.083}\right)}\right] = £43\,480$$

4.6   Yields

Returning to Eq. (4.22), the yield (more precisely, the income yield) y is the ratio of periodic (usually annual) income to value or price. It represents the rate of return that an investment provides or yields. The income yield can be calculated at any time during the life on an investment and may be referred to as a *running yield*. The *initial yield* is a particular type of income yield and is the current rent divided by purchase price. It is a common market measure of investment performance because it represents the yield accepted by an investor at acquisition. The fact that initial yields from similar types of property investment are similar demonstrates that they typically sell for a certain multiplier of income. For example, if a property recently let at a MR of £100 000 per annum and the investment was purchased for £1 667 000 the initial yield is £100 000 divided by £1 667 000 or 6%.

In practice valuers usually work with the *net* initial yield, where the purchase price is adjusted to include purchase costs. These costs typically include transfer tax, brokerage and legal fees. For example, an investor pays £1 000 000 for a property that is let at an MR of £80 000 per annum.[4] Purchase costs are estimated to be 6.5% of the purchase price. So, whereas the gross initial yield is 80 000/1 000 000 or 8.00%, the net initial yield is calculated as follows:

Net initial yield = MR/(price × 1.065) = 80 000/(1 000 000 × 1.065) = 7.51%
To recap, the process for calculating a net initial yield is:

[1] Increase the purchase price to include purchase costs
[2] Divide the MR by the purchase price plus costs

This net initial yield can then be used to value comparable property investments. First, obtain (or estimate if vacant) the MR. Then use Eq. (4.22) where y represents the net initial yield. For example, a property investment is let at an MR of £70 000 per annum. To estimate its market value (MV):

$$MV(\text{including purchase costs}) = 70\ 000 / 0.0751 = £932\ 091$$

Then, to obtain the MV net of purchase costs:

$$MV(\text{excluding purchase costs}) = £932\ 091 / 1.065 = £875\ 203$$

We could get the same answer by dividing £70 000 by the gross initial yield of 8%, but valuers prefer to report evidence in the form of net initial yields. This is because purchase costs can vary over time, perhaps due to changes in transfer tax rates, so gross yields would be influenced by these changes to purchase costs, thus distorting valuations.

If there are two property investments, both producing the same income, but one sells for more than the other, the investment with the higher price has the lower initial yield. In other words, more attractive property investments offer lower yields. This might seem counterintuitive, but it occurs because investors bid up the price of the more attractive investment. In this way, yields reflect the attractiveness of investments and are used by valuers to compare one investment against another. Although, note that, because it is a simple ratio, a yield cannot distinguish between investments of different sizes.

The price paid for an investment is affected by many other characteristics of the investment in addition to current income level. These include future expectations of income and capital growth and perceived risk, which are, in turn, determined by a range of factors such as location, age, use, condition of the property, the financial standing of the tenant and so on. Attention would also be paid to the returns obtainable from other investments and, of these, government bonds often form an important reference point because they are expected to provide a 'risk-free' return.[5] As far as property investments are concerned, the initial yield is usually lower than the rate of return that will be obtained over the life of the investment because the purchaser is paying a sum that prices in expectations of future growth in the rent paid by a tenant and the capital value of the property.

4.7   Rates of return

Whereas a yield provides a measure of *ex-poste* return from an investment, a target *rate of return* signifies the *ex-ante* rate of return that an investor would like to see. Using this terminology, simple investment decision rules can be devised that compare investment yields with an investor's target rate of return, and, if the yield is above the target return, then an investment looks good.

A target rate of return depends on a range of factors and these, along with supply-side factors, determine the price that will be paid and the resultant net initial yield that will be obtained. Fisher (1930) argued that the total return expected from an investment may be made up of three economic variables. First, the prevailing market rate of interest, as this determines the cost of acquiring the capital to invest and sets a minimum level of return that could be obtained if funds were placed in a savings account – a measure of opportunity cost or loss of *liquidity*. Second, the anticipated rate of inflation. If inflation is expected to rise, then the target rate should increase to compensate. Third, a risk premium to compensate for the chance of incurring a financial loss and the uncertainty surrounding expected future benefits. Investors expect a reward for taking risk; the greater the perceived risk, the greater the return necessary to attract investment. Risk may be categorised as market risk or as property risk. Market risk refers to events that might affect the return on all property investments such as shifts in supply and demand, unexpected inflation, availability and cost of equity and debt finance, liquidity problems and returns available from other types of investment. Property risk might be added to reflect specific risks associated with the location, type of tenant, property use and condition and how this might impact on depreciation of capital and rental value, and management costs. The amount added to a target rate as risk premium will vary for each investor[6] and investment, and each type of risk can influence separately or in combination.

Obtaining rates to reflect these three components of total return allowed Fisher (1930) to construct an equation so that the nominal target rate of return r required by an investor may be expressed as:

$$r = (1+i)(1+\Delta)(1+RP) - 1 \tag{4.29}$$

where i is the prevailing interest rate, Δ is the rate of inflation, and RP is the risk premium.

Government bonds are regarded as a risk-free investment (except for the risk of *un*expected inflation), so investors expect a return that adequately compensates them in terms of opportunity cost of capital and expected inflation. Therefore, the rate of return that investors expect from government bonds provides a useful combined measure of i and Δ. The rate of return (or gross redemption yield) on short- and medium-dated government bonds is used as a benchmark risk-free rate on which to build target rates of return for other types of investment. Property investments tend to be held for periods of five to seven years so it may be more appropriate to consider using medium-term gilt yields. Mathematically the risk-free rate, RFR, required from government bonds may be expressed as:

$$RFR = (1+i)(1+\Delta) - 1 \tag{4.30}$$

So, the RFR can now be inserted into Eq. (4.29) as follows:

$$r = (1+RFR)(1+RP) - 1 \tag{4.31}$$

As Baum and Crosby (1995) note, an approximation of this is given by:

$$r = RFR + RP \tag{4.32}$$

Often, an investor's choice of target rate of return will be affected by the actual returns that have been achieved within a sector of the property investment market, central London offices for example. An important point to remember is that if the target rate is set too high, good investments will be rejected; if it is set too low, uneconomic investments will be accepted.

For investments where income is expected to grow over time, investors are prepared to accept a lower initial yield. Gordon (1962) argued that the initial yield y from an investment (a perpetual annuity in this case) can be related to the target rate of return r in terms of the annual growth g in net income that is anticipated:

$$y = r - g \tag{4.33}$$

Lease terms can distort this simple relationship. If rent increases at intervals greater than annually, a higher rate of g is required to reconcile y and r. If the growth pattern is uneven, perhaps the lease specifies rent reviews every five years, but the valuation takes place in between rent reviews – then the initial cash flow must be assessed separately until the even rental growth pattern resumes. This matter will be considered in more depth in Chapter 7.

So, combining Fisher and Gordon, i.e. Eqs. (4.32) and (4.33):

$$y = RFR + RP - g \tag{4.34}$$

Ball et al. (1998) extend this model to include an annual rate of property depreciation d:

$$y = RFR + RP - g + d \tag{4.35}$$

This provides a means of deriving y from the components of r. For example, if the RFR is 4%, RP is 3%, g is 2%, and d is 1%, then y will be 6%. Although the construction of y in this way can clarify its return components, it should not be considered as a replacement means of deriving a market y for use in valuation. Analysis of yields obtained from comparable investments is the best way to estimate a market y for a valuing a property investment. That said, the model can be helpful in determining whether the market is correctly pricing property investments. Rearranging Eq. (4.35):

$$RFR + RP = y + g - d \tag{4.36}$$

For example, an investor's target rate of return comprises a risk-free rate of 2% and a risk premium of 4%. An investment opportunity with an asking price of £5 m was recently let at £250 000 per annum. Annual rental growth net is expected to be 1.5% and an annual depreciation rate is estimated to be 0.5%. Should the investor purchase this investment?

$$2\% + 4\% = 5\% + 1.5\% - 0.5\%$$

Given that the investor's target return is the same as the initial yield plus growth and net of depreciation, the answer is yes. Of course, the future return is uncertain, but it is possible to examine past performance as a guide.

When dealing with inflation in investment appraisal, a distinction is made between real (or effective) rates of return and nominal rates of return. Income that does not take inflation into account is known as a nominal cash flow, whereas income that does take inflation into account is known as a real cash flow. A nominal rate of return does not reward an investor for bearing inflation whereas real rate does. For example, an index-linked government bond is an example of a risk-free real cash flow and should therefore be discounted at a real risk-free rate of return. Whereas a conventional government bond is a nominal cash flow and should be discounted at a nominal rate of return.

Finally, regarding rates of return, it is worth noting how annual rates can be converted to monthly or quarterly rates, and *vice versa*. To convert between an annual rate of return r_a and a monthly equivalent r_m:

$$r_a = \left(1+r_m\right)^{12} - 1 \quad \text{and} \quad r_m = \left(1+r_a\right)^{1/12} - 1$$

And between an annual rate r_a and a quarterly rate r_q:

$$r_a = \left(1+r_q\right)^{4} - 1 \quad \text{and} \quad r_q = \left(1+r_a\right)^{1/4} - 1$$

Key points

- Property ownership and occupation are often separate interests, and the capital amount paid for a property is therefore a function of its income-producing potential.
- Even when occupiers buy property for their own occupation, they often consider the opportunity cost of the capital and the financial return the asset may produce.
- The inverse of the yield is known as the years purchase; a multiplier used to compare different investments by stating how many years need to pass until the income received equals the capital value.
- Valuation is the estimation of the future financial benefits derived from the ownership expressed in terms of their PV.
- The size of the PV will depend upon the duration of the investment and the discount rate.
- Differences in PVs are more pronounced in the short term than in the long term.
- So, in valuation terms, rents receivable in the early years largely dictate the value of an interest, unless a substantial reversionary value is expected. For example, a landlord is able to regain possession after a long lease and perhaps redevelop the property.
- In choosing a yield, a valuer must be mindful of many factors, including type of property, status of the tenant, nature of the lease (particularly the rent review), yield on alternative investments, anticipated rental growth, when income is first received, and its frequency.
- The yield is a frequently quoted valuation metric derived from property transactions. It implies rather than reveals return and growth expectations.
- Yields are used in pricing models to estimate market value.

4.A   Appendix 4A

Alternative means of deriving PV:

$$PV = \frac{1}{\left(1+r\right)} + \frac{1}{\left(1+r\right)^2} + \frac{1}{\left(1+r\right)^3} + \cdots + \frac{1}{\left(1+r\right)^{n-1}} + \frac{1}{\left(1+r\right)^n} \tag{A.1}$$

Multiply both sides by $(1 + r)$:

$$PV(1+r) = 1 + \frac{(1+r)}{(1+r)} + \frac{(1+r)}{(1+r)^2} + \frac{(1+r)}{(1+r)^3} + \ldots + \frac{(1+r)}{(1+r)^{n-1}} + \frac{(1+r)}{(1+r)^n}$$

$$= 1 + \frac{1}{(1+r)} + \frac{1}{(1+r)^2} + \frac{1}{(1+r)^3} + \ldots + \frac{1}{(1+r)^{n-2}} + \frac{1}{(1+r)^{n-1}} \qquad \text{(A.2)}$$

Subtract Eq. (A.1) from Eq. (A.2)"

$$PV(1+r) - PV = 1 - \frac{1}{(1+r)^n}$$

$$PV + PVr - PV = 1 - \frac{1}{(1+r)^n}$$

$$PVr = 1 - \frac{1}{(1+r)^n}$$

So,

$$PV = \frac{1 - \dfrac{1}{(1+r)^n}}{r} = \frac{1 - (1+r)^{-n}}{r} \qquad \text{(A.3)}$$

4.B   Appendix 4B

Unlike a building society account or bond investment where the capital invested remains, the capital invested in an annuity is not paid back. Instead, the return from an annuity is partly a return *on* capital (at *r*) and partly a return *of* capital in the form of a sinking fund, which must recoup the capital originally invested by the end of *n* periods. To correctly calculate the PV of an annuity, the *PV* formula must include a sinking fund so that capital is recovered by the end of the investment period (the return of capital) while, at the same time, a return on capital is maintained at *r*. The formula therefore comprises these two parts, *r* and *SF*:

$$PV = PMT \left[\frac{1}{r + SF} \right] = PMT \left[\frac{1}{r + \left(\dfrac{r}{(1+r)^{n-1}} \right)} \right] \qquad \text{(B.1)}$$

So, there are two formulae for calculating the *PV of a level annuity* (Eqs. 4.19 and B.1). For example, what is the PV of an investment that offers an annual income of £10 000 over the next four years at a return of 5% per annum? Using Eq. (4.19):

$$PV = \pounds 10\,000 \times \frac{1-(1+0.05)^{-4}}{0.05} = \pounds 10\,000 \times 3.5460 = \pounds 35\,460$$

Using Eq. (B.1):

$$PV = \pounds 10\,000 \times \frac{1}{0.05 + \left[\dfrac{0.05}{(1+0.05)^4 - 1}\right]} = \pounds 35\,460$$

Table 4.2 shows the returns on and of capital broken down year by- year.

Table 4.2 Return on and of capital.

Year	Capital outstanding	Income	Return on capital	Return of capital (sinking fund)
1	35460	10000	1773	8227
2	27233	10000	1362	8638
3	18595	10000	930	9070
4	9525	10000	476	9524
				35460

The income provides for a return on capital at the remunerative rate (5% per annum) and a return of capital at the accumulative rate (also at 5% per annum). The sinking fund invests income at the accumulative rate to recover the original capital outlay of £35460. Because the sinking fund is returning some of the capital at the end of each year the amount of capital outstanding reduces, causing the return on capital to reduce too, leading to more of the fixed income being available for return of capital, and so on. Because the accumulative and remunerative rates are the same, the annuity and PV formula in Eq. (4.19) is known as *single rate*; the sinking fund is, in effect, a hypothetical one. The other version (Eq. B.1) is known as 'dual rate' and is used when the remunerative rate r and the accumulative rate SF (or s for short) are different. Note that $r + s$ becomes r when the period over which income is received is really long because the annual amount that needs to be invested in a sinking fund becomes negligible as n gets bigger, so s tends to 0 and the formula simplifies.

Notes

1. Property cash flows can take other forms, capital profit from a completed development or capital payments such as premiums for example, but let us keep things simple at this stage and just think about rental income.
2. An alternative approach to deriving PV is shown in Appendix 4A.
3. See also Appendix 4B.
4. Given that rent might be paid quarterly or monthly in advance, it might seem a mistake to calculate a yield assuming income is received annually in arrears. But if all

valuers analyse comparable evidence in the same way, then the yield works as a unit of comparison, which is its primary function.

5. How these factors might be expressed mathematically in a rate of return is discussed in Section 4.7.

6. It is important to remember that a *market* value is being estimated so factors considered relevant to investors should be considered only if they reflect market sentiment. Specific investor requirements can be considered in an appraisal of worth rather than a market valuation.

References

Ball, M., Lizieri, C., and MacGregor, B. (1998). *The Economics of Commercial Property Markets*. London, UK: Routledge.

Baum, A. and Crosby, N. (1995). *Property Investment Appraisal*, 2e. London, UK: Routledge.

Fisher, I. (1930). *The Theory of Interest*. Philadelphia, PA: Porcupine Press.

Gordon, M. (1962). *The Investment Financing and Valuation of Corporations*. Homeward, Illinois: Richard D Irwin, Inc.

Questions

Single-Sum Payments

[1] *Future value of a single sum*

a)   A and B have been given £5000 each. B will receive the money immediately while A must wait for one year. Who has received the most favourable treatment?

b)   To treat A and B equally but still have a gap of one year between the endowments, how much should be paid to A, assuming money can earn interest at 10% per annum.

c)   If A's gift is payable in five years' time, how much should be paid to make both equal value?

d)   An investor bought an investment for £50 000 and sold it one year later for £55 000. What yield did the investment show?

e)   An investor purchased an antique for £3000 and sold it nine years later for £5400. What yield did the investment produce?

f)   Your tenant owes £10 000 rent but wants to postpone payment for one year. You are willing but only for a 15% return. How much will your tenant have to pay?

g)   A roof will need replacing in four years' time. The current cost is £25 000. If building costs are forecast to increase at 3.5% per annum, what will the cost be in four years?

h)   You can buy some land for £1 m. You think you will be able to sell it to a developer in five years for twice that amount. You think an investment with this much risk requires an annual rate of return of 20%. Should you buy the land?

i) A property was let three years ago on a 10-year lease with a rent review in year five. The current rent passing is £100 000 per annum and is forecast to grow at 3% per annum. What is the rent expected to be at the rent review?

j) How much will £100 be worth if it grows at 8% per annum for five years with interest payable annually?

k) How much will £100 be worth if it grows at 2% per quarter for five years?

[2] *Present value of a single sum*

a) Referring to Question 1a, suppose A's endowment was £5000 and was payable in one year. What sum should be paid as an endowment to B today to equalise the sums?

b) If money can be invested and receive an annual return of 4%, how much capital should be invested now to meet an estimated future expenditure calculated in Question 1g?

c) You are interested in acquiring a property you think will be worth £1 000 000 in five years' time. After taking advice, you revise this forecast to £900 000. How much does this reduce what you are willing to pay for the property today, assuming your required return is 15% per annum?

d) You manage a block of four flats. In five years, the central heating system will need upgrading at an estimated cost then of £15 000. You propose to set aside money now to make provision for this outlay. Assuming this money will earn interest at 4% per annum and all tenants will contribute an equal share, how much should each contribute?

e) If the PV of £12 000 in 4 years' time is £7000, what is the annual discount rate?

Multi-Sum Payments

[3] *Future value of a level annuity*
There are major repair works planned in eight years' time for an industrial estate in your investment portfolio. Assuming you can invest money at an average rate of return of 6.5% per annum, how much will accrue if you invest £500 000 at the end of each year for the next eight years?

[4] *Payments on a level annuity (sinking fund)*
Rather than set aside a capital amount now for the roof repair from Question 1g, you decide to set aside equal annual instalments. What should these instalments be, assuming you can invest money at a rate of return of 4% per annum?

[5] *Present value of a level annuity of fixed duration*

a) Your client proposes to purchase a property which will produce a rent of £5000 per annum for the next five years, receivable annually in arrears. Assuming a discount rate (yield) of 10%, what is the capital value of these rents?

b) Which of the following would you choose, assuming an investment return rate of 5% per annum: £100 000 in five years' time or £10 000 at the end of each of the next 10 years?
   How much would you pay for the right to receive £50 000 per annum over the next 15 years assuming average investment returns of 8% per annum? What if the £50 000 per annum in c was receivable over the next 60 years? Or over 1000 years? What do you notice?

c) Your client receives a rent of £1000 per annum for the next three years, the first income being received in one year's time (in arrears). Assuming a discount rate of 8% per annum, what is the capital value of these incomes?

d) A property has just been let on a five-year lease at a rent of £40 000 per annum, payable annually in arrears. Using a discount rate of 5%, calculate the PV of the rent received during the lease.

[6] *Present value of a stepped annuity of fixed duration*
   a) A property has a current rent of £30 000 per annum, payable annually in arrears. After three years the rent is reviewed to a new rent, which will then be fixed for the following three years. The new rent will be based upon an anticipated growth of 3% per annum over the next three years. Assuming a yield of 5%, calculate the PV of the six years of rent.

   b) A property has just been let on a five-year lease at a rent of £40 000 per annum receivable annually in arrears. It is anticipated that £25 000 will need to be spent in three years' time to renew the roof. Assuming a yield of 7.5%, calculate the PV of the investment during this lease.

   c) A property is expected to produce a rent of £1000 per annum for three years followed by £1500 per annum for the following three years followed by £2000 per annum for the final three years of a nine-year lease. The rents are paid annually in arrears. Using a yield of 8%, value the rent from this lease.

   d) A property has just been let at a rent of £50 000 per annum. It is let on a nine-year lease with three-year rent reviews (i.e. at the end of year three and year six). Assuming that rents are increasing at an average rate of 4% per annum, calculate the PV of this rental income over the lease. Assume a yield of 6% and rents are paid annually in arrears.

[7] *Present value of a constant growth annuity*
   a) What is the PV of an eight-year lease that has an annual rent of £200 per square metre in arrears, increasing by 1% each year? The required return is 8%.

   b) A landlord has offered a tenant a six-year lease with an annual rent of £250 per square metre in arrears. The tenant has asked the landlord for a lower initial rent in return for annual increases of 2% per annum. Assuming a yield of 10%, what should the initial rent be?

[8] *Present value of a perpetual level annuity*
   a) What is the PV of the right to receive £100 000 per annum in arrears in perpetuity assuming a discount rate of 5%?

b) A freehold shop investment is for sale and currently produces an annual rent of £80000 per annum. Assuming a yield of 5%, what is the PV of this investment?

c) What is the PV of the right to receive £100000 in perpetuity annually in arrears starting in three years' time (so that the first payment is in four years' time) assuming a yield of 7.5%?

d) What is the PV of the right to receive £100000 annually in arrears for the next three years plus £150000 annually in arrears in perpetuity starting in three years' time assuming a yield of 10%?

[9] *Timing of receipts*

a) What is the PV of the right to receive £100000 per annum in advance in perpetuity assuming a discount rate of 8%?

b) What is the PV of the right to receive £25000 per quarter in advance in perpetuity assuming a discount rate of 8%?

c) Calculate the PV of the right to receive £100 every year for five years payable annually in advance assuming a discount rate of 8%

d) A property has just been let on a five-year lease at a rent of £40000 per annum, payable annually in advance. Using a yield of 5%, calculate the PV of the rents to be received during the lease.

e) How much will £100 be worth if it grows at 2% per quarter for five years with interest payable annually in arrears?

f) How much will £100 be worth if it grows at 2% per quarter for five years with interest payable quarterly in arrears?

[10] *Yields*

a) A property has just sold for £500000 and was recently let at £40000 per annum. What is the initial yield?

b) An investment pays a gross yield of 11.5% per annum. What annual income is produced from an initial capital deposit of £90000?

c) An investor wishes to receive an income of £7000 per annum from a property currently showing a yield of 5%. What capital sum must be invested?

d) What is the PV of a property producing an income of £300000 per annum assuming a yield of (i) 5% and (ii) 5.5%?

e) A property is about to be auctioned. It is currently let a rent of £120000 per annum. The bidding starts at £1000000 and proceeds in £100000 steps to the eventual sale figure of £1800000. Tabulate the initial yield at each bid and explain what you see.

[11] *Net initial yields*

a) Assuming purchase costs at 6.5%, what net initial yield is reflected in the sale of a property for £2116000 where the rent received is £100000 per annum?

b) A property has just been sold at a net initial yield of 7.5%. The rent received is £1058000 per annum. Assuming purchase costs of 6.5%, what was the price paid?

c) A property has just been let at a rent of £100000 per annum. Market transactions suggest that similar properties are selling at yields of 6.5%.

Assuming purchase costs of 6.5%, what is the market value of the property?

d) A property has recently let at a rent of £250000 per annum and has just sold to an investor for £3 m. Assuming purchase costs at 6.5%, what is the net initial yield?

[12] *Equivalent rates of return*

a) An investment offers a monthly return of 2%. What is the equivalent annual return?

b) An investment offers an annual return of 24%. What is the equivalent (i) monthly return and (ii) daily return?

c) If your borrowing rate is 8% per annum but your payments are quarterly, what is the quarterly rate that is used to calculate your payments?

d) If your borrowing rate is 12% per annum but your payments are monthly, what is the monthly rate that is used to calculate your payments?

e) (i) If a monthly rate is 1.5%, what is the equivalent annual rate? (ii) If an annual rate is 20%, what is the equivalent monthly rate? (iii) If a daily rate is 0.1%, what is the equivalent annual rate?

f) What is the monthly equivalent of an annual interest rate of 15%?

g) If the effective annual rate is 12%, what is the effective quarterly rate?

Nominal and Real Rates of Return

[13]

a) What is the value of the right to receive £100000 per annum for three years receivable annually in arrears and growing at a growth rate of 3% per annum? Assume a nominal discount rate of 10% per annum.

b) What is the value of the right to receive £100000 per annum for three years receivable annually in arrears assuming a discount rate of 6.80%?

c) What do you notice?

[14]

a) Assuming a discount rate of 10%, what is the PV of the right to receive £100000 per annum for five years, receivable annually in arrears, and growing at 4% per annum?

b) What is the value of the cash flow, but this time keep the £100000 annual income static and capitalise it at a discount rate of 5.7692%.

Answers

[1] *Future value of a single sum*

a) B (time value of money)

b) £5500

c) $£5000 \times (1+0.1)^5 = £8053$

d) $(£55000 - £50000)/£50000 = 10\%$

e) $£3000 \times (1+r)^9 = £5400$, so $r = 0.0675$ or 6.75%
f) $£10\,000 \times (1+0.15)^1 = £11\,500$
g) $£25\,000 \times (1+0.035)^4 = £28\,688$
h) No. $£1\,000\,000 \times (1+r)^5 = £2\,000\,000$. So, $r = (2\,000\,000/1\,000\,000)^{1/5} - 1 = 14.87\%$
i) $£100\,000 \times (1+0.03)^2 = £106\,090$
j) $£100 \times (1+0.08)^5 = £146.93$
k) $£100 \times (1.02)^{20} = £148.59$

[2] *Present value of a single sum*
a) $£5000 \times (1+0.1)^{-1} = £4545$
b) $£28\,688 \times (1+0.04)^{-4} = £24\,523$
c) $£1\,000\,000 \times (1+0.15)^{-5} = £497\,177$ and $£900\,000 \times (1+0.15)^{-5} = £447\,459$. $£497\,177 - £447\,459 = £49\,718$ so reduce to offer to $£950\,282$.
d) $£15\,000 \times (1+0.04)^{-5} = £12\,329$, so each tenant contributes $£3082$
e) $£12\,000 \times (1+r)^{-4} = £7000$ so $r = (7000/12000)^{-1/4} - 1 = 14.42\%$

[3] *Future value of a level annuity*
$£500\,000 \times [(1+0.065)^8 - 1] / 0.065 = £5\,038\,428$

[4] *Payments on a level annuity (sinking fund)*
$£28\,688 \times [0.04/(1+0.04)^4 - 1] = £6756$. This amount should be invested at the start of each of the next four years to accrue $£28\,688$ assuming an interest rate of 4% per annum paid annually in arrears. This can be checked using the *FV of a level annuity* formula to calculate the FV of $£6756$ invested in each of the next four years at 4% per annum. The answer should be $£28\,688$.

[5] *Present value of a level annuity of fixed duration*
a) $£5000 \times \{[1 - (1+0.10)^{-5}]/0.10\} = £18\,954$
b) PV $£100\,000 = £100\,000 \times (1+0.05)^{-5} = £78\,353$
PV $£10\,000$ p.a. $= £10\,000 \times \{[1 - (1+0.05)^{-10}]/0.05\} = £77\,217$
c) $£50\,000 \times \{[1 - (1+0.08)^{-15}]/0.08\} = £427\,974$
$£50\,000 \times \{[1 - (1+0.08)^{-60}]/0.08\} = £618\,828$
$£50\,000 \times \{[1 - (1+0.08)^{-1000}]/0.08\} = £625\,000$
As $n > 60$ years, the formula simplifies to $1/r$.
d) $£1000 \times \{[1 - (1+0.08)^{-3}]/0.08\} = £2577$
e) $£40\,000 \times \{[1 - (1+0.05)^{-5}]/0.05\} = £173\,179$

[6] *Present value of a stepped annuity of fixed duration – cash-flow model*
a) PV of first three years' rent: $£30\,000 \times \{[1-(1+0.05)^{-3}]/0.05\} = £81\,697$
New rent in three years' time $= £30\,000 \times (1.03)^3 = £32\,782$
PV of second three years' rent: $£32\,782 \times \{[1-(1+0.05)^{-3}]/0.05\} = £89\,274$
Defer PV of second three years' rent by three years: $£89\,274 \times (1+0.05)^{-3} = £77\,118$
So, the total PV is $£81\,697 + £77\,118 = £158\,815$
b) PV of rent: $£40\,000 \times \{[1 - (1+0.075)^{-5}]/0.075\} = £161\,835$
PV of repairs: $£25\,000 \times (1+0.075)^{-3} = £20\,124$
So, PV $= £161\,835 - £20\,124 = £141\,711$
There may be some debate about whether expenditure should be discounted at the yield.

c) PV of first three years: £1000 × {[1 − (1+0.08)$^{-3}$]/0.08} = £2577
PV of second three years, deferred three years: £1500 × {[1 − (1+0.08)$^{-3}$]/0.08} × (1+0.08)$^{-3}$ = £3069
PV of third three years, deferred six years: £2000 × {[1 − (1+0.08)$^{-3}$]/0.08} × (1+0.08)$^{-6}$ = £3248
So, total PV = £2577 + £3069 + £3248 = £8894.

d) PV of first three years: £50000 × {[1 − (1+0.06)$^{-3}$]/0.06} = £133651
PV of second three years, deferred three years: £50000 × (1+0.04)^3 × {[1 − (1+0.06)$^{-3}$]/0.06} × (1+0.06)$^{-3}$ = £126227
PV of third three years, deferred six years: £50000 × (1+0.04)^6 × {[1 − (1+0.06)$^{-3}$]/0.06} × (1+0.06)$^{-6}$ = £119216
So, total PV = £133651 + £126227 + £119216 = £379094.

[7] *Present value of a constant growth annuity*

a)
$$PV = £200 \times \left(\frac{1 - \left(\dfrac{1+0.01}{1+0.08}\right)^8}{0.10 - 0.02} \right) = £1{,}186$$

b) First, solve the level annuity PV: £250 × {[1 − (1+0.10)$^{-6}$]/0.10} = £1089. Then use this in the growth annuity formula and invert to solve for the initial rent:

$$\text{Rent} = £1{,}089 \times \left(\frac{0.10 - 0.02}{1 - \left(\dfrac{1+0.02}{1+0.10}\right)^6} \right) = £239$$

The tenant pays an initial rent of £239/m² instead of £250/m², but the landlord receives the same PV from the lease because the rent grows.

[8] *Present value of a perpetual level annuity*
a) PV = £100000/0.05 = £2000000
b) PV = £80000/0.05 = £1600000
c) PV = ((£100000/0.075) × (1+0.075)$^{-3}$ = £1073281
d) PV of £100000 per annum for first three years: PV = £100000 × {[1 − (1+0.10)$^{-3}$]/0.10} = £248685
PV of £150000 per annum in perpetuity starting in three years: (£150000/0.10) × (1+0.10)$^{-3}$ = £1126972
So, the total PV is £248685 + £1126972 = £1375657.

[9] *Timing of receipts*
a) PV = (£100000/0.08) + £100000 = £1350000
b) PV = (£25000/(1+0.08)$^{0.25}$ − 1) + £25000 = £1311899
c) PV = £100 × {[1 − (1+0.08)$^{-5}$]/0.08} × (1+0.08) = £431
d) PV = £40000 × {[1 − (1+0.05)$^{-5}$]/0.05} × (1+0.05) = £181838
e) FV = £100 × (1.08)^5 = £147
f) FV = £100 × (1+ 0.02)^{20} = £149

[10] *Yields*
a) Yield: £40000/£500000 = 8%
b) Income: £90000 × 0.115 = £10350 p.a.

c) Value: £7000/0.05 = £140 000
d) (i) £300 000/0.05 = £6 000 000, (ii) £300 000/0.055 = £5 454 545
e)

Bid (£)	Initial yield
1 000 000	12.00%
1 100 000	10.91%
1 200 000	10.00%
1 300 000	9.23%
1 400 000	8.57%
1 500 000	8.00%
1 600 000	7.50%
1 700 000	7.06%
1 800 000	6.67%

At each bid, the yield produced by the investment dropped until it reached a level that represented the lowest yield the market bidders were prepared to accept.

[11] Net initial yields
a) Net initial yield: £100 000 / (£2 116 000 × 1.065) = 4.44%
b) Price paid: (£1 058 000/0.075)/1.065 = £13 245 696
c) Market value: (£100 000/0.065)/1.065 = £1 444 565
d) Net initial yield: £250 000/(£3 000 000 × 1.065) = 7.82%

[12] *Equivalent rates of return*
a) EAR = $(1+0.02)^{12} - 1 = 0.2682$ or 26.82%
b) (i) $r_m = \sqrt[12]{(1+0.24)} - 1 = 0.01809$ or 1.81%,
(ii) $r_d = \sqrt[365]{(1+0.24)} - 1 = 0.0005895$ or 0.06%
c) $r_q = (1+0.08)^{0.25} - 1 = 1.94\%$
d) $r_m = (1+0.12)^{1/12} - 1 = 0.95\%$
e) (i) EAR = $(1+0.015)^{12} - 1 = 19.56\%$, (ii) $r_m = (1+0.20)^{1/12} - 1 = 1.53\%$,
(iii) EAR = $(1+0.001)^{365} - 1 = 44.03\%$
f) $r_m = (1+0.15)^{1/12} - 1 = 1.17\%$
g) $r_m = (1+0.12)^{0.25} - 1 = 2.87\%$

Nominal and real rates of return

[13]
a)

Year	Nominal cash flow (£)	PV £1 @ 10%	PV of cash flow (£)
1	103 000	0.9091	93 637
2	106 090	0.8264	87 673
3	109 273	0.7513	82 097
Value			263 407

b)

Year	Real cash flow (£)	PV £1 @ 6.80%	PV of cash flow (£)
1	100 000	0.9363	93 630
2	100 000	0.8767	87 670
3	100 000	0.8209	82 090
Value			263 390

c)   They have the same value, barring rounding. To value a cash flow where you expect growth, you can do two things: either create a nominal cash flow by applying the growth rate and discount at a nominal discount rate or discount the real cash flow at a real discount rate.

[14]

a)   Changing the cash flow

Year	Nominal cash flow (£)	PV £1 @ 10%	PV of cash flow (£)
1	104 000	0.9091	94 546
2	108 160	0.8264	89 383
3	112 486	0.7513	84 511
4	116 986	0.6830	79 901
5	121 665	0.6209	75 542
Value			423 883

b)   Changing the discount rate
Using the PV of £1 per annum formula:

$$£100\,000 \times \frac{\left(1-\left(1+0.57692\right)\right)^{-5}}{0.057692} = £423\,894$$

The values are the same. Future growth can be valued by changing the cash flow or by changing the discount rate. The 5.7692% is calculated using the following formula:

$$\frac{1+\text{target rate}}{1+\text{growth rate}} - 1$$

This is a growth-adjusted discount rate. Roughly, if you require a target rate of return of 10% and you expect growth to be 4%, you can discount the current income by about 6% (approximately the difference between them). The precise number is a function of timing of receipt of growth in income.

Chapter 5
Valuation Process and Governance

5.1 Valuation process

The term 'valuation' can mean two things, the act of preparing an estimate of value and the estimate of value itself. Valuation is, therefore, a term used to describe both a process and an outcome. As a process, valuation is based on an analysis of market information under certain assumptions, supported by experience and knowledge. The valuation process should be objective and impartial or 'client-neutral'. The aim is to provide a consistent approach, credible and consistent valuations, independence, objectivity and transparency, clear terms of engagement, a clear basis of value and clear reporting.

The process of valuation involves the following steps, which can be codified in valuation *standards* as a way of promoting consistency in approach:

- Confirm instruction and agree terms of engagement
- Inspect the property
- Gather and analyse comparable evidence
- Establish basis of value
- Make assumptions and special assumptions as appropriate
- Select the appropriate valuation method(s) and undertake the valuation
- Produce the valuation report

5.1.1 Confirm instruction and agree terms of engagement

5.1.1.1 Understanding and agreeing the valuation task

Valuations in the public sector – land and property taxation or expropriation for example – are usually in response to statutory requirements and detailed instructions

Property Valuation, Third Edition. Peter Wyatt.
© 2023 John Wiley & Sons Ltd. Published 2023 by John Wiley & Sons Ltd.
Companion website: www.wiley.com/go/wyatt/propertyvaluation3e

are often prescribed in laws and regulations. Private-sector valuations, on the other hand, are likely to be motivated by market need. The valuation process and output are not set by statute, so it is vital that clients understand what they are getting when they request a valuation and have redress for malpractice should the need arise.

Before confirming a valuation instruction, it is important to understand the precise nature of the land and property interests to be valued and the purpose of the valuation, summarised in Table 5.1. This helps determine the size of the task, the type of valuation required, and the method or methods likely to be used. It also ensures that the valuer is suitably qualified to do the job and helps detect any conflict of interest.

It is also important to determine which international and national valuation standards apply and whether the valuation will be affected by legislation or state regulations. For example, in some countries, only qualified registered valuers are legally permitted to undertake property tax valuations.

Tasks undertaken at this initial stage include the following:

- Establish the purpose and subject of the valuation and the date at which value is to be estimated.
- Agree 'terms of engagement'.
- Ensure the valuer individual/firm has appropriate knowledge and experience and is free of conflict.
- Agree appropriate basis/bases of value and valuation assumptions.

5.1.1.2   Terms of engagement

The terms of engagement should be confirmed before the valuation report is issued. If the valuation is one in which the public has an interest or upon which third parties may rely, the terms should disclose any previous involvement that the valuer may have had with either the property to be valued or the client commissioning the valuation. This reduces the potential for conflicts of interest. Cherry (2006) lists some of the more likely conflicts of interest that may arise:

- The valuer acts for both buyer and seller of a property in the same transaction
- Valuing for two or more parties competing for an opportunity
- Valuing for a lender where advice is being provided to the borrower
- Valuing a property previously valued for another client
- Valuing both parties' interests in a leasehold transaction

Table 5.1   Reasons for commissioning a property valuation.

- To provide price discovery proxies to support trading of property rights
- To inform decisions about whether to invest in or develop land
- To support lending decisions, insurance risk assessment and the reporting of fair value of land and property assets in financial statements
- In relation to taxation, it may be necessary to value very large numbers of land and property holdings on a regular basis
- Regarding expropriation of tenure rights, assessments of market and non-market value will be required on an individual basis
- For planning policy and development control decisions

- Valuing for a third party where the valuer's firm has substantial fee-earning relationships with the commissioning client
- Recurring valuations of the same asset unless there are controls to minimise the risk of self-review
- Requests for valuer to act as advocate and expert

Should a conflict arise, the valuer must decide whether to accept the instruction depending on the specific circumstances. If the instruction is accepted, the valuer should:

- Disclose to the client(s) the possibility and nature of the conflict, the circumstances surrounding it and any other relevant facts;
- Advise the client(s) in writing to seek independent advice on the conflict;
- Inform client(s) in writing that the valuer or valuer's firm is not prepared to accept the instruction unless either the client(s) request(s) the valuer to do so unconditionally or it is subject to specified conditions that the valuer has put in place as well as arrangements for handling the conflict, which the client has in writing approved as acceptable (Cherry 2006).

In addition, any assumptions, special assumptions, reservations, special instructions or departures, consent to or restrictions on publication and any limits or exclusion of liability to parties other than client should be noted. The fee basis and complaints handling procedure or reference thereto will also be set out. As a minimum, terms of engagement should identify the items set out in Table 5.2.

When received from the client, terms of engagement become the first element in the documentation and must be on file.

5.1.2   Inspect the property

A physical inspection usually precedes a valuation but is not mandatory. For example, mass valuation for taxation purposes might well dispense with physical inspections of each taxable unit. The extent of inspections and investigations would have been agreed in the terms of engagement and must be properly documented. The

Table 5.2   Minimum items for terms of engagement.

- The client
- The subject of the valuation, nature and extent of tenure rights, and type of asset to be valued
- Purpose, basis, date and currency of valuation, including details of any anticipated or actual marketing constraint
- Affiliation, experience and qualifications of the valuer
- Status of valuer (internal, external, independent), any managed conflicts of interest
- Source and nature of information relied upon, scope of information supplied by client, extent of inspections, investigations, assumptions and reservations
- Any consent to or restrictions on publication of the valuation report
- Any limits or exclusions of liability to parties other than client
- Confirmation that the valuation will be undertaken in accordance with standards
- Fee basis
- Availability of complaints handling procedure

inspection draws attention to the characteristics of the locality (including the availability of infrastructure communications and other facilities that affect value) and the physical nature of the property (including the dimensions and areas of land and buildings, age and construction of buildings, use[s] of land and buildings, a description of accommodation, installations, amenities, services, fixtures, fittings, improvements, any plant and machinery that would normally form an integral part of the building). The assessment of physical factors does not involve a structural survey but a record of the repair and condition of the premises including the decorative order, whether the property has been adequately maintained and any basic defects. Care is needed when valuing buildings of non-traditional construction.

Floor areas are calculated in accordance with the International Property Measurement Standards (IPMS) or nationally adopted measurement standards, and if drawings are supplied, they should be sample checked on site. Plant and machinery items that would normally be sold with the property are included in the valuation. Trade fixtures and fittings are normally excluded from a valuation unless the property is being valued as part of an operational entity.

It is acceptable to revalue a property without inspection so long as the client has confirmed that no material changes to the property or area have occurred, and the valuation should state this as an assumption. Market practice suggests an inspection every three years for investment properties, but this will vary (Cherry 2006). When valuing a development property, the valuation should reflect the stage of construction that has been reached.

The nature of the legal interest must also be ascertained including details of any leases or subleases, easements and other legal rights, restrictions on, say, use or further development and any improvements that may have been made to the premises by a tenant. If the property rights are registered, then searches of the land register will verify ownership, and if subordinate interests such as easements and rights of way are also registered with the local authority, then these can be checked too. Because of the complexity and diversity of property interests, apparently minor legal or physical details can have a significant effect on value, such as an overly restrictive user clause in the lease or non-compliance with a fire regulation. When there is a separation of ownership and occupation, via a lease for example, liability for running costs, liability for repair, subsidence, flooding, and/or other risks should be identified. Table 5.3 provides a typical inspection checklist.

Planning and environmental issues such as abnormal ground conditions, historic mining or quarrying, coastal erosion, flood risks, proximity of high-voltage electrical equipment, contamination (potentially hazardous or harmful substances in the land or buildings), hazardous materials (potentially harmful material that has not yet contaminated land or buildings) and deleterious materials (building materials that degrade with age, causing structural problems) must also be raised and are of paramount importance if the property is to be (re)developed, as are potential alternative uses.

5.1.3   Gather and analyse comparable evidence

It is important to identify any potential comparable evidence, noting rents and prices achieved together with physical, legal and locational attributes of the properties. Specifically, the information on comparable evidence should include the

Table 5.3   Typical inspection and information requirements.

Desk-top Information:
- Cadastral and ownership records, tenure, details of any leases[a], sub-leases, licences and any other legal documents
- Property address, floor plans, site plans, satellite imagery and aerial photography; topographic maps, soil maps, land-use maps
- Details of any sales and lettings; previous valuations, surveys and inspections
- Contact details and access arrangements
- Planning information, legality of use
- Land and property tax details
- Buildings insurance details

Inspection details:
- Date of inspection, arrival and departure times
- Occupied/unoccupied. If occupied, was occupier met at the subject property?
- Weather
- Limitations on inspection

Site:
- Size, topography, stability, flood risk, drainage, trees, buildings
- Abnormal ground conditions (coastal erosion, mines, quarries or other underground works, services, cables)
- Contamination, filled land, hazardous or deleterious materials
- Location in relation to utilities and services (including broadband)
- Boundaries (definition, responsibility, evidence of unauthorised access and encroachment, stability of adjacent buildings, light and support)
- Potential for alternative uses(s)

Exterior of building(s):
- Age, use and any previous use of building(s), protected building(s)
- Type of construction (converted or purpose built, floor loading for industrial/warehousing)
- Condition (rot, structural movement)
- Evidence of extension(s), refurbishment
- Services (water, gas, electricity, heating and drainage)
- Standard of maintenance

Interior of building(s):
- Number of floors, floor levels, floor areas and dimensions
- Layout of each floor, configuration and use of each room
- Use, fit-out specification, condition, lift (goods, passenger), escalator, ventilation
- Fire and security systems, disabled access

Communications:
- Roads (made/unmade), rights of way or easements
- Public transport
- Loading/access, including any weight/height restrictions, parking facilities

Locality:
- Adjacent land uses
- Character of locality, presence of protected areas (e.g. conservation area)
- High-voltage cables or substations, telecommunications masts

[a] Lease details (if applicable): rent, lease length, repair and insurance obligations, break options, rent review terms, alienation, sub-letting and assignment terms, obligations to refurbish, use clause.

address and nature of the property, and the details of the transaction. This would include the date of transaction, the quality and availability of pre-sale information, probity of bidding process, whether the transfer was open market, lease details where appropriate and transaction costs. Information on the wider market context is also important, including:

- Market size and type;
- Stakeholders: Occupiers, investors and dominant tenure pattern (state-owned land, number of parcels by type, number registered, etc.);
- Transaction activity (number of transactions per annum and as a percentage of stock);
- Type of transactions Capital or rental, land use;
- Prices: Rents and capital values;
- Development activity (new supply, take-up, removal of old stock);
- Vacancy levels;
- Lending activity (percentage of stock subject to a mortgage, loan-to-value ratios, finance rates) and
- Supply and demand: Economic and social influences, and competition.

Information should be as timely and as detailed as possible. This is easier stated than achieved, even in countries with the most developed property markets. Valuers should take reasonable steps to verify any information relied upon. It is helpful in this respect if there is readily accessible information about the provenance of secondary data including its source, date collected, frequency of update and so on. Client information that is not in the public domain and that is obtained while valuing a property must be treated confidentially.

The analysis of comparable evidence is a key part of the valuation process and should be recorded on file. This may include handwritten notes, notes of telephone calls, copies of sales particulars and so on. This information is not passed to the client as part of the valuation report but is retained by the valuer for future reference as required.

5.1.4   Establish basis of value

The precise definition of value on which a valuation is based is referred to as the *basis* of value. It is usually accompanied by a set of assumptions that refine the definition. To avoid ambiguity, international valuation standards (IVS) provide a conceptual framework and explain the meaning of valuation bases in precise terms. The standards ensure that consistent bases of value and valuation assumptions are applied. Other definitions of value may be prescribed legally at the national level. The basis on which value is estimated should be made clear in the valuation. IVS define several bases of value, and jurisdictions might define additional ones. The main ones are Market Value, Market Rent and Investment Value.

5.1.4.1   Market value

Market Value is the basis that corresponds to the concept of value in exchange and is the most widely adopted. It is defined as:

'the estimated amount for which a property should exchange on the date of valuation between a willing buyer and a willing seller in an arm's length transaction after property marketing wherein the parties had each acted knowledgeably, prudently and without compulsion'. (RICS 2019, p. 55).

The Red Book (RICS 2019), as the UK valuation standards are colloquially referred to, provides explanatory notes on the conceptual framework for this definition. In essence, Market Value is measured as the most probable price reasonably obtainable in the market reflecting the optimum use that is legally permissible and financially feasible.

The estimate must not include any element of *special value*, an amount that reflects attributes of the property that would only be of value to a special purchaser over and above value to the market in general. This might include, for example, price inflated or deflated by special circumstances such as unusual financing arrangements, *synergistic value* (which arises from the merger of two or more physical properties or two or more legal interests within the same property) or a special relationship between the parties to the transaction. Special Value might be reported, but it must be separate from the Market Value estimate.

Market Value can include value that might arise from expectations of changing circumstances surrounding the property such as development potential (even if there is no planning permission at the time of the valuation). The Market Value basis will normally be accompanied by assumptions to apply it in the correct context, such as vacant possession, subject to a lease, functioning as an operational entity or as a surplus property for removal.

5.1.4.2   Market rent

Because property values can be capital and rental, a definition of *market rent* is also published. It is:

'the estimated amount for which a property, or space within a property, should lease on the date of valuation between a willing lessor and a willing lessee on appropriate lease terms, in an arm's-length transaction, after proper marketing wherein the parties had each acted knowledgeably, prudently and without compulsion'. (RICS 2019, p. 56).

'Appropriate lease terms' should be stated in the valuation and usually cover repair liability, lease duration, rent review pattern and incentives. If the property is vacant, then lease terms will be notional but typical for the type of property being valued.

5.1.4.3   Investment value

Investment Value is defined as:

'the value of property to a particular owner, investor, or class of investors for identified investment or operational objectives'. (RICS 2019, p. 57).

It represents value to the owner or prospective owner. It reflects the benefits received from holding a property and, therefore, does not involve a presumed exchange.

In developed economies at least, Investment Value is usually measured in financial terms, representing the monetary worth of a property either as an investment or as part of an operational entity. It is an estimate of value-in-use or worth and usually involves consideration of specific circumstances of an investor or group of investors, such as a particular attitude to risk, a financing strategy or tax position. An investment valuation often involves forecasts of costs and values in a cash flow discounted at an investor's target rate of return. Specific circumstances may also relate to the way in which the property is to be managed if it is to be held as an investment (a small-scale niche investor may wish to manage the property much more actively than a large institutional investor) or the way in which the property might be used by an occupying business. Wider considerations relating to the investment portfolio of an investor or the property estate of an occupier may also be considered.

In a perfect market, where buyers and sellers have instant access to market information, their economic requirements are identical and properties are homogeneous. It could be assumed that market participants would arrive at similar decisions and thus investment values would converge on a market value. However, property markets are decentralised and imperfect, properties are heterogeneous, as are buyers and sellers, and there are many typological and geographically distinct sub-markets. Differences between the *investment value* of a property and its *market value* provide the motivation for buyers or sellers to enter the marketplace.

In some countries, it can be difficult to use market value as a basis because the trading of property rights is restricted in certain ways. For example, it may not be possible to sell tenure rights and so market value may be impossible to determine. Use rights may be restricted, so it is not possible to determine the optimum use. Such property rights could be valued on an Investment Value basis instead. This is in line with a recommendation from the UN that:

> 'Policies and laws related to valuation should strive to ensure that valuation systems take into account non-market values, such as social, cultural, religious, spiritual and environmental values where applicable'. (UN FAO 2017, p. 22).

Non-market values refer to the attributes of property that are seldom, if ever, transacted in a market context. As such, they fall within the scope of Investment Value rather than Market Value. However, they do not lend themselves to monetary measurement in the way that investment valuation techniques presume. Non-market value reflects social and environmental benefits associated with holding property rights. Acknowledgement of this non-market value is essential when assessing compensation for expropriated property rights because the affected party is not a willing seller and therefore market value (an estimate of value in exchange) does not fully reflect value-in-use.[1]

In many societies, holders of property rights are custodians of a scarce resource. Market value is only one – economic – approach to measuring the value of that resource, and because social, cultural and environmental values are not transacted in the way that economic values are, Market Value fails to measure these values. For example, if property rights are to be expropriated from an owner, the relevant legal framework may state that the amount to be offered should be based on Market Value. The owner, however, may have a notion of Investment Value that is more than Market Value. If this Investment Value encompasses social, cultural and environmental values, measurement in monetary terms is difficult, and if the values are of benefit to a wider community,[2] then these positive externalities will not be reflected.

The problem is illustrated by Small and Sheehan (2008). Some communities consider land, its ownership and transfer of ownership to be part of a spiritual or cultural matrix of rights, obligations and relationships. The land might be regarded as collectively owned by all members of the community – past, present and future; the living are mere custodians. In this situation, value is likely to represent something more than the present value of future economic benefits and so sale at that price would not only under-value the land but also disenfranchise future members of the community. Essentially, while it may be possible to estimate the economic rent that might be charged for a fixed duration of occupation of the community land (or part thereof), it is not possible to reliably estimate the capital value of its outright and permanent sale.

The Food and Agriculture Organization of the United Nations (UN FAO 2017) recognises that there are values for which and with which people may trade but that some values are not traded at all. These social, cultural, religious, spiritual and environmental values constitute an important part of a person or community's identity. The basis of Investment Value should be extended to encompass social, cultural and environmental values as well as economic value. Herein lies a difficult task, how to identify and quantify non-market values: a task made harder when property rights are customary and informal. For example, a valuer may be asked to estimate the 'value' of trees or other natural features that hold spiritual significance to a community. The features may possess market value in terms of their fruit-bearing or material-providing capacity but their non-market social, cultural, and environmental values to the local community may be much greater.

The difficulty in measuring non-market value is exacerbated by the time-value basis of investment valuation models, whereby benefits received in the future are valued less than equivalent benefits received now. Given that social, cultural, and environmental values can involve long time horizons, these valuation models fail to adequately account for non-market values. It is, essentially, an example of market failure. Many countries recognise this and implement environmental and land-use regulations as a means of attempting to account for non-market value.

5.1.5   Make assumptions and special assumptions as appropriate

In addition to agreeing the basis of value, it may also be necessary to agree certain *assumptions* to set the valuation in the appropriate context. Assumptions are suppositions taken to be true without the need for verification. They are made where it is reasonable for valuer to accept something as fact without specific investigation. Typical valuation assumptions are that the property is in good condition, services are operational, there are no deleterious materials, structural defects or hazardous materials present and statutory requirements relating to construction have been met. Regarding the site, it is usually assumed that it is capable of development or redevelopment with no unusual costs, that there are no archaeological remains and that there is no pollution, contamination or risk of flooding. It is normally assumed that the current use has planning consent and that no compulsory purchase powers are proposed.

Assumptions may need to be added to the basis when estimating the market value of certain types of property. Trade-related properties, for example, are designed or adapted for specific uses and they often transfer as part of an operational business. Consequently, such properties tend not to be valued separately from the business as a whole and include the value of personal property. Often a separate valuation of plant and equipment is required, particularly for industrial premises where such assets represent a significant component of the tangible assets of a company. Plant and equipment may be valued in its working place or for removal from the premises at the expense of the purchaser (RICS 2019, p. 94). If a property includes land that is mineral-bearing or is suitable for use as a waste-management facility, an assumption may be necessary to reflect the potential for such uses in the valuation.

Whereas assumptions might be presumed to be true at the date of the valuation, but have not been verified, special assumptions are presumptions that differ from those that exist at the valuation date or would not normally be made by a purchaser in the market. Examples include: that a development or refurbishment is finished when in fact it is still under way, that a property has been let on specified terms when it is actually vacant (and vice versa), that planning consent has been, or will be, granted for development, that there is a restricted period in which to sell the property, or that the exchange takes place between parties where one or more has a special interest and that additional value, or *synergistic value*, is created as a result of the merger of the interests. Also, where a property has been damaged, special assumptions may include treating the property as reinstated, as a cleared site with planning permission assumed for the existing use or refurbished/redeveloped for a use for which there is a prospect of obtaining planning permission. For example, when valuing a piece of development land for lending purposes, it may be practical to assume that the proposed development is complete. Specific lending criteria would then determine how and when money is released to fund the development.

Special assumptions must be clearly stated together with a note of the effect on value. Assumptions and special assumptions need to be considered and explained to client before proceeding. There is no such thing as a standard assumption, which can be implied without being stated.

5.1.6   Select valuation approach(es) and method(s) and undertake the valuation

Chapters 6–8 describe the valuation approaches, methods and techniques. Whichever method of valuation is employed, it should reflect the behaviour of participants and affected parties. The use of more than one method is widely recommended as it provides a cross-check. Calculations may be handwritten, with working notes retained on file, or they may be undertaken on a spreadsheet or other computer software, in which case a printout should be kept on file.

5.1.7   Produce a valuation report

The minimum content for a valuation report is set out in Table 5.4. Conditions might include the handling of taxation, expenses, transaction costs, goodwill, fixtures and fittings. When reporting the value of a portfolio of properties, if it is

Table 5.4   Minimum valuation report content.

- Identification of the client and any other intended user(s)
- Consent to or restrictions on use, distribution and publication
- Identification and status of valuer, disclosure of any previous involvement
- Purpose of the valuation
- Identification of the subject property, including type, use and property rights
- Source and nature of information relied upon, details of inspections (including dates) and investigations
- Basis/bases of value
- Assumptions and special assumptions
- Valuation approach and reasoning
- Valuation amount and currency (stated in figures and words)
- Signature and date (signatory must be an individual)
- Confirmation that valuation accords with relevant standards
- Basis on which the fee is calculated
- Conditions, reservations, special instructions and departures
- Limits or exclusions of liability
- Complaints handling procedure or reference thereto

Section A

suspected that the value of the portfolio is different from the sum of individual property values then this should be mentioned in the report. Also, negative values[3] must be reported separately. Any changes made to provisional valuations and draft reports should be recorded.

The output from the valuation process will be determined by the purpose of the valuation. For many public sector valuations – for taxation or expropriation purposes for example – a simple figure might suffice. Indeed, in the case of tax assessment, many thousands of valuations may be required and valuation figures, along with basic land and property details, are usually the extent of output.

In the private sector, valuations often form part of wider advice. A valuation for lending purposes may be accompanied by a risk assessment, and valuations for accounts may include a valuation of a property in its existing use and its potential development value. To this end, a valuation figure is usually incorporated into a report. Being the result of a contractual relationship between valuer and client, the content of a valuation report carries legal liability. Mistakes, errors and omissions can lead to negligence claims and, in some cases, the losses incurred can be substantial.

Given the above, and the importance of a valuation report to many financial and business decisions, many professional valuer associations set out the requirements for valuation reports so that there is consistency of output and potential for omitting items is minimised. Indeed, indemnifiers of valuers will usually insist on adherence to prescribed reporting requirements as a condition of insurance.

A valuation report can take several forms, prescribed formats from say mortgage lenders, internal memoranda or full written reports. Some reports are likely to be very detailed and others much briefer. A detailed report might include a schedule of comparable evidence and an accompanying analysis of that evidence. There may be a commentary on current market conditions, trading activity and future development proposals. Some clients require fully annotated valuation calculations; others might only be interested in the final valuation figure.

The nature and extent of these elements of a valuation report should be agreed before the valuation is undertaken. The calculations that underpin a valuation are likely to be in the form of handwritten notes or computer files. These should be retained but need not form part of the report nor shared with the client. White and Curtis (2018) recommend the following information be included in the valuation file:

- Engagement letter and acceptance of terms of engagement
- Records of key decisions and events
- Evidence of research undertaken, especially collation and analysis of comparable evidence
- Correspondence between valuer, client, third parties and any relevant internal correspondence
- Evidence of any internal peer review of the valuation

5.2   Valuation governance

Governance of valuation is essential and encompasses policy, legal and regulatory structure, regulation of valuers, regulation of valuations and education. Continuously monitored and regularly revised international standards govern valuer responsibilities and ethics. National standards can provide guidance on valuation methods.

Governance refers to all processes of governing whether undertaken by the State, a market, professional association, community or other social network. The processes might take the form of laws and regulations, codes of practice or accepted norms. The degree to which valuation is self-regulated (by a valuers' association for example) or statutorily regulated needs to be considered. It may be preferable to impose statutory regulation of valuers either in place of or alongside self-regulation. There are examples of both self-regulating professional associations of valuers and of statutory regulation of valuations and valuers around the world.

Governance takes the form of 'rules' by which valuers are expected to conform and regulates responsibility for incorrect valuations and resolving disputes. Usually, these rules are codified into valuation standards that address conduct and process. In terms of conduct, these standards ensure objectivity, transparency and accountability. They also establish minimum competency levels via education and training requirements, monitor conduct and impose disciplinary procedures as appropriate. In terms of process, standards set mandatory requirements and provide guidance and information. This helps achieve a more consistent and reliable interpretation of valuation principles and their application in practice and assists clients' understanding of what is being valued, assumptions made and limitations that apply.

Principles and objectives should be codified, and a common terminology developed to clearly communicate value and valuation concepts. Failure to meet the standards could lead to a negligence claim, and professional indemnity insurance is essential. In some countries, valuation standards exist but are not sufficiently enforced. Professional bodies and institutions may be under-resourced both financially and in terms of human resources. Due to poor enforcement, bogus and unqualified practitioners masquerading as valuers take advantage of this situation and this tarnishes the image and credibility of the profession. Also, corruption is very damaging to the valuation profession. It takes many forms including influence on valuations.

5.2.1   Standards of conduct

Conduct may seem a rather nebulous concept, and indeed it does cover a range of aspects of ethical behaviour, including (IVSC 2011; IESC 2016; RICS 2020):

- *Integrity*. Be honest and straightforward. Treat others with respect, be courteous, polite, considerate and never discriminate. For example, do not take advantage of a client, third party or anyone to whom a duty of care is owed; do not offer or accept gifts, hospitality or services, which might suggest an improper obligation; do not knowingly associate with a valuation that contains false or misleading information or that is made recklessly or omits or obscures information where it would be misleading.
- *Objectivity*. Base advice on relevant, valid and objective evidence. Disclose and then either avoid or manage conflicts of interest as appropriate. Be aware of potential causes of valuation bias including client influence. Valuers should be free from conflict of interest or identify and manage conflicts of interest when they arise. This might involve rotating valuers so that the same valuer does not repeatedly value the same property or value properties of the same client. It might also involve limiting the extent and duration of a business relationship between a valuer's firm and a particular client. This is particularly important when the fee-earning relationship is a significant one. Any conflict management must be disclosed in the valuation report.
- *Accountability*. Take responsibility for valuation services, and recognise and respect client, third party and stakeholder rights and interests.
- *Confidentiality*. Do not disclose confidential or proprietary information without permission unless required by law or regulations.
- *Competency*. Attain and maintain knowledge and skill required and act in accordance with technical and professional standards. Act within scope of competence and observe legal requirements. Qualifications are a way of ensuring competence, that valuers exercise due skill and have regard for the technical standards expected of them. To pass the qualification, a valuer will need to demonstrate a range of skills that are usually obtained through a mixture of academic education and practice-based training and experience.
- *Professionalism*. Avoid action that discredits the profession. Give due attention to social and environmental considerations. Avoid actions or situations that are inconsistent with professional obligations. Regularly reflect on standards and evaluate services to ensure consistency with evolving ethical principles and professional standards. Act in a way that promotes trust in the profession and recognise how practice and conduct bear upon public trust and confidence.
- *Service*. Be clear about service required and offered, communicate effectively, and deliver and pay for services within agreed timescales.

- *Accessibility*. Do not mislead, misinform or withhold information regarding products and terms of service. Present relevant documentary or other material in plain and intelligible language. Be transparent about fees and costs.
- *Responsive*. Respond to complaints appropriately and professionally; question things that do not seem right.

Threats to a valuer's ability to comply with appropriate standards of conduct can come from the valuer, from the organisation that employs a valuer or from clients. A valuer may be conflicted by a personal interest in a transaction to which the valuation relates, a firm may instruct one of its valuers to provide a loan-security valuation brokered by the same firm or a client may attempt to influence, pressurise, intimidate or even threaten a valuer to behave unethically. For example, a client may exert pressure on a valuer to arrive at a valuation figure that might not otherwise be the case. This kind of influence is not necessarily explicit. Baum (2000) identified examples of the ways in which valuers may be indirectly influenced:

- If a valuer's fee is contingent upon completion of a transaction, then there is an incentive to confirm the agreed transaction price.
- If a valuer's fee is calculated as a percentage of a transaction price, then there is an incentive to maximise this price.
- A valuer may be motivated to keep a particular client to improve career prospects or firm's success.
- When valuing assets in an investment fund, a valuer may be influenced to lower valuations by a new fund manager so that higher valuations subsequently reward his or her performance.

Client influence can come in different guises. There is the more obvious reward power and coercive power. Then there is the more difficult-to-detect expert power and information power. For example, when a valuer presents a draft valuation for review, the client may introduce new information about a property that hitherto was not available to the valuer. The primary factors affecting the degree to which clients influence valuations are the:

- Type of client,
- Characteristics of valuer and valuation firm,
- Purpose of the valuation and
- Information endowments of client and valuer.

It is important that valuers do not allow undue influence or bias to override professional judgements and obligations.

To ensure valuers conduct themselves in a professional manner, safeguards might take the following forms:

- Statutes and regulations at the jurisdiction level. There may be legislation that sets out corporate structures and governance, such as external auditing for example. More specifically, statutory licensing of valuers for certain valuations might be required, and these might set out minimum requirements for education, training and experience requirements.
- Rules of conduct at the professional level. Members of a valuation professional organisation will be required to comply with professional standards. This

would likely include monitoring of compliance and disciplinary procedures in the event of rule breaches.

- Working procedures and quality control at the firm level. This might involve separating parts of a firm's business using 'Chinese walls' to prevent conflicts of interest from arising. It might require staff to register any personal interests that may be material to the work undertaken by a firm, placing controls on the acceptance of gifts and hospitality. Assets that are required to be valued on a regular basis, perhaps for auditing purposes or for investment performance monitoring, may see a rotation of valuers to combat potential familiarity bias.

5.2.2   Valuation process standards

As well as conduct, there is a need to regulate the valuation process to ensure due diligence and due process, reduce the potential for professional negligence, maintain and enhance valuer education and training, and benchmark performance. Section 5.1 described the stages of the valuation process and demonstrated how many of these are regulated using international and national standards. These standards codify valuation principles and objectives and provide a common terminology and definitions to clearly communicate valuation concepts. The standards seek to ensure that valuers clearly explain what is to be done and what has been done. This helps the end-users understand what they are getting and helps the valuer whose reputation would suffer if users were dissatisfied with the service they receive.

Valuation, as a process, is a judgement based on facts, assumptions, experience and knowledge. It is a mixture of technical skill and professional judgement. Judgement, although professionally rooted, is a subjective act and is therefore susceptible to behavioural influences, including the use of heuristics or rules of thumb, reliance on intuition or 'gut feel', and anchoring to certain information such as a recent sale or an agreed price, also known as confirmation bias.

5.2.3   International valuation standards (IVS)

There is a desire to harmonise or standardise valuation practice around the world because of increasing globalisation and associated international trade and investment, international accounting standards and banking regulation. IVS, published by the International Valuation Standards Council or IVSC, set out core principles of valuation at an international level (IVSC 2022) Valuers should:

- Follow the ethical principles of integrity, objectivity, impartiality, confidentiality, competence and professionalism to promote and preserve public trust.
- Have the technical skills and knowledge required to appropriately complete the valuation assignment.
- Identify what is being valued, use appropriate information and data in a clear and transparent manner and use the appropriate valuation methodologies to develop a credible valuation.

Section A

- Communicate the analyses, opinions and conclusions of the valuation to the intended user(s), typically via a valuation report, in which:
  - Valuers report the valuation standards used and comply with those standards.
  - The basis (or bases) of value are defined or cited.
  - Valuers report the date of the valuation.
  - Valuers disclose significant assumptions and conditions related to the valuation.
  - Valuers report the intended use and user(s) of the valuation.
  - Valuers determine, perform and report an appropriate scope of work.
- Keep a copy of the valuation and a record of the valuation work performed for an appropriate period after completion of the assignment.

Paragraph 50.1 helpfully summarises the competence requirement of valuers:

'Valuations must be prepared by an individual or individuals within an entity, regardless of whether employed (internal) or engaged (contracted/external) possessing the necessary qualifications, ability and experience to execute a valuation in an objective, unbiased, ethical and competent manner and having the appropriate technical skills, experience and knowledge of the subject of the valuation, the market(s) in which it trades and the purpose of the valuation' (IVSC 2022, p. 11).

5.2.4   National valuation standards

National valuation standards fall into two main groups. First, technical or methods-driven standards, which ensure uniformity of practice, and, second, principles-based standards, which place more emphasis on skills and valuer judgement. National standards in developed economies have moved towards a principles-based approach, whereas in developing economies, a more detailed rule-based approach is followed, particularly where core valuation skills are not widely available. Regardless of the approach, the development of and adherence to national standards that align with international standards help avoid inconsistencies in definitions of value, valuation approaches and methods.

National standards can set out more detailed protocols, including legislative or regulatory requirements that must be accommodated within national standards. National valuation standards might contain information on the relevant codes and requirements that must be adhered to when undertaking certain valuations. These might include protocols agreed with certain client groups such as mortgage lenders, perhaps requiring more detailed risk assessment for example. Or they might stipulate protocols for valuations for accounting purposes. The more technical and specific nature of national valuation standards means that they are likely to need updating more regularly than the IVS.

It is important that valuation standards are consistently and rigorously enforced. To this end, some countries may operate a licensing or registration scheme or certification programme to ensure compliance as well as regulating the technical competencies. For example, the Royal Institution of Chartered Surveyors (RICS) is the valuation standards setter for the United Kingdom. The standards regulate process, promote the use of consistent bases and other definitions, provide a regulatory framework for valuation advice, require ethical and transparent approaches, and assist clients in understanding what is being valued, the assumptions made and the limitations that apply. Procedurally the standards establish the purpose and subject of the valuation and ensure the valuer has appropriate knowledge and experience, is free of conflict and has confirmed that appropriate terms and assumptions have been agreed with the client. They also set out the reporting requirements for the valuation.

The RICS valuation standards are contained within the 'Red Book', which, in fact, now comprises two publications: RICS Valuation – Global Standards (RICS 2019) and a UK National Supplement (RICS 2017). It might seem odd that a national standards setter has a set of global standards, but they apply to RICS valuers working outside of the United Kingdom. In the global standards, the highest level of regulation is mandatory and includes Professional Statements, which set out compliance procedures and ethical requirements, and Valuation Technical and Performance Standards, which are concerned with the valuation process and valuation reporting. The next part of the Red Book, Valuation Practice Guidance – Applications, is advisory and comprises guidance on valuations for specific applications, such as financial statements or secured lending, and guidance on valuing certain types of asset, such as businesses, plant and equipment, and personal property. Similarly, in the UK national supplement, the UK professional and valuation standards are mandatory and the UK valuation practice guidance on applications is advisory.

Valuations for certain purposes may be subject to additional standards. For example, the RICS stipulate additional standards for valuations that are to be included in financial statements and other regulated purposes such as incorporation or reference in stock market listing particulars and for takeovers and mergers, collective investment schemes, unregulated property unit trusts, financial statements of pension schemes and assets of insurance companies for purposes of calculating their margin of solvency. Valuations for commercial secured lending are undertaken in accordance with the protocol agreed between the RICS and the British Bankers Association, which requires detailed commentary on market trends and risks and extends the general rule on disclosing conflicts to disclosure of past involvement too.

The RICS also publishes a wide range of guidance notes and information papers that relate to valuation practice. The guidance notes provide valuers with recommendations or approaches for accepted good practice as followed by competent and conscientious practitioners. The information papers provide valuers with the latest technical information, knowledge and findings from regulatory reviews.

5.3 Valuation systems

It is important to place valuations into a context of land policy. By providing estimates of transaction prices in land and property markets, assessing the value of property rights used as loan security, underpinning tax assessments, helping to optimise the use of state land, resolving disputes and analysing the value implications of spatial planning and infrastructure development, valuations play a vital role in land-policy decisions.

The relationship between land-policy drivers and valuation is symbiotic. Active land and property markets and high levels of lending activity generate much-needed market-pricing information. The existence of a land registry, together with accurate, complete and accessible records of ownership of tenure rights, is of paramount importance: state guarantee of the accuracy of land registry records or the availability of title insurance help reduce risk. The existence, predictability and enforcement of land-use rules and zoning are especially important to valuation, as are the existence and enforcement of building codes and safety standards for buildings.

Similarly, business law, specifically legislation that relates to contracts, agency, and tort, will be highly influential on values and valuation practice. Taxation policy and related legislation are also relevant to valuation practice as they establish the framework and rules for tax assessment.

The degree to which these policies and laws influence valuation practice will depend on specific drafting. A precise definition of value may be stipulated, together with assumptions as to which factors must be disregarded or assumed, for example in the case of tax valuations or valuations for expropriation. Alternatively, definitions might be more loosely defined and the detail left for the courts to decide. It might be preferable for definitions to be left open to interpretation so that market participants can decide on precise terms that are pertinent at the time they are negotiated. Courts may intervene when clarification is required or in cases of disagreement.

Land and property markets – whether development, occupier, investment or finance markets – can be a major contributor to an economy. It is advisable, therefore, to have valuer representation at a senior level within government. Sometimes, the more technical requirements for valuation are divided among a finance ministry (often responsible for valuations associated with real-estate taxation), central bank (valuations for lending) and land-related ministry or agency (which may be responsible for valuations for expropriation, state land management and so on). Valuer representation at a policy level, which can oversee these areas, is recommended. This enables cross-cutting issues that affect values, such as a market downturn, currency fluctuations or political unrest, to be understood.

> 'States should ensure that appropriate systems are used for the fair and timely valuation of tenure rights for specific purposes, such as operation of markets, security for loans, transactions in tenure rights as a result of investments, expropriation and taxation' (UN FAO 2017, p. 77).

There are three key components of effective valuation systems (Figure 5.1): first, access to information on the nature and extent of the land and property to be valued, together with comparable evidence and information on the wider market; second, a sufficiently qualified and adequately resourced valuation profession; and third, robust governance of that profession.

Valuation systems and processes should be capable of managing complex structures of formal and informal property rights. A valuation system does not succeed in isolation; it needs functioning land and property markets, land and property information, industry support and political support from the State.

Equal and effective access to land and property information is an essential requirement for fair and open negotiations and transactions in relation to property rights. This information is therefore of great importance to valuers.

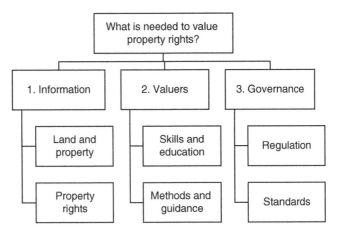

Figure 5.1   Key components of valuation systems.

In many countries, trading of property rights may be infrequent, mechanisms might not be in place to share transaction information, or it might be difficult to identify the rights that are being traded or the buyers and sellers. This has implications for holders of those rights, particularly if they are vulnerable members of society. Potential purchasers may have power or access to resources (skills and knowledge as well as information) that may not be available to holders of legitimate property rights.

These asymmetric trading positions can lead to the acquisition of property rights at prices that are not a true reflection of their market value, let alone their non-market value. If these acquisitions of property rights are compulsory – for expropriation purposes perhaps – then this combination of information paucity and power asymmetry can be contentious. Such transactions are therefore to be disregarded as evidence of market value as they do not sufficiently comply with the IVSC definition.

Disputes are particularly evident in locations undergoing change. In rapidly urbanising locations, on a city fringe perhaps, rural and urban markets interact and pressure to convert less-valuable farmland to more-valuable urban land intensifies. Without adequate access to market information by all parties, together with a valuation profession capable of providing objective interpretation of that information, it is possible for holders of agricultural property rights to be inadequately remunerated when their rights are acquired for development. In circumstances where extensive tracts of land are acquired, this problem is exacerbated.

It is only in transparent markets that rising values are readily observable. If market trading is not visible, it is neither possible to monitor price information nor estimate land value. As a result, transaction prices may not fully reflect the optimum use of the land.

5.3.1   Information systems

Many tasks related to the administration of land and property benefit enormously from the sharing of parcel-level information on ownership, use and price. If this information is kept up to date and accessible to all, then stakeholders are able make informed decisions about the future use and development of land and property.

The responsibility for the development and maintenance of a comprehensive land and property information system usually lies with the State. This is because the State often initiates the formation of nationwide systems for recording and registering ownership and price information in support of planning, taxation and other government duties.

Many countries have a land and property tax, and this motivates the creation of dedicated databases, sometimes referred to as fiscal cadastres, to record location or address information together with ownership, use and value details in relation to individual and communal tenure rights. Such databases can be useful for registration purposes, essentially forming the basis for a multipurpose cadastre or land and property information system. It is vital that, when maintaining a land and property tax base, information is shared between the valuation and taxation department and other land administration departments, including the land registry, planning, surveys and utilities. This ensures that new taxable units are added to the valuation roll and records any changes to those already on the list.

Tenure registration systems and other land and property information systems should be developed, maintained and adequately resourced. It is likely that coordination will be necessary when the different systems are developed and maintained in separate government departments. Common referencing and unique identifiers are essential in these circumstances, particularly with the growing move towards integrated land information systems.

In many countries, information on customary and informal property rights is not readily available and this makes markets in these rights riskier. Financial investors will place a higher risk-adjusted discount rate on potential revenue and impose more-stringent lending criteria, thus devaluing property rights. Potential buyers will need to spend more time and resources on due diligence, either trying to obtain the information they require to make a decision or arrange insurance cover for the risk associated with the transaction. Maintaining information on unregistered, customary and informal tenure rights is difficult but very helpful when valuing those rights.

A key task of a valuer is to obtain data on transaction prices and, together with details of property rights, land and property and wider contextual information, apply this to a valuation. To assist this task, information relating to all transactions involving transfers of property rights should be recorded and made publicly accessible. This should be the case for capital and rental transactions, for State-owned and private land, and for formal and informal markets. States should decide what 'land and property' information should be publicly accessible and what 'personal' information should remain secret. For example, the address, date and price of sale might be released but the name of the owner may not be. In some countries, the private sector might be motivated to collect land and property information if it sees a commercial advantage in doing so. For example, large firms may use the information to conduct market analysis, and smaller firms working in specific locations or market sectors can become valuable sources of specialised information. State valuers might have responsibility for land and property tax assessments and for transfer tax administration. Collecting and recording price information is an important part of the information base for these tasks.

A State may also deal with land directly – sales and leasing of State-owned land for example – and this is an important source of price information.

High transfer taxes are detrimental to market transparency because they create an incentive to under-declare prices to reduce tax liability. Low transaction taxes reduce this incentive and can therefore improve the reliability and accuracy of transaction data recorded by the State. Making transaction data accessible will improve the ability to detect errors and false declarations of price paid, and the imposition of a capital gains tax can encourage more accurate reporting of acquisition prices because it encourages the regular and truthful reporting of the value of land and property.

5.3.2   Valuation capacity

Effective land administration requires the establishment of strong institutional frameworks to support cadastral mapping, spatial planning, registration, valuation, taxation and billing, and collection functions. These rely upon highly skilled professionals, who in turn require an enabling environment of academic training, regulated professions and ethical conduct to deliver fair, equitable, transparent and sustainable services.

In many countries, inadequate budgetary resources, inefficient administration and lack of regulation have led to poor and even corrupt service delivery. In many developing countries and countries in transition, government agencies are typified by a lack of qualified staff and low levels of pay, while private-sector professionals often operate in poorly regulated environments. States are therefore encouraged to provide adequate, often significant, resources to ensure that capacity, including valuation capacity, is developed and maintained and to provide a suitable enabling environment for the development of professional associations.

Administration of valuations (developing valuation techniques, managing mass appraisals, constructing and maintaining databases) in the public sector might be undertaken by a central valuation agency. This has the advantage of offering a single service to those parts of government that require valuations for taxation, expropriation and state land leasing. The creation of a single valuation service should reduce duplication of resources, promote data sharing and benefit from economies of scale in terms of data handling. It should also encourage development of skills and knowledge in specialist areas of valuation such as specialised premises, infrastructure, utilities and plant and machinery.

The extent of a public sector valuation service provision will be a matter to consider in the light of a trade-off between the need for market data and analysis (where the private sector often performs more favourably) and legal, technical and administrative experience (at which the public sector often outperforms). For example, a government seeking to implement an *ad valorem* land and property tax may require thousands of valuations. This can be expensive to administer, particularly if the tax policy requires regular revaluations. State valuers may possess the requisite skills to undertake valuations for tax purposes but may lack knowledge of market conditions. The result can be a mechanistic or 'market unaware' approach to valuation and a tax base that is seldom updated to reflect extant market conditions.

Section A

A similar set of issues can arise when undertaking valuations for expropriation purposes. Without sufficient understanding of local market conditions, including demand for and supply of land for different uses and the development potential of land, compensation paid to holders of tenure rights may be based on cost replacement of existing use rather than fair value.

It may be that the public sector takes the role of facilitator, compiling land use, ownership and transaction data, and the private sector combines this with its market trading knowledge. A public–private partnership of this type can form the basis for the development of a valuation profession that is able to apply its skill set to market data and ensure that statutory valuations are grounded in market evidence.

In the private sector, there is growing demand for valuations from individuals, groups and corporations not only in response to statutory valuations (appealing tax assessments or contesting compensation for expropriated tenure rights for example) but also valuations for loan security and valuations of land and property (see Chapter 3). In many countries, there is insufficient capacity for these valuations to be undertaken using local resources. Instead, overseas valuers may undertake valuations, particularly in the case of land and property operated by large multinational businesses. It is important to be mindful of potential information and knowledge asymmetry when negotiating with local holders of tenure rights.

5.3.3   Professional valuers associations

The establishment of an association that represents valuers can be a key facilitator in ensuring enough valuers exist to meet demand. A valuers association can:

- Oversee education and skills development, including preparation and update of guidance on valuation processes and methods.
- Regulate the conduct of valuers, including ethical considerations and the provision of liability assurance for valuation advice through indemnity insurance.
- Provide an affordable and accessible means of dispute resolution.
- Finally, and perhaps most importantly, it will have a governance structure that includes a set of valuation standards to help ensure objectivity of valuations and integrity of valuers.

5.3.3.1   Education and skills development

An essential component of an effective valuation system is an educational faculty that can deliver appropriate training and education. It will have a distinct body of knowledge, centred on valuation methods. Experiences in some post-Soviet transition countries provide a note of caution here. Despite a strong need for market-based valuations to facilitate transition from command economies, valuation professions were created in a non-market environment and valuers were recruited from engineering and land-surveying professions. These groups often managed to 'capture' the valuation profession and overemphasise technical and legal aspects of valuation at the expense of economic and financial aspects. This was then reflected in education and training curricula and ultimately in valuation methodology.

Valuation curricula will vary according to the political, legal and economic environment in which valuers operate but there are some fundamental areas of knowledge. These are the economics of land and finance, and the law of property, contract, tort, and administration (specifically land-use planning and tax).

Usually, the suitability of a curriculum is monitored through a process of professional accreditation, so the concurrent development of an education system and professional association is common. Sometimes, a valuation profession develops in isolation from market trading and in such cases the profession can become overly technical. Professional representation and training that encompass both statutory and market valuations, together with strong links to agency, brokerage, and transactional responsibilities, can reduce this type of detachment.

Institutions with regulatory functions over the valuation profession should ensure adequate opportunity for its development. It is vital that educational and professional development is adequately resourced and maintained.

To build valuation capacity, it is important for potential valuers to be able to develop skills in property market issues and valuation techniques through professionally accredited training and education including degree courses, apprenticeships and continuing professional development. Although international valuation bodies can accredit education programmes, there should be sufficient local capacity to train and educate valuers.

Oversight of the content of training and education curricula is essential and is unlikely to be satisfactorily achieved by the State alone. A professional body is best placed to provide thought leadership in respect of educational requirements for entry into the profession, the skills and training that are required and how they can be provided.

5.3.3.2 Regulation of conduct

Valuer regulation should centre on the creation and adoption of professional codes of conduct and ethics. Governments should support such activity and encourage openness and transparency in the valuation process. The aim is to build trust in valuer activities by ensuring valuers act with independence, integrity and objectivity, have sufficient knowledge of the relevant tenure rights and land and property assets, and possess appropriate skills and experience to undertake the valuation competently. It also helps valuers withstand external pressures, client influence and the potential for corruption.

Quality assurance and procedural standards are also important. These protect users, so they understand what they are getting, and protect valuers, whose reputation might suffer if users are dissatisfied with their service. Therefore, procedural rules require valuers to explain what is to be done (terms of engagement or service agreement) and what has been done (valuation output). They also facilitate the appeal and review of valuations via tribunals and courts.

If confidence in valuations is to be assured then they should be transparent, coherent and consistent, and they should be undertaken by honest, impartial and competent valuers.

The dangers of ineffective valuer regulation:
- Increased levels of valuation variance, where valuers are unable to agree with one another to an acceptable degree
- Valuation inaccuracy, where valuations are unacceptably different from market prices
- Susceptibility to client influence, perhaps because of over-reliance on a single client's fee income for example
- Conflicts of interest, where the impartial representation of a stakeholder's interests may be open to question
- Valuation negligence and, most seriously,
- Fraud and corruption

Valuers producing valuations for financial reporting and lending purposes can be subjected to intense client pressure and they should work with regulatory authorities to combat this. In some countries, valuers may rely heavily on a small number of very large clients, state-owned enterprises for example. The ability to withstand client influence and pressure will need to be particularly robust in these circumstances.

If valuers cannot be held liable for their errors, omissions and other mistakes, then there is little incentive for them to avoid malpractice that might benefit them in some way. Some countries impose obligatory insurance on valuers, which engages insurance companies in assuring quality control on the profession. The existence and use of professional indemnity insurance are important.

5.3.3.3   Dispute resolution

Market and non-market values are socially created so dispute resolution should include representation from valuers who are well versed in the local context. They should also be transparent and impartial in their decision-making processes.

This is particularly so in cases of valuation for expropriation, as this can involve parties that are very different in terms of size, power and access to resources. It must be emphasised that the IVSC definition of market value is the basis on which value is determined, rather than value to a particular seller or buyer; they are addressed under separate heads of compensation. Their power differences are disregarded as matters for consideration under the 'without compulsion' aspect of the definition. When it comes to non-market values, similar principles apply according to the relevant definition.

While the courts must remain the final arbiters in valuation disputes, deliberative and inclusionary processes (DIPs) may be adopted prior to any court engagement and their results made available to the courts. DIPs might include citizen juries charged with deciding values or valuation workshops. Some jurisdictions consider that valuation determinations made by tribunals comprising, for example, a professional valuer, a magistrate and an eminent local may be accepted by the courts.

In all such DIPs or more formal independent valuation tribunals, it is essential that the affected parties can present evidence and see it transparently and accountably considered. In cases of expropriation, unless the courts later determine that claims were baseless and vexatious, the cost of expert representation should be considered a legitimate claim for disturbance. Therefore, subject to the laws of the relevant jurisdiction, a preferred order of dispute resolution in cases of expropriation would be:

- Arm's length negotiations between the parties;
- Continued negotiation, with appropriate expertise available to the parties, including mediators if appropriate;
- DIPs
- Tribunals
- The court system.

The earlier that values can be agreed upon, the better for all concerned.

5.4   Conclusion

Valuations fill the information gaps in markets that are often opaque and shrouded in secrecy. They inform stakeholders and market participants about the value of their tenure rights, and this can be vital in ensuring fairness when dealing with exchanges and acquisitions of these rights. Ready access to price and value information in relation to tenure rights also have a stabilising influence on this large and sometimes volatile sector of the economy.

Valuations are important for planning: value is a very good measure of the success or failure of land-use planning policy. Achieving the right mix of land uses, and the right amount of infrastructure and service provision in the places where it is needed, will stimulate economic activity. Spatial planning that optimises value generates more revenue from land and property taxation. This creates a virtuous circle as tax revenue is used to improve infrastructure and service provision, leading to further increases in value. Because value is such a useful indicator of land-policy success, it can be very helpful for governments to consult with valuers on land-policy matters.

To conclude, valuations are an important part of land-policy formulation. They reveal the value inherent in legitimate tenure rights, form the essential information base for negotiations over transfers of those rights and, by extension, provide a foundation for land and property markets. In short, valuations:

- Ensure the equity of tax assessments and fairness of compensation for expropriation;
- Ensure both market and non-market values are considered when buying, selling and leasing formal, customary, communal and informal tenure rights;
- Provide objective price information to holders of tenure rights in situations where transactions are scarce, such as privatisation of State assets or the acquisition of large tracts of land;
- Are a principal component of risk assessment when making loans secured by tenure rights;
- Check the value implications of planning policy and development decisions;
- Help reduce the risk associated with occupying, developing and investment in land and property; and
- Provide reassurance to individuals and businesses regarding the fairness of transactions in tenure rights and help reduce disputes.

Key points

- Valuations should be transparent, coherent and consistent and be undertaken by honest, impartial and competent valuers.
- A sufficient number of valuers are required, with requisite education, qualifications, skills and experience, and a valuers association can oversee education and skills development, regulate the conduct of valuers and provide an affordable and accessible means of dispute resolution.
- Valuer regulation should centre on the creation and adoption of professional codes of conduct and ethics. Governments should support such activity and encourage openness and transparency in the valuation process.
- Successful valuation practice requires well-drafted standards that are effectively enforced. International standards should govern valuer responsibilities and ethics, and national standards should operationalise approach and method. All valuation standards should be continuously monitored and revised on a regular basis.
- Valuation procedures are regulated in the United Kingdom at the national and international levels by a long-established set of standards. These standards are continuously monitored by professional bodies and are revised on a regular basis. It is essential therefore for valuers to keep themselves up to date.
- The valuation standards do not concern themselves with method but regulate the procedures surrounding the initial instruction, terms of engagement, valuation preparation and reporting. Specific valuations standards regulate certain valuation applications or purposes.
- Property rights and associated land and property are in a constant state of flux; economic activity, political involvement and changing social and environmental needs mean that a profession tasked with valuing tenure rights needs to continuously update guidance, standards and governance procedures.
- Tenure registration systems and other land and property information systems should be developed, maintained and adequately resourced. These systems should record ownership and use and value details in relation to individual and communal tenure rights. It is likely that coordination will be necessary when the different systems are developed in separate government departments. Common referencing and unique identifiers are essential in these circumstances.
- It is important to record information on customary and informal tenure rights as well as formal rights. Valuation systems and processes should be capable of managing complex structures of formal and informal tenure rights.
- Transaction details for state-owned and private land, for formal and informal tenure rights, and for capital and rental transactions, should be publicly accessible, with appropriate safeguards on personal information. Transaction taxes should be kept low.

Notes

1. Developed economies with established markets in private property rights may set standard formulae for quantifying compensation in respect of non-market value for expropriated property rights. This may be acceptable because property rights are capable of being unambiguously identified and money is a recognised and generally accepted form of compensation for the loss of these rights. For example, in the United

Kingdom, in addition to receiving the market value of expropriated property rights, owners and occupiers of a 'dwelling' may claim a 'home loss payment'.

2. Non-market values associated with property rights often benefit communities beyond the holders of those rights. For example, benefits associated with forests (timber and non-timber products, climate regulation, carbon sequestration, watershed services, soil stabilisation/erosion control, air quality, biodiversity, recreation, tourism, etc.) are more in the nature of public services rather than personal benefits.

3. Negative values might arise where expenditure exceeds income – a leasehold interest where the head-rent is higher than the sub-rent for example.

References

Baum, A. (2000). *Commercial Real Estate Investment*. UK: Chandos.

Cherry, A. (2006). *A Valuer's Guide to the Red Book*. London, UK: RICS Books.

Hemphill, L., Lim, J., Adair, A. et al. (2014). *The Role of International and Local Valuation Standards in Influencing Valuation Practice in Emerging and Established Markets*. Royal Institution of Chartered Surveyors.

IESC (2016). *International Ethical Standards: An Ethical Framework for the Global Property Market*. International Ethical Standards Coalition.

IVSC (2011). *Code of Ethical Principles for Professional Valuers*. International Valuation Standards Council.

IVSC (2022). *International Valuation Standards (IVS)*. International Valuation Standards Council.

RICS (2017). *RICS Valuation – Global Standards 2017: UK National Supplement*. London, UK: Royal Institution of Chartered Surveyors.

RICS (2019). *RICS Valuation – Global Standards: Incorporating the IVSC International Valuation Standards*. London, UK: Royal Institution of Chartered Surveyors.

RICS (2020). *Rules of Conduct for Members, Version 7 with Effect from 2 March 2020*. London, UK: Royal Institution of Chartered Surveyors.

Small, G. and Sheehan, J. (2008). The metaphysics of indigenous ownership: why indigenous ownership is incomparable to Western concepts of property value. In: *Indigenous Peoples and Real Estate Valuation, Research Issues in Real Estate*, Chapter 6 (ed. R. Simons, R. Malmgren and G. Small). New York: Springer.

UN FAO (2017). *Valuing Land Tenure Rights: A Technical Guide on Valuing Land Tenure Rights in Line with the Voluntary Guidelines on the Responsible Governance of Tenure of Land, Fisheries and Forests in the Context of National Food Security, Governance of Tenure Technical Guide*. Rome: United Nations Food and Agriculture Organisation.

White, T. and Curtis, C. (2018). File charges. *RICS Prop. J., Royal Institution of Chartered Surveyors* October/November: 24–25.

Section B
Valuation Approaches and Methods

Chapter 6
Market Approach

6.1　Introduction

If properties are to be exchanged, then buyers and sellers must agree prices for the property rights they are acquiring. In a competitive market, suppliers to, users of and investors in property agree on exchange prices. These exchange prices take the form of rents and prices. Rents (periodic payments) provide evidence of the cost of occupation. Prices (capitalised sums) provide evidence of either (i) the cost of owner occupation or (ii) the cost of an investment (ownership without occupation but with receipt of rent from the occupier instead). Often, agreeing an exchange price is easier said than done due to infrequent trading and market opacity leading to information paucity. Valuations fill this information gap.

Although each property interest is unique, for the more common uses such as residential, commercial, industrial and retail space, it may be possible to group properties into relatively homogeneous market sectors defined by land use and location. This means that it should be possible to analyse exchange prices within these groups to help estimate market values. This *market approach* is central to valuation. The economic concept that underpins the approach is *substitution*, that a knowledgeable and prudent person would not pay more for a property than the cost of acquiring an equally satisfactory substitute. This implies that, within a suitable time frame, the values of properties that are close substitutes in terms of location, utility and desirability will tend to be similar and the lowest price of the best alternative tends to establish market value. The market approach is all about identifying comparable evidence and adjusting it to support an opinion of value.

There are two methods under the umbrella heading of the market approach, the comparison method and the hedonic regression method. Both methods are revealed preference methods, but the first is mechanical or algorithmic and the second is statistical.

Property Valuation, Third Edition. Peter Wyatt.
© 2023 John Wiley & Sons Ltd. Published 2023 by John Wiley & Sons Ltd.
Companion website: www.wiley.com/go/wyatt/propertyvaluation3e

6.2   The comparison method

Value is estimated by examining the prices of comparable property rights that have recently transacted in the market. The principle of comparison underpins all valuation methods. Prices paid by owner-occupiers of property will provide evidence of the capital value of freehold or long-leasehold interests and, if enough capital transactions can be obtained, this would be good comparable evidence of capital values. Purchasers of property investments usually concentrate on the property's income-producing characteristics. Therefore, rental value and yield comparisons are essential for valuing property investments. The comparison method is also used to help value specialised trading property and is useful for valuing auxiliary facilities such as car parking spaces and land uses that are ancillary to the main business accommodation such as storage land. Procedurally, the comparison method involves the following steps:

[1]   Collect comparable evidence of market transactions.
[2]   Identify the value-significant characteristics.
[3]   Adjust comparable evidence to provide a consistent estimate of rent, price or yield.

6.2.1   Collect comparable evidence of market transactions

Valuers collect evidence of transactions and eliminate those not conducted at *arm's length* (between parent and subsidiary companies for example). Properties yet to transact or beyond a suitable time frame should be given less weight or disregarded. The RICS sets out a *hierarchy of evidence* shown in Table 6.1.

Transactions can be sales (which reveal evidence of capital values and yields) or new lettings (which reveal evidence of rental values). Other types of transaction can also be used to provide evidence of market rents. These include renewals of existing leases, rent-review settlements and assignments of leases. These transactions

Table 6.1   Hierarchy of evidence (based on RICS, 2019).

A: Direct comparable evidence
Contemporary, completed transactions of near-identical properties for which full and accurate information is available (may include subject property)
As above but for similar properties
As above but where full data may not be available
As above but where properties are being marketed and offers made

B: General market data
Published sources, commercial databases
Other indirect evidence such as indices (useful for dealing with historical transactions)
Historic evidence
Demand and supply data for rent, owner-occupation, investment

C: Other sources
Transactions from other real-estate types and locations
Other background data (interest rates, stock market movements and returns)

---

may not be as good as open market transactions, but they are useful. Sayce et al. (2006) provide a useful ranking of the usefulness of these sources of comparable evidence for valuation purposes: the best evidence is obtained from open market lettings that are conducted at arm's length, then lease renewals (from which the tenant can walk away although the significant costs in doing so should be borne in mind) and then rent reviews (where both parties have contractual obligations under the lease). If the comparable is a lease renewal, this is usually negotiated by professionals and agreed on similar terms to the previous lease, but it is important to note whether the value of any improvements that the tenant may have made to the property was disregarded when setting the level of rent under the new lease. Under legislation that will be discussed later, the value of a tenant's improvements may be disregarded for certain valuations. The rent agreed at review will reflect the terms of the rent-review clause in the lease and it is important to consider these terms in detail. Chief concerns are:

- The timescale for operation of the rent review and the precise terms on which it should take place, including the interval between each review (the rent-review period),
- Whether the review of the rent is upward-only,
- Whether there is an assumption that the property is vacant and to let for the purposes of determining the rent,
- Assumptions regarding the user clause (a lease term that may restrict the use of the premises),
- Assignment and sub-letting (*alienation*) provisions and
- Whether the value of tenant's improvements should be disregarded.

The rent-review clause in the lease will also state how disputes over the amount of reviewed rent should be resolved. Sub-lettings are secondary evidence as they are usually contracted out and affected by the head rent and other terms in the head lease. Assignments, where the current tenant sells (assigns) the lease to a new tenant, do not involve a reassessment of the rent passing but may involve a premium if there is a profit rent (the rent passing is below market rent) or a reverse premium if the property is over-rented (the rent passing is above market rent). Consequently, they are regarded as secondary evidence of market rents. If a rent at lease renewal or at a rent review cannot be agreed by the two parties and is determined by a third party, then this provides relatively weak evidence of market rent. At arbitration, the arbitrator must weigh up the evidence supplied by expert advisors who are appointed by the parties to the dispute. Contrastingly, if an independent expert is called in to resolve the dispute, more reliance may be placed on the judgement. Disputes that end up in a law court often do so to resolve a legal matter or require an interpretation of a point of law and are far removed from the open market.

Gathering several comparable transactions can help cross-reference and check whether each transaction is consistent with local market practice. It is not always possible to meet these aspirations for the evidence base because markets can be opaque and 'thin' (few properties coupled with few bidders with specific or non-market circumstances).

Surveying firms or property consultants offer consultancy and agency services and can provide an up-to-date and readily available source of transaction information for valuers working in the same firm. Moreover, valuers and their agency colleagues tend to share transaction information on an informal basis. Much of this information is not released into the public domain at the transaction level by the surveying firms themselves. Instead, they prefer to release aggregate information, usually on a quarterly, biannual or annual basis. Typically, the sort of information that is published includes supply, demand and resultant take-up figures, yields and rents across the main urban areas. There is no single definitive source and firms often publish information relating to specific sectors of the markets such as warehouse space or out-of-town retailing. Some of the larger surveying firms with offices in many countries publish international data. Surveyors specialising in auctioning commercial property and transaction results from auctions can provide very useful information on, typically, secondary property, as it is this type of property that tends to be sold at auction. Although surveying firms do not publish individual transaction details, publishers and specialist data providers do compile details of individual market transactions.

6.2.2   Identification of value-significant characteristics

Comparable transactions may be selected based on their value-significant characteristics, which can be classified as property-specific or market-related.

6.2.2.1   Property-specific characteristics

In terms of the site itself, location is key. In Chapter 2, the influence of location on property value was considered at the scale of the urban area and it was argued that accessibility was a key determinant of the location for a business. In short, the importance of accessibility is dependent upon the use to which the property is put and the various needs for accessibility result in a process of competitive bidding between different land uses and a property rent pattern emerges that is positively correlated with the pattern of accessibility. This usually means that the highest rents are paid in the centre of an urban area, but there are an increasing number of exceptions to this simple assertion. Nevertheless, the theory is sound and empirical evidence supports it.

Other value-significant site characteristics include size and layout and use(s) of the land, proximity to transport networks (both public and private transport), parking (covered/uncovered), compatibility of neighbouring uses, occupiers, development activity, soil quality, improvements, planning situation and potential for alternative uses. The geographic extent from which comparable evidence can be selected depends on the type of property and the state of the market.

Specification and condition of building(s) and space are important physical characteristics. This includes size, construction type (walls, roof, structure, external and internal finishes), date constructed or refurbished, number of floors in the building and the range of floors occupied, condition, external appearance (including aspect and visibility), internal specification, configuration, and services (heating, ventilation, lifts, fire safety, lighting, toilets). For retail property, dimensions of sales space, display frontages, escalators (and floors served), delivery access and staff area are important considerations. For offices, the type of entrance (sole/shared/prestige/standard), partitioning, suspended ceilings, natural light, floor height, raised floors, windows, security and food facilities are relevant. For industrial space and warehousing, eaves height, wall/roof lining, floor loading, partitioning, mezzanine, food facilities, loading facilities, storage (open/covered), access to road and public transport networks are important.

There is evidence that buyers pay a premium for new-build dwellings and, although it is possible to disregard this new-build premium by way of a special assumption (for example, for loan security valuations as it does not exist beyond the initial sale), when it is a consideration, it is important to note the following market dynamics (RICS 2012):

- Transactions that are pre-development resales and other special purchases
- Bulk purchases by investors as these may not reflect the market price of individual dwellings
- Buy-to-let purchases, which can be above owner-occupier prices in the short term but can drop in value in the medium to long term as social mix is altered and dwellings may be vacant
- Builders and developers purchasing first homes (using third parties) to establish headline prices

Running costs such as repairs and maintenance of common parts are important considerations, as is the adaptability of the premises in the face of changing production methods, technological advances or a rapidly expanding or contracting market. Flexibility for change of use will enhance the marketability of the property should the current occupier wish to move. Occupiers increasingly demand adaptable internal space so that it can meet their changing business requirements without having to move premises. Design considerations and corporate image are important to occupiers who may be using the premises as a headquarters or a use that requires regular client contact. These characteristics help the property combat obsolescence.

Of increasing concern to occupiers and owners is the energy efficiency and environmental performance of properties. Features include:

- Renewable energy generation such as wind power and photovoltaic cells
- Daylight-orientated design. Dark, south-facing surfaces
- Thermal and radiant heating for water and climate control
- Elevator counterweights that capture and store potential energy
- Recycled and sustainably harvested materials
- Fans, opening windows, daylight sensing and timers, sunshades and programmable louvers
- Healthier paints, finishes and adhesives, carpets with recycled fibres and eco-friendly dyes
- Low energy lighting, low water fixtures, rainwater collection and re-use

There has been a great deal of research into whether these characteristics affect capital and rental transaction prices. For example, Chegut et al. (2020) analysed valuations of rental housing undertaken by international firms in England and the Netherlands and found that valuers were reducing the values of dwellings with poor Energy Performance Certificates (EPCs) and in the Netherlands they found higher valuations for dwellings with higher energy efficiency. Most of the studies have focused on residential properties, but some have focused on commercial property transactions. Dalton and Fuerst (2018) published a meta-analysis of 42 studies that examined the relationship between energy efficiency and property prices. They found an average rental premium of 6.02% and a sales premium of 7.61%, but they found considerable statistical heterogeneity and potential publication bias in the results.

Legal factors can have a significant impact on value. If the legal interest is a freehold, then it is important to consider any easements or other statutory rights and obligations (such as restrictive covenants) over the land, the nature and extent of permitted use(s), potential for change of use and proposed development plans. If the freehold is held as an investment and let to an occupying tenant, then for the landlord the quality of that tenant is a primary concern, not only in terms of an ability to keep paying rent but also in complying with other lease terms such as repair and maintenance. If the property is let to more than one tenant, then the mix of tenants is important. User restrictions are sometimes inserted into each lease contract to protect the landlord's balance of lettings. For example, if the landlord owns a large shopping mall, then it would be wise to ensure that there is a variety of shops. To do this, the landlord and each tenant agree what limitations are to be placed on the trade that can occur in a particular shop unit. The landlord will need to ensure that potential tenants are financially able to meet terms of lease and that they are of a sufficient standing so as not to harm the investment value of the shopping mall as a whole; references and guarantees are often taken up.

The terms of the lease will be value significant to both the landlord and the tenant. Of overriding concern is the length of the lease and the ability or legal right to renew the lease at the end of the term. Also important is the responsibility for internal and external repairs and insurance as well as other regular expenditure such as a service charge and property taxes. But there are many other issues and lease provisions that the tenant must be mindful of: any restrictions on use and the ability to make changes to the premises, sub-letting or assignment; the nature and frequency of any rent reviews and options to renew or terminate the lease, known as break options; the nature of any incentives offered by the landlord (such as a rent-free period) or by the tenant (such as a premium) and the remedies for breach of lease terms.

It is important to consider how much rent is left after all expenditure has been accounted for. This *net rent* is usually calculated by deducting the cost of insurance, management and maintenance from the gross rent. Usually, the precise amounts of expenditure are not known and percentage deductions from the gross rent are estimated instead (a 2.5% deduction to cover the cost of insurance, 10% for management costs for example). Ideally, investors want leases that oblige the tenant to be responsible for repairs and insurance. This (partly) explains why leasehold investments are less attractive; the additional repair and management responsibilities, the wasting nature of the asset and a lack of reversionary value (redevelopment potential perhaps) are not attractive characteristics to an investor. A primary concern of the landlord is the security of rent in real terms so the negotiation of a new rent at rent review or lease renewal is of great importance. If rent reviews were not inserted into the lease contract, then the rent that the landlord receives would be eroded by inflation over the duration of the lease. Rent reviews ensure that the landlord receives an inflation-proof income.

6.2.2.2   Market-related characteristics

The wider market factors form part of the cognitive background that valuers bring to a valuation, including market knowledge and an awareness of the current legislative framework, environmental policy and economic activity.

The principal macroeconomic influences on property values include national output (measured using gross domestic product or GDP), inflation, household disposable income, consumer spending and retail sales, employment, construction activity, net household formation, production costs (including wage levels) and the cost and availability of finance. Changes in the size and demographic profile of the population can affect demand for goods and services as well as the availability and cost of the workforce used to produce them. Economic factors that affect the value of retail property in particular centre on the propensity to attract custom; for example, purchasing power (credit restrictions), consumer behaviour (spending habits, changes in tastes or fashion) and population density. Office property value may be influenced by the period of establishment in a region. But, regardless of the property type, the valuer tries to ascertain market strengths and weaknesses, assess the likely supply of and demand for properties comparable to the one being valued and determine the factors likely to influence value. Important local market characteristics include stock availability, rental growth rates, yields, rents, capital values, take-up rate, vacancy rate and the development pipeline. As a way of obtaining a mixture of macroeconomic information and market information, valuers can obtain summary statistics relating to the urban and regional location in which the property is located. The extent to which a valuer is concerned with national and regional economy depends on the size and type of property being valued; a large regional shopping centre or car assembly plant would require a great deal of market analysis at the national level whereas the valuation of a doctor's surgery or suburban shop would require analysis primarily at the local level.

Social factors include tastes of consumers and clients and changes in those tastes. For example, a wholesale shift towards the purchase of organic produce, to working at home or internet-based retailing will impact various sectors of the property market including shops, warehousing, offices and transport logistics. Important data include demographic, household and employment data, economic data and estimates of floor space for the main commercial property market sectors (offices, shops and restaurants, industrial and warehousing). Data are readily available at the national, regional and town/city levels as well as market-defined sub-locations such as the West End, Mid-Town and City of London. Location-specific market reports from the main property agents typically consist of some headlines and then report the availability (in terms of floor space) of new, refurbished and second-hand business space and space under construction, the level of take-up (also measured in terms of floor space), asking prices and quoted rents for new and second-hand space and the amount of vacant floor space. Table 6.2 summarises the kind of market data that might be collected at the town/city level.

Table 6.2 Urban area data.

- Infrastructure, details of road, rail and air communications
- Name and population of neighbouring urban centres
- Population and number of households in the urban area and its environs
- Resident population classified by gender, age, sociodemographic profile
- Car ownership
- Household tenure mix
- Employment profile (% employed full-time and part-time work, self-employed, unemployed, retired, studying, looking after the home, permanently disabled)
- Employment sectors in which the working population is employed
- Name, activity and number of staff of the largest employers
- Rent and price levels for commercial, industrial and residential property
- Vacancy rates for the main property types (retail, office, industrial, warehousing, residential)
- Proportion of top retailers present in the town/city and the names of those not present
- The names of the main shopping streets and main commercial developments
- Annual spend on comparison and core convenience goods within the catchment area of the town/city

Finally, the property market is a market for a tangible product that has influences and implications beyond its straightforward economic use as a factor of production or as an investment asset. The aesthetic and architectural qualities of individual properties are there for all to see. Similarly, the layout and design of property in its collective sense – across an urban area – impose a skyline that influences not only how we feel about a place but also how we work, reside, interact with others, and spend our leisure time. The 'invisible hand' of free trade is not always able to optimise these 'public' benefits and can sometimes impose unacceptable public or social costs on society. It is therefore the role of government to intervene. The main way that government intervention affects property values is through development control and land-use regulation or planning, but other activities can also have a significant impact including compulsory purchase of real estate, legislation that may protect certain rights of occupiers (security of tenure for example) and regulations that may affect revenue such as Sunday trading and gambling laws.

To summarise, the key value-significant characteristics that form the basis for the selection of market transactions in the comparison method of valuation are:

- Location and size of the land;
- Building size, age, specification, condition, layout, efficiency and adaptability;
- Tenure and legal rights;
- Land use regulation and other planning controls and
- Transaction date.

Comparable transactions may be ranked or weighted, depending on the detail of the information available and the number or size of the monetary or percentage adjustments applied to the sale or rental price of each comparable property. The valuer estimates the degree of similarity between the property to be valued and comparable sales or lettings by considering value-significant characteristics.

For example, transactions that took place near the property being valued would be of interest, whereas those further away might be considered less relevant. The geographic extent from which comparable transactions are selected depends on the type of property and the state of the market. The date of each transaction would be considered on a similar basis, while having regard market conditions at the time of each sale or letting. Similarly, the degree of similarity in terms of the nature and extent of the property rights held, physical attributes, current and potential land uses would also form the basis for ranking and weighting.

Table 6.3 shows comparable evidence being considered in respect of a rental valuation of an office suite of 350 square metres on the third floor of 28 King's Road. The building comprises an eight-storey office building in the heart of the business district of a large city. The locations of the comparables are shown in Figure 6.1, along with the subject property, shown as a star.

It may be helpful to rank the comparables in order of usefulness in relation to the subject property. There is no 'correct' outcome to this process, and it is important not to equate mathematical process to defensible output, but it can help identify relative differences between the comparable transactions. Table 6.4 illustrates one approach using rankings of one to four, with one being the most comparable.

6.2.3 Adjustment of value-significant characteristics

The value-significant characteristics described above can be used to adjust transaction prices or rents so that they provide an evidence base for estimating the value of a similar property for which the price information is not known.

Table 6.3 Comparable evidence schedule.

Comparable	Date (months ago)	Address	Size (m²)	Transaction type	Formality
A	2	6th Floor, 30 King's Road	300	New lease	Documented
B	1	3rd Floor 34 King's Road	500	Lease renewal	By agreement and documented
C	3	2nd Floor, 18 Abbey Street (no lift)	600	New lease	Telephoned agent
D	3	1st Floor, 16 King's Road (no lift)	250	Rent review	By arbitration and documented
E	4	5th Floor 11 Abbey Square	350	Rent review	By agreement and documented
F	1	Ground Floor, 8 Abbey Square	700	Asking rent	Letting details
G	2	4th Floor 15 King's Road	650	New lease	Documented
H	6	7th Floor 9 King's Road	800	Rent review	By arbitration and documented

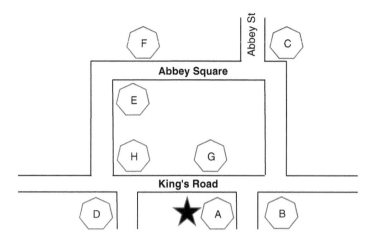

Figure 6.1   Location of subject property and comparable properties.

Table 6.4   Weighting of comparable evidence.

Comparable	Location	Date	Floor	Size	Transaction type	Formality	Weighted average
A	1	2	4	1	1	2	1.65
B	2	1	1	3	3	1	1.90
C	4	3	2	3	1	3	2.60
D	2	3	3	2	2	4	2.55
E	3	3	3	1	2	1	2.35
F	4	1	3	4	4	3	3.05
G	2	2	2	4	1	2	1.95
H	2	4	4	4	2	4	3.10
Weight	20%	25%	10%	10%	25%	10%	

6.2.3.1   Selecting a unit of comparison

The first task is to select the unit of comparison. This is usually sale price, rent or yield. The latter is not a transaction amount as such; it is a ratio of sale price to rent. It is, therefore, an important unit of comparison for investment transactions. Initial yields tend to be comparable for similar property investments in similar locations because their income growth prospects and risk to capital and income will tend to be similar. Yet, remembering that small differences in yield can result in big differences in value, adjustments may be made to initial yields from recent comparable transactions to reflect any differences between them and the property being valued. When deciding on whether to adjust the yield obtained from comparable evidence, consideration would be given to circumstances specific to the subject property. For example, in the case of an investment interest:

Circumstance	Outcome	+ or −
The property is let to a first-class tenant	Good	−
The property has been recently modernised	Good	−
The tenants want to take a very short lease	Bad	+
There is a public right of way running across part of the rear service yard	Bad	+

For comparison purposes, prices and rents are usually expressed as amounts per unit of area. This normalises differences in the size of site or the property, although selection of comparable properties might be confined to those within a similar size range. The price of a parcel of land is usually expressed as an amount per hectare, per acre or per square metre. The land area metrics that are widely used for planning and development purposes are (RICS 2021):

- Land ownership area: Land area held in a single legal interest or title by one or more legal owners
- Site area: Land area on which development authorisation is sought

- Net development area: Extent of site area that can be built on

- Plot ratio: Ratio of development floor area (either gross external or gross internal area) to site area. Floor area ratio and floor space ratio are similar to plot ratio
- Site coverage: Ratio of gross floor area (gross external area) to site area, expressed as a percentage

For building floor space, measurement is a little more complicated, although in recent years there has been a concerted effort to establish international building measurement standards, culminating in the publication of International Property Measurement Standards (IPMS) for dwellings, offices, retail and industrial buildings. These standards define three metrics, IPMS1, IPMS2 and IPMS3.

IPMS1 is used for measuring the area, on a floor-by-floor basis, of the external envelope of buildings. The metric is useful for planning purposes or for cost estimation. Measurement includes basements, balconies, covered galleries, generally accessible rooftop terraces (but stated separately). It excludes open light wells or upper-level voids of an atrium, open external stairways, ground-level patios and decks, external car parking, equipment yards, cooling equipment and refuse areas, other ground level areas not fully enclosed, but these may be stated separately. Space with limited use should be separately stated. These include areas with limited height or limited natural light. IPMS2 and 3 measure internal areas within the dimensions of IPMS and their definitions vary according to land use.

For offices, IPMSC (2014) states that IPMS2 is the sum of each floor measured to the internal dominant face. It includes interior walls, columns and enclosed walkways or passages between separate buildings. It is used for measuring interior areas and categorising the use of space. Covered void areas such as atria are only included at their lowest floor level. Balconies, covered galleries and generally accessible rooftop terraces are included, stated separately and measured to their inner faces. IPMS2 categorises areas by function. IPMS3 is used for measuring occupation of floor areas in exclusive use. It is the floor area available on an exclusive basis to an occupier excluding standard facilities[1] and shared circulation areas.

It is calculated on an occupier-by-occupier or floor-by-floor basis for each building. Internal walls and columns in an occupant's exclusive area are included. Floor area is taken to the internal dominant face and, where there is a common wall with an adjacent tenant, to the centre line of the common wall. Balconies, covered galleries and generally accessible rooftop terraces are stated separately and measured to their inner faces.

For residential dwellings, IPMSC (2016) states that IPMS2 is measured to the internal dominant face for external construction features, otherwise to the finished surface. This is done floor by floor and component by component if required. For multi-use buildings, areas in exclusive use can be stated separately from common facilities. The measured area includes internal walls, columns, enclosed walkways or passages between buildings, covered void areas (e.g. atria) at their lowest level. Balconies, internal catwalks, covered galleries, internal loading bays, internal permanent mezzanines and verandas are also included but stated separately. Excluded are temporary mezzanines, open light wells and upper-level voids of an atrium, any ground-level structures or areas beyond the external walls such as sheltered areas, external catwalks and external loading bays. IPMS3 is used for measuring occupation floor areas in exclusive use. There are variations of IPMS3 that either include or exclude internal walls and columns. Floor area occupied by stairs is only included at the lowest level. Vertical penetrations or voids over 0.25 square metres (including the enclosing wall) are excluded. Room dimensions and areas are measured to internal dominant faces or finished surfaces as appropriate.

For industrial buildings, IPMSC (2018) states that IPMS2 is a whole building measurement used to measure the interior boundary area of the building. It includes internal areas, walls and columns, enclosed void areas at the lowest floor level. Balconies, internal loading bays, mezzanines, enclosed walkways or passages between separate buildings are stated separately. Excluded but stated separately are areas outside external walls. IPMS3 is used for measuring occupation or floor areas in exclusive use.

For retail properties, IPMSC (2019) states that IPMS2 includes areas within the internal dominant face (including internal walls and pillars), and covered voids at their lowest level. Internal loading bays, internal mezzanines, enclosed passageways between separate buildings and external floor areas measured to inner face of balustrade are included (but stated separately). Excluded (but stated separately if measured) are areas outside external wall, other areas that are defined as being within the shop line (presumed boundary) and sheltered areas. IPMS3 has three variations: 3A, 3B and 3C, but all are used for measuring floor areas in exclusive occupation.

For commercial properties that are let, rents are expressed as an annual figure per square metre. Consider an industrial property with an IPMS3 area of 325 square metres. A comparable property has an IPMS3 area of 350 square metres and was recently let on a new 15-year lease with five-yearly, upward-only rent reviews at a rent of £12 200 per annum. Analysis of the comparable property reveals that the rent paid was equivalent to £34.86 per square metre. This information can be used to estimate the rental value of the subject property as follows:

Area (m^2)	325	
× Rent (£/m^2)	34.86	
Estimated rental value (£)		11 330

It may be appropriate to apply different rates for different locations but perhaps maintain relativities between values per unit area for different uses. Similarly, the value per unit area might vary according to the use of the property. For properties where there is a mix of uses, comparable evidence may reveal a value relationship between the uses. For example, Hayward (2009) shows how, in estimating the rental value of a car showroom and ancillary accommodation, the rent on ancillary office space is proportionate to the rent for the main showroom space. These rents for office and workshop space could be cross-checked against comparable evidence in the area. Car-parking spaces may either be separately valued on a unit rent per space basis or their value might be implied in the overall rent per square metre that is applied to the main floor space. The rental valuation might be set out as follows:

Use	Area (m²)	Rent (£/m²)	Annual rent (£ p.a.)
Showroom	300	120	36 000
Sales office	30	60 (1/2 of showroom rent)	1800
General office	100	60 (1/2 of showroom rent)	6000
Reception area	20	40 (1/3 of showroom rent)	800
Workshops	600	30 (1/4 of showroom rent)	18 000
Parts store	20	30 (1/4 of showroom rent)	600
Mezzanine floor	50	15 (1/8 of showroom rent)	+750
Rental value of buildings			63 950
Uncovered car stances	No. 25	400	10 000
Car-parking spaces	No. 40	100	+4000
Estimated rental value (£)			77 950

It may be appropriate to determine a typical size for a particular use and alter value per unit area for smaller sites and larger properties or adopt base prices for properties within certain sizes bands. In some markets, gross rent (before any deductions for outgoings) may be used as a unit of comparison, whereas in other markets net rent is used. The valuer must define the basis of comparison.

Take another example. A 30-year-old empty factory includes 1000 square metres of factory area and 100 square metres of offices, and an estimate of the market rent is required. The following comparable evidence has been gathered:

Factory	Details	Analysis
A	2000 m² including 150 m² of offices, modern building recently let at £55 000 per annum on a 15-year lease with 5-year rent reviews.	New building. Larger than the subject property. Also has office space. £27.50/m².
B	2500 m² plus 300 m² of office space above part, 40 years old, currently let on a 15-year lease and recently reviewed to £65 000 per annum for the next 5 years.	Significantly larger than the subject property. Similar age but slightly newer than the subject property. £26.00/m². Factory with office space.
C	750 m² without office accommodation, built 80 years ago on a cramped site and let at a rent of £11 250 per annum.	A much older unit of a smaller size. Poor site. Factory only. £15/m².

Factory	Details	Analysis
D	Recently completed basic factory units of 1500 m² being marketed at £28/m². The developers are prepared to provide the area of office space required on demand at a rent of £60/m². Net initial yields are expected to be 7.5%.	New buildings. No sales yet – asking rents only. Similarly, the yield evidence is a forecast rather than achieved. Provides some evidence of demand for 1500 m² units with the option for office space to be provided.
E	An old factory with a floor area of 3000 m² in good structural condition, recently let at £16/m² and sold immediately afterwards to a local investor for £500 000.	Old but sound, £48 000 per annum. Factory only. The sale indicates it is investible and the net initial yield is 9%.

Given more information on lease terms, quality of building, site area and so on, it would be possible to tabulate information for ease of comparison and perhaps rank the evidence too. There is an indication from comparable D that the rental value of office and factory space could be separated. However, the argument could equally be made that an overall rent for factory and office space is warranted. Adopting the former approach, the valuation might proceed as follows:

Factory space	1000 m² @ £19/m²	£19 000
Office space	100 m² @ £25/m²	£2 500
Market rent		£21 500

Shops are a little different because a higher value per unit area is assigned to the front of a shop compared to the rear and to the ground floor compared to other floors. This means that not only is area an important metric, but so are the dimensions of the floor space, in particular, the length of the front face or *frontage*, which is typically used for window display. The gross frontage of a shop is the overall external measurement in a straight line across the front of the building, from the outside of external walls or from the centre line of party walls. The net frontage is the overall external frontage on the shop line measured between the internal face of the external walls, or the internal face of support columns including the display window frame and shop entrance but excluding recesses, doorways or access to other accommodation.

We know from Ricardian rent theory that the rent that a tenant can afford to pay depends upon the level of trade and hence profit that can be produced. The level of trade in a shop is influenced by the ability to display items in the window to attract passing trade. The more that can be displayed, the more trade will be generated, the more the profit, the higher the rent. Consider two shops of identical size but different shapes, shown in Figure 6.2.

Even though they are the same size, shop 2 would be regarded as more valuable because its wider frontage will allow more goods to be displayed. A technique known as *zoning* is used to divide up the sales area of standard shop units. It is a means of reflecting the fact that the trading area nearest front of shop is most valuable. The ground-floor sales area is divided into zones parallel to the frontage

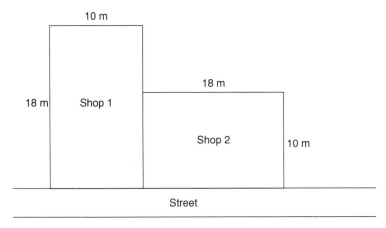

Figure 6.2   Shop shapes.

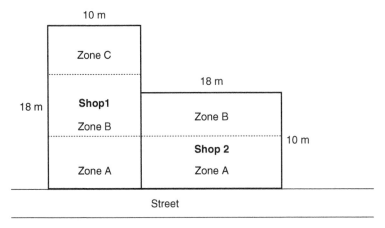

Figure 6.3   Zoning shops.

and to a depth of 6.1 metres (20 feet). Zone A is always at the front and a maximum of three zones is usual with a 'remainder' area encompassing all that is left over[2]. Figure 6.3 illustrates how the shops in Figure 6.2 would be zoned.

The areas of the zones are as follows:

Shop 1	Shop 2
Zone A = 6.1 m × 10.0 m = 61.0 m²	Zone A = 6.1 m × 18.0 m = 109.8 m²
Zone B = 6.1 m × 10.0 m = 61.0 m²	Zone B = 3.9 m × 18.0 m = 70.2 m²
Zone C = 5.8 m × 10.0 m = 58.0 m²	

To weight the space at the front of the shop more highly, the area of zone A is kept the same, but the area of each subsequent zone is 'halved back'. This process derives an area 'in terms of zone A space' or *ITZA* for short. Looking at the shops in Figure 6.3, the area ITZA for zone A is the actual area, the area of zone B is halved, and the area of zone C is halved again (i.e. quartered). Any remaining space beyond zone C might be halved again (i.e. divided by eight), but the magnitude of this fraction may vary depending on any special features of the remaining area. For shop 1, the ground floor (GF) would be divided into zones as follows:

Space	Width (m²)	Depth (m²)	Area (m²)	% Zone A	Area ITZA (m²)
GF zone A	10.00	6.10	61.00	100.0%	61.00
GF zone B	10.00	6.10	61.00	50.0%	30.50
GF zone C	10.00	5.80	58.00	25.0%	14.50
Total		18.00	180.00		106.00

For shop 2, the area ITZA would be:

Space	Width (m²)	Depth (m²)	Area (m²)	% Zone A	Area ITZA (m²)
GF zone A	18.00	6.10	109.80	100.0%	109.80
GF zone B	18.00	3.90	70.20	50.0%	35.10
Total		10.00	180.00		144.90

Zoning calculates a larger area ITZA for shop 2 due to its wider frontage and consequent larger zone A, the most valuable sales space. Assuming these two shops are good comparables and shop 1 recently let for £35 700 per annum, this equates to £337 for each square metre of shop space when it is expressed ITZA. Similarly, shop 2 recently let for £48 500 per annum, so this equates to £335 per square metre ITZA. In this way, the ITZA areas allow ITZA rents per square metre to be a standard metric for comparison purposes. If it is assumed that an ITZA rent of £336 per square metre represents the market rent from these two shops, this figure can be multiplied by the ITZA area of a comparable shop to estimate its market rent.

Sales space on floors other than the ground floor is considered to be less valuable and is expressed as a small fraction of the area ITZA, perhaps a sixth or a tenth. There is not much demand for sales space above first-floor level in a standard shop unit, but the value applied will depend upon the ease with which the other floors can be reached by customers (facilitated, perhaps, by escalators and lifts or stairs at the front of the shop) and the ease with which goods can be transported to these floors. Ancillary space such as storage is even less valuable and may be expressed as a smaller fraction of the area ITZA or as a nominal rent per square metre. Office space that is ancillary to the sales area may also be expressed as a fraction of the area ITZA or may be related to rents for similar office space in the locality.

Of course, not all shops are 'standard'. A typical frontage-to-depth ratio is 1: 2.5 or 1: 3 (shop 1 in Figure 6.4). Shops with a much higher ratio, shop 2 for instance, may warrant a reduction to the valuation because, although the zone A space gives display prominence, there is relatively limited space for the retailer to stock goods for sale. Similar adjustments may be made if the shop is unusually large or an unusual shape such as shop 3, which has a masked area towards the rear of the premises (a masked area is an area made less prominent by, say, an L-shaped layout or features such as split levels or pillars getting in the way of displays). If the

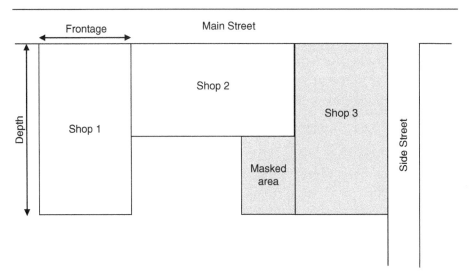

Figure 6.4   Shop shapes.

shop has a return frontage (where a shop is positioned on a corner and fronts two roads or pedestrian flows, shop 3) it is usual to either zone from both frontages if both provide good pedestrian flow, or zone from the prominent frontage and make an end allowance (say a 5–15% addition to the zone A rent) if warranted, bearing in mind that an excessively long return frontage could adversely affect the layout of sales space. The size of the end allowance will depend on the nature of the return frontage; is it a quiet street, can the property be accessed from it, what is the security like? Some shops, in a shopping centre for example, may have frontages on two floors. The valuation below is an example of a shop with complex floor areas and illustrates how this might be handled when estimating a rental value. The valuation may be subject to end allowances including deductions for abnormal size.

Space	Width (m²)	Depth (m²)	Area (m²)	% Zone A	Area ITZA (m²)
Ground-floor zone A	9.00	6.10	54.90	100.0%	54.90
Ground-floor zone B	9.00	6.10	54.90	50.0%	27.45
Ground-floor zone C	9.00	6.10	54.90	25.0%	13.73
Ground-floor remainder	9.00	4.70	42.30	10.0%	4.23
End allowance (as % of ground floor area ITZA)				5.0%	5.02
Other sales space			100.00	20.0%	20.00
Storage area			80.00	5.0%	4.00
Total		23.00	387.00		129.32

Because shoppers appreciate the convenience of a well laid-out and easily accessible shopping area, the rents that retailers are prepared to pay decline quite rapidly with increasing distance from the prime (most accessible) shopping location in an urban area. In valuation terms, the prime location is often referred to as the 100% prime position and zone A rents of neighbouring shops may be related to this position by expressing them as a percentage of 'retail prime'.

For large shops it is important to consider their location in relation to the retail centre, their prominence, quality of fit-out (including ventilation), vertical accessibility (number of floors, escalators, lifts), compartmentalisation and horizontal circulation, car parking, varying floor levels, ceiling heights, second entrances, loading and sorting facilities.

One final point regarding shops is that they are often let as *shells*. In other words, their internal fittings are excluded, and the landlord grants a short (say, three-month) rent-free period to enable the incoming tenant to fit out the shop. Care must be taken to ensure that any measurements taken when the property was a shell are suitably adjusted or re-measured when calculating the IPMS3 area for valuation purposes. Furthermore, when selecting comparables, it is important to ensure these fit-out periods are not confused with longer rent-free periods that may be granted as an incentive to take occupation.

Certain types of leisure property, which are normally valued with regard to their trading potential, may be compared using specific units of comparison too. For these types of property, size may not be such an important factor and different comparison metrics may be used. For example, if sufficient comparable evidence is available, a capital or rental value per hotel room (inclusive of dining and conference facilities), per cinema seat, per tent or caravan pitch might be determined.

6.2.3.2   Adjustments to comparison metrics

The next stage is to adjust market prices, market rents and net initial yields from comparable evidence, as appropriate. This is done so that they provide a consistent estimate of market rent, price or yield, before using them to value a subject property. These adjustments can be undertaken qualitatively by the valuer, who would have experience and knowledge of the local market, or a quantitative technique can be used to weight comparable properties, isolate differences in the elements, quantify these differences and adjust the values accordingly. Quantitative adjustments are more transparent whereas qualitative adjustments may be necessary when comparable evidence is scarce. Both approaches have their place and often a combination of qualitative and quantitative approaches would be employed. Adjustments to price and rent are often more quantitative, whereas yield adjustments are often more qualitative. The aim is to ensure that any adjustments made to the comparable evidence reflect the likely reactions of market participants.

A quantitative approach might compare two or more transactions to derive the size of the adjustment for a single value-influencing characteristic. Ideally, two sales will be identical apart from the characteristic being measured, but this is rare and usually a series of 'paired' comparisons are made to isolate the effect of a single characteristic. Each comparable may be weighted depending on the number of adjustments applied, the total adjustment in absolute terms, the difference between positive and negative adjustments, any large adjustments made or any other factors that suggest weight should be applied. The aim is to arrive at an adjusted or net *effective* rent, price or yield that takes account of differences between the comparables and therefore provides a set of consistent evidence. Mathematically, the

Table 6.5 Adjustments to elements using the comparison method of valuation.

	Comparable a	Comparable b	Comparable c	Comparable n
Rent (£/m²)	R_a	R_b	R_c	R_n
Elements:				
▪ Location	±	±	±	±
▪ Physical description	±	±	±	±
▪ Sale date	±	±	±	±
▪ Sale conditions	±	±	±	±
▪ Lease terms	±	±	±	±
▪ etc.	±	±	±	±
Net adjustment (£/m²)	$±_a$	$±_b$	$±_c$	$±_c$
Adjusted rent (£/m²)	$R_a + (±)_a$	$R_b + (±)_b$	$R_c + (±)_c$	$R_n + (±)_n$

adjustment process is presented in Table 6.5 for establishing market rent, but the process would be the same for estimating a market price or yield.

In the absence of sufficient data to allow a quantitative approach, value-significant characteristics may be expressed in qualitative terms such as 'inferior' or 'superior'. Reconciliation involves consideration of the strengths and weaknesses of each characteristic. The valuer uses judgement to determine the direction and magnitude of the effect that each characteristic has on value and assesses its relative importance. When this has been done for each characteristic and for every comparable, the net adjustment for each is resolved.

A qualitative approach is popular because it reflects the imperfect nature of the property market, but it is usual to combine quantitative and qualitative approaches when using the comparison method. Table 6.6 provides an example of how this

Table 6.6 Comparison valuation using quantitative and qualitative approaches.

	Comp A	Comp B	Comp C	Comp D	Comp E	Subject property
Market rent (MR) (£ p.a.)	67 000	75 000	66 000	80 000	83 200	—
IPMS3 area (m²)	100	90	95	115	130	125
MR (£/m²)	670	830	694	609	640	
Management costs (% MR)	—	−5%	−5%	—	−5%	
Repair liability (% MR)	—	−5%	−10%	—	−5%	
Insurance liability (% MR)	—	−2.5%	—	—	−2.5%	
Age allowance (% MR)	+5%	−5%	—	—	−5%	
Net quantitative adjustment to MR	+5%	−17.5%	−15%	—	−17.5%	
Adjusted MR (£/m²)	704	685	590	609	528	
Condition	Average	Average	Average	Average	Average	Average
Ratio of parking space to IPMS3 area	Average	Average	Average	Poor	Good	Good
Location	Superior	Inferior	Average	Superior	Inferior	Superior
Net qualitative adjustment	−ve	+ve	−ve	−ve	+ve	—

might be done when estimating a market rent. The inclusion of IPMS3 area helps determine comparability in terms of size so it is best not to calculate market rent per square metre straightaway. Comparables A and D appear to be very strong and may attract the greatest weight when reconciling these comparables to derive an estimate of market rent for the subject property.

The comparison method is predicated on comprehensive and up-to-date records of transactions and is therefore a reliable method in an active market where recent evidence is available. It is important, therefore, to maintain records of inspections of any comparables, investigations of supporting data (such as planning and legal information) and the analysis of the evidence. Reliability of the comparison method declines when market conditions are volatile or when valuing specialised land and property for which there is little market evidence, and the uniqueness of each property precludes attempts at meaningful comparison. In these cases, it is necessary to look more closely at the financial decisions that underpin the prices agreed. As an example, some specialised types of property are valued by quantifying the contribution of the property to business profit. Yet it may be possible to compare properties based on broad value determinants such as land quality, availability of infrastructure, and accessibility as well as building size, type, age and quality.

Wiltshaw (1991) argues that the comparison method is statistically flawed, primarily because of the small number of comparable transactions used in many valuations and, as the number of comparables decreases relative to the number of comparison elements to be adjusted, it increases the likelihood of statistical insignificance. Nevertheless, the principle of comparison is central to property valuation. If sufficient transaction data were available, it would be possible to use a hedonic regression method and indeed such techniques are used for valuing residential property.

6.3   Hedonic regression method

Computer-based algorithms have been developed to automate the comparison method described above. Generally, they consist of two steps: select comparables from a database according to value-significant characteristics and then estimate the value of the subject property. Known as automated valuation models or AVMs, these point to the future of valuation practice. For markets where transactions are frequent and properties are relatively comparable, such as residential apartments in a city, it is possible to employ statistical methods such as hedonic regression, and that is the method that will be described in this section. Other automated methods are emerging, primarily based on machine-learning algorithms and neural networks. Because these automated methods lack the granularity of comparables-based methods, their main applications are for the construction of price indices, for portfolio valuations and for mass appraisal (high-volume valuations, usually for property taxation purposes) (European AVM Alliance 2019).

The premise of hedonic regression is that a statistical relationship can be identified between value-significant characteristics of properties and the prices that have been paid for them. This relationship, encapsulated in a mathematical equation, can then be used to estimate the values of properties where the characteristics are known but the prices are not. A key difference between this statistical approach and conventional valuation is the way in which data are used: hedonic regression relies on a large quantity of data to produce a mathematical equation that can be used to predict property value. To explain how the method works, it is best to start with a simple linear regression.

6.3.1   Simple linear regression

Simple linear regression is a mathematical model of a linear[3] relationship between the values of two variables. The model can be used to estimate values of a response variable given values of a predictor variable. If this relationship were to be plotted on a graph, it should be possible to draw a 'best fit' line through the data points, the slope of which is the average predicted change in the response variable per unit of change in the predictor variable.

But to be more precise, we could derive a best-fit line that minimises the deviation (measured in terms of the sum of squared errors or residual term) of each observed value from the line, known as the *least-squares* principle. This best-fit line is called the *regression line* (of response variable y on predictor variable x). In this way, any linear function involving two variables can be expressed in the form:

$$y_i = b_0 + b_1 x_i + u_i \tag{6.1}$$

where y_i is an estimate of the average value of the response variable y corresponding to a given value of x, x is the actual value of the predictor variable, b_0 is an estimate of y when $x = 0$ (the intercept of the regression line), b_1 is an estimate of the gradient of the regression line and u_i is a random component (residual error term).

Using the least-squares principle, the regression line can be derived by solving for b_1 and b_0 using the variance of x and the covariance between x and y. The expression from which b_1 can be calculated is:

$$b_i = \frac{\text{cov}_{xy}}{\text{Var}_x} = \frac{s_{xy}^2}{s_x^2} = \frac{\sum_{i=1}^{n}(x_i - \bar{x})(y_i - \bar{y})}{\sum_{i=1}^{n}(x_i - \bar{x})^2} \tag{6.2}$$

where Cov_{xy} is the covariance between x and y, Var_x is the variance of x, $\bar{y}$ is the mean of y, $\bar{x}$ is the mean of x and $i = 1, \ldots, n$ (where n is the number of observations).

For b_0, the expression is:

$$b_0 = \frac{\sum y - b_1 \sum x}{n} = \bar{y} - b_1 \bar{x} \tag{6.3}$$

The mean value of the response variable y (sale price for example) would appear as a straight line on a scatterplot as it would be the same for all values of a predictor variable x (say, floor space). This is illustrated in Figure 6.5.

Total variation SS_T of each value of y about the mean value of y is calculated by taking the sum of the squared differences between observed values of y and the mean value of y:

$$SS_T = \sum(y_i - \bar{y})^2 \tag{6.4}$$

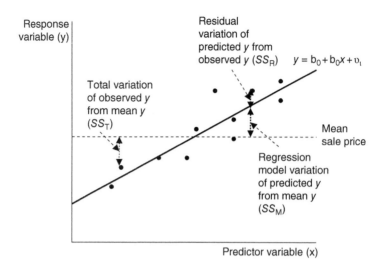

Figure 6.5 Regression line of y on x.

Each point on the regression line (which slopes) varies from the mean value of y, and this regression model variation SS_M can be calculated as the sum of the squared differences between mean value of y and the regression line:

$$SS_M = \Sigma(\hat{y}_i - \bar{y})^2 \qquad (6.5)$$

where $\hat{y}_i$ is modelled sale price of property i.

Finally, residual variation SS_R (variation unexplained by the regression model) can be calculated as the sum of the squared differences between observed values of y and the regression line:

$$SS_R = \Sigma(y_i - \hat{y}_i)^2 \qquad (6.6)$$

We would expect the total variation to comprise variation explained by the regression model plus residual variation:

$$SS_T = SS_M + SS_R \qquad (6.7)$$

6.3.1.1   Interpreting the model

6.3.1.1.1   Model performance
As a measure of the relationship between the two variables, we can calculate the amount of variation in the values of the dependent variable SS_T, which is explained by the model SS_M, i.e. explained variation divided by total variation:

$$R^2 = \frac{SS_M}{SS_T} = \frac{\Sigma(\hat{y}_i - \bar{y})^2}{\Sigma(y_i - \bar{y})^2} \qquad (6.8)$$

This is known as the *coefficient of determination, R^2*, and it ranges from 0 to 1. It provides a measure of the size of the relationship between the two variables. The square root of R^2 is *Pearson's correlation coefficient, R*, which provides an estimate of the overall fit of the regression model.

The F-ratio measures how much the model has improved the prediction of the response variable compared to the level of inaccuracy in the model. It is the mean sum of squares for the model MS_M divided by the mean residual sum of squares MS_R:

$$F = \frac{MS_M}{MS_R} \tag{6.9}$$

The means are calculated by dividing the sums of squares by the degrees of freedom. For MS_M the degrees of freedom is the number of variables and for MS_R the degrees of freedom is the sample size n minus the number of parameters being estimated p (i.e. the number of beta coefficients plus the constant). A good model will have a high F-ratio.

The coefficient of variation (CoV) is a measure of the average squared error or variance of the regression model, calculated by dividing the sum of the squared errors by the mean value of y:

$$CoV = \frac{SEE}{\bar{y}} \tag{6.10}$$

where SEE is the standard error of the estimate, a measure of the *amount* of deviation between actual and predicted values of the response variable:

$$SEE = s_{y,x} = \sqrt{\frac{\sum(y - \hat{y})^2}{n - p - 1}} \tag{6.11}$$

It is a measure of the average squared error or variance of the regression model, calculated by dividing the sum of the squared errors by the degrees of freedom. Statistical software can calculate standard errors and confidence intervals for individual predicted values. These values are a function of overall SEE and the individual characteristics of the observation; the more typical the characteristics (closer to average), the lower the standard error and confidence intervals about the predicted value.

The CoV is a standardised SEE and is analogous to the conventional CoV (standard deviation divided by mean). It can be regarded as the standard deviation of the regression errors. So, if the errors are normally distributed, two thirds of actual values of y fall within one SEE of the predicted values, 95% within two SEEs and so on. The result provides confidence intervals around the regression line. Unlike R^2, which evaluates seriousness of errors indirectly by comparing them with variation in observed values of y, SEE evaluates them directly in units of y.

6.3.1.1.2   Model Parameters

If x significantly predicts y, then it should have a b_1 (beta coefficient) significantly different from zero[4]. This hypothesis is tested using a t-test. The t-statistic measures the significance of a predictor variable in explaining differences in the response variable. It is the ratio of the beta coefficient of the predictor variable to its standard error s[5]:

$$t = \frac{b}{s} \tag{6.12}$$

The larger t is and the smaller s is, the greater the contribution of that predictor[6]. Degrees of freedom are $n - p - 1$. Generally, for samples with at least 60 observations (plus one additional observation for each parameter to be estimated) a predictor variable with a t-statistic $\geq \pm 1.96$ indicates 95% confidence that b does not equal 0 and therefore x is significant in predicting y. If $\geq \pm 2.58$, then 99% confident. Statistical software usually computes the probability p that the observed value of t would occur if b was 0, and if $p < 0.05$ then b is significantly different from 0.

6.3.1.1.3 Residuals

Standardised residuals (difference between observed and predicted outcomes) should be normally distributed about the predicted responses with a mean of zero. A normal P–P plot of regression standardised residuals is a check on normality. Plotted points should follow a straight line. When the model fit is appropriate, a scatterplot of standardised residuals (residual divided by standard deviation) against predicted responses should be random, centred on the line of zero standard residual value. Standardised residuals with z-scores $> \pm 3$ are outliers and therefore concerning. If more than 1% standardised residuals have z-score over 2.5, the error in the model is unacceptable. If more than 5% of standardised residuals have z-score over 2, this is also evidence that the model poorly represents the data. The variance of the residuals about the predicted responses should be the same for all predicted responses (homoscedastic). When standardised residuals are plotted against standardised predicted residuals, random dots show homoscedasticity, a funnel indicates heteroscedasticity and a pattern is a violation of linearity assumption. If the dots are more spread out at some points than others, this indicates a violation of homogeneity of variance and linearity assumptions.

6.3.1.2 Example

The sale prices and monthly rents have been recorded for 60 residential investment properties. Table 6.7 shows an extract of the observations.

The sale price will be the response variable and the histogram in Figure 6.6 shows that the frequency distribution of sale prices is slightly positively skewed. The mean sale price was £301 860 with a standard deviation of £65 140 and the median was £297 000. Monthly rent will be the predictor variable.

The coefficients are: $b_0 = 69.59$ and $b_1 = 0.9239$. Both are significantly different from 0 at the 0.01% level. The model is therefore: $y = 69.59 + 0.9239x$. So that is £69 590 plus 0.9239 times the observed monthly rent. This simple linear relationship is illustrated in Figure 6.7.

The R^2 is 0.7962 or 79.62% and R is 0.8923. The F-ratio is 226.52 (199 336/880) and this is significantly different from zero at the 1% level. The standard error of the estimate is:

$$SEE = \sqrt{\frac{51033}{58}} = 29.66$$

The CoV is therefore $SEE/\bar{y} = 29.66/302 = 0.0982$ or 9.82%.

Statistical software can be used to calculate the outputs more easily. Using SPSS, model coefficients are shown in Table 6.8, with slight differences due to rounding.

Table 6.7   Extract of sample of data points.

ID	Price (£000) (y)	Rent (£/month) (x)
1	341	268
2	242	130
3	242	130
4	297	253
5	297	211
6	396	343
7	270	134
8	462	378
9	176	157
10	176	157
. . .	. . .	. . .
50	407	317
51	231	191
52	231	191
53	226	178
54	215	178
55	220	178
56	209	178
57	220	178
58	264	211
59	330	297
60	341	303
Average	$\bar{y} = 302$	$\bar{x} = 251$

Section B

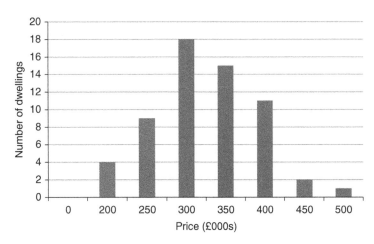

Figure 6.6   Frequency distribution of sale prices.

Statistics for the residuals are shown in Table 6.9. The standard residual has a mean of zero and a standard deviation of one which is good, but Figure 6.8 shows that the distribution is not particularly normal.

As a check on normality, the normal P–P plot in Figure 6.9 shows that the standardised residuals follow a relatively straight line.

Table 6.8 Coefficients calculated using SPSS.

Model[a]	Unstandardised coefficients		Standardised coefficients			95.0% confidence interval for B	
	B	Std. error	Beta	t	Sig.	Lower bound	Upper bound
1 (Constant)	69.593	15.897		4.378	0.000	37.772	101.415
Rent (£/ month)	0.924	0.061	0.892	15.054	0.000	0.801	1.047

[a] Dependent variable: price (£000s).

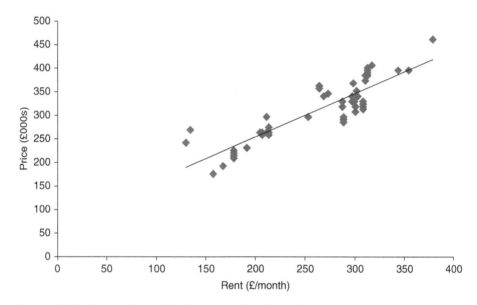

Figure 6.7   Linear relationship between sale price and rent.

Table 6.9   Residuals statistics.

Model[a]	Minimum	Maximum	Mean	Std. deviation	N
Predicted value	189.520172	419.210388	301.858333	58.1256071	60
Residual	−49.8714561	75.9145203	0.0000000	29.4059720	60
Std. predicted value	−1.933	2.019	0.000	1.000	60
Std. residual	−1.682	2.560	0.000	0.991	60

[a] Dependent variable: price_000s.

As a check on model fit, the scatterplot of standardised residuals against predicted responses in Figure 6.10 is relatively random and centred on 0. There were no outliers.

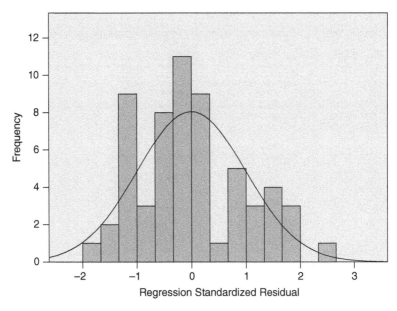

Figure 6.8   Histogram of dependent variable (Price £000).

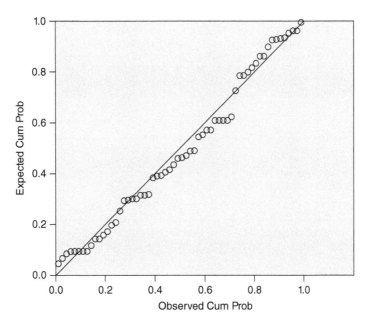

Figure 6.9   Normal P–P plot of regression standardised residuals for dependent variable (Price £000).

So rent is a pretty good predictor of price. This is unsurprising as investors pay prices that bear a relationship (expressed as a yield or multiple) to the rent.

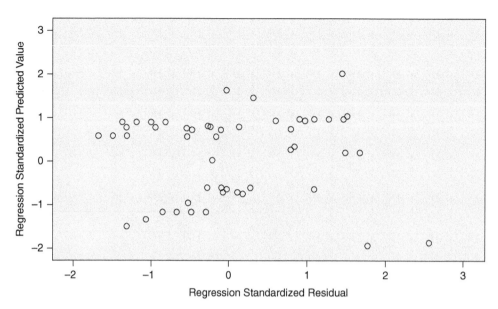

Figure 6.10   Scatterplot of standardised residuals against predicted values of response variable.

6.3.2   Multiple linear regression

Multiple linear regression seeks a linear combination of two or more predictor variables that correlate maximally with a response variable. The model looks like this:

$$y_i = b_0 + b_1 x_1 + b_2 x_2 + \cdots + b_k x_k + u \tag{6.13}$$

The model requires certain assumptions to be met if it is to perform effectively:

- All predictor variables are continuous or categorical with only two categories (coded as dummy variables) and the response variable is continuous and unbounded.
- Predictors should not have variances of zero and at each level of the predictor variable, the variance of the residual terms should be homoscedastic (have the same variance).
- Predictors should not correlate too highly (display multi-collinearity). This is investigated by scanning the correlation matrix of predictors to see if any have correlations above 0.80–0.90 (statistical software has multi-collinearity diagnostics).
- Predictors are uncorrelated with 'external variables' (omitted variable bias)
- All values of the response variable are independent.

Commercial, income-producing properties may require separate models for different property types or dummy variables to distinguish types. In additive models the dependent variable should be value per unit area, but multiplicative models are probably preferable due to wide variation in prices.

6.3.2.1   Preparing the predictor variables

Untransformed data directly describe attributes such as age and floor area. Some qualitative attribute data may need to be transformed into quantitative data before it can be included in the model. Category data can be transformed into sets of binary or 'dummy' (yes/no) variables using the most typical category as the benchmark and then coefficients of other categories are interpreted relative to the benchmark, for example how much more would someone pay for a detached dwelling relative to a benchmark terraced house? Qualitative scale variables (such as age bands) can also be transformed into quantitative ones. Scale variables should be centred on zero in additive models and one in multiplicative models. Some predictor variables may explain variation in the response variable non-linearly. Also, combinations of predictor variables might be included. An example would be two variables multiplied together such as floor area (a continuous variable where values get bigger as floor area increases) and quality of space (perhaps measured as a scale variable where poor is one, average is two and good is three). The resultant multiplicative variable should capture interactive effects. Another more commonly used example is a quotient variable whereby one variable, such as floor area, is divided by another, number of rooms for example, to produce average room size.

The minimum ratio of observations to predictor variables is 5:1 but if the response variable is skewed, many more observations may be needed. Univariate outliers can be identified using boxplots and multivariate outliers using Mahalanobis distances or residual scatterplots.

6.3.2.2   Interpreting multiple linear regression

6.3.2.2.1   Multi-collinearity between predictor variables

Using a matrix of correlation coefficients r between each pair of predictor variables, check whether correlations are >0.9. Note that r requires values of each variable to be normally distributed so if the independent variable is dichotomous (a dummy variable for example), then the correlation between it and a continuous variable is called a serial point correlation and equivalent to the independent sample t test. Therefore, do not put too much weight on these observations.

6.3.2.2.2   Model performance

Examine whether the change to R^2 resulting from adding each predictor is significant. Remember that as more predictor variables are added to the model, R^2 can only increase or stay the same and this can overstate goodness of fit when insignificant variables are included, or the number of variables is large compared to the sample size[7]. The significance of R^2 can be tested at each stage of model building using an F-ratio, which represents the ratio of improvement in model prediction relative to inaccuracy that remains:

$$F = \frac{(n-k-1)R^2}{k(1-R^2)} \tag{6.14}$$

where n is the number of cases and k is the number of predictors in the model.

6.3.2.2.3   Model parameters

In multiple regression analysis, the objective is to examine the relative importance of each predictor variable. Beta coefficients quantify the degree to which each predictor variable explains y if all the others are held constant. As in simple linear regression, the associated standard error indicates extent to which beta coefficients would vary across samples and is used to determine extent to which they are significantly different from zero. Because standardised beta coefficients are all in units of standard deviation, relative importance of each predictor can be compared[8]. t-statistics and corresponding p-values allow comparisons of explanatory importance to be made. The t-statistic measures the significance of a predictor in explaining differences in the dependent variable.

6.3.2.3　Residuals

6.3.2.3.1　Multi-collinearity

For any two observations, the residual terms should be uncorrelated (independent); in other words, they should lack autocorrelation. The Durbin–Watson test checks for serial correlation between the residuals – it tests whether adjacent residuals are correlated or independent – the test statistic varies between zero and four with two meaning residuals are uncorrelated. A value close to two is preferred; a value less than one or greater than three is cause for concern.

If the largest *variance inflation factor* (VIF) is >10, there is cause for concern. If the average VIF is substantially >1, the regression may be biased. Tolerance <0.1 is a serious problem and <0.2 is a potential problem. Check *eigenvalues* (how many distinct dimensions there are among independent variables). If several are close to zero, variables are highly inter-correlated. *Condition indices* are the square roots of the ratios of largest eigenvalue to each successive eigenvalue. A condition index >15 is a possible problem and > 30 is a serious problem. *Variance proportions* are the proportions of the variance of the estimate accounted for by each principal component associated with each of the eigenvalues. Collinearity is a problem when a component associated with a high condition index contributes substantially to the variance of more than two variables.

6.3.2.3.2　Influence of observations

Standardised residuals should be random, normally distributed and with a mean of zero. The observations should be checked to see if more than 5% have standardised residuals greater than ±2. Also, Cook's distance checks whether individual observations have an undue influence on the model. If the statistic is greater than one, then this is cause for concern. Leverage gauges influence of the observed value of the response variable over the predicted values. If no cases exert undue influence over the model, then leverage values should all be close to average ($[k + 1]/n$). It would be cause for concern if a case is more than two or three times the average. Mahalanobis distances measure the distance of cases from mean(s) of the predictor(s). For a large sample ($n > 500$) and 5 predictors, values greater than 25 are concerning; for smaller samples ($n = 100$) and three predictors, values greater than 15 would be cause for concern. DFBeta statistics show whether any case would have a large influence on regression parameters. An absolute value greater than one is a problem.

6.3.2.4　Example

The simple example described above can be expanded by introducing additional predictor variables. These are type of dwelling, type of heating and number of rooms. The expanded data set consists of the fields described in Table 6.10.

Table 6.10   Expanded data set.

Name of variable	Description of variable	Type of variable	Sub-type	Values
ID	Identification number	Quantitative	Category	Unique identifiers
Type	Type of dwelling	Qualitative	Category	D – Detached SD – Semi-detached ET – End-terrace MT – Mid-terrace
Rooms	Number of rooms	Quantitative	Interval	Ranges from three to eight rooms
Heating	Type of heating	Qualitative	Category	G – Gas AD – Air duct E – Electricity SF – Solid fuel O – Oil
Price	Capital value	Quantitative	Continuous	Capital value (£'000s)
Rent	Rental value	Quantitative	Continuous	Rental value (£/month)

None of these additional variables are continuous measures; two are categorical and one is ordinal (number of rooms), so they need to be converted to dummy variables. To do this, a baseline dwelling is established with the following attributes: detached, gas central heating and eight rooms, and dummy variables are created for other categories of these variables.

Frequencies of observations in each category are shown in Table 6.11. Some of the numbers are low – the end and mid-terraced dwellings for example but, as this is a hypothetical example, we will not worry too much about that. In practice, it will be necessary to collect a much bigger sample.

To look for multi-collinearity between predictor variables, a correlation matrix is shown in Table 6.12. None of the correlations are higher than 0.9, but this should not be given too much weight as the variables are dichotomous (dummy).

Model performance is reported in Table 6.13. Inclusion of the heating variables adds little explanatory power, so the tables hereafter relate to model number three.

Statistics for model parameters (coefficients) are shown in Table 6.14. The variable indicating whether the property is mid-terraced (TypeDumMT) is not a statistically significant predictor of price nor is the variable indicating seven bedrooms (Rms7).

Table 6.11   Frequency of observations.

Number of rooms	Frequency	Heating type	Frequency	Type of dwelling	Frequency
3	2	AD	20	D	21
4	11	E	8	ET	9
5	18	G	18	MT	8
6	18	O	7	SD	22
7	8	SF	7		
8	3				

Table 6.12 Correlation matrix.

Pearson correlation

	RENTpmth	Type SD	Type ET	Type MT	Rms3	Rms4	Rms5	Rms6	Rms7	HtgSF	HtgO	HtgE	HtgAD
RENTpmth	1.000	-0.033	-0.541	0.360	-0.280	-0.620	-0.220	0.390	0.393	0.387	-0.102	-0.400	-0.050
TypeDumSD	-0.033	1.000	-0.320	-0.298	-0.141	-0.182	0.257	0.257	-0.298	-0.277	-0.061	-0.298	0.783
TypeDumET	-0.541	-0.320	1.000	-0.165	0.442	0.645	-0.275	-0.275	-0.165	-0.153	0.138	0.796	-0.297
TypeDumMT	0.360	-0.298	-0.165	1.000	-0.073	-0.186	-0.257	-0.043	0.712	0.774	-0.143	-0.010	-0.277
Rms3	-0.280	-0.141	0.442	-0.073	1.000	-0.088	-0.122	-0.122	-0.073	-0.067	-0.067	0.473	-0.131
Rms4	-0.620	-0.182	0.645	-0.186	-0.088	1.000	-0.310	-0.310	-0.186	-0.172	0.365	0.448	-0.152
Rms5	-0.220	0.257	-0.275	-0.257	-0.122	-0.310	1.000	-0.429	-0.257	-0.238	-0.238	-0.257	0.154
Rms6	0.390	0.257	-0.275	-0.043	-0.122	-0.310	-0.429	1.000	-0.257	-0.238	-0.011	-0.150	0.309
Rms7	0.393	-0.298	-0.165	0.712	-0.073	-0.186	-0.257	-0.257	1.000	0.774	-0.143	-0.154	-0.277
HtgSF	0.387	-0.277	-0.153	0.774	-0.067	-0.172	-0.238	-0.238	0.774	1.000	-0.132	-0.143	-0.257
HtgO	-0.102	-0.061	0.138	-0.143	-0.067	0.365	-0.238	-0.011	-0.143	-0.132	1.000	-0.143	-0.257
HtgE	-0.400	-0.298	0.796	-0.010	0.473	0.448	-0.257	-0.150	-0.154	-0.143	-0.143	1.000	-0.277
HtgAD	-0.050	0.783	-0.297	-0.277	-0.131	-0.152	0.154	0.309	-0.277	-0.257	-0.257	-0.277	1.000

Sig. (1-tailed)

	RENTpmth	Type SD	Type ET	Type MT	Rms3	Rms4	Rms5	Rms6	Rms7	HtgSF	HtgO	HtgE	HtgAD
RENTpmth		0.402	0.000	0.002	0.015	0.000	0.045	0.001	0.001	0.001	0.218	0.001	0.351
TypeDumSD	0.402		0.006	0.010	0.141	0.082	0.024	0.024	0.010	0.016	0.322	0.010	0.000
TypeDumET	0.000	0.006		0.104	0.000	0.000	0.017	0.017	0.104	0.122	0.146	0.000	0.011
TypeDumMT	0.002	0.010	0.104		0.290	0.078	0.024	0.373	0.000	0.000	0.139	0.471	0.016
Rms3	0.015	0.141	0.000	0.290		0.252	0.177	0.177	0.290	0.304	0.304	0.000	0.159
Rms4	0.000	0.082	0.000	0.078	0.252		0.008	0.008	0.078	0.094	0.002	0.000	0.123
Rms5	0.045	0.024	0.017	0.024	0.177	0.008		0.000	0.024	0.034	0.034	0.024	0.120
Rms6	0.001	0.024	0.017	0.373	0.177	0.008	0.000		0.024	0.034	0.466	0.127	0.008
Rms7	0.001	0.010	0.104	0.000	0.290	0.078	0.024	0.024		0.000	0.139	0.120	0.016
HtgSF	0.001	0.016	0.122	0.000	0.304	0.094	0.034	0.034	0.000		0.157	0.139	0.024
HtgO	0.218	0.322	0.146	0.139	0.304	0.002	0.034	0.466	0.139	0.157		0.139	0.024
HtgE	0.001	0.010	0.000	0.471	0.000	0.000	0.024	0.127	0.120	0.139	0.139		0.016
HtgAD	0.351	0.000	0.011	0.016	0.159	0.123	0.120	0.008	0.016	0.024	0.024	0.016	

Table 6.13 Model summary.

Model[a]	R	R^2	Adjusted R^2	Std. error of the estimate	Change statistics					Durbin–Watson
					R^2 change	F change	df1	df2	Sig. F change	
1	0.892[b]	0.796	0.793	29.6583884	0.796	226.617	1	58	0.000	
2	0.961[c]	0.924	0.919	18.5888583	0.128	30.882	3	55	0.000	
3	0.981[d]	0.963	0.956	13.6630103	0.039	10.361	5	50	0.000	
4	0.982[e]	0.964	0.954	13.9581541	0.001	0.477	4	46	0.752	1.652

[a] Dependent variable: price_000s.
[b] Predictors: (Constant), RENTpmth.
[c] Predictors: (Constant), RENTpmth, TypeDumSD, TypeDumMT, TypeDumET.
[d] Predictors: (Constant), RENTpmth, TypeDumSD, TypeDumMT, TypeDumET, Rms3, Rms6, Rms5, Rms7, Rms4.
[e] Predictors: (Constant), RENTpmth, TypeDumSD, TypeDumMT, TypeDumET, Rms3, Rms6, Rms5, Rms7, Rms4, HtgO, HtgAD, HtgSF, HtgE.

Table 6.14 Coefficients.

Model[a]	Unstandardised coefficients		Standardised Co-efficients			95.0% confidence interval for B		Correlations			Collinearity statistics	
	B	Std. error	Beta	t	Sig.	Lower bound	Upper bound	Zero-order	Partial	Part	Tolerance	VIF
3 (Constant)	227.285	23.023		9.872	0.000	181.042	273.528					
RENTpmth	0.556	0.062	0.537	8.987	0.000	0.431	0.680	0.892	0.786	0.245	0.209	4.783
TypeSD	−33.163	4.436	−0.247	−7.476	0.000	−42.073	−24.252	−0.225	−0.726	−0.204	0.681	1.469
TypeET	−50.544	8.880	−0.279	−5.692	0.000	−68.379	−32.708	−0.644	−0.627	−0.155	0.309	3.231
TypeMT	1.726	7.695	0.009	0.224	0.823	−13.730	17.182	0.446	0.032	0.006	0.455	2.199
Rms3	−88.152	19.865	−0.245	−4.438	0.000	−128.052	−48.252	−0.362	−0.532	−0.121	0.245	4.087
Rms4	−63.448	15.747	−0.380	−4.029	0.000	−95.077	−31.820	−0.600	−0.495	−0.110	0.084	11.933
Rms5	−43.554	11.458	−0.309	−3.801	0.000	−66.568	−20.540	−0.086	−0.473	−0.104	0.113	8.861
Rms6	−52.818	9.679	−0.375	−5.457	0.000	−72.259	−33.377	0.149	−0.611	−0.149	0.158	6.323
Rms7	−15.956	11.140	−0.084	−1.432	0.158	−38.331	6.419	0.517	−0.199	−0.039	0.217	4.609

[a] Dependent variable: price_000s.

In terms of the residual statistics reported in Table 6.15, there are no Cook's distances greater than one, and the leverage values are fairly closely packed around the mean. There are two data points with quite high Mahalanobis distances (28.517).

Only three observations had standardised residuals greater than plus or minus two standard deviations adrift, shown in Table 6.16, well below the 5% threshold.

The scatterplot of standardised residuals against standardised predicted values in Figure 6.11 appears random, indicating homoscedasticity.

Figure 6.12 shows that the distribution is closer to normal than was the case with the simple linear regression.

The normal P–P plot in Figure 6.13 shows that that the standardised residuals follow a relatively straight line.

Table 6.15   Residuals statistics.

Statistic[a]	Minimum	Maximum	Mean	Std. deviation	N
Predicted value	176.000000	437.559479	301.858333	63.9147543	60
Standard predicted value	−1.969	2.123	0.000	1.000	60
Standard error of predicted value	3.786	9.661	5.348	1.599	60
Adjusted predicted value	176.000000	438.192657	302.101918	64.1676400	60
Residual	−28.1117153	32.5655594	0.0000000	12.5778205	60
Standard residual	−2.058	2.383	0.000	0.921	60
Mahal. distance	3.546	28.517	8.850	6.281	60
Cook's distance	0.000	0.318	0.024	0.056	60
Centred leverage value	0.060	0.483	0.150	0.106	60

[a] Dependent variable: price_000s.

Table 6.16   Case-wise diagnostics (dependent variable: price_000s).

Case number	Std. residual	Price_000s	Predicted value	Residual
2	−2.058	396.0000	424.111715	−28.1117150
27	−2.037	264.0000	291.829285	−27.8292851
38	2.383	363.0000	330.434440	32.5655604

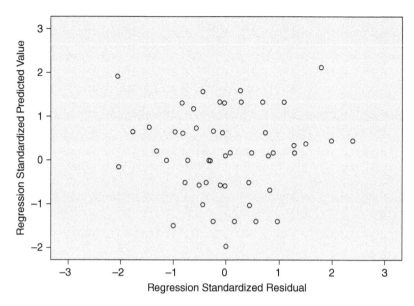

Figure 6.11   Scatterplot (dependent variable: price).

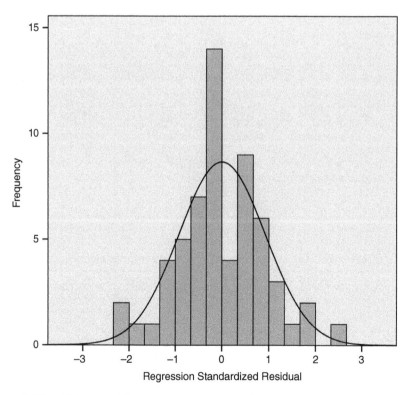

Figure 6.12   Histogram (dependent variable: price).

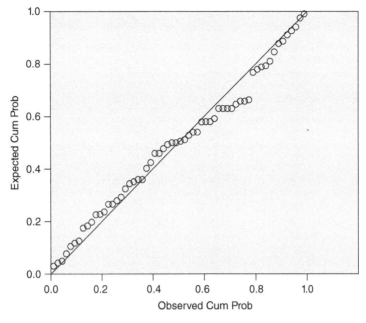

Figure 6.13   Normal P–P plot of regression standardised residual (response variable: price).

Key points

- The comparison method utilises transaction data generated by the market and is based on a rational approach that compares characteristics and adjusts for any differences. The approach is less reliable when data are scarce.
- Complex income-producing properties are harder to analyse due to the possible existence of special circumstances. For example, a landlord may accept a lower rent from a tenant who renews a lease. Also, incentives offered by the landlord such as a rent-free period and incentives offered by tenants such as a premium must be handled carefully to ensure a rational and defensible adjustment is made. Other dangers include transactions that are not at arm's length.
- The principle of comparison is fundamental to all methods used to value properties: estimates of market rents, yields, expenses, land values, construction costs and depreciation may be derived using comparison techniques.
- The comparison method relies on transaction prices and rents generated by trading activity and transactions to provide the evidence on which to base an estimate of market value. Of course, no two properties are the same so a valuer must adjust for differences in tenure rights, land and property characteristics or transaction date. The approach relies on comprehensive and up-to-date records of trading activity.
- Utilising a large data set of rents or prices, along with a range of value-significant attributes, the hedonic regression method uses a mathematical model to estimate property values. The method is best suited to valuing relatively homogeneous properties.
- The hedonic regression method can be used to value large numbers of properties quickly and at low cost, so long as data are available. It is often used to value properties for taxation purposes, where it is referred to as mass appraisal.

Notes

1. Stairs, lifts, motor rooms, toilets, cleaners' cupboards, plant rooms, fire refuge areas and maintenance rooms.
2. In some parts of the United Kingdom, deeper zones are used. For example, in parts of Oxford Street in London, where retail units are larger, 9.14 metres (30 feet) zones are used.
3. It is possible to model non-linear relationships between two variables by transforming one or both sets of values using, say, reciprocal, exponential, logarithmic or power functions. For example, the relationship between price and distance to a town centre may be non-linear; price might fall at a decreasing rate. Transforming one of the variables would allow the relationship to be included in the linear model.
4. In other words, significantly different from the mean. As the mean is a horizontal line, its b (gradient) is zero.
5. The standard error measures the error associated with using b as an estimator of the true, but unknown, relationship between predictor variable x and response variable y.
6. When t is small, one cannot reject H_0 that b equals 0 and that the predictor is unimportant in explaining y. This does not mean that the predictor is not correlated with y: the t-value measures the marginal contribution of a dependent variable in predicting y when all other variables in the equation are held constant. Because of multi-collinearity, some variables that duplicate information provided by others may be highly correlated with y but insignificant predictors as indicated by their t-values. Conversely, some variables might predict y in combination with others but individually none may be highly correlated with y.
7. Adjusted $\bar{R}^2 = 1 - \dfrac{(n-1)SS_R}{(n-p-1)SS_T}$, where n is the sample size (number of observations) and p is the number of independent variables, adjusts for the number of explanatory terms by limiting the degrees of freedom and increases only if the new term improves the model more than would be expected by chance. $\bar{R}^2$ can be negative and will always be less than or equal to R^2. It is relevant when data sets are small (<30 observations) in relation to the number of explanatory variables.
8. To interpret βs literally requires standard deviations of the variables. For example, if a predictor variable's $\beta = 0.5$ and standard deviation was 10 and the standard deviation of the response variable was 15 then, as the predictor value increases by 10, the response variable increases by 7.5 (with other predictors held constant).

References

Chegut, A., Eichholtz, P., Holtermans, R., and Palacios, J. (2020). Energy efficiency information and valuation practices in rental housing. *J. Real Estate Financ. Econ.* 60: 181–204.

Dalton, B. and Fuerst, F. (2018). The 'green value' proposition in real estate: a meta analysis. In: *Routledge Handbook of Sustainable Real Estate* (ed. S. Wilkinson, T. Dixon, N. Miller and S. Sayce), 177–200. London: Routledge.

European AVM Alliance (2019). *European Standards for Statistical Valuation Methods for Residential Properties*, 2e.

Hayward, R. (2009). *Valuation: Principles into Practice*, 6e. London: Estates Gazette.

IPMSC (2014) International Property Measurement Standards: Office Buildings, International Property Measurement Standards Coalition, November 2014.

IPMSC (2016) International Property Measurement Standards: Residential Buildings, International Property Measurement Standards Coalition, September 2016.

IPMSC (2018) International Property Measurement Standards: Retail Buildings, International Property Measurement Standards Coalition.

IPMSC (2019) International Property Measurement Standards: Office Buildings, International Property Measurement Standards Coalition, November 2014.

RICS (2012). *Valuation of Individual New-Build Homes: Guidance Note*, 2e. RICS Professional Guidance UK: Royal Institution of Chartered Surveyors.

RICS (2019). *Comparable Evidence in Property Valuation, Guidance Note*, 1e. Royal Institution of Chartered Surveyors.

RICS (2021). *Land Measurement for Planning and Development Purposes, Guidance Note*, 1e, May 2021. Royal Institution of Chartered Surveyors.

Sayce, S., Smith, J., Cooper, R., and Venmore-Rowland, P. (2006). *Real Estate Appraisal: From Value to Worth*. Oxford: Blackwell Publishers.

Wiltshaw, D. (1991). Valuation by comparable sales and linear algebra. *Land Dev. Stud.* 8 (1): 3. Spring.

Questions

Comparison method

[1]
- a) What are the typical characteristics of a prime property?
- b) Excluding rent and value, list the ways in which properties may differ.
- c) What problems do confidentiality, thin trading and heterogeneity poses for valuers?

[2] An office suite of 1000 square metres has just been let at £15 000 per annum. The landlord is liable for maintaining the structure and the common parts and the insurance of the building. A service charge covers the cost of heating and lighting. The landlord's liability amounts to £3 per square metre. What is the net rent?

[3] Calculate the MR of a retail property that comprises a ground floor only, with a frontage of 3 metres and a depth of 26 metres. Use a zone A rate of £1400 per square metre, 6-metre deep zones and value any 'remainder' at A/8.

[4] Shop 2 has just been let for £63 000 per annum. It has a frontage of 4 metres and a depth of 12 metres. Using 6-metre zones, value shop 1, which has a frontage of 9 metres and a depth of 18 metres.

[5] A shop just let at £100 000 per annum. Its ground floor has a frontage length of 8 metres and a depth of 20 metres, and the first-floor measures 8 metres by 15 metres. A second shop measures 7 metres frontage by 24 metres depth on ground floor and 7 metres by 15 metres on first floor. What is market rent of the second shop? Assume 6-metre zones, 'remainder' valued at 20% of zone A (i.e. A/5), and first-floor space valued at 12.5% of zone A (i.e. A/8).

Hedonic regression method

[1] The data below give the selling price (£'000s) and floor space (m²) of seven houses:

Size (x)	Price (y)
74	52
91	79
126	137
152	165
108	85
89	96
97	84

a) Draw a scatterplot with size (the explanatory variable) on the x axis and price (the response variable) on the y axis.

b) Draw a 'best-fit' line through the points on the graph by eye (the slope of which is the average predicted change in y per unit of change in x).

c) Using a spreadsheet, estimate the regression line of y on x.

d) Report the r^2 of this regression line.

[2] For a survey of overcrowding in cities, the following data were collected for eight districts:

Distance from city centre (km)	Population density (000s/km²)
5	27.5
10	20.8
16	19.1
21	15.0
25	10.0
28	9.0
32	6.2
38	4.9

a) Using formulae, calculate the regression equation.

b) For every kilometre away from the city centre, what is the average decline in population density?

c) How accurately does this equation fit the data?

Answers

Comparison method

[1]

a) A good location, occupied by a financially sound tenant, an 'institutional' lease, i.e. not too long or short with regular upward-only rent reviews, modern flexibly designed building with no structural problems.

b) Location and legal factors: Location, tenant, lease terms, tenure, use. The nature of the building: Age, condition, size, construction, services (aircon, heating, lighting systems, etc.), configuration.

c) The key issue is that valuers rely on information about market transactions to inform them of market conditions, so confidentiality restricts its availability of information, heterogeneity limits its relevance, and 'thin' trading reduces its reliability.

[2]

Rent on internal repairing and insuring basis	£15 000
Less landlord's outgoings @ £3/m²	£3 000
Rent on full repair and insuring (FRI) terms	£12 000

Rent on FRI terms is sometimes called net rent, and this is the rental value normally used in the market valuation of freehold property using income-based valuation methods. It is a net rent because it is the rent for the space only and excludes repairs and insurance costs.

[3]

Space	Width (m²)	Depth (m²)	Area (m²)	% Zone A	Area ITZA (m²)
GF zone A	3.00	6.00	18.00	100.0%	18.00
GF zone B	3.00	6.00	18.00	50.0%	9.00
GF zone C	3.00	6.00	18.00	25.0%	4.50
GF remainder	3.00	8.00	24.00	12.5%	3.00
End allowance (as % of GF area ITZA)				0.0%	0.00
Other sales space			0.00	0.0%	0.00
Storage area			0.00	0.0%	0.00
Totals		26.00	78.00		34.50

So, $34.50 \, m^2 \times £1400/m^2 = £48\,300$ p.a.

[4]

Analyse shop 2

Zone A	4 m × 6 m	24 m²
Zone B	(4 m × 6 m)/2	12 m²
Area ITZA		36 m²
Zone A rent/m² = MR/area ITZA = £63 000/36 m² = £1750/m² p.a.		

Value shop 1

Zone A	9 m × 6 m	54 m²
Zone B	(9 m × 6 m)/2	27 m²
Zone C	(9 m × 6 m)/4	13.5 m²
Area ITZA		94.5 m²
MR = 94.5 m² × £1750/m² = £165 375 p.a.		

[5]

Analysis of comparable

Zone A	8 m × 6 m	48 m²
Zone B	(8 m × 6 m)/2	24 m²
Zone C	(8 m × 6 m)/4	12 m²
Remainder	(8 m × 2 m)/5	3.2 m²
First Floor	(8 m × 15 m)/8	15 m²
Area ITZA		102.2 m²
Zone A rent/m² = £100 000/102.2 m² = £978/m²		

Valuation of subject

Zone A	7 m × 6 m	42 m²
Zone B	(7 m × 6 m)/2	21 m²
Zone C	(7 m × 6 m)/4	10.5 m²
Remainder	(7 m × 6 m)/5	8.4 m²
First floor	(7 m × 15 m)/8	13.1 m²
Area ITZA		95 m²
MR = 95 m² x £978/m² = £92 910 p.a.		

Hedonic regression method

[1]

a) and b)

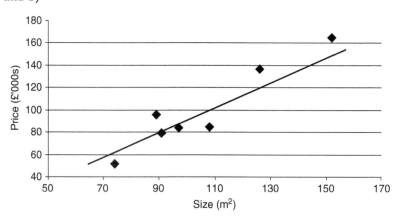

c) $y = 1.39x - 46.97$ (price increases at a rate of £1390/m²)

d) 0.9085

[2]

a) $y = 29.34 - 0.6984x$

b) For every km away from the centre, average population density declines by –698.4/km².

c) $r^2 = 0.9652$

Chapter 7
Income Approach

7.1   Introduction

Land and property can be held as investments where the holder of tenure rights passes some of those rights to another party in return for regular payments. The most common arrangement is where an owner leases occupation rights to a tenant. The tenant pays rent to the owner and the level of rent is determined by the supply of and demand for that type of property in the occupier market.

To the owner, rent represents the income return on the investment and the present value of the property may be determined by capitalising the rent at a suitable yield or capitalisation rate. The yield is usually derived from analysis of recent transactions involving comparable properties. Alfred Marshall was the first to expound methods of capitalising urban rental income as a means of pricing property investments. He focused on the scenario whereby landowners let sites on long ground leases, for 99 years say, and stated that the

> capitalized value of any plot of land is the actuarial 'discounted' value of all the net incomes which it is likely to afford, allowance being made on the one hand for all incidental expenses, including those of collecting the rents, and on the other for its mineral wealth, its capabilities of development for any kind of business, and its advantages, material, social and æsthetic, for the purposes of residence (Marshall 1920: Book Five, Chapter 11).

The cash flows of some income-producing properties can be complex, such as an office building or a shopping centre let to many tenants with different rents, lease terms and rent-review dates. In these cases, the *income capitalisation* method may not be appropriate. Instead, a cash flow may be constructed where inflows and outflows are dealt with in more detail. At the end of this period, if the investment is a perpetual annuity such as a freehold interest, a sale may be assumed. This

Property Valuation, Third Edition. Peter Wyatt.
© 2023 John Wiley & Sons Ltd. Published 2023 by John Wiley & Sons Ltd.
Companion website: www.wiley.com/go/wyatt/propertyvaluation3e

discounted cash flow (DCF) method requires estimates of the cash-flow period, net income on a period-by-period basis, a discount rate for capitalising the net income, and a sale value at the end of the cash flow.

Certain types of properties are linked to the businesses that operate from them. These *trade-related* properties might be purpose-built or have some monopoly value due to their legal status or planning consent. Examples include mining rights, hotels, casinos and bars. Other businesses such as fisheries, safari parks and fuel stations may be linked to a specific location. Trade-related properties can be valued using the income approach by capitalising estimated future trading potential (rather than capitalising estimated rental income). There is a heavy reliance on accounts and other financial information about the business and also reliance on expertise to value the goodwill element of the business, such as advance bookings. Personal property may need to be valued in conjunction with real property and a special assumption would be required to reflect this.

This chapter is structured as follows:

- Income capitalisation method: For valuing perpetuities (freehold property investments) and annuities (leasehold property investments)
- DCF method: For valuing complex property investments
- Profits method: For valuing trade-related property investments

7.2   Income capitalisation method

Lean and Goodall (1966) and Fraser (1993) provide excellent summaries of the investment characteristics of the main types of freehold property investment. A freehold in possession (the interest of an owner-occupier where there are no sub-interests) is a pure equity interest, which affords the owner a perpetual right to the full benefits of the property. For a business, this is the right to the profit obtainable from undertaking business activity on the premises without the liability to pay rent. The notional annual return from this interest, known as *imputed rent*, is the market rent (MR) of the property. For a freehold acquired as an investment (where the property is let) the equity extent of the freehold depends on the lease terms and, in particular, the frequency with which the rent is reviewed to MR. For example, a long lease without review is a fixed-income investment whereas an annually reviewed turnover rent is an equity investment. Rent reviews offer a degree of income security in real terms.

A fixed-income freehold property investment, such as a long (ground) lease without rent reviews, tends to have a higher yield than a freehold property investment let on a shorter lease with lots of rent reviews. This is due to the lack of growth potential in the ground lease, and the yield might be similar to yields on undated bonds but somewhat above to reflect their comparative illiquidity. As the end of a long ground lease approaches, the yield falls in anticipation of (often very high) reversionary value. When the value of the freehold ground rent for the remaining term plus the value of the reversion exceed the value of the freehold ground rent in perpetuity, the reversion is affecting value. From this point, the investment will exhibit equity investment characteristics because the reversionary MR can be affected by rental growth prospects and the like. In more recent years, rent-review clauses have been inserted into freehold ground leases and these introduce a further equity element. Freehold investments let on shorter leases or leases with regular rent reviews are equity-type investments.

The actual rent specified in the lease and currently paid is known as the rent passing or *contract rent* and the rent that a property would normally command in the open market as indicated by rents paid for comparable space near to the valuation date is known as the *market rent* (MR). Whether a contract rent or MR, the valuer must determine the *net* income (receivable after deductions for any repairs, insurance, services, rates, head rents and other rent charges) and the period for which it will be received.

Income capitalisation involves capitalising the net income[1] at an appropriate yield derived from comparable evidence of similar investment transactions. Valuers analyse the current and anticipated supply and demand for properties similar to the one being valued, analyse rents and prices of comparable investment transactions, calculate their yields, derive a suitable yield for the subject property and use it to capitalise its actual or estimated rent. Any future growth in economic benefits (either rental income or capital value) is accounted for in (implied by) the choice of yield. The approach is therefore *growth implicit* in that it does not explicitly project the rent into the future. This contrasts with the DCF method where rent is projected or forecast, and the discount rate is *growth explicit*.

An investor may be willing to pay more than market value if the property satisfies requirements specific to that investor (a gap in an investment portfolio for example) but if this sort of decision-making is not reflected in the market, then it should not influence an opinion of market value. Instead, this is *investment value*, discussed later.

7.2.1   Perpetual annuities (freeholds)

Depending on the timing of the investment acquisition, a freehold property investment might be *rack-rented* or *reversionary*. A rack-rented property is one which is let at the current MR while a reversionary freehold property investment is one where the property is let below MR but with a reversion (usually at a rent review or lease renewal) to MR in the future.

7.2.1.1   Valuation of rack-rented freehold property investments

For a property to be rack-rented at the valuation date, it must have either been let or been subject to a lease renewal or rent review so recently that the contract rent is assumed to be the MR. If the property is vacant at the valuation date it is common practice to assume an MR possibly subject to a letting period or an adjustment to the yield to reflect the fact that the property is vacant.

For example, value the freehold interest in a property that was recently let at a net MR of £100 000 per annum. Analysis of recent transactions for similar premises reveals that an appropriate net initial yield (NIY) is 8%. The MR is receivable in perpetuity[2] and this can be capitalised at 8% using Eq. (4.24) from Chapter 4 (in which y is the NIY):

$$V = \frac{MR}{NIY} = \frac{£100000}{0.08} = £1250000$$

The inverse of the yield is a multiplier known as the years purchase (YP), so called because it represents the number of years over which the net income must be received in order to recoup the present value. Mathematically the YP is the equivalent of the *present value of £1 pa* and, conventionally, is multiplied by the net rent (from now on, we will dispense with the word 'net') to determine the total present value or, simply, the value of the property. The valuation would therefore be set out as follows:

MR (£)	100 000
× YP in perpetuity @ 8%	12.5
Valuation before PCs (£)	1 250 000

As a final step, purchase costs (PCs) should be deducted from the valuation. These comprise stamp duty or transfer taxes plus any legal fees and broker fees.

The formula above can be rearranged to derive NIYs from comparable evidence where the MR and price paid (P) are known:

$$NIY = \frac{MR}{P}$$

For example, a property was recently let at a rent of £150 000 per annum and the freehold interest has just been sold for £2 250 000 (including purchase costs). The yield from this investment is calculated as follows:

$$NIY = \frac{£150\,000}{£2\,250\,000} = 6.67\%$$

The income profile of a typical rack-rented property investment is illustrated in Figure 7.1. At the beginning of a new lease the property is let at MR or, if the property is empty, an estimate of MR is derived from comparable evidence. Over time the MR of equivalent new properties increases (the solid line) if there is growth in the economy, but the MR of the subject property, which is getting older,

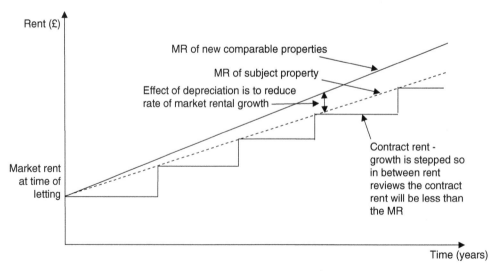

Figure 7.1   Income profile of a rack-rented property investment.

will not keep pace (the dashed line). The actual rent received by the investor rises in steps under a typical UK lease arrangement to the MR of the subject property every five years (the stepped solid line).

7.2.1.2   *Valuation of reversionary freehold property investments*

Often the contract rent is not the current MR because it was agreed some time ago, usually when the lease began or at the last rent review, or perhaps a premium was paid, and the rent was reduced to reflect this. The income from a reversionary freehold investment comprises a contract rent secured by the lease contract and a potential uplift or reversion to a higher MR at the next rent review or lease renewal. The value of this potential reversion should be reflected in the price the investor pays. Theoretically, according to Baum and Crosby (1995), the growth potential of reversionary investments where the term is less than the normal rent-review period of five years is greater than for a rack-rented property because the first rent review will be in less than five years' time. However, it is rare for reversionary freehold investments to be valued more highly than rack-rented equivalents because the reversion comes with a degree of uncertainty. Fraser (1993) argues that, for a reversionary investment, the value impact of the reversion becomes greater as it draws nearer; immediately after a rent review the capital value growth rate tends to be less than the rental growth rate but as the reversion draws nearer it tends to exceed it. Thus, investors purchasing reversionary investments anticipate three elements of return: current income, capital gain deriving from rental growth and capital gain deriving from the passage of time to reversion uplift. The latter is in effect rental growth from earlier years (not yet received because of five-year rent reviews) being stored up and released as capital gain as the reversion approaches.

Two techniques are used to value reversionary investments: term and reversion, and core and top slice.

7.2.1.2.1   Term and reversion

The contract rent (also known as the term rent or rent passing) is capitalised until the point at which it reverts to MR. Then the MR is capitalised in perpetuity, but this capital value is deferred from now until the point at which it is received. These two capital values are then added together. This is shown diagrammatically in Figure 7.2.

Mathematically, the income streams are valued as follows:

$$V = (t \times \text{YP for term}) + (m \times \text{YP in perpetuity} \times \text{PV for term})$$

$$= \left[t \times \left(\frac{1 - (1 + y_t)^{-n}}{y_t} \right) \right] + \left[m \times \frac{1}{y_r} \times \frac{1}{(1 + y_r)^n} \right] \qquad (7.1)$$

where V = value

t = contract rent for term

YP = years purchase (*PV £1 pa*)

m = market rent

n = period to rent revision

y_t = term yield

y_r = reversion yield

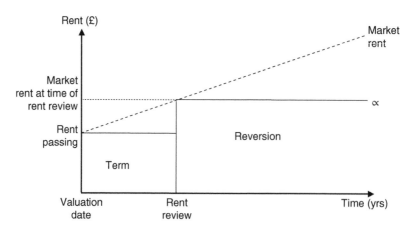

Figure 7.2   Term and reversion valuation.

For example, a property is currently let at £250000 per annum on a lease with four years unexpired. The MR is £300000 per annum and the NIY is estimated to be 8%. A valuation of this property investment is set out below.

Term rent (£)	250 000	
YP 4 years @ 8%	3.3121	
		828 025
Reversion to MR (£)	300 000	
YP perpetuity @ 8%	12.5000	
Deferred 4 years (PV £1 for 4 years @ 8%)	0.7350	
		2 756 250
Valuation before PCs (£)		3 584 275

For some investments, where the rent is fixed for a long period before the reversion, a higher yield might be appropriate to reflect the delay in receiving rental growth. The term and reversion technique is useful for valuing property investments let on short leases and with break clauses because it is straightforward to insert a gap or 'void' between the two tranches of rental income that represents a re-letting period.

7.2.1.2.2   Core and top slice

The contract rent (the core) is capitalised in perpetuity and the top-slice, which is the difference between the MR and the contract rent, is also capitalised in perpetuity but deferred until the rent review or lease renewal. These capital values are then added together. This is shown diagrammatically in Figure 7.3.

Mathematically, the valuation would be as follows:

$$V = \left(c \times \text{YP into perpetuity} \right) + \left((m - c) \times \text{YP in perpetuity} \times \text{PV for term} \right)$$

$$= \left[c \times \frac{1}{y_c} \right] + \left[(m - c) \times \frac{1}{y_t} \times \frac{1}{(1 + y_t)^n} \right] \tag{7.2}$$

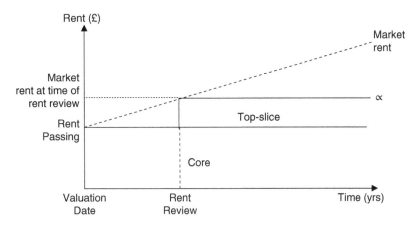

Figure 7.3   Core and top-slice valuation.

where c = contract (core) rent in perpetuity
y_c = core yield
y_t = top-slice yield and the other variables are as defined for Eq. (7.1).
Using the same example:

Core: contract rent (£)	250 000	
YP in perpetuity @ 8%	12.5000	
		3 125 000
Top slice: uplift to MR (£)	50 000	
YP perpetuity @ 8%	12.5000	
Deferred 4 years (PV £1 for 4 years @ 8%)	0.7350	
		459 375
Valuation before PCs (£)		3 584 375

The rationale for dividing the income into these two layers is that the core rent is assumed to extend into perpetuity on the basis that there is little likelihood of the rent falling below the rent passing because of upward-only rent reviews and rental growth prospects. The contract rent and top slice are capitalised at a yield based on comparable evidence. Sometimes, if it is felt that the top-slice element is riskier, it might be capitalised at a slightly higher yield.

The core and top-slice technique can be adapted to value over-rented properties where the contract rent is higher than the current estimate of MR. The core is the MR and the top slice (known as *overage* in this case) is the uplift from MR to the contract rent. A higher yield may be used for the overage if it is considered to be at higher risk.

7.2.2   Annuities with a term certain (leaseholds)

7.2.2.1   Valuation of leasehold property investments

The freehold investments described above are capable of producing income for as long as the land is capable of economic use and the capital value may be realised at any time at least equivalent to the value of the site but usually enhanced by whatever buildings have been constructed. Leaseholds, on the other hand, are a more diverse group of property interests. There must be at least two legal interests in a property to create a leasehold investment; perhaps a head tenant leases the property from the freeholder via a head-lease, paying a head rent, and sublets to a sub-tenant by a sub-lease and receives a sub-rent. Assuming the rent received from the sub-tenant is greater than the rent paid to the landlord, the head tenant receives a *profit rent* and if the head-lease is assignable this profit rent may have a market value.

Generally speaking, leaseholds are a less popular form of property investment than freeholds. A leasehold investment is terminable, so the investment return – the profit rent – is in the form of income only. Being the difference between the rent paid and the rent received, changes in either of these will be magnified in the profit rent. In other words, the profit rent is contingent upon and geared to these rents in a way that 'clean' rent to a freeholder is not. If, for example, the sub-tenant stops paying rent, the head tenant (i.e. the leasehold investor) will still have to pay the head rent to the landlord. As the head-lease nears the end of its term, it will be harder to sub-let the property. Also, complex patterns of profit rent can occur if revisions to the rent received and rent paid are at different times. It is even possible for the profit rent to become negative when the contract rent exceeds the MR and, particularly in these situations but in general also, the quality of the tenant, and especially the ability to pay rent, is critical in determining risk. A leasehold investor may have repair, insurance and other liabilities under the terms of the lease as well as restrictions over the way that the interest may be transferred. These constraints can be inconvenient and costly, and consequently affect value. Some or all of these liabilities can be passed on to the sub-tenant under the terms of the sub-lease, but this is still a management cost, and the constraints remain nonetheless. All of this add to the costs and risk associated with leasehold investment.

Due to the relative unattractiveness of leasehold investments, valuations are less frequently undertaken but where they are required, they are more difficult than freeholds. Lease terms and termination dates, the gearing characteristics caused by the size of the profit rent compared to the head rent, and repair, insurance and other liabilities under the terms of the lease will all vary. This means that any yields that can be obtained from comparable evidence will be diverse and adjustments difficult to justify. The conventional solution to this problem is to capitalise the profit rent at a yield derived from freehold investments, which tend to be more commonplace and homogeneous than leasehold investments. The yield may be increased to reflect additional risk and reduced attractiveness of a leasehold investment compared to a freehold investment. An alternative is to use yields derived from non-property investments such as bonds (Baum and Crosby 1995). Fraser (1993) argues that if the MR and head rent (and hence the profit rent) are fixed for the whole term, the yield would be similar to long-dated gilts plus a risk premium to reflect default risk and terminable nature of the interest (remember that gilts return the capital invested).

For valuation purposes, it is useful to divide leasehold property investments into two types: those with fixed-profit rents and those with variable-profit rents.

7.2.2.1.1   Valuation of fixed-profit rents

These tend to take the form of short periods of profit rent between rent reviews or where the lease is short. As with freehold investment valuations, before the profit rent is capitalised all irrecoverable expenses must be deducted to arrive at a net profit rent.

For example, value the net profit rent of £25 000 per annum from a head-lease that has a remaining term of four years. The NIY is assumed to be 9%. The valuation may be set out in the conventional format:

Profit rent (£)	25 000	
YP for 4 years @ 9%	3.2397	
Valuation before PCs (£)		80 993

7.2.2.1.2   Valuation of variable profit rents

A variable-profit rent would arise if the head rent and sub-rent do not move in unison, perhaps because the sub-lease contains rent reviews, and the head rent is fixed (essentially the head rent is a fixed deduction from a growth income) or both head-lease and sub-lease contain reviews but at different times. Variable-profit rents tend to be for longer periods of say 10 or more years and can be for very long periods. Referring back to the freehold ground rents described at the beginning of this chapter, the head-leasehold interest in such an arrangement would take the form of a variable-profit rent. The head tenant could develop the site and let the property at an occupation rent (containing rent reviews) far in excess of the fixed ground rent.

A long ground lease (more than 50 years remaining) can be very similar in its income growth characteristics to a freehold investment over its early life. Figure 7.4 illustrates this. The lines track the capital values of two investments over a period of 50 years; the upper line is a freehold with a current MR of £100 000 per annum and a rental income growth rate of 5% per annum. The rent is projected every five years and capitalised at an NIY of 8%. So, in year 0 the current MR of £100 000 per annum is capitalised at 8% giving a capital value of £1 250 000 and in year 25 it is £100 000 compounded at 5% per annum for 25 years capitalised at 8% giving £4 232 944. Because this investment is a freehold, the capital value will keep rising as long as the growth rate and yield assumptions hold. The leasehold investment takes the form of a long (50 year) head-lease where the head rent is fixed at £10 000 per annum for the whole term and is sub-let at an MR of £100 000 per annum. Like the freehold, this sub-rent is predicted to grow at 5% per annum. So, the value of the long leasehold now is £100 000 less £10 000 giving a profit rent of £90 000 per annum capitalised at 8% (assume same as freehold yield for simplicity) for a fixed term of 50 years giving a capital value of £1 101 014. In year 25, the profit rent would have grown to £328 635 (£100 000 compounded at 5% per annum over 25 years less fixed ground rent of £10 000) and this is capitalised at 8% over the remaining 25 years of the lease giving a capital value of £3 508 110. However, towards the end of the lease term the capital value of the long lease drops dramatically.

For varying profit rent like this, a yield can be used to capitalise the net profit rent subject to the same caveats mentioned above. Using this approach, changes to (and growth in) the profit rent are implicitly handled by the yield. For example,

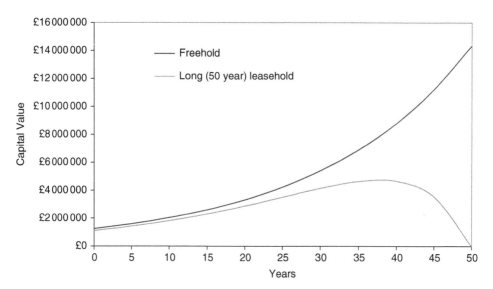

Figure 7.4   Freehold and leasehold capital values over time.

a property is held on a head-lease with 12 years unexpired at a fixed rent of £10 000 per annum with no further rent reviews. The property is sublet for the remainder of the head-lease term (less one day) at a current rent of £30 000 per annum with five-year rent reviews. The MR is £35 000 per annum. NIYs on comparable properties let on a freehold basis are 6%. This yield is increased by 2% to reflect the investment characteristics of the leasehold compared to freehold.

Rent received (£)	30 000	
Less rent paid (£)	(10 000)	
Profit rent (£)	20 000	
YP for 2 years @ 8%	1.7833	
Term value (£)		35 666
Reversion to MR (£)	35 000	
Less rent paid (£)	(10 000)	
Profit rent (£)	25 000	
YP for 10 years 8%	6.7101	
PV for 2 years @ 8%	0.8573	
Reversion value (£)		143 814
Valuation before PCs (£)		179 480

An alternative approach is to calculate the capital value of the ground rent separately and then deduct this amount from the capital value of the tenant's profit rent (from which the ground rent had not been deducted).

Fraser (1977) argued that it is wrong to compare yields from freehold investments with those from leaseholds with geared terminable profit rents because the income growth patterns will be different. For the profit rent from a leasehold to be comparable to the MR from a freehold investment, it needs to be for a long term and the head rent should be fixed and significantly below MR. Alternatively, the head-lease should contain reviews to a small fraction of the market rent payable under the sub-lease which should contain regular rent reviews to market levels. The yield that could be used to capitalise this type of investment might then be comparable to freehold yields but higher to reflect terminable nature of the investment and increased management and possible maintenance liability that the landlord might face. As the lease nears termination increasing capital depreciation (see the rapid decline in capital value of the long leasehold interest in Figure 6.7) means a leasehold investment bears little comparison to other types of investment and any supposed relationship between freehold and leasehold yields becomes tenuous. If valuing variable profits using a yield derived from freehold investments is considered to be too simplistic, the valuer might consider using the discounted cash-flow method, where the profit rent is forecast over the lease term.

7.3   Discounted cash-flow method

A yield is simply a ratio between income and capital value. Because income and capital value are expected to change (usually grow) over the life of an investment, investors are often prepared to accept a lower return (initial yield) at the start of the investment in expectation of higher returns later on. Rather than attempt to predict how income and value might change in the future, the income capitalisation method described above capitalises current estimates of rent at yields derived from comparable evidence. This means that these yields are often lower than the rate of return that an investor expects to receive because they imply future rental income and capital growth expectations. The gap between the yield and expected rate of return represents the expected or implied rental growth hidden in the valuation. Consequently, the assumed static cash flow is not the expected cash flow, and the yield is not the expected rate of return from the investment. This might sound bad but remember that income capitalisation is simply a method of valuation that uses market-derived evidence wherever possible, and this evidence takes the form of MRs and NIYs.

The income capitalisation method relies on comparison to justify adjustments to NIYs obtained from comparable investment transactions. These adjustments reflect differences in income and capital risk and growth potential, as well as economic and property-specific factors including macro-economic conditions, property market and sub-sector activity, the financial standing of individual tenants, property depreciation and changes in planning, taxation, landlord and tenant legislation. Yield adjustments must reflect all of these factors and this means that the resulting valuation is sensitive to small adjustments to the yield.

Income capitalisation is appropriate where there are sufficient comparable market transactions to provide evidence of yields and MRs. But there are circumstances when evidence is scarce, either because market activity is slow or because the property is infrequently traded. Also, there might be greater variability in investments, meaning more variables must be accounted for in the yield. For example, secondary properties are generally more variable in terms of location, physical quality, condition or covenant and are therefore riskier. Similarly, shorter leases can cause greater diversity in property investment cash flows, often with gaps in rental income.

The DCF method focuses more explicitly on the expected flow of income, expenditure and capital growth that might be expected from an investment. The method requires an explicit forecast of the cash flow, which may consist of net rental income plus a reversion or resale value, and this is discounted at a target rate of return.

7.3.1   A discounted Cash-Flow valuation model

The income capitalisation method relies on analysis of prices and rents achieved on recent comparable transactions to estimate a yield for the subject property. This growth-implicit yield is then used to capitalise the current contract rent and/ or current MR. In other words, the rent is not projected or forecast. The DCF method, on the other hand, capitalises projected rent at the investor's *target rate of return*. The method thus requires two additional assumptions compared to the income capitalisation method; a target rate of return (which should cover the opportunity cost of investment capital plus perceived risk) and an expected rental growth rate.

7.3.1.1   Constructing a DCF valuation model

The relationship between the income capitalisation method and the DCF method can be represented by Eq. (4.33) from Chapter 4:

$$y = r - g$$

where y is the yield, r is the investor's target return and g is the annual rental growth rate.

The left side of the equation represents the income capitalisation method and the right side represents the DCF method. The DCF method separates the yield into two parts: a target rate of return and a rental growth rate. In other words, the yield implies the rental growth that the investor expects in order to achieve the target rate of return. An investor accepting a relatively low initial yield from a property investment when higher yields might be available from fixed-income investments implies an expectation of future growth. For example, an investor with a target rate of 8% who purchases a property investment for a price that reflects an initial yield of 5% would require a 3% annual growth to achieve the target rate. This simple relationship is made more complex in the United Kingdom because rental income from property investments is normally reviewed every five years instead of annually. This means that a slightly higher annual growth rate will be required to meet the investor's annual target rate of return. Provided the growth rate, target return and rent-review period in the DCF method are mathematically consistent with the yield adopted in the income capitalisation method, the valuation will be the same. The following explains why.

Starting with the income capitalisation method, the present (capital) value, V, of an income stream from a rack-rented freehold property investment is the PV £1 pa or YP (see Eq. 4.19 in Chapter 4) multiplied by the annual income or MR:

$$V = MR \cdot \frac{1-(1+y)^{-n}}{y} \tag{7.3}$$

where y is the growth-implicit yield and n is the number of years for which the rent is received.If the rent is receivable in perpetuity, i.e. a freehold property investment, the above formula simplifies to Eq. (4.23) from Chapter 4:

$$V = \frac{MR}{y}$$

In other words, the present value is equivalent to a constant annual income capitalised at (divided by) the yield. In a DCF, the rent is discounted at the investor's target rate of return r rather than the yield. So, the present value of a rack-rented freehold property investment, which consists of a constant (i.e. non-growth) annual MR receivable in perpetuity annually in arrears, can be expressed as follows:

$$V = \frac{MR}{r} \tag{7.4}$$

But because the DCF method is explicit about rental growth, it must be stated separately in the valuation model. To begin with, assume rent is receivable in perpetuity and there are annual rent reviews that inflate the MR at an estimated long-term average annual rental growth rate g. As long as $r > g$, rental growth can be incorporated as follows:

$$V = \frac{MR}{r-g} \tag{7.5}$$

For rent that is reviewed at intervals of greater than annually, this equation must be altered. The following equation represents a rack-rented freehold property investment let at MR in perpetuity with rent reviews every three years:

$$V = \frac{MR}{(1+r)} + \frac{MR(1+g)}{(1+r)^2} + \frac{MR(1+g)}{(1+r)^3} + \frac{MR(1+g)^3}{(1+r)^4} + \frac{MR(1+g)^3}{(1+r)^5} + \frac{MR(1+g)^3}{(1+r)^6} + \frac{MR(1+g)^6}{(1+r)^7} + \cdots \infty$$

The above expression (which takes form of a geometric progression) simplifies to (see Appendix 7.B):

$$V = \frac{MR}{r - r \left(\dfrac{(1+g)^3 - 1}{(1+r)^3 - 1} \right)} \tag{7.6}$$

Rearranging Eq. (4.24) we can show that $\dfrac{MR}{V} = y$ and, substituting these variables into Eq. (7.6), the relationship between the income capitalisation method and the DCF method can be shown by:

$$y = r - r \left(\frac{(1+g)^p - 1}{(1+r)^p - 1} \right) \tag{7.7}$$

This is the property yield equation derived by Fraser (1993) and based on a rack-rented freehold property investment. It shows that y is determined by the investor's target rate of return r, the annual rental growth rate g, and the number of years between each rent review (the rent-review period) p. This equation is the same as Eq. (4.33) except that the annual rental growth rate g has been increased to compensate for the fact that rental growth is not actually received until each non-annual rent review.

If the property to be valued is rack-rented and the rent and review period are known, using the income capitalisation method, the valuer only has one variable, the yield, to predict in order to value the property. If sufficient evidence is available, this is straightforward. With the DCF method, there are two unknowns: the investor's target rate of return and the rental growth rate. To predict the rental growth rate, yields on recently let comparable freehold properties can be compared with an estimate of the investor's target return for those properties. Armed with this information and rearranging Eq. 7.7, an average annual growth rate can be implied as follows:

$$g = \left[\frac{(r-y)(1+r)^p + y}{r} \right]^{\frac{1}{p}} - 1 \tag{7.8}$$

If reviews were annual, the growth rate would be the target rate minus the initial yield on a rack-rented freehold property, i.e. $g = r - y$. For example, if an investor accepts an initial yield of 4% but requires an overall return of 7%, then the income must grow by 3% over the year, but with five-year rent reviews the implied rental growth rate would be:

$$g = \left[\frac{(0.07-0.04)(1+0.07)^5 + 0.04}{0.07} \right]^{\frac{1}{5}} - 1 = 3.23\%$$

So, an investor accepting an initial yield of 4% would require 3.23% per annum growth in the income, on average (compounded at each review), to achieve the target return. Figure 7.5 illustrates this.

Equation (7.8) is referred to as the implied rental growth rate formula. The higher the target rate of return relative to the yield, the higher the rental growth rate must be to achieve the target return. The implied growth rate formula is constructed assuming that the property is rack-rented. g represents the market's expectations of future growth and is an average growth rate. In fact, it is a discounted growth rate into perpetuity, so g is influenced by expectations in the near future more than ones further away (Fraser 1993). As an alternative, it is possible to derive an explicit growth rate from direct analysis of rental growth rates prevalent in various market sectors, regions and towns. Some argue that the assumption of a stable and constant growth rate is simplistic, but it can be taken to be an adequate reflection of the decision-making process of most investors. Before looking at the application of the DCF method, the next section will look at the input variables in more detail.

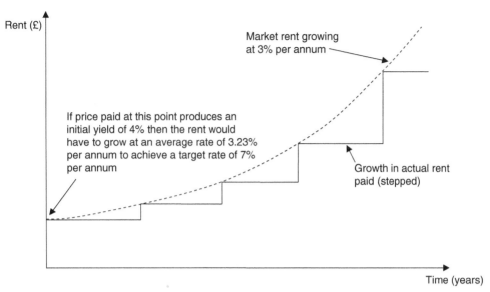

Rent (£)

Market rent growing
at 3% per annum

If price paid at this point produces an
initial yield of 4% then the rent would
have to grow at an average rate of 3.23%
per annum to achieve a target rate of 7%
per annum

Growth in actual rent
paid (stepped)

Time (years)

NB. Assumes rent received in perpetuity

Figure 7.5   Rental growth.

7.3.1.2   Key variables in the DCF valuation model

The key variables in the DCF model are rent, rental growth rate, target rate of return and, in cases of long-term and perpetual annuities, exit value.

7.3.1.2.1   Rent and rental growth
The estimation of MR is undertaken in the same way as for the income capitalisation method. Rental growth can be separated into two components: growth in line with inflation and real growth in excess of inflation. Depreciation is the rate at which the MR of an existing property falls away from the MR of a property that is comparable in all respects except that it is (hypothetically) permanently new. So, assuming constant rental growth, an annual rate of rental growth must be net of an average annual rate of depreciation. As these two components are interacting growth rates, their mathematical relationship with g is (Fraser 1993):

$$g = g_m - d - dg_m \tag{7.9}$$

where g is the average annual rental growth rate of actual property, g_m is the average annual rental growth rate of permanently new property and d is the average annual rate of depreciation. As dg_m is usually very small, the equation can be simplified to:

$$g = g_m - d \tag{7.10}$$

It is possible to conduct or commission forecasts of rents and rental growth rates, and the models that underpin these forecasts are usually based on analysis of past performance. It is important to determine whether the forecast is based on actual properties (and therefore include depreciation) or new properties (and therefore ignore effect of depreciation). Simple models might take the form of a historic time series of rents and capital values from which a moving average or exponentially smoothed set of values for future years might be predicted. More complex regression-based models will produce equations, which identify independent variables such as gross domestic product (GDP) or other output measures, expenditure, employment, stock, vacancy, absorption and development pipeline, and measure their effect on a dependent variable such as rental growth or yield (Baum 2000). Forecasts, although not at the individual property level, provide useful information on rental growth performance across the main investment sectors and locations in the United Kingdom and allow an implied rental growth rate to be verified against growth rates achieved in the market. It must be remembered, though, that rents can be volatile in the short term and very little is known about depreciation rates and their effect on rental growth prospects in the long term.

7.3.1.2.2 Target rate of return

The *target rate of return* (or *discount rate* because it is the rate at which cash flows are discounted to present value) should adequately compensate an investor for the opportunity cost of capital plus the risk that the investor expects to be exposed to. It is therefore a function of *a risk-free rate of return* and a *risk premium*: a higher risk premium (and thus higher target rate) would be used to discount the future cash flow of a riskier property investment and cause its present value to reduce accordingly. It is difficult to obtain evidence of the target rate from the market, but the baseline is the return from a risk-free investment such as the gross redemption yield on medium-dated government bonds. A risk premium is then added to this risk-free rate, which should cover (Baum and Crosby 1995):

- *Tenant risk*: Risk of default on lease terms, particularly payment of rent but also repair and other obligations, risk of tenant exercising a break option or not renewing lease (higher risk if the lease is short). The level of tenant risk will depend to an extent on the type of tenant; a public sector organisation may be considered less likely to default than a fledgling private sector company.
- *Physical property risk*: Management costs (e.g. rent collection, rent reviews and lease renewal) and depreciation. This type of risk is less acute in the case of prime retail premises because land value is a high proportion of total value, but the reverse is true for, say, small industrial units. A certain amount of physical property risk can be passed on to the tenant via lease terms.
- *Property market risk*: Illiquidity caused by high transaction costs, complexity of arranging finance and accentuated by the large lot size of property investments
- *Macroeconomic risk*: Fluctuating interest rate, inflation, GDP, etc., all affect occupier and investment markets in terms of rental and capital values and potential for letting voids.
- *Legal risk*: In the main, this refers to planning policy and development control. For example, presumption against out-of-town retailing, promotion of mixed-use, developments on previously developed land.

Baum and Crosby (1995) point out that, for valuation, it is not usually feasible to quantify all of these components of risk as this would need to be done for each comparable. Instead, the valuer might subjectively choose and adjust a target rate not at the individual property level but by grouping various property investments and examining the risk characteristics of each. By far the most frequently encountered investment type is a rack-rented freehold. Regular rent reviews mean that this is an equity-type investment that benefits from income and capital growth just as equities do, albeit with less-frequent income growth participation. Whereas the return from an investment in company shares relies on the continued existence and profitability of that company, a property investment will remain even if the occupying company fails. Unlike share dividends, rent is a contractual obligation paid quarterly in advance and is a priority payment in the event of bankruptcy. After a likely rent void, the premises can be re-let and perhaps used for a different purpose, subject to location, design and planning considerations. This reduces the reliance of the investment on a single business occupier, helps underpin the value of the investment and reduces risk. A freehold let on fixed ground rent has a risk profile similar to undated gilts as it generates a fixed income from a head tenant who is very unlikely to default on what will probably be a significant profit rent. Consequently, this type of property investment is very secure, and risk will derive from changes in the level of long-term interest rate and inflation rather than property or tenant-specific factors (Fraser 1993).

Some of the more general 'market' risks, such as illiquidity, tenant covenant and yield movement, are best incorporated by adjusting the target rate of return. Other, property-specific risks such as regular deductions from gross rent, a depreciation rate slowing rental growth, voids and management costs can be reflected in adjustments to the cash flow. In this way, properties of the same type can be grouped together to help estimate a risk premium for a particular sector or subsector of the market such as high street shops or secondary industrials on the basis that properties within each sector have similar tenant risks or lease structures. Any remaining costs (fees, management, dilapidations, etc.) can be incorporated by making adjustments to the cash flow.

Risks that relate to income can be handled by adjusting the cash flow. Risks that relate to the security of the land and property as an investment should be handled by adjusting the discount rate – the higher the perceived risk, the higher the discount rate or yield, all else equal. In emerging markets, there may be additional risks resulting from: high inflation and other macroeconomic volatility; capital and other regulatory controls; political changes, war, civil unrest; poorly defined or enforced contracts; lax accounting controls and corruption.

It is worth noting that, because the rental growth rate is not part of the target rate of return (as it is with the yield), the valuation is less sensitive to the choice of target rate than is the case for yield.

7.3.1.2.3 Exit value

A property is a durable, long-term investment asset and in order to avoid trying to estimate cash flows far off into the future, a *holding period* of between 5 and 10 years is normally specified, after which a notional sale may be assumed. The length of the holding period can be influenced by the length of the lease, by the timing of a break clause or by the physical nature of the property, perhaps timed to coincide with a redevelopment opportunity.

The notional sale value or *exit value* is usually calculated by capitalising the estimated rent at the end of the holding period at a yield. When a yield is used to estimate an exit value, it is called an *exit yield* and is usually higher than initial yields on comparable but new and recently let property investments because it must reflect the reduction in remaining economic life of the property and the higher risk of estimating cash- flow at the end of the holding period. Where an allowance has been made for refurbishment in the cash flow during the holding period, the exit yield should reflect the anticipated state of the property. The exit yield may reflect land values if demolition is anticipated. If the holding period is less than 20 years, the exit yield has a significant impact on the valuation figure.

7.3.2   Perpetual annuities

7.3.2.1   Valuation of rack-rented freehold property investments

A freehold property investment was let recently at an MR of £100 000 per annum on a 15-year FRI lease with 5-year rent reviews. Assuming a target return of 7%, an implied annual rental growth rate (calculated above) of 3.23%, a holding period of 10 years after which a sale is assumed at an exit yield of 4%, the valuation of this property is shown below:

Years 1–5 MR (£)	100 000		
YP 5 years @ 7%	4.1002		
		410 020	
Years 6–10 MR (£)	117 228		$(100\,000 \times 1.0323^5)$
YP 5 years @ 7%	4.1002		
PV £1 5 years @ 7%	0.7130		
		342 701	
Exit value in year 10 MR (£)	137 423		$(100\,000 \times 1.0323^{10})$
YP in perpetuity at 4%	25.0000		
PV £1 10 years @ 7%	0.5083		
		1 746 472	
Valuation before PCs (£)		2 499 192	

The net income in each period is discounted at the target rate of return to a present value and these are totalled to obtain a total present value or valuation of the subject property. Because no growth is implied in the target rate, the rental income must be inflated at the appropriate times (rent reviews) over the term of the investment to account for growth. At the end of the holding period, a notional sale is assumed so the projected rent of £137 423 is capitalised at an exit yield of 4%.

Checking this answer against the income capitalisation method, because the rental growth rate has been obtained mathematically from the relationship between the target rate and the yield, the answers will be the same.

MR (£)	100 000	
YP in perpetuity @ 4%	25.0000	
Valuation before PCs (£)		2 500 000

The advantage of the DCF technique is that more information is presented, use of a target rate enables cross-investment comparisons and specific cash-flow problems such as voids and refurbishment expenditure can be incorporated.

7.3.2.2 Valuation of reversionary freehold property investments

The valuation of a freehold reversionary interest in a property let at £100 000 per annum on a 15-year lease with 3 years until the next rent review and a 5-year rent-review pattern is shown below. The MR is estimated to be £125 000 per annum. The investor's target rate of return is 8% and the holding period is until the second rent review in eight years' time. The exit yield is estimated to be 6%.

Using the implied growth rate formula, the annual growth rate implied by a target rate of 8% and an initial yield of 6% assuming five-year rent reviews is 2.24% per annum. The DCF valuation is as follows:

Years 1–3 MR (£)	100 000	
YP 3 years @ 8%	2.5771	
	257 710	
Years 4–8 MR (£)	133 590	$(125\,000 \times 1.0224^3)$
YP 5 years @ 8%	3.9927	
PV £1 3 years @ 8%	0.7938	
	423 418	
Exit value in year 8 MR (£)	149 237	$(125\,000 \times 1.0224^8)$
YP in perpetuity at 6%	16.6667	
PV £1 8 years @ 8%	0.5403	
	1 343 803	
Valuation before PCs (£)	2 024 930	

This valuation can be compared to income capitalisation methods shown below. Term and reversion technique:

Term (contract) rent passing	100 000	
YP for 3 years @ 6%	2.6730	
	267 301	
Reversion to estimated MR	125 000	
YP perpetuity @ 6%	16.6667	
PV £1 for 3 years @ 6%	0.8396	
	1 749 207	
Valuation before PCs (£)	2 016 508	

Core and top-slice technique:

Core rent	100 000	
YP in perpetuity @ 6%	16.6667	
		1 666 667
Top slice: uplift to estimated MR	25 000	
YP in perpetuity @ 6%	16.6667	
PV £1 for 3 years @ 6%	0.8396	
		349 841
Valuation before PCs (£)		2 016 508

When using the same yield for all tranches of rental income, the valuations for both income capitalisation techniques will be the equivalent, any small differences in the two valuations will be due to rounding. Note how these valuations are slightly different to the DCF valuation. The difference is small but worthy of explanation. Taking the core and top-slice valuation above, mathematically the valuation proceeds as shown in Eq. (6.3):

$$V = \frac{c}{y_c} + \frac{m-c}{y_t(1+y_t)^n} = \frac{100000}{0.06} + \frac{125000 - 100000}{0.06(1+0.06)^3}$$
$$= 1666667 + 349841 = 2016508$$

In this income capitalisation technique, the yield implies rental growth and therefore the rent is not explicitly projected at a growth rate. The DCF method does project rent at g but, unlike a rack-rented property with five-year rent reviews where the implied rental growth rate is 2.24% per annum, here there are two periods to incorporate into the calculation: one that lasts until the first rent review and then the five-year rent-review period thereafter (Brown and Matysiak 2000):

$$V = \frac{c\left(1-(1+r)^{-n}\right)}{r} + \frac{m(1+g)^n}{r(1+r)^n}\left[\frac{(1+r)^p - 1}{(1+r)^p - (1+g)^p}\right] \qquad (7.11)$$

where n is the period to the next rent revision and p is the rent-review period thereafter. If we assume that the valuations from the income capitalisation model and the DCF model should produce the same valuation, we can calculate the implied growth rate for a reversionary property investment. Using spreadsheet iteration in the final stage, g can be calculated as follows:

$$2016508 = \frac{100000\left(1-(1+0.08)^{-3}\right)}{0.08} + \frac{125000(1+g)^3}{0.08(1+0.08)^3}$$
$$\left[\frac{(1+0.08)^5 - 1}{(1+0.08)^5 - (1+g)^5}\right] \therefore g = 2.22\%$$

The implied growth rate from this reversionary property is slightly lower than from the rack-rented equivalent because the rental growth will arrive sooner due to the rent review in two years' time rather than in five years. Diagrammatically, the situation is illustrated in Figure 7.6.

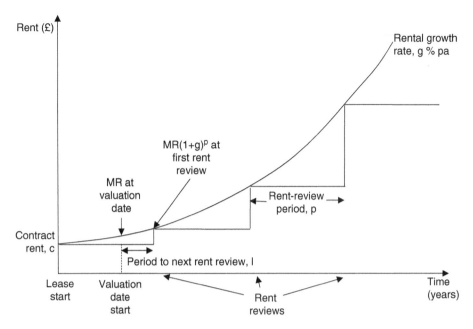

Figure 7.6   Rental growth between rent reviews.

7.3.3   Annuities with a term certain

7.3.3.1   Valuation of leasehold property investments

Baum and Crosby (1995) argue that a leasehold property investment producing a fixed-profit rent over its entire term produces a risk that is almost entirely dependent upon the quality of the sub-tenant: a cash flow from a good-quality tenant is similar to the return from a fixed-income bond plus a suitable risk premium. The target rate used to discount a fixed-profit rent is therefore likely to be derived from comparison to other fixed-income investments such as government bonds with similar maturity dates. This approach is more logical and is not based on questionable comparisons with the freehold investment market.

If the profit rent is variable, then there is a gearing effect. Basically, if a fixed head rent is deducted from a sub-rent, which includes rent reviews, the resultant profit rent will vary by an amount greater than the variation in the sub-rent itself. The magnitude of this variability depends on the size of the fixed deduction of head rent from the variable sub-rent and can be expressed as the income-gearing ratio.

To illustrate this, consider three property investments; a freehold, leasehold A where the head rent is very similar to the sub-rent and leasehold B where the sub-rent is very much larger than the head rent. All three investments generate an initial income of £100000 per annum subject to annual rent reviews and rental growth is estimated to be 5% per annum. As can be seen from Table 7.1, the income from the freehold grows at the rental growth rate of 5% per annum. Leasehold A receives a £900000 per annum sub-rent and pays a £800000 per annum head rent, leaving £100000 per annum profit rent. Leasehold B receives a £110000 per annum sub-rent and pays a £10000 per annum head rent, leaving £100000 per annum profit rent.

Table 7.1 Geared leasehold profit rents.

Year	Freehold initial net income (£)	Freehold income growth (%)	Leasehold A initial net income (£)	Leasehold A income growth (%)	Leasehold B initial net income (£)	Leasehold B income growth (%)
0	100000	—	100000	—	100000	—
1	105000	5.00%	145000	45.00%	105500	5.50%
2	110250	5.00%	192250	32.59%	111275	5.47%
3	115763	5.00%	241863	25.81%	117339	5.45%
4	121551	5.00%	293956	21.54%	123706	5.43%
5	127628	5.00%	348653	18.61%	130391	5.40%
6	134010	5.00%	406086	16.47%	137411	5.38%
7	140710	5.00%	466390	14.85%	144781	5.36%
8	147746	5.00%	529710	13.58%	152520	5.35%
9	155133	5.00%	596195	12.55%	160646	5.33%
10	162889	5.00%	666005	11.71%	169178	5.31%
...						
40	703999	5.00%	5535990	5.76%	764399	5.07%
41	739199	5.00%	5852789	5.72%	803119	5.07%
42	776159	5.00%	6185429	5.68%	843775	5.06%
43	814967	5.00%	6534700	5.65%	886463	5.06%
44	855715	5.00%	6901435	5.61%	931287	5.06%
45	898501	5.00%	7286507	5.58%	978351	5.05%
46	943426	5.00%	7690832	5.55%	1027768	5.05%
47	990597	5.00%	8115374	5.52%	1079657	5.05%
48	1040127	5.00%	8561143	5.49%	1134140	5.05%
49	1092133	5.00%	9029200	5.47%	1191347	5.04%
50	1146740	5.00%	9520660	5.44%	1251414	5.04%

Except where the head rent is a peppercorn (very low) rent, rental growth for a leasehold profit rent is greater than the rental growth on an equivalent freehold. The growth rate diminishes at each subsequent rent review and tends towards the market rental growth rate in perpetuity (Baum and Crosby 1995). The income-gearing ratio for the first leasehold is 89% and for second it is 9%, and the way that a profit rent might be expected to grow depends on this ratio. Use of income capitalisation to value varying profit rents is hard to justify because of heterogeneity of interests and potential complexity profit rent cash flows.

That leaves the DCF method, but identifying a market target rate of return for leaseholds with variable and geared profit rents is difficult as each investment opportunity will have unique ratios between head rent and sub-rent, leading to individual profit rent cash flows and gearing circumstances. Furthermore, there will also be differences in terms of tenant quality and remaining lease term. The leasehold target rate must relate to the lease structure and any profit rent gearing and Baum and Crosby (1995) suggest that attention should focus on the choice of risk premium when moving from a freehold to a leasehold target rate. Other cash-flow variables such as the head rent, rent reviews and so on can be incorporated in the cash flow.

Freehold investment transactions can be analysed to derive a suitable rental growth rate, which can be applied to the leasehold investment cash flow and this should be done in preference to estimating a growth rate that is implied by the relationship between target rate and yield on a leasehold investment because of the heterogeneity of cash flows from leasehold investments (Baum and Crosby 1995). If the leasehold includes a head rent and sub-rent both with rent reviews at the same time and both rents are assumed to grow at the same rate, then the profit rent would grow at the same rate as the growth in MR for a free-hold. But in cases where rent reviews in the sub-lease (say every 5 years) are different to those in the head lease (say every 15 years) the complexities are best handled by presenting a full DCF. For example, the leasehold investment described above will be valued again but this time using a DCF. Assuming a target rate of 10% and an yield of 6% for freehold property, this implies rental growth of 4.47% per annum. But the target rate at which the cash flow from a leasehold investment is discounted must be adjusted to reflect additional risk. Here the adjustment is from 10 to 12%.

Years	Rent received (£)	Growth @ 4.47% pa	Inflated rent (£)	*Less* rent paid (£)	Profit rent (£)	PV @ 12%	PV (£)
1	30 000	1	30 000	(10 000)	20 000	0.8929	17 857
2	30 000	1	30 000	(10 000)	20 000	0.7972	15 944
3	35 000	1.0914	38 199	(10 000)	28 199	0.7118	20 071
4	35 000	1.0914	38 199	(10 000)	28 199	0.6355	17 921
5	35 000	1.0914	38 199	(10 000)	28 199	0.5674	16 001
6	35 000	1.0914	38 199	(10 000)	28 199	0.5066	14 286
7	35 000	1.0914	38 199	(10 000)	28 199	0.4523	12 756
8	35 000	1.3581	47 535	(10 000)	37 535	0.4039	15 160
9	35 000	1.3581	47 535	(10 000)	37 535	0.3606	13 535
10	35 000	1.3581	47 535	(10 000)	37 535	0.3220	12 085
11	35 000	1.3581	47 535	(10 000)	37 535	0.2875	10 790
12	35 000	1.3581	47 535	(10 000)	37 535	0.2567	9634
Valuation before PCs (£)							176 041

Two criticisms of the DCF approach are that growth is unlikely to be constant over the life of the investment and the target rate of return is subjectively estimated and possibly different for the lease term and reversion parts of the cash flow.

7.4 Profits method

Certain types of property are inextricably linked to the businesses that operate from them: special characteristics of the property itself are central to the capacity of the business to generate profit. Such trade-related properties might be purpose built or have some monopoly value due to their unique location, legal status or planning permission. Consequently, the attributes of the property with regard to the business operating therein are more important than the flexibility of the property for change of use. For example, a property or the proprietor may be licenced to conduct a particular type of business, to sell wines, beers and spirits in a public house, restaurant or hotel say. Such licences may, in combination with the location of the property, make for strong trading potential such as a petrol station on a busy road into town. Other businesses may trade well simply on the basis of their location alone: hotels, garden centres, theme parks, cinemas, theatres, car parks and so on.

With non-specialised property (shops, offices, factories, warehouses, etc.), there is normally sufficient trading activity and comparability of asset within each market sector to observe prices and rents. Trade-related properties, on the other hand, are more heterogeneous and there are fewer transactions, so use of comparison is more difficult. Even when comparable evidence is available, properties may be sold as part of a portfolio or business and individual property values are difficult to isolate. Trade-related properties are not usually held on a leasehold basis because of the significant investment in fixtures, fitting, furniture and equipment, so there is not much rental evidence. Consequently, the valuer needs to employ a method that examines the economic fundamentals of the business.

Because the business and the property are closely linked and tend to be bought and sold as operational entities rather than with vacant possession, these premises are valued by capitalising their estimated future trading potential (as opposed capitalising estimated rental income).

Some trade-related properties are more commonplace than others; pubs, restaurants, hotels and entertainment complexes for example, so it may be possible to make comparisons with similar trades on a wider geographical scale, perhaps examining profit made per hotel bedroom or nightclub floor for example. Leisure properties tend to not make a standard amount of profit per square foot, so comparison metrics based on unit floor areas are not particularly useful. It is important to consider differences between properties such as lease length (typically longer for leisure properties) and user clauses (often more restrictive than those found in standard commercial leases due to specific planning permission or licencing for the use). Improvements and fit-outs are often more expensive and more frequently undertaken than on standard commercial premises and therefore it is particularly important to check the handling of tenants' improvements at rent review. Some small businesses such as a hotel, guest house or pub might attract purchasers willing to pay a price that includes a non-pecuniary return as well as capitalised financial income because it represents a lifestyle or location that they desire.

7.4.1   Method

The principle behind the method is that price or rent paid for a trade-related property derives from its trading potential as utilised by a hypothetical *reasonably efficient operator* (REO). The method consists of the following steps:

Estimate	Fair maintainable turnover (FMT)
Deduct	Cost of sales (purchases and adjustment for change in value of stock)
=	Gross profit
Deduct	Operating costs and wage costs
=	Net profit
Deduct	Remuneration to operator
Deduct	Interest on capital invested, stock and consumables
=	Fair maintainable operating profit (FMOP)
Then either	(1) Capital valuation: capitalise adjusted net profit at an appropriate freehold or leasehold yield
	(2) Rental valuation: apportion adjusted net annual profit between rent and profit

Trade-related properties are valued on the basis of their potential net profit adjusted to reflect the trading of an REO. This adjusted net annual profit is then either (i) capitalised at an appropriate capitalisation rate to arrive at a capital value from an occupier's perspective or (ii) divided into two, one portion of which is assumed to be available as rent for the premises in which the business takes place and the other portion is a residual profit for the operator of the business. The rent portion may be capitalised at an appropriate yield to arrive at a capital value from an investor's perspective. The method is therefore based on two economic assumptions, that the business makes a profit and that rent is a surplus paid out of this profit. It is also assumed that the current trading activity represents the optimum use of the property and that the business is efficiently run.

The property valuation is undertaken assuming that the business will at all times be effectively and competently managed, operated and promoted and that it is properly staffed, stocked and capitalised. The valuation includes land, buildings, trade fixtures, fittings, furnishings and equipment associated with the business and assumes that they are working and owned outright. It also includes market perception of inherent trading potential including transferable goodwill (i.e. attached to the property) and assumes that advanced bookings and order books can be transferred and that existing licences, consents, permits, certificates and registrations can be obtained and renewed. Moreover, freeholds are often offered for sale with the benefit of a trade inventory. The valuation excludes personal goodwill, wet and dry consumable stock and any badged items and the value attached to brand name should be separately identified. Value-significant factors for trade-related properties are shown in Table 7.2

The first step is to estimate the fair maintainable turnover (FMT) that could be produced at the premises by an REO on a year-by-year basis. FMT is defined as the '. . .level of trade that an REO would expect to achieve on the assumption that the property is properly equipped, repaired, maintained and decorated' (RICS 2019, p. 85). It excludes any trade that can be attributed to personal

Table 7.2 Value-significant factors for trade-related properties.

Property
- Efficiency of layout
- Level of comfort afforded in trading areas (e.g. floor space per cover in a restaurant)
- Number and quality of rooms
- Quality of owners accommodation
- External facilities
- Repair

Business
- Licences, agreements, leases – look at detail such as restrictions on live entertainment, beer supply agreements, preferential loans for refits, etc. (perhaps paid back by discounting beer supply). Check whether such arrangements are cheaper than going to market for discount supply.
- Compliance with Environmental Health and Fire Authority regulations
- Customer profile (age, type – demographic)
- Opening hours and peak trading periods (loss-making during week?)
- Staffing: Costs, efficiencies of layout (valuations usually assume two-person proprietary team so record any variation from this assumption)
- Tariffs (needed to assess gross yields): Beer and wine tariff (compare to local area, last price increase?), catering, accommodation – discounting, occupation levels, average room rate, ratio of rack room rate to average room rate
- Expenses; purchasing arrangements, promotion, functions, entertainments, methods of payment
- Trading information: Accounts (purchase invoices are last resort), VAT returns, net sales per quarter, stock-takers reports and records, weekly/monthly records, management accounts

goodwill of a specific operator. This definition means that the use of actual trading accounts and the strength of the occupying business are not always relevant.

The second step is to deduct purchases to arrive at an estimate of annual gross profit resulting from the FMT. Operating costs (including energy costs, business rates, insurance, maintenance and repair costs[3], accountant's fees, marketing, interest in operator's capital[4], printing, stationery, etc.) and wage costs (including national insurance and holiday pay) are then deducted from gross profit to arrive at net profit. From this an estimate is made of fair maintainable operating profit (FMOP). This is annual profit before depreciation and mortgage payments (and rent if leasehold) have been deducted.

The estimate of profit will refer to income, expenditure and the operator's capital. Typically, these figures will be reported in the company's annual accounts, the previous three to five years of which should be analysed to identify whether the profit is maintainable over a period of time. It is advisable to examine profit over several years because profit in any one year may be due to exceptional circumstances. Audited accounts are to be preferred but should not necessarily be accepted at face value. It should be borne in mind that profit and loss accounts may be prepared for various purposes and when using them to estimate FMOP, it is important to consider whether the business has more than one property. This is because consolidated accounts may not apportion expenditure on marketing, training, accountancy, depreciation, cyclical repairs or management expenses for head office premises between each property. When analysing accounts information, particular attention should be paid to:

- Costs associated with a head office (if relevant)
- Accounts of an owner-occupied business may not show management salaries or director's remuneration
- Depreciation policies vary and valuers should 'add back' depreciation that has been deducted when assessing maintainable profit
- Adjustments to reflect repair and maintenance as well as refitting and re-equipping
- Treatment of tips and effect on stated turnover and wage costs
- Business mix, profit potential and risk attached to each component
- For licenced properties, the gross profit should reflect actual and projected terms of any supply tie (where operator is tied to the landlord for all or part of supplies)
- If the interest is leasehold, adjust to reflect MR for the business rather than rent passing at the valuation date

An inspection of the business identifies likely sources and amounts of income and expenditure and provides a basis for comparison with the accounts. The valuer should look for any unusual items and conditions such as cash sales and purchases, multiple bank accounts or additional revenue such as tips in the case of licenced premises.

When assessing whether the profit is maintainable, it is important to consider impact of competition in terms of its degree (volume, local/regional/national), type (level and style) and the extent to which it is detrimental or indeed beneficial.

It must be stressed that the valuation of trade-related properties is a specialist area and requires a thorough understanding of the business for which the property is being used. Enquiries should be made into the background and history of the business: how long has the current operator owned it for, how much was paid for the business, was a mortgage taken out to help fund the acquisition and, if so, is the mortgage valuation available for inspection? Operators of small-scale and family run businesses may use some personal items and capital to run the business and these costs may not be reflected in the annual accounts; these costs should be identified and added to the working expenses. Such items might include salary or remuneration to the proprietor, interest payments on personal loans used to support the business, depreciation of property and reduced wages to family members. However, if the market for the business in question generally encompasses family run employees then this should be acknowledged (Hayward 2009). If the information in the accounts does not contain sufficient detail the company should provide trading information on a property-by-property basis for current and previous years, perhaps including receipts if necessary and percentages of gross turnover allocated to individual income and expenditure sources.

It is useful to calculate gross and net yields, wage ratios and plot trends in key figures such as staff costs, turnover and so on over several years. If there are any peculiarities from year to year or season to season which distort the pattern of trade, then these ratios will help to indicate these. The profits method assumes the business is operated at maximum capacity, but some operators choose to under-trade or indeed generate an extraordinary level of trade perhaps resulting from excessive levels of personal goodwill. It is useful, therefore, to assess the physical capacity of premises and compare this to the actual turnover data. This is more difficult in the case of new asset classes since the relationship between business and underlying property asset is unknown.

If the property is to be valued assuming that it is a fully equipped operational entity, then care must be taken regarding how any trade inventory is handled. Some plant, machinery and equipment (tangible assets) may not be owned outright – they may be leased. Assumptions will need to clearly state how these things are handled. It may be assumed that leasing agreements can be transferred on sale. If they cannot, then the impact on valuation should be considered. The valuer also needs to consider how licences and other statutory consents can be transferred or renewed. The valuation may be subject to certain Special Assumptions such as (RICS 2019, p. 87):

a) On the basis that trade has ceased and no trading records are available to prospective purchasers or tenants
b) On the same basis as (a) but also assuming the trade inventory has been removed
c) As a fully equipped operational entity that has yet to trade (a 'day one' valuation)
d) Subject to stated trade projections, assuming they are proven (appropriate when considering development of the property)

In licenced premises, a landlord may collect a share of some income (from amusement machines for example) at source in addition to property rent. For businesses operating from standard retail units, it may be appropriate to value on a floor area rate basis but with regard to lease provisions, comparables and fit-out obligations.

Once the valuer is satisfied with the estimate of FMOP, the market valuation can take two forms: an estimate of capital value or an estimate of rental value. To calculate capital value, FMOP is capitalised at a rate of return that reflects risk and rewards of the property and its trading potential. Where possible, comparable evidence would be used to select an appropriate rate. To estimate MR, a deduction is made from FMOP to reflect a return on any tenant's capital invested in the business[5] (such as stock and working capital). The remaining amount is referred to as the *divisible balance,* which is apportioned between landlord and tenant, having regard to the risks and rewards of each, with the landlord's proportion representing the annual rent. The rent portion is conventionally between 40 and 50% with a more precise figure derived from comparison to similar businesses. It is this rent portion that is capitalised at an investment yield to arrive at an investment value for the business.

It should be borne in mind that specialised trading properties can be of interest to conventional property investors and to business operators and the rate of return that these groups of purchasers require may be different. For the former it may relate to the perceived risk of the market and specific asset and the return that can be achieved on alternative investment assets. For the latter the rate of return may relate to the return required from the business as a whole, taking into account any mortgage and equity requirements for the type of property being valued. Consequently, the yield at which rent or the capitalisation rate and which profit is capitalised should be chosen with these distinct markets in mind.

To see how the profits method might be applied to a specific type of trade-related property, consider a 50-bed hotel that has an average annual occupancy of 50% and charges, on average, £70 per room per night. To value this property, first, use the comparison method to check whether £70 is a reasonable price and whether the occupancy rate is satisfactory compared to other hotels in the area. The data in Table 7.3 has been extracted from the accounts and the next step is to ensure that all sources of revenue are accounted for. The premises are held on a freehold basis and the hotel is part of a small chain and must contribute towards head office overheads.

Assuming the gross turnover to be a reasonable estimate of FMT, purchases and depreciation in the value of stock are deducted[6] to arrive at gross profit. Operating costs and wage costs are then deducted to arrive at a net profit. From this figure, operator's remuneration and interest on operator's capital (assumed to be charged at 10% per annum in this example) are deducted to arrive at an adjusted net profit. This figure is capitalised at appropriate yield, 10% in this case.

Table 7.3   Information extracted from hotel accounts.

Accounts information	£
Gross turnover:	
▪ Accommodation (£70 × 365 days × 50 rooms × 50% occupancy)	638 750
▪ Bar	45 000
▪ Restaurant	25 000
Value of fixtures, fittings, furniture and equipment	250 000
Value of stock at year start	105 000
Value of stock at year end	95 000
Cash (say 1 month's working expenses)	21 000
Value of operator's capital	271 000
Purchases	45 000
Wage costs	200 000
Operator's remuneration	50 000
Operating costs:	
▪ Utilities	3500
▪ Rates	36 000
▪ Building insurance	1000
▪ Annual repair and maintenance	2500
▪ Laundry and cleaning	2500
▪ Marketing	1000
▪ Contribution to head office costs	2000
▪ Contents insurance	1250
Repairs to building[a]	1500
Mortgage[b]	1250

[a] These are regarded as one-off repairs and not an annual expenditure. It may signify that the annual repair and maintenance allowance need increasing, though.

[b] Not regarded as a typical business expenditure and therefore excluded.

FMT:

Accommodation (£)		638 750	
■ Bar (£)		45 000	
■ Restaurant (£)		25 000	
			708 750
Purchases (£)		(45 000)	
Adjustment for depreciation in value of stock (£)	1 Jan: £105 000		
	31 Dec: (10 000)		
	£95 000		
		(55 000)	
Gross profit (£)			653 750

Operating costs:

■ Utilities (£)		(3500)	
■ Laundry and cleaning (£)		(2500)	
■ Rates (£)		(36000)	
■ Advertising, stationery, telephone, postage, etc. (£)		(1000)	
■ Contents insurance (£)		(1250)	
■ Annual sinking fund for repairs and renewals (£)		(2500)	
■ Building insurance (£)		(1000)	
■ Contribution to HQ overheads (£)		(2000)	
■ Wages (£)		(200 000)	
			(249 750)
Net profit (£)			404 000
Operator's remuneration (£)			(50 000)

Interest on operator's capital (£)

■ Furniture, fixtures, fitting and equipment	(250 000)		
■ Stock (average)	(100 000)		
■ Cash (1 month's working expenses)	(21 000)		
Operator's capital		(371 000)	
Annualised @ 10%		0.10	(37 100)
FMOP (£)			316 900
YP in perpetuity @ 10%			10
Property Valuation (£)			3 169 000

If the value of the business as a whole is also required, then the value of the inventory, stock and cash float can be added to the value of the premises, as follows.

Capital value of property (£)	3 169 000
Value of inventory (£)	250 000
Average value of stock for year (£)	100 000
Cash (£)	21 000
Capital value of business (£)	3 540 000

Because of the specialised nature of the businesses concerned, valuers tend to specialise in the valuation of properties used for particular trades. Some valuers may concentrate on the valuation of licenced premises such as pubs, clubs, restaurants and casinos; others may specialise in the valuation of hotels, guest houses or care homes. The overriding requirement for any valuer agreeing to value a specialised trading property is to have adequate knowledge and experience of the relevant business sector operating from the property.

Key points

- The yield describes ratio of income to capital value and is used to compare investments because yields are often comparable for similar types of property in the same area. The unit of comparison for rack-rented freeholds is the current rental income yield (initial yield) and for reversionary investments it is the equivalent yield (see Chapter 9). A *running yield* follows changes in income as a result of rent reviews, rent steps, cost changes, etc.
- The capital value of an investment depends on expected rental income and the yield and is very sensitive to changes in the latter. Using the income capitalisation method, the yield used to capitalise property investments is based on initial yields and equivalent yields derived from the analysis of recent transactions of comparable property investments.
- Freehold property investments are perpetual and have a base land value whereas leasehold investments decline to zero value at lease end. A leasehold interest can only have market value if it produces income via a profit rent and is assignable. The conventional method of valuing leasehold investments was to convert the terminable interest (mathematically at least) into the equivalent of a freehold investment; the use of yields derived from freehold investments could then be justified. Valuers now tend to look much more closely at the nature of the cash flow from a leasehold investment before applying a yield or yields.
- The income capitalisation method does not reveal the total return that an investor expects; instead, future rental income is discounted (capitalised) at a rate that implies that the investor expects the income to grow in order to achieve a target rate of return whereas the DCF method involves selecting a suitable holding period, forecasting the cash flow over this period and selecting an appropriate target rate and exit yield. All of these assumptions should reflect market behaviour.
- The value of an investment can be considered to be a multiple of the current rent where the multiplier is the reciprocal of the investor's required yield (income capitalisation method) or the present value of the expected future cash-flow (DCF method) (Fraser 1993). Techniques vary depending on the extent to which assumptions are made explicit. For example, a valuer may wish to include an explicit growth rate forecast rather than imply a long-term average from analysis of comparable evidence, or depreciation may be explicitly accounted for in the cash flow.
- The DCF method is better at isolating factors affecting future income flow from those that affect the target rate of return required by the investor, thus allowing direct comparison with other investment opportunities. It can also deal with complexity in cash flows and reveal assumptions about income growth, depreciation, holding period, timing of income and expenditure and the target rate of return.

- Choice of method is a matter of availability of evidence and complexity of the property interest being valued: use the income capitalisation method for investments with a standard pattern of income and rent reviews and use the DCF method for non-standard cash flows such as varying profit rents.
- The valuation of trade-related properties requires specialist skill. There is heavy reliance on accounts and other financial information about the business and also on expertise to value the goodwill element of the business. Attention should be focused on two things. first, the adjustment of the costs to bring net profit back to a point where there is no regard to the individual operator – the business is assumed to be run by an averagely competent operator, and, second, the selection of an appropriate capitalisation rate.

Notes

1. Rent net of any regular management and maintenance expenditure.
2. It is assumed that, even though freehold property investments are usually let on leases of fixed terms, the property will re-let on expiry of current lease. Therefore, rent can be regarded as perpetual. This assumption might be altered if redevelopment or the like is planned, in which case a DCF is more able to handle these sorts of circumstances.
3. Which may take the form of an annual sinking fund to cover periodic replacement of items.
4. This includes working capital, furniture, fixtures, fittings and equipment, and stock. These are capital sums, and the application of an interest rate converts them into annualised deductions. The rate should be based on the cost of borrowing or a reasonable rate of return, as appropriate.
5. This is personal and not transferable.
6. Any appreciation in stock value would be added.

References

Baum, A. (2000). *Commercial Real Estate Investment*. Oxford: Chandos.

Baum, A. and Crosby, N. (1995). *Property Investment Appraisal*, 2e. London: Routledge.

Brown, G. and Matysiak, G. (2000). *Real Estate Investment: A Capital Market Approach*. Harlow, UK: FT Prentice Hall.

Fraser, W. (1977). The valuation and analysis of leasehold investments in times of inflation. *Estates Gazette* (244): 197–203.

Fraser, W. (1993). *Principles of Property Investment and Pricing*, 2e. Macmillan.

Hayward, R. (2009). *Valuation: Principles into Practice*, 6e. London: Estates Gazette.

Lean, W. and Goodall, B. (1966). *Aspects of Land Economics*. London: Estates Gazette Ltd.

Marshall, A. (1920). *Principles of Economics*, 8e (first published in 1890). London: Macmillan and Co. Ltd.

McAllister, P. (2001). Pricing short leases and break clauses using simulation methodology. *Journal of Property Investment & Finance* 19 (4): 361–374.

McAllister, P. and Loizou, P. (2007). *The Appraisal of Data Centres: Deconstructing the Cash Flow, Working Papers in Real Estate & Planning, 04/07*. University of Reading.

RICS (2004). *Capital and Rental Valuations Of Hotels in the UK*, Valuation Information Paper 6. London: Royal Institution of Chartered Surveyors.

RICS (2006). *Capital and Rental Valuation of Restaurants, Bars, Public Houses and Nightclubs in England and Wales*, Valuation Information Paper 2, 2e. London: Royal Institution of Chartered Surveyors.

RICS (2019). *RICS Valuation – Global Standards: Incorporating the IVSC International Valuation Standards*. London, UK: Royal Institution of Chartered Surveyors.

Sidwell, A. (1991). The valuation of nursing and residential homes. *Journal of Property Valuation and Investment* 9 (4): 352–356.

Questions

Freehold interests

[1]
  a) Value a property just let at £50000 per annum on a 10-year FRI lease with an upward-only rent review in year five. A similar property in the locality let at £75000 per annum has sold for £1500000. Assume PCs are 6.5% of the sale price.
  b) Using the term and reversion technique, value the freehold interest in a shop let on FRI terms at £50000 per annum. The lease has four years unexpired, the current MR is £60000 per annum and the NIY is estimated to be 6%.

[2] A 5000 square metre office building is let to two tenants occupying 2500 square metres each. The details of the tenancies are:

Tenant	Lease start	Rent review period (years)	Current rent (£)	Market rent (£)	Lease term (years)	FRI
1	7 years ago	5	500000	750000	10	Yes
2	8 years ago	5	450000	750000	10	Yes

Assuming an NIY of 6%, value the freehold interest in the office building.

[3] A shop is let on a 10-year FRI lease, which has 2 years unexpired at a current rent of £60000 per annum. You are aware of a comparable shop next door, which has just sold for £1000000. It had been let a few days before the sale on an FRI lease for 10 years with a 5-year upward-only review at an MR of £80000 per annum. Making appropriate assumptions, value the freehold interest in the first shop.

[4] Two years and nine months ago, a shop was let on a 15-year FRI lease with upward-only rent reviews every 5 years. Its internal dimensions are 5 metres (frontage) by 22 metres (depth). This includes an area of 4 metres by 5 metres at the rear of the shop, used for storage. The current rent is £107000 per annum. Recent evidence indicates that similar properties are letting at £2500 per square metre zone A and selling at NIYs of 6%. Assuming zone depths of 6 metres, value the freehold interest in the shop.

[5] A retail property at 24 High Street (with a frontage length of 8 metres and a depth of 15 metres) was let 7 years ago on a 25-year lease. Rent reviews are every five years and at the last review, the rent was agreed at £16000 per annum. 28 High

Street (frontage 7 metres, depth 17 metres), on the same side of the road as the subject property, recently let on standard lease terms for £19050 per annum with five-year rent reviews, no incentives and was subsequently sold for £317500. Assuming zone depth of 6 metres, value the freehold interest at 24 High Street.

[6] An industrial unit with a gross internal area (GIA) of 2500 square metres is located on a well-established industrial estate with good road connections. It was let 22 years ago and let on a 25-year lease to an engineering firm on FRI terms, with 5-year upward-only rent reviews. The passing rent is £100000 per annum. The following comparable evidence of properties on the same industrial estate is available:

Comparable A	A unit of 1500 m² GIA, built 10 years ago, has recently been let on a 10-year FRI lease to a company fabricating window units at a rent of £67500 per annum.
Comparable B	An older unit, let to firm manufacturing automotive components, has recently been sold to a local property investor for £800000. A net rent of £80000 per annum was recently set at rent review.

Value the freehold interest in the industrial premises, making any assumptions that you think are appropriate.

Leasehold interests

[1] A shop is let on an FRI lease at £10000 per annum with eight years unexpired and there are no rent reviews in the remaining term. The tenant has sub-let the premises on the same lease terms at £12500 per annum and the sub-lease expires just before the end of the head-lease. Value the leasehold and sub-leasehold interests, assuming a leasehold yield of 8% and a sub-leasehold yield of 8.5%. The current MR of the property is £16000 per annum.

[2] Paragon Pension Co. is contemplating purchasing the head-leasehold interest of a factory from Urban Property Investments. The head lease was granted to Urban Property 40 years ago for a term of 50 years at a fixed net rent of £10000 per annum. Urban Property sub-let the factory on a 20-year FRI lease with 5-yearly rent reviews to Make-a-Splash Ltd. 10 years ago. Make-a-Splash currently occupies the factory. The rent passing is £223000 per annum and the next rent review is due in one month. Market evidence suggests that freehold investments of similar industrial premises are selling at 9% NIY and rental growth has been 2% per annum for the last few years. Assuming a target rate of return of 11%, value the head-leasehold interest using a DCF.

[3] A company owns the head-leasehold interest of a large industrial unit. The head lease was granted 50 years ago for a term of 70 years at a head rent of £15000 per annum without review. The company sub-let the unit 5 years ago on a 25-year FRI lease. The rent was reviewed last month to £345000 per annum. Using a DCF, value the head lessee's interest assuming a leasehold target rate of 14% and a rental growth rate of 1.5% per annum.

[4] An investor owns the head lease of an office building. The head lease was granted for a term of 99 years and has 18 years left to run at a fixed head rent of £5000 per annum. The office is sub-let in its entirety on a 25-year lease granted

7 years ago with 5-year upward-only rent reviews at a passing rent of £180 000 per annum. The MR is £200 000 per annum. Using a DCF, value the head lessee's interest assuming a leasehold target rate of 10% and a rental growth rate of 2% per annum.

[5] A head lease has 15 years to run at a fixed rent of £10 000 per annum on FRI terms. The property has just been sublet on a lease due to expire one day before the end of head lease and this sub-lease contains rent reviews in years 5 and 10. The current MR is £15 000 per annum on FRI terms. You are aware that freehold interests in similar properties are being valued using NIYs of 5% and you are aware that investors for these properties are seeking a target rate of return of 11%. You assume that the rental growth rate at the rent reviews in the sub-lease will be the same as that on similar rack-rented freeholds. Value the head-leasehold interest.

[6] An office property is currently let on an FRI lease with seven years unexpired at a fixed rent of £60 000 per annum. The estimated MR based on an FRI lease with five-year rent reviews is £100 000 per annum. An identical property, currently let on an FRI lease with three years unexpired at £85 000 per annum with an estimated MR of £100 000 per annum, has just been sold for £1.6 million. The tenant sub-let the property on an FRI lease, which also has seven years unexpired. The rent passing is £90 000 per annum.

a) Value the freehold interest using the income capitalisation method.

a) Assuming a target rate of 12% and a rental growth rate of 2.49% per annum, value the head-leasehold interest using a DCF.

Profits method

[1] The freehold interest in a café is being offered for sale as a going concern. Details from the profit and loss account for the last financial year are:

Turnover (£)	345 000
Purchases (£)	(78 600)

Operating expenses (£):

■ Staff	(82 000)
■ Electricity and gas	(12 000)
■ Business rates	(3 250)
■ Repairs and maintenance	(1 900)

Value the café on the basis that an REO will continue to operate the business and generate a similar level of turnover and profit. Assume a capitalisation rate of 6.5%.

[2] A car park has 100 spaces. The accounts show that, in the last year, the car park has averaged 70% occupancy, with an average income per space per day of £4.50. Ticketing costs are £5000 per annum, business rates are £25 000 and an annual repairs allowance amounts to £8500. The operator charges 7% of turnover as a management fee. Assuming a capitalisation rate of 8%, value the car park.

[3] A privately owned 300-bed student hall of residence is 98% let during term time, at a rent of £100 per week. There are 30 term weeks in a calendar year with the remainder being vacation weeks. During the vacation periods, the premises are 70% occupied on average and let at £70 per week. Additional

income from sundry services amount to approximately £45 000 per annum. Annual expenditure includes £100 000 for maintenance, £150 000 for cleaning, £75 000 for wages and £45 000 for utilities. Assuming a capitalisation rate of 7.25%, value the property.

[4] You have been asked to value a marina. Summary details of the last five years of revenue and expenditure are shown in the table below. In addition, £500 000 of replacement pontoons is required as well as £500 000 of sea wall defence work. A yield of 9% and a capitalised earnings technique is regarded as appropriate for this type of investment. Your answer should include advice on what factors you might think would affect this yield.

Year ending	2018	2019	2020	2021	2022
Revenue (£)	787 000	802 000	783 000	771 000	787 000
Expenditure including operating and wage costs (£)	368 000	338 000	321 000	331 000	360 000

[5] You have been asked to value the freehold interest in a town centre restaurant. The business occupies the premises under a lease. There is also a pavement eating area held on an annually renewable licence at £525 per annum from the local authority. One owner, who is also the chef, operates the business on a full-time basis. The most recent accounting information is shown below. Assuming a capitalisation rate of 12.5%, value the restaurant.

Sales (£)	550 000
Purchases (£)	(121 000)
Depreciation in value of stock (£)	(2500)
Operating costs and wage costs (£):	
▪ Wages	(150 000)
▪ Director's emoluments	(35 000)
▪ Director's pension	(10 000)
▪ Rent	(41 025)
▪ Licence fee	(525)
▪ Rates, water and environmental charges	(13 500)
▪ Heating and lighting	(10 500)
▪ Telephone	(3000)
▪ Insurance	(5250)
▪ Repairs and maintenance	(6000)
▪ Printing, postage and stationery	(2500)
▪ Promotion	(3000)
▪ Amortisation of goodwill	(2410)
▪ Accountancy and professional fees	(3500)
▪ Transport	(2000)
▪ Laundry, cleaning and linen hire	(1500)
▪ Depreciation	(5035)
▪ Entertainment	(1250)
▪ Hire purchase, leasing and rental agreements	(5250)
▪ Credit and charge card commissions	(3000)
▪ Profit from sale of an asset	(3012)
▪ Sundries	(1268)

[6]

a) The operator of Plaza Café owns the freehold interest in the property. The accounts show a steady increase in both turnover and profit over the past three years, and the latest set of accounts shows receipts of £260 000 and a net profit to the business of £67 150. The café (business and property) is being offered for sale. Details from the profit and loss account for the last financial year are:

Sales (£)	260 000
Cost of sales (£)	(98 800)

Operating costs (£):

Wages	(80 000)
Electricity and gas	(10 000)
Business rates	(2250)
Repairs and maintenance	(1800)

You have been asked to produce a valuation on the basis that an REO will continue to operate the café and generate a similar level of turnover and profit. Assume an investment yield of 7% for this type of business.

b) Using the premises described in (a) but this time assume the operator is a tenant (rather than an owner-occupier). Estimate the current MR of the café. The tenant is currently paying a rent of £12 000 per annum and has invested the following sums in the business.

Furniture, fixtures, fittings and equipment (£)	(20 000)
Stock (£)	(2500)
Cash and working capital (£)	(500)

Assume the tenant's required return on this capital is 5% but make all other assumptions as necessary.

[7] Your client operates a hotel held under a 25-year FRI lease granted 2 years ago and that is still considered to be rack-rented. Your client purchased the previous tenant's fixtures and fittings from the landlord at the time the lease was granted. You have been asked to value the lease, assuming a capitalisation rate of 10%. The following accounts have been provided.

Sales (£):	
Drinks	120 000
Food	140 000
Accommodation	150 000
Own consumption	(500)
Cost of sales:	
Wet	(55 000)
Dry	(35 000)
Gross profit:	319 500
Other income:	
Deposit account interest	5000
Operating expenses:	
Wages and national insurance	(95 000)
Director's remuneration	(25 000)
Telephone	(2000)

Motor	(2500)
Licences and memberships	(1500)
Hardware and crockery	(500)
Leasing and machine rental	(1500)
Equipment repairs and renewals	(6000)
Cleaning and laundry	(8000)
Print, stationery and ads	(2000)
Sundry expenses	(2000)
Accountancy	(1500)
Bookkeeping	(1500)
Legal and professional	(2000)
Stocktaking fees	(500)
Rent	(45000)
Rates	(12000)
Insurance	(3000)
Light, heat and water	(12000)
Repairs to property	(3000)
Bank loan interest	(20000)
Bank charges and interest	(2000)
Credit card charges	(3500)
Depreciation of fixtures and fittings	(10000)
Depreciation motor	(500)
Total operating expenses:	(262500)
Net profit:	62000

Answers

Freehold interests

[1]

a) Rack-rented freehold valuation
Analysis of comparable:

$$£75000/(£1500000 \times 1.065) = 4.69\%(NIY)$$

Valuation:

Market rent (£)	50000	
YP in perpetuity @ 4.69%	21.3220	
Valuation gross of PCs (£)		1066099
Valuation net of PCs at 6.5% (£)		1000000

b) Reversionary freehold valuation

Term rent (£)	50000	
YP 4 years @ 6%	3.4651	
		173255
Reversion to MR (£)	60000	
YP in perpetuity @ 6%	16.6667	
PV £1 4 years @ 6%	0.7921	
		792102

Valuation gross of PCs (£)		965 357
Valuation net of PCs		906 438
@ 6.5% (£)		

[2]

Tenant 1

Term rent (£)	500 000	
YP 3 years @ 6%	2.6730	
		1 336 500
Reversion to MR (£)	750 000	
YP in perpetuity @ 6%	16.6667	
PV £1 3 years @ 6%	0.8396	
		10 495 241
Valuation gross of PCs (£)		11 831 741
Valuation net of PCs @ 6.5% (£)		11 109 616

Tenant 2

Term rent (£)	450 000	
YP 2 years @ 6%	1.8334	
		825 030
Reversion to MR (£)	750 000	
YP in perpetuity @ 6%	16.6667	
PV £1 2 years @ 6%	0.8900	
		11 124 956
		11 949 986
		11 220 644

Valuation of building net of PCs @ 6.5% (£) 22 330 260.

[3]   Analysis of comparable evidence:

$$NIY = 80000 / 1065000 = 7.51\%$$

Valuation:

Term: rent passing (£)	60 000	
YP 2 years @ 7.51%	1.7953	
		107 718
Reversion to MR (£)	80 000	
YP perp @ 7.51%	13.3156	
PV 2 years @ 7.51%	0.8652	
		921 653
Valuation gross of PCs (£)		1 029 371
Valuation net of PCs (£) @ 6.5%		966 546

[4]   Analysis of comparable evidence:

Zone A	5 m x 6 m	30 m²
Zone B	(5 m x 6 m)/2	15 m²
Zone C	(5 m x 6 m)/4	7.5 m²
Storage area	(5 m x 4 m)/20	1 m²
Area ITZA		53.5 m²

MR/m² ZA is £2500/m² so the MR of the subject property is $53.5\,m^2 \times £2500/m^2 = £133\,750$.

Valuation:

Term rent (£)	107000	
YP 2.25 years @ 6%	2.0479	
		219126
Reversion to MR (£)	133750	
YP in perp @ 6%	16.6667	
PV £1 2.25 years @ 6%	0.8771	
		1955259
Valuation gross of PCs (£)		2174385
Valuation net of PCs @ 6.5% (£)		2041676

[5] Analysis of comparable evidence:

Zone A	7 m × 6 m	42 m²
Zone B	(7 m × 6 m)/2	21 m²
Zone C	(7 m × 5 m)/4	8.75 m²
Area ITZA		71.75 m²

MR ITZA = £19050/71.75 m² = £265.50/m²
NIY is £19050/(£317 500 × 1.065) = 5.63%.

Valuation:

Zone A	8 m × 6 m	48 m²
Zone B	(8 m × 6 m)/2	24 m²
Zone C	(8 m × 3 m)/4	6 m²
Area ITZA		78 m²

MR = £265.50/m² × 78 m² = £20 709 p.a.

Term (contract) rent passing (£)	16000	
YP for 3 years @ 5.63%	2.6914	
		43063
Reversion to estimated MR (£)	20709	
YP perpetuity @ 5.63%	17.7620	
PV £1 for 3 years @ 5.63%	0.8485	
		312107
Valuation before PCs (£)		355170
Valuation net of PCs @ 6.5% (£)		333493

[6] Analysis of comparable evidence:

- Comparable A: £45/m² p.a. FRI. This is a more modern unit and so the rent could be adjusted downwards before being applied to the subject property. If no adjustment is deemed necessary, then the MR of the subject property is 2500 m² GIA × £45/m² = £112 500 p.a.
- Comparable B: NIY = 80 000/(800000 × 1.065) = 9.39%
- The subject property and the two comparable properties are all let to heavy industrial users so there is no need to adjust for differences there.

Valuation:

Term rent (£)	100 000	
YP for 3 years @9.39%	2.5138	
		251 380
Reversion to estimated MR (£)	112 500	
YP perpetuity @ 9.39%	10.6496	
PV £1 for 3 years @ 9.39%	0.7640	
		915 280
Valuation before PCs (£)		1 166 661
Valuation net of PCs (£) @ 6.5%		1 095 456

Leasehold valuations

[1]

Head-leasehold		
Rent received (£)	12 500	
Less rent paid (£)	10 000	
Profit rent (£)	2500	
YP 8 years @ 8%	5.7466	
Valuation before PCs (£)		14 367

Sub-leasehold		
Market rent (£)	16 000	
Less rent paid (£)	12 500	
Profit rent (£)	3500	
YP 8 years @ 8.5%	5.6392	
Valuation before PCs (£)		19 737

[2]   Since the rent review is in one month, grow the current rent for five years at 2% p.a. to determine likely MR. There is one more review due in 5 years' time, and both leases will expire in 10 years' time. Therefore, the projected future rental income from sub-lease is:

- Projected rent at forthcoming review: $£223\,000 \times (1.02)^5 = £246\,210$
- Projected rent at subsequent review (in five years): $£223\,000 \times (1.02)^{10} = £271\,816$

Years	Rental growth @ 2% p.a.	Projected rent received (£)	Head rent (£)	Profit rent (£)	YP 5 years @11%	PV £1 @ 11%	PV (£)
1–5	1.1041	246 210	(10000)	236 210	3.6959	1	862 167
6–10	1.2190	271 816	(10000)	261 816	3.6959	0.5934	554 264
Valuation before PCs (£)							1 416 431

[3] The head-lease expires in 20 years' time, the sub-lease has 20 years unexpired, and the rent was recently reviewed, so assume MR.

Years	Rental growth @ 1.5%	Projected rent received (£)	Head rent (£)	Profit rent (£)	YP @ 14%	PV £1 @ 14%	PV (£)
1–5	1	345 000	(15 000)	330 000	3.433	1	1 132 890
6–10	1.0773	371 669	(15 000)	356 669	3.433	0.519	635 487
11–15	1.1605	400 373	(15 000)	385 398	3.433	0.270	357 229
19–20	1.2502	431 319	(15 000)	416 349	3.433	0.140	200 106
Valuation before PCs (£)							2 325 712

[4]

Years	Rental growth @ 2% p.a.	Projected rent received (£)	Head rent (£)	Profit rent (£)	YP @ 10%	PV £1 @ 10%	PV (£)
1–3	1	180 000	(5000)	175 000	2.4869	1	435 207
4–8	1.0612	212 240	(5000)	207 240	3.7908	0.7513	590 225
9–13	1.1717	234 340	(5000)	229 340	3.7908	0.4665	405 567
14–18	1.2936	258 720	(5000)	253 720	3.7908	0.2897	278 634
Valuation before PCs							1 709 633

[5]

Years	Rental growth @ 6.55% p.a.	Projected rent received (£)	Rent paid (£)	Profit rent (£)	YP @ 11%	PV £1 @ 11%	PV (£)
1–5	1	15 000	(10 000)	5000	3.6959	1	18 480
6–10	1.3733	20 605	(10 000)	10 605	3.6959	0.5934	23 258
11–15	1.8859	28 304	(10 000)	18 304	3.6959	0.3522	23 826
Valuation before PCs (£)							65 564

Calculation of rental growth rate implied from a target rate of return of 11% and a NIY of 5%:

$$g = \left[\frac{(r-y)(1+r)^p + y}{r} \right]^{\frac{1}{p}} - 1$$

$$g = \left[\frac{(0.11 - 0.05)(1 + 0.11)^5 + 0.05}{0.11} \right]^{\frac{1}{5}} - 1 = 6.55\%$$

[6] Analysis of comparable property:

Equivalent yield = 5.77 %

a)   Valuation of freehold interest using income capitalisation

Adjust equivalent yield upwards for abnormal seven-year unexpired term compared to three years unexpired in comparable, so, using a yield of say 5.85%:

Rent passing (£)	60 000	
YP 7 years @ 5.85%	5.6123	
		336 736
Reversion to MR	100 000	
YP perp @ 5.85%	17.0940	
PV £1 7 years @ 5.85%	0.6717	
		1 148 175
Valuation before PCs (£)		1 484 911
Valuation net of PCs (£)		1 404 171

b)   Valuation of leasehold interest

Years	Rent received (£)	Rent paid (£)	Profit rent (£)	YP @ 12%	PV £1 @ 12%	PV (£)
0–2	90 000	(60 000)	30 000	1.6901	1	50 703
3–7	105 042[a]	(60 000)	45 042	3.6048	0.7972	129 439
Valuation before PCs (£)						180 142

[a] £100 000 × (1.0249)².

Profits method

[1]

Turnover -> FMT (£)	345 000	
Purchases (£)	(78 600)	
Wages (£)	(82 000)	
Utilities (£)	(12 000)	
Business rates (£)	(3250)	
Repairs and maintenance (£)	(1900)	
FMOP (£)	167 250	
YP in perpetuity @ 6.50%	15.3846	
Valuation before PCs (£)		2 573 077

[2]

Number of spaces	100	
Average occupancy	70%	
Average income per day (£)	4.50	
Number of days	365	
FMT (£)		114 975
Ticketing costs (£)		(5000)
Management costs @ 7% of total income (£)		(8048)
Business rates (£)		(25 000)
Repairs allowance (£)		(8500)
FMOP (£)		68 427
Capitalised @ 8%		12.5
Valuation before PCs (£)		855 338

[3]

Number of bedrooms	300
Number of term weeks	30
Rent (£ per week)	100
Occupancy rate	98%
Total term income (£)	882 000
Number of bedrooms	300
Number of vacation weeks	22
Rent per week	70
Occupancy rate	70%
Total vacation income (£)	323 400
Other income (£)	45 000
FMT (£)	1 250 400
Maintenance (£)	(100 000)
Cleaning (£)	(150 000)
Wages (£)	(75 000)
Utilities (£)	(45 000)
FMOP (£)	880 400
Capitalised @ 7.25%	13.7931
Valuation before PCs (£)	12 143 448

[4] Estimate of FMT: Here, simple averages are used but in practice more intuition might be employed.

Years	2018	2019	2020	2021	2022	FMT
Revenue (£)	787 000	802 000	783 000	771 000	787 000	786 000
Expenditure (£)	368 000	338 000	321 000	331 000	360 000	343 600
Net profit (£)	419 000	464 000	462 000	440 000	427 000	442 400
Net profit as % of turnover	53%	58%	59%	57%	54%	56%

Valuation:

FMOP (£)	442 400
YP in perpetuity @ 9%	11.1111
	4 915 551
Capital expenditure deductions:	
Replacement pontoons (£)	(500 000)
Sea wall repairs (£)	(500 000)
	(1 000 000)
Valuation before PCs (£)	3 915 551

Yield influences:

- Location and specification
- Income security
- Future growth prospects
- Allowances for irregular or one-off income/expenditure

[5] For valuation purposes, interest received from capital employed in the business is ignored as it is not a trade-related revenue stream. The business is considered to be one most suited to a sole proprietor, so the wage cost has been adjusted and director's emoluments and pension costs have not been deducted. The rent paid under the lease is not deducted nor is amortisation of goodwill as it assumed that it is non-transferable. Depreciation and the cost of hire purchase, leasing and other rental agreements are also not deducted nor is the profit realised on the sale of asset. The valuation therefore proceeds as follows:

Sales -> FMT (£)		550 000
Purchases (£)		(121 000)
Depreciation in value of stock (£)		(2500)
Gross profit (£)		426 500
Operating costs and wage costs (£):		
Wage cost (adjusted)	(135 000)	
Licence fee	(525)	
Rates, water and environmental charges	(13 500)	
Heating and lighting	(10 500)	
Telephone	(3000)	
Insurance	(5250)	
Repairs and maintenance	(6000)	
Printing, postage and stationery	(2500)	
Promotion	(3000)	
Accountancy and professional fees	(3500)	
Transport	(2000)	
Laundry, cleaning and linen hire	(1500)	
Entertainment	(1250)	
Credit and charge card commissions	(3000)	
Sundries	(1268)	
		(191 793)
FMOP (£)		234 707
YP in perpetuity at a yield of 12.5%		8.0000
Valuation before PCs (£)		1 877 656

[6]
a) The operator is an owner-occupier:

Sales -> FMT (£)	260 000	
Cost of sales (£)	(98 800)	
Wages	(80 000)	
Electricity and gas	(10 000)	
Business rates	(2250)	
Repairs and maintenance	(1800)	
FMOP (£)	67 150	
Capitalised at 7%	14.2857	
Valuation before PCs (£)		959 286

b) The operator is a tenant:

FMOP as before (£)		67 150
Allowances for tenant's investment in business:		
Furniture, fixtures, fittings and equipment (£)	(20 000)	
Stock (£)	(2500)	

Cash and working capital (£)	(500)
Total (£)	(23 000)
Return on capital invested @ 5%	×0.05
	(1150)
Divisible balance (£)	66 000

In the absence of any information to the contrary, assume this balance is split between landlord and tenant on a 50:50 basis. So, the MR is estimated to be £33 000 per annum. Note that the rent that the tenant is actually paying is ignored because the valuation is an estimate of market rent.

[7]   Adjust profit and loss account by disregarding deductions for director's remuneration, rent, bank loan interest, credit card charges and depreciation for fixtures and fittings and vehicles.

Sales (£):		
▪ Wet	120 000	
▪ Dry	140 000	
▪ Other	150 000	
▪ Own consumption	(500)	
Total (£)	409 500	
Cost of sales (£):		
▪ Wet	(55 000)	
▪ Dry	(35 000)	
Gross profit:	319 500	
Other income:		
▪ Deposit account interest	5000	
FMT (£)		324 500
▪ Operating costs and wage costs:		
▪ Wages and NI	(95 000)	
▪ Telephone	(2000)	
▪ Motor	(2500)	
▪ Licences and memberships	(1500)	
▪ Hardware and crockery	(500)	
▪ Leasing and machine rental	(1500)	
▪ Equipment repairs and renewals	(6000)	
▪ Cleaning and laundry	(8000)	
▪ Print, stationery and ads	(2000)	
▪ Sundry expenses	(2000)	
▪ Accountancy	(1500)	
▪ Bookkeeping	(1500)	
▪ Legal and professional	(2000)	
▪ Stocktaking fees	(500)	
▪ Rates	(12 000)	
▪ Insurance	(3000)	
▪ Light, heat and water	(12 000)	
▪ Repairs to property	(3000)	
▪ Bank charges and interest	(2000)	
Total expenditure (£)		(158 500)
Adjusted net profit -> FMOP (£)		(166 000)
YP perpetuity at 10%		10.0000
Valuation before PCs (£)		1 660 000

Appendix

7.A   Valuation of trade-related properties using the profits method

Trade-related properties are usually valued by capitalising estimated FMOP. The profit is estimated with reference to past performance and the key source of information is the accounts. The capitalisation rate should reflect the risk, growth and income security associated with the business and its estimation is reliant upon good comparable evidence. Such evidence is also important in determining whether gross profit margins, wages levels and other costs and revenues are in line with market expectations.

7.A.1   *Hotels, guest houses, bed and breakfast and self-catering accommodation*

Whenever possible, valuers would seek to use the comparison method of valuation, perhaps using metrics such as price per 'double-bed unit' for small hotels, guest houses and bed-and-breakfast accommodation (based on individual properties and dependent upon size, type and location). Price per double-bed unit (which includes double or twin room with ensuite facilities) is usually calculated by looking at annual rents but can be based on capital values too, and this might be the case for trophy or privately owned lifestyle hotels. The value of rooms that are not double-bed units is adjusted using a formula based on size and ensuite facilities. The value of common areas of guest houses and smaller hotels is reflected in overall valuations of letting rooms. The value of bar/restaurant facilities open to non-residents is assessed separately. In the absence of suitable comparable evidence, normally in the case of large hotels, it is necessary to capitalise an estimate of FMT and use the profits method. The capitalisation rate should reflect risk, growth and income security and is heavily reliant on comparable evidence (which is also important in judging gross profit levels, wages and other costs). Most three-star and some four-star hotels are valued using this method. A discounted cash-flow method might be used to value four-star and five-star hotels where purchasers may include investors, in which case it is necessary to consider holding period, exit yield, target rate of return and whether to inflate figures in the cash flow over the holding period.

The profits method and discounted cash-flow method require analysis of several years' trading performance plus projections until trading has stabilised. Changes in supply and demand, as well as changes in the legislative and regulatory environment, may affect trading performance and these factors should be reflected in the cap rate and adjusted net profit in the case of the profits method and in the target rate of return and cash flow in the case of a discounted cash flow (RICS 2004).

As a comparison-based short-cut to the profits method, the valuation of self-catering holiday accommodation can be based on the number of single bed spaces. The price per bed space can be estimated by looking at the profit levels of a range of comparable properties and making adjustment for size, type and location as appropriate. The level of profit is estimated by looking at the total income (excluding VAT) and deducting costs of maintenance, utilities and services, TV licences and depreciation of fixtures. The estimation should not include the cost of any loan or mortgage used to buy the property.

For example, a modern purpose-built four-star provincial hotel has 125 double bedrooms with well-planned, flexible accommodation including a restaurant, bar, conference rooms and leisure club with 200 members. The hotel is easily accessible, just off a motorway junction and with good car-parking facilities. It is reliant on corporate business and conference trade during the week with some leisure-based trade at weekends. The advertised tariff is £95 per night for a double and £75 per night for a single (including VAT). Overall room occupancy averages 75%, comprising 35% double occupancy and 65% single occupancy. The average achieved room rate was 70% of the advertised tariff. Other sources of income (and expenditure) include:

Item	Revenue (£)	Expenditure (£)
Rooms	See above. . .	470 000
Food	1 260 000	850 000
Beverages	468 000	200 000
Phone	72 000	30 000
Other	179 500	80 000

Other operating expenses include:

Administration (£)	340 000
Marketing (£)	90 000
Property management (£)	130 000
Energy (£)	120 000
Business rates (£)	135 000
Insurance (£)	50 000
Renewal fund: £10 000 per room @ 10% per annum (£)	125 000

The valuation might proceed as follows:

Total rooms available per annum: 125 rooms × 365 days	45 625 room days
75% room occupancy	34 219 let room days
Double occupancy @ 35%	11 977 let room days
Single occupancy @ 65%	22 242 let room days

Averaged achieved room rates:

Double room @ £95.00 @ 70% = £66.50	£55.42 net of 20% VAT
Single room @ £75.00 @ 70% = £52.50	£43.75 net of 20% VAT

Room revenue:

11977 doubles @ £55.42 (£)	663765
22242 singles @ £43.75 (£)	973088

Other revenue:

Food, beverages, phones, other (from above) (£)	1979500
Total	3616353

Running expenses:

Expenditure (from above) (£)	(1630000)	
Other operating expenses (from above) (£)	(990000)	
Adjusted net profit -> FMOP (£)	996353	
YP perpetuity @ 10%	10	
Valuation before PCs (£)		9963530

A 10-year investigation into collapse of international hotel chain 'Queens Moat Houses', undertaken by DTI and published in 2004, strongly criticised hotel valuation methods. Volatility of valuations depend on the valuer's view of maintainable profit, which itself depends on how many years' trading are factored in, as well as economic circumstances of the time and adjustments made for factors such as competition. The report argues that hotel value depended on a hybrid of tangible and intangible assets and based on many subjective judgements. If valuations of this kind are to be incorporated into balance sheets, the bases and assumptions on which they have been calculated should be clearly stated in the accounts. The report noted that hotel valuations should be done by, or in conjunction with, someone possessing considerable accounting skills. In summary, the investigators questioned whether hotel values should be included in balance sheets without detailed explanations.

7.A.2   Restaurants, pubs and nightclubs

The valuation of restaurants, pubs and clubs involves a similar approach to that described above for hotels. Capital valuations of freehold premises are a capitalisation of the estimated FMOP whereas, for leasehold premises, FMOP becomes a divisible balance which is split into a rent portion and a residual profit portion. The rent can be capitalised at a suitable property investment yield and the profit can be capitalised at a suitable business cap rate. Normally, company accounts are used to derive the FMOP. If the net profit figure from the accounts includes a deduction for depreciation or interest on the operator's capital, then this should be added back. The reason for this is that the figures may not be representative of the sector. Also, if the net profit included a deduction for rent, then this should be added back too as this is a component of the divisible balance being calculated. Once an adjusted net profit is estimated, a 'market-derived' interest on tenant's capital in the business can be deducted.

Public houses are labour-intensive, intricate businesses and subject to the demands of a changeable clientele. The value-significant attributes of pubs are likely to include:

- Location: Passing trade, catchment, demographics, competition
- Bar length and room size
- Sales area and drinking space
- Display fittings
- Surveillance of bar by landlord
- Temperature of cellar
- Internal or external toilets
- Catering facilities
- Car parking
- Whether the pub is 'tied' to a brewery or a free house

There are several types of purchasers including breweries, caterers and retailers, investors and owner-occupiers. Variations of the profits method are often used to value pubs depending on the type of likely purchaser. For a pub that is owned by an investor (a brewery for example) and operated by a tenant publican, income is typically generated by three revenue streams; wholesale (beer, liquor and maybe food), retail (food and other sales) and the 'tied' rent. For a freehold pub (a free house), profits are derived from essentially two sources: retail sales (beer, liquor and food) and machine income. By analysing the income and expenditure streams, a net-adjusted profit can be determined, which is then divided between remuneration for the tenant and rent to the landlord. The rent can then be capitalised to determine a capital value. A pub may also be valued by reference to 'barrelage', and this enables the valuer to estimate likely turnover and profit without recourse to a full accounts approach. If a pub lease is terminating in the short term, it may be worth considering the reversion to capital value rather than a revised lease rent.

The profits method is also used to value nightclubs and the approach is similar to pubs but with an additional risk premium on the required rate of return due to the threat of licence revocation and the fickle nature of the market. If the pub, club or other licenced premises is leased to a tenant with a strong covenant it may be possible to value the property as an investment by capitalising the MR. The difficulty is in establishing the appropriate yield at which to do so. The term rent may be regarded as relatively secure, but the reversion may require a little more thought in terms of alternative-use value, flexibility of the space, quality of the building and location. For some types of licenced premises in certain locations, an investment market is firmly established but the profits method remains primary valuation method.

Valuations of licenced properties should report (RICS 2006):

- Market activity at the national, regional and local scale as relevant
- The size and layout of the premises, noting any internal areas incapable of efficient use and any outside areas that may enhance value (seating, etc.)
- Availability of services, refuse collection information
- Means of escape
- Ownership arrangements (owned, leased, hired) and age of plant, fittings, furnishings and equipment

- Existence and duration of licences, and planning details (consents, conservation areas, listed buildings, etc.)
- Competition (which may, it should be noted, also serve as comparable evidence)
- Proximity of transport and other entertainment facilities
- Proximity to sensitive neighbours
- Lease details
- Contamination

The following two examples illustrate the application of the profits method to the valuation of a restaurant and a pub.

You have been asked to estimate the capital value of the freehold interest in a town centre restaurant. The business is currently leasehold and also has a pavement eating area held on an annually renewable licence at £525 per annum from the local authority. The business is operated on a full-time basis by one owner, who is also the chef. The most recent accounting information is shown below.

Revenue:		
Sales (£)		550 000
Interest received on capital (£)		1075
Expenditure:		
Purchases (£)		(121 000)
Opening stock value (£)	10 000	
Closing stock value (£)	7500	
Change in value of stock (£)		(2500)
Operating costs:		
Wages and salaries (£)	(150 000)	
Director's emoluments (£)	(35 000)	
Director's pension (£)	(10 000)	
Rent (£)	(41 025)	
Rates, water and environmental charges (£)	(13 500)	
Heating and lighting (£)	(10 500)	
Telephone (£)	(3000)	
Insurance (£)	(5250)	
Repairs and maintenance (£)	(6000)	
Printing, postage and stationery (£)	(2500)	
Promotion (£)	(3000)	
Accountancy and professional fees (£)	(3500)	
Transport (£)	(2000)	
Amortisation of goodwill (£)	(2410)	
Laundry, cleaning and linen hire (£)	(1500)	
Depreciation (£)	(5035)	
Entertainment (£)	(1250)	
Hire purchase, leasing and rental agreements (£)	(5250)	
Credit and charge card commissions (£)	(3000)	
Sundries including licence fees (£)	(1268)	
Profit/loss on sale of asset (£)	(3012)	
		(308 000)
Net profit/loss (£)		119 575

For valuation purposes, interest received from capital employed in the business is ignored as it is not a trade-related revenue stream. The business is one most suited to a sole proprietor and therefore the net profit figure has been adjusted so that director's emoluments and pension costs are not deducted, and the salaries amount has also been adjusted. The rent paid under the lease is ignored because it forms the basis of the adjusted net profit which is capitalised. Amortisation of goodwill is also ignored as it assumed that it is non-transferable. Depreciation and the cost of hire purchase, leasing and other rental agreements are also ignored, as is the profit realised on the sale of an asset. The valuation would therefore proceed as follows:

FMT (£)		550 000	
Less purchases (£)		(121 000)	
Depreciation in value of stock (£)		(2500)	
Gross profit (£)		426 500	
Running expenses:			
Wages and salaries (£)	(135 000)		
Licence fee (£)	(525)		
Rates, water and environmental charges (£)	(13 500)		
Heating and lighting (£)	(10 500)		
Telephone (£)	(3000)		
Insurance (£)	(5250)		
Repairs and maintenance (£)	(6000)		
Printing, postage and stationery (£)	(2500)		
Promotion (£)	(3000)		
Accountancy and professional fees (£)	(3500)		
Transport (£)	(2000)		
Laundry, cleaning and linen hire (£)	(1500)		
Entertainment (£)	(1250)		
Credit and charge card commissions (£)	(3000)		
Sundries (£)	(1268)		
TOTAL (£)		(191 793)	
Adjusted net profit -> FMOP (£)		234 707	
YP in perpetuity at a yield of 12.5%		8.0000	
Valuation before PCs (£)			1 877 656

It is also useful to consider some of the figures as percentages of turnover. Gross profit is 78%, wages and salaries (before adjustment) are 27% and adjusted net profit is 43%. These metrics are helpful when comparing the trade figures of the subject property to comparable businesses.

You have been asked to estimate the capital value of the leasehold interest in a public house. The landlord is a brewery, which granted a new 30-year lease last year on FRI terms with 5-year upward-only rent reviews. The lease is assignable, subject to landlord's approval. There is a 'tie' for beers, wines and spirits.

Revenue (£):			
Sales receipts		364 082	
Net machine receipts		4 312	
Less Costs (£):			
Purchases		(150 108)	
Opening stock value	6 671		
Closing stock value	7 953		
Increase / Reduction in stock value		1 282	
		(148 826)	
Gross Profit (£)		219 568	
Operating Expenses (£):			
Wages and salaries		(73 210)	(20% sales)
Postage, stationery and advertising		(1 508)	
Telephone		(837)	
Accountancy and book-keeping		(1 362)	
Cleaning		(4 115)	
Sundry		(6 367)	
Motor expenses		(3 706)	
Tied rent[1]		(42 000)	(11.5% sales)
Rates and Water		(11 637)	
Insurance		(2 165)	
Light and heat		(6 753)	
Repairs and renewals		(9 268)	(2.5% sales)
Total		(162 928)	
Adjusted Net Profit -> FMOP (£)		56 640	

[1] The beer tie is relevant in that without it the adjusted net profit would be higher

The adjusted net profit can then be capitalized in the normal way at an appropriate market yield.

7.A.3   Care homes

Care homes are registered to provide personal care and possibly nursing care. The profits method is also used to value care homes (see Sidwell 1991 for example) with the comparison method (recent sales or per registered bed multiplier in the locality) as a check. The main revenue stream is the occupancy fees, and the main costs are staff. Comparison metrics are helpful in determining whether occupancy levels, fee rates, staff and non-staff costs are reasonable for the sector and locality. An adjusted net profit of around 25–38% of revenue for a nursing home and 38–41% for a less staff intensive residential or personal care home would be expected (Hayward 2009). The selection of yield at which the adjusted net profit is capitalised is matter of judgement based on experience in the sector and evidence from previous sales.

The yield will depend on the location and quality of the home, and it is particularly important to look at the quality of the catering facilities, staff costs, agency fees and medical charges.

For example, you have been asked to value a 24-registration residential care home built 12 years ago. You have limited accounting information. The home is in a town with a higher-than-average proportion of residents aged 75 years or over. Total day-space extends to 96 square metres. Residents are located on ground-floor and first-floor levels and there is a lift (shared by residents and staff) between the two floors. There are 20 single and two twin rooms. 16 have ensuite provision. All single rooms exceed 10 square metres and the twin rooms exceed 16 square metres. (So the proportion of total space allocated to single rooms is 83%.) There are four bathrooms (a ratio of 1:3). At the time of the last inspection, there are 22 residents in occupation and both twin rooms were being used as singles. There were no requirements following this inspection. The 22 residents exclude two beds contracted for respite care to the local authority. Including these, the average fee is £379.91 per resident per week. The home is run under management with the manager supernumerary to the staffing rota. The manager receives £23 500 per annum. Senior carers are paid £6.35 per hour and care assistants £5.65. The cook is paid £5.85 per hour, as is the activities organiser, while the housekeeper receives £5.35 per hour. The home operates the following rota:

Hours	Staff designation	Number
07:00–13:30	Senior care assistant	1
	Care assistant	1
13:30–20:00	Senior care assistant	1
	Care assistant	2
20:00–07:00	Care assistant	2

Domestic staff contribute 101 hours per week while the activities organiser works 15 hours per week. The management accounts for the 121-day period to the 31 December record a total gross wage bill of £65 450. The management figures show fee income for the 121-day period to 31 December of £158 023. Before analysing the figures, the following assumptions are made, that the number of registrations is maintained at 24, the average weekly fee is £379.91 and the overall occupancy rate is 98.5%.

Calculation of wages and salaries

Staff position	No. staff	Hours	Days per week	Rate (£/hour)	Total per week (£)
Senior care assistant	1	13	7	6.35	577.85
Care assistant	2	13	7	5.65	1028.30
Junior care assistant	2	11	7	5.65	870.10
Activities organiser	1	15	1	5.85	87.75
Cook	1	40	1	5.85	234.00
Housekeeper	1	61	1	5.85	326.35

Sub-total	3124.35		
Number of weeks per year plus Holiday weeks	56		
NI (part-time) inflation factor	1.07		
		187211	
Plus manager's salary	23 500		
13/12 to reflect holiday cover	1.0833		
NI (full-time) inflation factor	1.09		
		27 750	
		214 961	
Income	(£379.91 × 52 weeks × 24 rooms × 98.5%)	465 000	

Operating costs (£):

Wages and salaries (calculated above)	(215 000)	(46.23% of income)
Provisions	(20 000)	
Heating and lighting	(10 000)	
Repairs and maintenance	(10 000)	
Insurance	(3500)	
Telephone	(2000)	
Printing and advertising	(1000)	
Professional fees	(3000)	
Transport	(1500)	
Laundry and cleaning	(4000)	
Residents' welfare	(2000)	
Staff training	(2000)	
Water and environmental charges	(3750)	
Sundries (including reg. fees)	(7250)	
	(285 000)	
Adjusted net profit -> FMOP (£)	180 000 (38.7% of income)	

This adjusted net profit can be capitalised in the normal way at an appropriate yield.

7.A.4 Petrol-filling stations

Petrol-filling stations may be attached to car dealers and motorway services, found on supermarket sites, along main trunk roads and in other urban and suburban locations. They can be broadly classified as those with large throughput (of *acquisition* interest to oil companies) and those with less throughput (of *supply* interest to oil companies). Outlets tend to be owned and operated by major oil companies, owned and operated by a dealer or retailer or owned by a major oil company (who also supplies fuel) and operated by tenant (who pays a 'tied' or low rent to the oil company). A valuer should classify the petrol station by throughput and tenure, and then analyse the capital values and throughput figures of comparable outlets to determine a scale of capital values per litre of throughput, effectively a comparative sales approach. Table 7A.1 provides an example.

Table 7A.1   Variation in capital values of petrol stations depending on throughput.

Annual throughput (litres)	Capital value (pence per litre)
2 273 000	16.50
2 727 600	19.80
3 182 200	24.20
3 636 800	26.40
4 091 400	26.40
4 546 000	26.40
5 682 500	27.50
13 638 000	29.70

If the valuer believes the petrol station is one that an oil company might be interested in acquiring, the valuer will capitalise the throughput at a standard rate using a scale such as the one in Table 7A.1 and capitalise the additional facilities, such as shop, carwash and so on separately. If the throughput is such that an oil company would only be interested in supplying fuel, then the calculations will differ. A detailed examination of factors that influence the ability to trade can be undertaken, such as the volume of passing traffic, average 'turn-in', size of average petrol purchase and so on. Care must be exercised in adjusting throughputs of comparable petrol stations when reconciling them with the subject property. The trading potential of a specific station may depend upon many factors in addition to petrol sales and it is important that these are considered. The retail element of the petrol station sales can be substantial on many sites.

A valuation of a petrol-filling station might proceed as follows:

- Forecourt
  - Obtain annual throughput for each fuel type (litres).
  - Make deductions as appropriate for long (e.g. 24) hours, customer accounts, etc.
  - Sum to adjusted maintainable throughput @ £/000 litres.
  - Adjust as appropriate for credit card/agency sales, full/partial/no canopy, car wash, etc.
- Shop sales space might be valued at a rent per square metre obtained from comparable evidence.
- Ancillary space might be valued in proportion to the sales space, for example office space at 80%, staff and kitchen area at 50% and storage space at 30%.

For example, consider a petrol station that is currently owned and occupied by an independent retailer and fuel is supplied by an oil company. The property is an owner-occupied, self-service petrol station located on a busy trunk road. The road has good visibility, a 40-mph speed limit and average traffic volumes of 30 000 vehicles per day. The station has a turn-in rate of 4% from the near-side average of 16 000 vehicles per day plus 120 vehicles per day crossover from the other side of the road. This produces an average of 640 customers from the nearside plus 120 'crossovers'. Estimating an average purchase of 20 litres, this equates to

15 200 litres per day or approximately 5 138 000 litres per annum on a 6.5-day week basis. Other facilities include a forecourt shop and a car wash. The petrol station is one that would attract acquisition interest from oil companies and is valued as if this class of purchaser would be in the market.

Forecourt (£):		
5 138 000 l per annum @ capital value of		1 387 260
say 27 pence per litre (see Table 7A.1)		
Shop (£): 50 m² @ £130 per m[a]	6500	
Car wash (£): @ one third of net profit[b]	13 000	
	19 500	
YP in perpetuity @ 10%[c]	10	
		195 000
Valuation before PCs (£)		1 582 260

[a] As shop size increases, the profit per unit of floor area decreases as the range of goods is extended to include items with lower profit margins (Hayward 2009).
[b] A fully equipped car wash is estimated to cost £75 000 to build and, with a gross return of £50 000 per annum and running costs of £10 000 per annum, this leaves a net return of £40 000 per annum. It is assumed here that an oil company landlord would probably estimate one third of the net profit as rent, equating to approximately £13 000 per annum.
[c] Oil companies are not institutional property investors so a yield typically between 7 and 10% is used to capitalise annual (non-fuel) income.

7.A.5   Serviced offices

Serviced offices provide fully equipped office space plus access to support staff via a licence agreement. There is no lease liability, but occupation terms are usually standard. The speed of set-up is particularly helpful for new starts and gives access to technology and staff at an all-inclusive cost, but the cost can be expensive. The price charged for serviced office accommodation is usually quoted on a per workstation basis. Most operators of serviced officed quote a range of prices to reflect different rents dependant on the level of natural light and the size of office. Prices are per person per month, and this is usually inclusive of rent, business rates, service charge, furniture, electricity, lighting, heating and use of reception facilities.

McAllister (2001) explores the issues relating to the valuation of properties where income is derived from the provision of services as well as floor-space. He argues that, because serviced offices comprise a property and a business asset, the valuer should consider the derivation and risk profile of each income flow. In practice, serviced offices are valued using the profits method or investment method depending on whether the assets are owned by the operator. If the operator has no property assets, using the profits method, the value would typically be 2.5–4x Earnings Before Interest, Tax, Depreciation and Amortisation (EBITDA), in other words a 25–40% return. An investment method can also be used because serviced office businesses occupy standard office accommodation. Where a landlord is more directly involved with the business, the distinction between property value and business revenue blurs (payments for service provision can act as a substitute for rent for example). If the operator owns property assets, the value is derived from a combination of business profitability and property value. The profits method can be used to value the business, having regard to trading potential, on an existing-use basis and including plant and machinery, fixtures, fittings, furniture and equipment and assuming the business is competently managed, properly staffed, stocked and capitalised. Capital value would typically be 8–12.5x EBITDA (net of MR), an 8–12.5% return. The investment method can be used to value property assets. The valuation may exceed conventional investment valuation because the business derives profit from its services and charges rents, which may diverge from market due to more regular reviews, but extra profitability needs to be weighed against lower security from very short leases. An alternative approach is to estimate market value based on existing use with vacant possession.

7.A.6   Data centres

Data centres are highly specified and configured buildings, which integrate infrastructure to provide a secure, controlled environment to house and operate IT equipment. Occupational structures are divided into two main groups: conventional real-estate leases (and licences) and leases with managed services. They are specialised assets that produce investment income and so should be appraised as investment rather than occupational assets, despite an absence of significant market trading (McAllister and Loizou 2007). Table 7A.2 summarises the value-significant characteristics of data centres.

Table 7A.2   Value attributes of data centres.

Characteristic	Conventional asset	Technical real estate
Typical contract length	10 years	5 years
Rental change	Rent reviews	Annual inflation-linked
Non-performance penalty payments	None	Potential for substantial payments
Construction cost	1x	2–6x
Building infrastructure cost (relative to shell and core)	Low	High
Site value (relative to construction cost)	High	Low
Depreciation risk:		
■ Capex on infrastructure	Low	High
■ Capex on building	High	Low
■ Locational	Low	Low
■ Aesthetic	High	Low
■ Technological	Low	High
Market characteristics:		
■ Liquidity	High	Low-ish
■ Maturity	High	Low
■ Transparency	High	Low
■ Demand and supply shocks	Major	Major

7.A.7   Student accommodation

Now an established property investment sector, student accommodation is valued using a combination of profits method and discounted cash flow techniques.

7.A.7.1   Example 1

A 300-bedroom student hall of residence opened four years ago and has a nomination agreement with a nearby university. The property is 98% let during term time at £100 per week. During the holiday period, the property is 70% let at £70 per week. Using the profits method (capitalised earnings technique) to value the property, it is assumed that the FMT is equivalent to the current term rent, but the vacation rent is £70 per week with 70% occupancy. A cap rate of 7.25% is assumed.

FMT (£)	Term income: 300 rooms for 30 weeks @ £100 per week and 98% occupancy	882 000
	Vacation income: 300 rooms for 22 weeks @ £70 per week and 70% occupancy	323 400
	Other income	45 000
COSTS (£)	Maintenance	(100 000)
	Cleaning	(150 000)
	Staff	(75 000)
	Utilities	(45 000)
Net income -> ANP		880 400
-> FMOP		
Capitalised at 7.25%		13.7931
Valuation before PCs (£)		12 142 448

Analysis (unit of comparison): value per student bed = £40 475

7.A.7.2   Example 2

Another 300-bedroom student hall of residence is 98% let during term time (30 weeks) at £100 per week. During the vacation periods (22 weeks), the property is about 65% let at £60 per week. Using the profits method (DCF technique) to value the property, it is assumed that:

- Term rental value will increase by 3% p.a.,
- Vacation rental value will increase by 4% p.a.,
- Vacation occupancy levels will slowly improve,
- Utility costs grow at 4% p.a. and maintenance costs grow (to reflect rising energy costs and depreciation) at 2% p.a,
- Target rate of return (discount rate) of 10% and
- Exit yield of 7%.

First a cash flow is constructed to estimate the revenue and costs. The net income is then discounted to present value. The output net present value can be analysed by calculating capital value per bed type and per square foot.

Year	1	2	3	4	5	6	7	8	9	10	Exit
REVENUE (£):											
Term rent per week	100	103	106	109	113	116	119	123	127	130	
Term occupancy rate	98%	98%	98%	98%	98%	98%	98%	98%	98%	98%	
Term income	882 000	908 460	935 714	963 785	992 699	1 022 480	1 053 154	1 084 749	1 117 291	1 150 810	
Vacation rent per week	60	62	65	67	70	73	76	79	82	85	
Vacation occupancy rate	65.0%	67.5%	70.0%	72.5%	72.5%	72.5%	72.5%	72.5%	72.5%	72.5%	
Vacation income	257 400	277 992	299 820	322 948	335 866	349 301	363 273	377 804	392 916	408 633	
Other income	45 000	45 000	45 000	45 000	45 000	45 000	45 000	45 000	45 000	45 000	
Total	1 184 400	1 231 452	1 280 533	1 331 734	1 373 565	1 416 781	1 461 427	1 507 553	1 555 207	1 604 443	
COSTS (£):											
Maintenance	(100 000)	(102 000)	(104 040)	(106 121)	(108 243)	(110 408)	(112 616)	(114 869)	(117 166)	(119 509)	
Cleaning	(150 000)	(150 000)	(150 000)	(150 000)	(150 000)	(150 000)	(150 000)	(150 000)	(150 000)	(150 000)	
Staff	(75 000)	(75 000)	(75 000)	(75 000)	(75 000)	(75 000)	(75 000)	(75 000)	(75 000)	(75 000)	
Utilities	(45 000)	(46 800)	(48 672)	(50 619)	(52 644)	(54 749)	(56 939)	(59 217)	(61 586)	(64 049)	
Total	(370 000)	(373 800)	(377 712)	(381 740)	(385 887)	(390 157)	(394 556)	(399 085)	(403 752)	(408 558)	
Net income (£)	814 400	857 652	902 821	949 994	987 678	1 026 623	1 066 872	1 108 467	1 151 456	1 195 884	1 195 884
PV @ TRR of 10%	0.9091	0.8264	0.7513	0.6830	0.6209	0.5645	0.5132	0.4665	0.4241	0.3855	0.3855
Capitalised at exit yield of 7%											14.2857
PV (£)	740 364	708 803	678 303	648 859	613 271	579 502	547 474	517 108	488 330	461 065	6 586 646
Valuation before PCs (£)											12 569 724

7.B   To estimate the value, *V*, of a freehold property let at market rent, *MR*, which is subject to rent reviews every five years

If we let target rate of return = *r* and annual rental growth rate = *g*:

$$V = \frac{MR}{(1+r)} + \frac{MR}{(1+r)^2} + \frac{MR}{(1+r)^3} + \frac{MR}{(1+r)^4} + \frac{MR}{(1+r)^5} + \frac{MR(1+g)^5}{(1+r)^6}$$
$$+ \frac{MR(1+g)^5}{(1+r)^7} + \frac{MR(1+g)^5}{(1+r)^8} + \frac{MR(1+g)^5}{(1+r)^9} + \frac{MR(1+g)^5}{(1+r)^{10}} + \frac{MR(1+g)^{10}}{(1+r)^{11}} + \cdots \infty$$

Let rent-review period = *p*:

$$V = \frac{MR(1-(1+r)^{-p})}{r} + \frac{MR(1+g)^p}{(1+r)^p} \cdot \frac{(1-(1+r)^{-p}}{r} \frac{MR(1+g)^{2p}}{(1+r)^{2p}} \cdot \frac{(1-(1+r)^{-p}}{r} + \cdots \infty$$

Factor out expression for an annuity, which is common at each rent review:

$$V = \frac{MR(1-(1+r)^{-p}}{r}\left[1 + \frac{(1+g)^p}{(1+r)^p} + \frac{MR(1+g)^{2p}}{(1+r)^{2p}} + \frac{MR(1+g)^{2p}}{(1+r)^{2p}} + \cdots \infty\right]$$

The expression in brackets is a geometric progression S_∞ and can be summed using the following formula:

$$S_\infty = \frac{A}{1-R_c}\text{ where the first term } A = 1\text{ and the common ratio } R_c = \frac{(1+g)^p}{(1+r)^p}.$$

So $S_\infty = \dfrac{1}{1-\dfrac{(1+g)^p}{(1+r)^p}}.$ Multiplying top and bottom by $(1+r)^p$, we get $\dfrac{(1+r)^p}{(1+r)^p - (1+g)^p}.$

Substitute this into the valuation formula:

$$V = \frac{MR\left(1-(1+r)^{-p}\right)}{r}\left[\frac{(1+r)^p}{(1+r)^p - (1+g)^p}\right]$$

In the numerator, $1-(1+r)^{-p}$ multiplied by $(1+r)^p$ is $(1+r)^p - 1$ so the equation simplifies to (subject to $r > g$):

$$V = \frac{MR}{r}\left[\frac{(1+r)^p - 1}{(1+r)^p - (1+g)^p}\right]$$

This can be transformed as follows:

$$V = \frac{MR}{\dfrac{r\left((1+r)^p - (1+g)^p\right)}{(1+r)^p - 1}} = \frac{MR}{\dfrac{r(1+r)^p - r - r(1+g)^p + r}{(1+r)^p - 1}} = \frac{MR}{r - \dfrac{r(1+g)^p}{(1+r)^p - 1} + \dfrac{r}{(1+r)^p - 1}}$$

Giving:

$$V = \frac{MR}{r - r\left(\dfrac{(1+g)^p - 1}{(1+r)^p - 1}\right)}$$

Chapter 8
Cost Approach

8.1   Introduction

The comparison and income approaches to estimating market value are predicated on the availability of market price information. For certain types of land and property, market trading is sparse or non-existent. If there is no existing market for the tenure rights, then it may be possible to 'create' one using an auction or tender process. This approach may be appropriate in the case of large-scale land acquisitions or the sale of unusual rights such as airwaves for mobile phone networks. If this is not possible, then a cost approach may be appropriate. Two cost approach methods are described in this chapter: the replacement cost method and the residual method.

The *replacement cost method* is used to value properties that rarely, if ever, trade on the open market and therefore there is little or no evidence of comparable market prices on which to base value estimates. Some properties are very use-specific: bespoke manufacturing plants such as chemical works and oil refineries; public administration facilities such as prisons, schools and colleges, hospitals, town halls, art galleries and court facilities; and transport infrastructure such as airports and railway buildings. The method is used to value properties for financial reporting, taxation, and expropriation purposes. It is also used to estimate building reinstatement costs for insurance purposes. It is important to note that, when these sorts of properties are offered for sale, perhaps because they are no longer required for their current use, the primary market is likely to be for alternative uses for which a different valuation method may be appropriate.

The *residual method* is used to value the development potential of land. If a valuer thinks land is not being used to its full potential, it may have development value. Obtaining comparable evidence of development land values can be very difficult; each site is different in terms of size, condition, potential use, permitted

Property Valuation, Third Edition. Peter Wyatt.
© 2023 John Wiley & Sons Ltd. Published 2023 by John Wiley & Sons Ltd.
Companion website: www.wiley.com/go/wyatt/propertyvaluation3e

density of development, restrictions and so on, adjusting a standard value per hectare almost impossible. The *residual method* is based on a simple economic concept: the development value of land can be calculated as a surplus or *residual* remaining after estimated development costs have been deducted from the estimated value of the completed development.

8.2   Replacement cost method

The replacement cost method is used to value property '. . .that is rarely, if ever, sold in the market, except by way of a sale of the business or entity of which it is a part, due to the uniqueness arising from its specialised nature and design, its configuration, size, location or otherwise' (RICS Global Glossary). So, the value of a specialised property is intrinsically linked to its use and is only appropriate if continued exiting use is envisaged.

The method does not actually calculate a market value. Instead, it calculates a replacement cost for the improvements that have been made to the land, typically in the form of buildings and ancillary man-made land uses such as car parks, storage facilities and so on. Because of an almost complete lack of comparable market transaction information, the method relies on the economic principle of substitution to regard replacement cost as a proxy for exchange price. In other words, the value is essentially a deprival value of the property to the owner by assuming that the value of the existing property is comparable to the cost of providing a modern replacement that offers an equivalent service potential.

There are two parts to the replacement cost method, a valuation of the land in its existing use and unimproved state and a valuation of the improvements[1] to the land. The land is usually valued using the sales comparison method. The improvements are valued by estimating the cost of constructing a new replacement and then applying a depreciation allowance to reflect any deterioration and obsolescence inherent in the existing property.

The three main components of the replacement cost method are considered below: the cost of replacing the building(s) and other site improvements as appropriate, the depreciation of this cost and the valuation of the site.

8.2.1   Replacement cost

The replacement building and site improvements should be functionally equivalent to the subject property. This means that the modern equivalent replacement may not be the same size, layout, design and specification as the actual property. When estimating replacement building costs, the valuer needs to decide what constitutes the building and what constitutes plant and machinery. Sometimes, relatively standard properties have specialised features and adaptations and a valuer may decide that comparable evidence is available for the property and the specialised features can be valued separately using the replacement cost method. Some specialised features may have no market value and could have a negative value if they are regarded as an encumbrance.

The replacement cost will cover all the usual costs associated with the construction of a building, including setup costs (planning fees, site preparation), building costs, professional fees, a contingency allowance, and finance (unless there is an assumption of 'instant build') but not developer's profit. If replacement is likely to take a long time, then it may be justifiable to forecast variations in costs.

Professional fees are usually estimated as a percentage of the building cost. Complex schemes or highly specialised buildings may require specialist cost professionals such as quantity surveyors, civil engineers, and structural engineers.

The cost of finance for the construction can be calculated in one of two ways, both of which average the drawdown finance requirement over the estimated length of the construction period. Either calculate compound interest on *half* of the construction costs at the finance rate for the *whole* construction period *or* calculate compound interest on *all* the construction at the finance rate for the *half* of the construction period.

If the premises are historic, the extra cost associated with their direct replacement is usually ignored if the service or output could be provided from modern equivalent premises. However, if the historic nature of the premises is intrinsic to the use (a museum for example), then the cost of reproduction would be appropriate. If reproduction is not feasible, the construction of a building with a similarly distinctive design and specification might be appropriate. Of course, some buildings may simply be irreplaceable.

Estimates of construction costs can be obtained from professional cost estimators (quantity surveyors), cost manuals, builders and contractors. The actual cost of constructing the original property may be useful evidence once it has been adjusted for inflation, but with due consideration for any variation that might be due to the existence of a prepared site, the need to reconstruct as quickly as possible, possible changes to planning policy, building regulations and so on.

8.2.2   Depreciation

Because the subject property already exists, and may have done so for some time, the cost of an equivalent new one should be written down or depreciated to reflect any diminution in value. This might be due to age (and estimated remaining economic life[2]), comparative efficiency, functionality and running costs. The building is valued as it is, taking account of any lack of repair and maintenance, but, looking ahead, an assumption of routine repair and maintenance is acceptable.

Throughout a building's life, its value will tend to depreciate for two principal reasons: deterioration and obsolescence. Deterioration results from wear and tear and is hastened by a lack of maintenance. It is usually measured by reference to the estimated economic life of the asset. It is not easy to generalise about the life of various building types, but prime shop units are much less prone to deterioration than industrial units due to the nature of the use of the building and the proportion of total value attributable to land. Usually, occupiers and owners will want to delay the onset of deterioration as much as possible and this is achieved through good design and construction and active property management. Sound maintenance and management policies help to identify, plan and budget for the onset of deterioration. But inevitably, as the building gets older, maintenance costs increase and the rental value falls because the building is no longer modern and attractive. Consequently, the value of the building declines relative to site value until it becomes economically viable to redevelop the site.

Obsolescence refers to a decline in value resulting from changes that are extraneous to the property. It is a decline in utility not directly related to physical usage or the passage of time. A good-quality, flexible design can combat obsolescence but, to a certain extent, matters are beyond the control of the property owner or occupier and management and maintenance will have little impact. A building may become obsolete for any number of reasons that rarely work in isolation. Common forms of obsolescence are:

- *Functional.* The property can no longer be used for its intended purpose, perhaps due to technological changes rendering layout, configuration or internal specification of the property obsolete, and adaptation is not economically viable. Similarly, a property may be adequate in terms of its physical characteristics but is in the wrong location.
- *Socio-economic changes* in the optimum use for a site due to market movements may render the existing use obsolete, the building may not depreciate but the development potential of the site appreciates due to changes in the social fabric of the locality or changes in consumer demand, working environment and so on.
- *Aesthetic.* Image and design requirements are constantly changing and a property that no longer projects the right image may become obsolete.
- *Regulatory.* This includes changes in planning policy, environmental regulations, health and safety legislation, and lease terms for example.

Whereas physical deterioration may be a continual but gradual process, obsolescence may strike at irregular intervals regardless of age. The responsibility for maintaining the physical condition of a property is usually passed on to the tenant when a commercial property is let on full repairing and insuring (FRI) lease terms, but the risk of obsolescence cannot be managed in this way and is ultimately borne by the owner. The onset of deterioration and obsolescence can be measured by looking at the depreciation in value of the building in relation to modern replacements and by looking at the development value of the land in comparison to its value in its existing state. A sudden switch in the relative magnitudes of development land value and existing use value may occur because of a 'trigger event' that presents an opportunity for a more valuable use such as the granting of planning permission. For example, suppose a small industrial estate located on the edge of a town is around 15 years old and the units are looking a little tired. The owner can fill the units with small businesses paying low rents. A by-pass has recently been constructed around the town and accessibility to the industrial estate is greatly improved. At the same time 'factory outlet shopping' has become popular and planning permission to allow an element of retail trade from the industrial units is forthcoming. The owner of the industrial estate anticipates being able to charge higher rents to the factory outlet traders and therefore decides to redevelop the site.

Depreciation can vary according to type of structure and obsolescence may affect different parts of building at different rates. To counter depreciation, there may be a trade-off between spending more money on the initial design and specification of the building, thus achieving relatively low future costs-in-use, or spending less at the start and instead spending relatively more to maintain the premises over its life. The time value of money affects this trade-off significantly. In quantifying the diminution in value that results from depreciation, the objective is to reflect the way the market would view the asset. When there is a group of buildings to be valued, it is important to consider alternative uses for the premises and their associated lifespans. It is reasonable to assume that routine servicing and repairs are undertaken when estimating lifespan but not significant refurbishments or replacement of components. If refurbishment takes place, then the economic life of the building might be extended. Sometimes, obsolescence can be absolute, other times, partial, perhaps rendering less-efficient service potential.

8.2.2.1   Estimating depreciation rates

A depreciation rate for physical deterioration is usually estimated by comparing the decrease in value of a building of similar age with the value of a new building. The valuer will estimate a lifespan for the property and its remaining economic life. These will be based on the lower of the property's estimated physical life or economic life, although these are often the same in practice. Physical lifespan assumes the property could be used for any purpose and economic lifespan assumes it is used only for its designed purpose. Both ignore replacement of parts, refurbishment or reconstruction but include routine repairs and maintenance.

A depreciation rate for obsolescence might reflect the cost of upgrading or it might reflect the financial consequences of reduced efficiency compared to a modern equivalent. The modern equivalent replacement may be cheaper to recreate than the actual property, in which case the replacement cost already reflects the most efficient property, so no additional adjustment is necessary. Alternatively, an all-encompassing depreciation rate may be used for both physical deterioration and obsolescence. If separate rates are used, it is important to avoid double-counting.

There are various techniques employed in practice to account for depreciation over the estimated lifespan of a property and they work by spreading the reduction in value in a regular pattern over the estimated remaining economic life of the premises. Here are four examples:

- *Straight line.* By far the most common technique, it applies a percentage deduction based on the proportion of estimated remaining economic life of the premises. In this way, it writes off the value of the improvements over their lifetime to zero or to a residual value. For example, a building purchased for £800 000 with an expected five-year useful life might be depreciated as follows:

Year	Value at start of year	Depreciation charge	Value at end of year
1	800 000	160 000	640 000
2	640 000	160 000	480 000
3	480 000	160 000	320 000
4	320 000	160 000	160 000
5	160 000	160 000	0

- *Reducing balance.* A fixed percentage depreciation rate is applied. Using this approach, the value is never completely written off to zero. Taking the building again but this time applying a depreciation rate of 20%:

Year	Value at start of year	Depreciation rate @ 20%	Value at end of year
1	800 000	160 000	640 000
2	640 000	128 000	512 000
3	512 000	102 400	409 600
4	409 600	81 920	327 680
5	327 680	65 536	262 144

- *S-curve.* A varying rate of depreciation is devised, usually in the shape of an s-curve to reflect a low rate in early years, accelerating depreciation in middle years but then tailing off in final years. It is important to base the variation on empirical evidence.
- *Sinking fund.* A sinking fund may be set up that requires an annual investment to replace the capital value of the building at the end of its estimated economic life.

8.2.3   Land value

When estimating land value, a modern equivalent site is assumed to have the appropriate characteristics to deliver the required service potential at least cost. It may not be the same size or location as the actual site. Indeed, the actual site may not be appropriate if the surroundings have changed. For example, it may be an old prison or a hospital in a city centre location, which is now surrounded by housing. It is also important to note that certain public buildings, such as a school or a health centre, need to be in certain localities, and this constrains site selection.

The land value is based on the size of the plot that is required for the use so, if a school is on a 2.5-ha site when 1.5 ha is sufficient, it would be valued based on 1.5 ha. The land value is estimated by referring to evidence of transactions of comparable size, tenure and location. If the actual site uses space inefficiently or even inappropriately given changes in technology or production, then modern equivalent sites should be considered. The fundamental principle is an economic one; a hypothetical buyer would purchase the least cost site that is suitable and appropriate for the use. The valuer should also consider whether the actual site location is now one that a modern equivalent would use and whether any vacant land at the actual site is still required for a modern replacement. These considerations can be somewhat subjective as vacant land may be held for expansion, for security or simply may be surplus to requirements.

Because of the specialised nature of the businesses and operations, finding comparable land values is difficult and the valuer may need to broaden the search to include a wider range of alternative site uses. For example, if the use is specialised industrial, then reference to general industrial land prices is usually acceptable. The aim is to select the lowest cost site for an equivalent operation in a relevant location. If the actual site is held on a lease, then the lease terms should be considered when estimating the land value.

The finance cost for the land is estimated by calculating compound interest on the land value over the whole construction period. In other words, the land is assumed to be purchased first and therefore finance is payable for the whole construction period. This finance cost is hypothetical in the replacement cost method because the land is already owned.

Bringing these three components of the replacement cost method together, an example valuation can be presented. A secondary school comprising 3500 square metres (gross) has 12 years of its estimated 50-year life remaining. Construction costs for a modern equivalent school are £1700 per square metre and the construction period is estimated at 1.5 years. Professional fees are assumed to be 12.5% of building costs. The total site area including playing fields is 11 000 square metres and the land is estimated to be £18 000 per hectare. The local authority, which owns the school, can secure finance for construction and land purchase at a finance rate of 2% per annum.

Building area (m²)	3500		
Building cost (£/m²)	1700		
Cost of modern equivalent building (£)		5 950 000	
Fees (% build cost)	13%	743 750	
Finance on build cost and fees over half build period @	2.00%	100 157	
Gross replacement cost of building (£)		6 793 907	
Depreciation allowance (age/estimated economic life)		76%	
Net replacement cost of building (£)			1 630 538
Size of site (ha)	1.10		
Land value (£/ha)	18 000		
Value of land (£)		19 800	
Finance on land over build period @	2.00%	597	
Cost of land (£)			20 397
Valuation (before PCs) (£)			1 650 935

The final stage of the valuation is to stand back and look – a final reconciliation to ensure that depreciation has not been double-counted or ignored and to check whether the characteristics of the property being valued might lead a buyer to bid more than a modern equivalent. When reporting a valuation that is based on the replacement cost method, this must be stated in the report, along with the assumption that the value is subject to the adequate profitability of the company if the property is held in the private sector or subject to the prospect and viability of continued occupation and use if it is a public sector property.

At some point, the property value will fall below the development land value and development becomes viable. This would occur at different times depending

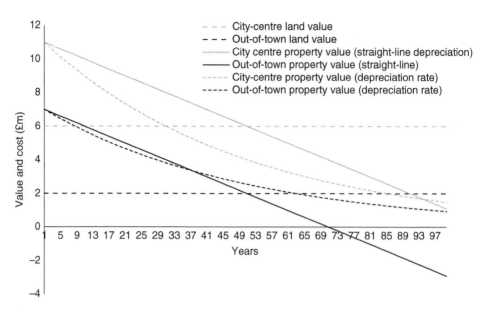

Figure 8.1   Accounting for depreciation.

on the ratio of land to building components of property value. Figure 8.1 provides an example where there are two sites: a city centre site with a land value of £6m and an out-of-town site with a land value of £2m. Building costs and estimated lifespans of the buildings are the same at £5m and 50 years respectively, and the property depreciation rate is 2% in both cases. Because the city centre site has a higher ratio, it has a shorter economic life. Accountants often simplify the process by assuming buildings have standard lifespans, typically 50 years, after which their value is written off, leaving just the land value. In economic terms this would be the point at which land value exceeds building value, signifying redevelopment. Figure 8.1 illustrates this arrangement and shows that now the two properties have the same economic life. This treatment of depreciation is forced by the 50-years' life-span assumption.

8.2.4   Application of the replacement cost method

The valuer may need to agree certain matters with the client as part of the valuation: the potential location of a replacement property, factors that may affect the estimated remaining economic life and details of capital expenditure on maintenance, repairs, improvements, refurbishments, and reconstruction. The valuer is reliant on information provided by the client to a greater extent than for non-specialised properties, particularly in relation to costs, design features and the performance of the property in relation to its use.

8.2.4.1   Valuation for accounts purposes

If an otherwise conventionally designed property has been specifically adapted (including the installation of plant and machinery), it may be valued subject to a special assumption that the adaptations do not exist and then treating the adaptation costs separately. The valuation should set out which adaptations are included with the property and which have been treated separately.

If the replacement cost value is significantly different from the market value of an alternative use for which planning permission is likely to be forthcoming, both should be reported in the accounts, but the latter need not take account of the costs associated with business closure or relocation. If appropriate, the valuer should report that the value of the premises would have a substantially lower value if the business ceased, but there is no requirement to report a figure.

The initial cost of the property, when first entered into the accounts, can include the purchase costs (such as legal costs and taxes) as well as the purchase price. On subsequent valuations, the reported amount should include purchase or sale costs but these can be stated separately.

8.2.4.2   Valuation for insurance purposes

These are also known as *reinstatement valuations* and are undertaken on behalf of lenders, normally in conjunction with a market valuation but are also undertaken for insurers and insurance brokers, property owners and occupiers. A reinstatement valuation provides for a similar property as at the date of the valuation or at the commencement of insurance policy cover and should be carried out at least every three years. In the case of insurance valuations, the site is assumed to continue in existence despite whatever disaster may have affected the buildings. Consequently, it does not include a valuation of the land. Furthermore, if the insurance policy provides for a replacement new property (a 'new-for-old' policy as it is known), then no deduction should be made to reflect deterioration and obsolescence.

8.2.4.3   Valuation for rating purposes

UK business rates are levied based on assessed rateable values, which are in turn based on rental valuations. For properties that are valued using the replacement cost method, the capital valuation must therefore be amortised at an appropriate yield (usually between 3 and 6%) to arrive at an estimate of annual rental value.

8.2.5   Issues arising from the application of the replacement cost method

It is important to remember that cost is a production-related concept whereas value is an exchange or use-related concept. Using cost as a proxy for value assumes that a property is worth its replacement cost rather than what someone is prepared to pay for it. For this reason, the method is often regarded as a last resort approach to estimating market value. The strong cost element and the necessity for extensive subjective input by the valuer throughout the valuation has led the courts to express 'considerable reservations about a basis which gives full rein to a valuer's judgement without offering any market evidence to support the opinion of value against which the results may be judged' (Scarrett 1991).

Although the resulting valuation from the method is a replacement cost, estimation of the inputs (land prices, build costs, depreciation allowances for example) can be based on market information when it is available. Adjustments may be made to these inputs to reflect differences in size, quality, utility, and so on in the same way as undertaken for the comparison approach. Comparable information is often available for build costs but less so for land prices, and particular difficulties may be encountered when trying to derive a depreciation rate. Valuers should observe economic lives of existing buildings and other improvements in comparison with new or recent replacements as a way of calculating depreciation rates. However, given the specialised nature of the properties concerned, it may be challenging to reconcile such diverse evidence.

Notwithstanding this challenge, depreciation rates may be all-encompassing or separated into physical deterioration, functional and economic obsolescence elements. Separation into component parts is only likely to be feasible when the body of comparable evidence allows. UK valuation guidance suggests that, for physical deterioration, costs of specific elements of rectification may be considered or direct unit value comparisons between properties in a similar condition may be undertaken. A further challenge is to identify changes in depreciation rates and remaining economic life estimates caused by market fluctuations.

Usually, it is not possible to obtain market data on which to base a measure of depreciation. Instead, valuers make assumptions about how improvements depreciate over time. This is typically a straight-line rate of depreciation based on the ratio of the estimated life of the building to its actual age at the valuation date. Because of the difficulty in putting a precise lifespan on a building, bands of say less than 20, 20–50 and over 50 years are often used.

The replacement cost method assumes that value is derived from an additive relationship between land value and depreciated building cost. This simple relationship is open to question: as Whipple (1995) argues, land and improvements 'merge to provide an undifferentiated stream of utility' so not only is it virtually impossible to determine the contribution to value made by each individual capital item, but their aggregate contribution is also highly unlikely to be a simple additive one.

The cost approach relies on the assumption of continuing existing use and so alternative uses of the land need to be considered separately. Unless instructed otherwise, a valuation of any alternative use, including redevelopment land value, is likely to take the form of a simple indication that the value of the site for a potential alternative use may be significantly higher than the replacement cost valuation.

In summary, because of a lack of comparable market transaction information, the method estimates replacement cost rather than exchange price. It does not produce a market valuation (an estimate of value-in-exchange) because cost relates to production rather than exchange and it is regarded as the method of last resort for this reason. It is a means of estimating the replacement cost or deprival value of company and public sector properties for which there is no market.

8.3   Residual method

The value of a piece of land (or site) will depend not only on its existing use but also on its potential for alternative uses, including its development potential – referred to as development land value. The residual method is used to estimate development land value. It involves estimating the value of the proposed completed development using the market approach or income approach, and then deducting all the costs of the development, including profit for the developer, leaving a residual development land value. The basic equation is:

Value of completed development – development costs (including developer's profit) = Development land value

For the development of a particular piece of land or site to be economically viable, the value of the completed development less all expenditure on land, construction and profit, must exceed existing use value.

Although widely used to estimate development land value, the residual method (and the equation above) can be adapted to estimate the level of potential profit. In this way, the residual method becomes a means of assessing development viability as well as estimating land value.

The need for development arises in three situations; where new buildings are to be created on previously undeveloped land (new development), where existing buildings on vacant/derelict sites are to be replaced by new structures (redevelopment) and where existing buildings are to be substantially converted or modernised (refurbishment). For the purposes of explaining the residual method, the generic term *development* will be used for all these situations. Redevelopment sites compete with new development sites for potential uses. New development sites may have the advantage of being clear of any previous development, but redevelopment sites often benefit from existing infrastructure and services.

Development activity is a highly visible, often intrusive process that is responsible for creating a landscape that influences the way that we interact with each other and with the built and natural environment. But here we focus on the financial economics of development because that is where valuation fits in to the process of development. Development land valuations differ markedly from other areas of valuation, principally because the subject of the valuation is a proposal for a development rather than an extant property. For this reason, obtaining comparable evidence of development land values can be very difficult. Each site will differ widely in terms of size, condition, potential use or uses, permitted density of development, restrictions and so on, making any adjustment to a standard value per hectare almost impossible. Instead, a valuer will frequently rely on comparable evidence to assess development value and costs.

The residual method is usually employed in two ways, a basic residual and a discounted cash flow. A basic residual is simple, quick to do and easy to interpret. A cash flow provides a detailed breakdown of income and expenditure as the scheme progresses from inception to fruition. Cash-flow techniques are useful because, once the initial feasibility has been established, a more detailed financial

Section B

appraisal is usually required not only by the developer but also by the lender (who may be financing the development) and the investor (who may be acquiring the scheme on completion). Being able to identify the cash flow at any point in time during a development project has obvious advantages over the 'snap-shot' estimate produced by a basic residual valuation.

8.3.1   Basic residual technique

The basic residual technique can be summarised as follows:

$$LV_0 = (1+i)^{-t}\left[\frac{DV_0}{(1+p)} - DC_0 - I\right]$$ (8.1)

where LV_0 = Present residual land value
i = Cost of finance (annual interest rate)
t = Development period (years)
DV_0 = Current estimate of development value
p = Profit, calculated as a percentage of DV
DC_0 = Current estimate of development costs
I = Finance costs (usually calculated over the construction phase of the development period only)

The model produces a simplified representation of the financial flows in development based on the following assumptions:

- The development value, expressed in current values, is received at the end of the development period.
- All development costs (land and construction) are debt financed with repayment in full at the end of the development period.
- Building costs are incurred evenly throughout the construction period and interest is calculated by halving the time over which interest accrues. The use of the finance rate effectively delays payment for the costs of development to the end of the development, placing them at the same date at which development value is received.
- Profit is deducted as a cash lump sum, taken as a proportion of total development costs or development value. As profit is also a cost to the development at the end of the development, all the income and outgoings are now placed at the end.
- Finally, the gross residual amount is the amount that can be paid for the site at the end of the development. But site value is a present value and therefore the residual surplus at completion of the development is discounted back from the end of the development to the beginning at the finance rate. Consequently, it is assumed that the land value is paid to the landowner at the commencement of the development and is funded entirely by debt.

An example of a basic residual valuation is shown below. This is an office scheme, and the relevant steps of the valuation are explained below the valuation.

Inputs

Areas

Gross internal area (GIA) (m²)	2000
Efficiency ratio (net/gross area)	85%
Net internal area (NIA) (m²)	1700

Time

Lead-in period (years)	0.25
Building period (years)	1.50
Void period (years)	0.25

Values

Estimated rental value (ERV) (£/m²)	200.00
Net initial yield	7.00%

Costs

Site preparation (£)	25 000
Building costs (£/m² GIA)	969
External costs (£)	120 000
Professional fees (% building costs and external works)	13.00%
Miscellaneous costs (£)	80 000
Contingency allowance (% construction costs)	3.00%
Planning fees (£)	5000
Building regulation fees (£)	20 000
Planning obligations (£)	0
Other fees, e.g. legal, loan, valuation (£)	95 238
Short-term finance rate (annual)	10.00%
Short-term finance rate (quarterly)	2.41%
Letting agent's fee (% ERV)	10.00%
Letting legal fee (% ERV)	5.00%
Marketing (£)	10 000
Sale costs (% NDV)	2.00%
Land purchase costs (% site purchase price)	6.50%
Developer's profit (% costs):	20.00%

Basic residual land valuation

Development value

Net internal area (NIA) (m²)	1700		
Estimated rental value (ERV) (£/m²)	200		
		340 000	
Net initial yield	7.00%	14.2857	
Gross development value (GDV) before sale costs (£)			4 857 143
Net development value (NDV) after sale costs (£)			4 761 905

Development costs

Site preparation (£)		(25 000)	
Building costs (£/m² GIA)	969	(1 938 000)	
External costs (£)		(120 000)	

Basic residual land valuation			
Professional fees (% building costs and external works)	13.00%	(267 540)	
Miscellaneous costs (£)		(80 000)	
Contingency allowance (% construction costs)	3.00%	(72 166)	
Planning fees (£)		(5000)	
Building regulation fees (£)		(20 000)	
Planning obligations (£)		0	
Other fees, e.g. legal, loan, valuation (£)		(95 238)	
Interest on costs and fees for half building period @	10.00%	(194 359)	
Interest on costs and finance for void period @	10.00%	(67 936)	
Letting agent's fee (% ERV)	10.00%	(34 000)	
Letting legal fee (% ERV)	5.00%	(17 000)	
Marketing (£)		(10 000)	
Developer's profit on total development costs (%):	20.00%	(589 248)	
Total development costs (TDC) (£)			(3 535 486)
NDV – TDC (£)			1 226 418
Land costs (£)			
Developer's profit on land costs (%)	20.00%	(204 403)	1 022 015
Finance on land costs over total development period	10.00%	2.00	0.8264
Residual land value before purchase costs (£)			844 641
Residual land value after purchase costs @ 6.5% (£)			793 090

8.3.1.1 Development value

Buildings should be measured in accordance with appropriate International Property Measurement Standards (IPMS). The gross internal area of the building to be developed (the area contained within the perimeter walls of the building) is the IPMS2 area (see Chapter 6). The net internal area, or IPMS3 area, is that part of the building on which rent can be charged and excludes corridors, plant rooms, lift lobbies, toilets, etc. Some properties such as supermarkets and industrial buildings are let on an IPMS2 basis while offices and shops are let on an IPMS3 basis, with shops being zoned to reflect the higher value attached to floor area (or sales space) nearer the front of the premises. The ratio of the IPMS2 to IPMS3 areas is called the *efficiency ratio*. The more efficient a building, the more space there is to charge rent on. Higher efficiency ratios lead to higher annual rentals per unit of constructed space. In practice, comparable properties would be examined to determine an appropriate efficiency ratio.

The market rent is estimated by considering rents that have been achieved on comparable properties. A net rent should be estimated that has been reduced to account for any regular expenditure such as management, repairs, or insurance. It is usual practice to estimate current rent rather than predict the rent that might be achieved when the development is complete.

The gross development value (GDV) is the price for which the completed development could be sold. For commercial property, GDV is calculated by undertaking an investment valuation based on the capitalisation of expected annual rent at an appropriate yield. For residential property, GDV would be based on estimated sale prices.

The price that an investor would be prepared to pay for the completed development would be net of any purchase costs such as legal fees and surveyor's fees. If the completed development is to be retained as an investment, it will usually need to be refinanced (perhaps by converting the short-term development loan into a long-term debt) and it is assumed that the lender will charge an arrangement fee together with the costs of a valuation of the investment. A percentage deduction is therefore made from GDV to reflect these costs and to arrive at a net development value (NDV).

8.3.1.2   Construction costs

An estimation of the construction costs, at the valuation date, is a major component in a residual valuation. In other than the most straightforward schemes, it is recommended that the costs be estimated with the assistance of an appropriately qualified expert. Care is to be taken to check that calculations provided by other professionals are on a consistent measurement basis, i.e. IPMS.

Site preparation costs can include:

- The cost of meeting environmental requirements;
- Remediation of contamination, noise abatement and emissions controls;
- Ground improvement works;
- Archaeological investigations;
- Diversion of essential services and highway works and other off-site infrastructure;
- Creating the site establishment and erection of hoardings;
- Conforming to health and safety regulations during the development;
- Realistic allowances for securing vacant possession, acquiring necessary interests in the subject site, extinguishing easements or removing restrictive covenants, rights of light compensation, party wall agreements, etc., reflecting that other parties expect to share in development value generated

Building costs and fees are usually estimated by a quantity surveyor, but an approximation can be gained by reference to recent contracts for similar developments or from building price books. It is usual to use current cost estimates and assume that cost inflation will match rental growth over the development period. Having said this, it is worth noting that construction contracts vary; they may be agreed on a 'rise and fall' or 'fixed-price' basis. A building contractor who agrees to a fixed-price contract is likely to charge a higher price because risk exposure is greater. Also, a fixed price contract is only fixed to the extent of the works outlined in the contract. A contractor can amend the pricing if any variations to the specification are made or unforeseen events occur.

External works might include demolition, access roads, car parking, landscaping, ground investigations or other costs associated with the development that are in addition to the unit price building cost estimated above.

Professional fees are usually agreed as a percentage of the construction costs but may be a fixed sum. Marshall and Kennedy (1993) found that a typical total for

Table 8.1 Typical professional fee levels.

Professional	Fee as a % of building costs
Architect	5–7.5%
Quantity surveyor	2–3%
Structural engineer	2.5–3%
Civil engineer	1–3%
Project manager	2+ %
Mechanical and electrical consultants	0.5–3%

fees averaged 14.5%; Table 8.1 shows a representative breakdown of these fees. The appropriate fee level depends on the type and location of the development. The following items may need consideration linked to the sale, letting, design, construction, and financing of the development (RICS 2019):

- Professional consultants to design, cost and project manage the development. A development team normally includes: an environmental and/or planning consultant, an architect, a quantity surveyor and a civil and/or structural engineer. Additional specialist services may be supplied as appropriate by mechanical and electrical engineers, landscape architects, traffic engineers, acoustic consultants, project managers and other disciplines depending on the nature of the development.
- Fees incurred in negotiating or conforming to statutory requirements (for example building consents) or any planning agreements.
- Costs related to the raising of development finance (including the lender's monitoring fees and legal fees).
- In some cases, the prospective tenant/purchaser may incur fees on monitoring the development (these may have to be reflected as an expense where they would normally be incurred by the developer).
- Lettings and sales expenses where the development is not pre-sold or fully pre-let. These expenses usually include incentives, promotion costs and agents' commissions. The cost of creating a 'show' unit in a residential development may also be appropriate.
- Incentives on letting such as fitting out periods, rent-free periods and capital payments to prospective tenants. These may be reflected by either continuing interest charges on the land and development costs until rent commencement or taking account of the costs in the valuation of the completed development.
- Legal advice and representation at any stage of the project.

In this example, professional fees of 13% of building costs and external works have been assumed, broken down as: Project Manager (2%), Quantity Surveyor (3%), Mechanical and Electrical Engineer (1%), Structural Engineer (1%) and Architect (6%). Miscellaneous costs are included as a catch-all for any other incidental expenditure such as insurance.

It is normal to include a contingency allowance for any unexpected increases in costs. The quantum, which is usually expressed as a percentage of building costs, is dependent upon the nature of the development, the procurement method and the perceived accuracy of the information obtained. However, whether a contingency allowance is appropriate is linked to the analysis of risk within development schemes. A contingency allowance could inadvertently double-count risk associated with development costs if that risk has allowed for in the developer's profit margin. Unforeseen increases in costs are an inherent risk in development and higher development returns are required to compensate for risks such as these. Therefore, a higher contingency allowance should be compensated by a relatively lower developer's profit.

The contingency allowance is a reserve fund to allow for any increase in costs or delays in construction. As construction costs are the single largest sum after land, any inflationary effect is likely to have a significant impact on costs. If the economy is particularly volatile, a cautionary approach is to apply the contingency allowance to all costs, including finance costs, but this will depend on the perceived risk of the project. Marshall and Kennedy (1993) found that a contingency fund is generally set at 3–5% of building costs and professional fees (and sometimes interest payments), but the figure varied depending on the nature of site (restrictive site, subsoil, etc.) and the development project itself. Generally, the longer the development period and the more complex the construction of the building, the higher the risk of unforeseen changes and, therefore, the higher the contingency allowance.

Regulatory fees might include planning fees, building regulation fees and the costs associated with legally binding conditions linked with the grant of development consent.

8.3.1.3   Cost of finance

Short-term finance is usually included in valuations of development land because development typically requires debt finance. The cost of finance depends on two factors, the interest rate and the duration of the loan. A lender will charge interest at the bank base rate for lending plus a return for risk. The magnitude of the risk premium will depend on the nature of the scheme, the status of developer, the size and length of loan and the amount of collateral the developer intends to contribute.

The duration of the loan depends on the estimated length of the development period. In simple terms, the development period comprises a lead-in period, the construction period, and a void period. The lead-in period allows time for obtaining planning consent, preparing drawings and so on. The void period sits between the end of the construction period and occupation by a tenant, including a possible rent-free period. During a void period, interest is payable on *all* costs so any extensions to this period will significantly increase the amount of loan finance incurred. In this example, a lead-in period of 3 months, a construction period of 18 months and a void period of 3 months has been assumed.

Interest payments on money borrowed to fund construction usually accrue monthly but are rolled up over the development period and paid back when the development is let or sold. In a basic residual valuation, finance is assumed at 100% of all construction and land-related costs. This is achieved by compounding interest on the construction costs over the construction period and by discounting the residual land value at the interest rate over the development period. Whereas finance is calculated on the land-related costs over the development period, because finance is not drawn on all construction costs at the start of the construction period, one of three techniques is usually employed to calculate the finance on the construction-related costs. The first is to set out the costs as a cash flow and determine the total interest payment. This technique requires a quarterly or monthly breakdown of construction costs and, for a basic residual valuation, this might seem too detailed. Therefore, an averaging technique is often employed by assuming that either interest accumulates on half the construction-related costs over the whole construction period, or that all construction costs are borrowed over half the construction period (the approach used in this example). This averaging of either cost or time reflects the fact that interest is not paid on the full amount over the entire building period. Usually costs start off low, peak in the middle and then tail off towards the end as illustrated in Figure 8.2. By averaging, a straight line rather than an s-shaped build-up of costs is assumed. Sometimes, interest on professional fees is calculated separately by compounding the amount over two thirds of the building period. This reflects the presumption that such fees tend to be incurred early in the development, during the planning and design phase and hence interest will be incurred for a longer period.

It is usual for interest to be treated as a development cost up to the assumed letting date of the last unit unless a forward sale agreement dictates otherwise. If an assumption is made that the completed development is held beyond the date of completion, the costs of holding that building during this void period must be added. These may include insurance, security, cleaning and energy costs. A proportion of the service charge on partially let properties may have to be included together with any potential liability for empty property taxes. Interest can then be accumulated in two parts; in the construction period as indicated above and then in the post construction void period, where the full costs of development can be

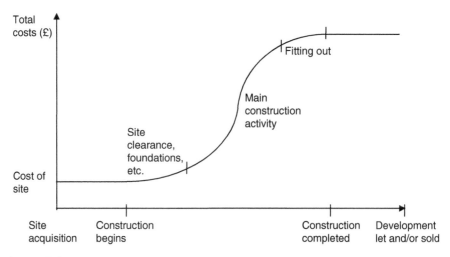

Figure 8.2   Build-up of costs over time.

included in the interest calculations. Interest accrued during the void period is calculated by compounding the total construction costs and interest rolled up during the construction period at the annual interest rate. A significant amount of interest can accrue during a void period. That is why it is very important to keep the length of any void period to a minimum. Detailed cash-flow projections are essential once the project is under way to incorporate changes in revenue and costs and particularly so for phased developments.

8.3.1.4   Letting and sale fees

The fees that agents and solicitors charge for either letting the completed development are usually calculated as a percentage of the estimated market rent. If the space is to be sold on a freehold or long leasehold basis (residential apartments for example), the fees are usually quoted as a percentage of sale price. The fee that agents charge will vary depending on whether they have been given sole or joint marketing rights. Marketing costs would cover items such as advertising, opening ceremony, brochure design and production and would obviously depend on the nature of the development.

8.3.1.5   Developer's profit

Developer's profit is the reward for initiating and facilitating the development and is dependent upon the size, length and type of development, the degree of competition for the site and whether it is pre-let or sold before construction is complete (a forward sale). Property development is perceived as riskier than investment in completed and let properties. Consequently, the required return will be higher than these 'standing' property investments. It is usual to express developer's profit in the basic residual valuation as a cash sum, expressed as a percentage of the development cost (including finance and land costs) or of development value.[3] A typical range of profit as a percentage of costs is 15–25%. Here, 20% has been assumed.

The developer also takes a profit margin on the residual land costs. The equation below shows how this is calculated.

$$\text{Developer's profit on residual land costs} = \text{Residual land cost} \times \left[1 - \left(1/(1+0.2)\right)\right]$$
$$= £1\,226\,418 \times \left[1 - (1/1.2)\right]$$
$$= £204\,403$$

8.3.1.6   Residual land costs

Once developer's profit has been deducted from the residual land costs, the remaining sum must cover the land price, purchase costs and finance on these costs. These are handled in reverse order. Regarding finance, assuming the site was purchased by the developer at the start of the development, interest on land costs will accrue over the total development period. To reflect this, the residual land costs are discounted at the short-term finance rate of 10% over the total development period to determine their present value.[4]

Acquisition costs are deducted to leave the net amount remaining for purchase of the land. These acquisition costs usually include legal costs, transfer tax or Stamp Duty agents' fees:

$$\text{Site value} = £844\,641 \div (1 + 6.50\%)$$
$$= £793\,090$$

The final figure is the residual land value and represents the maximum amount that should be paid for the site if the proposed development was to proceed and all the valuation assumptions held true.

8.3.2   Basic residual profit appraisal

Given all the value and cost inputs, including a developer's profit margin, the output from the residual method is usually a land valuation. However, if a land price is one of the cost inputs, then the method can be used to estimate the amount of developer's profit.

In a basic residual, if the land price or value is known, it becomes a cost to the development. It occurs at the beginning of the development and interest accrues on it over the development period. This cost is added to the other development costs and then deducted from development value to leave an estimate of residual profit at the end of the development period.

The basic equation for the residual land valuation can be transposed to determine the level of profit achieved given construction and site costs. The equation would look like this:

Value of completed development – development costs (including developer's
    profit) = Development land value
Less development costs, including site price
Equals developer's profit

Using the example above, the estimation of developer's profit would proceed as follows.

Basic residual profit appraisal			
Development value			
Net internal area (NIA) (m²)	1700		
Estimated rental value (ERV) (£/m²)	200		
		340 000	
Net initial yield	7.00%	14.2857	
Gross development value (GDV) before sale costs (£)			4 857 143
Net development value (NDV) after sale costs (£)			4 761 905
Development costs			
Land price (£)		(793 090)	
Land purchase costs (% land price)	6.50%	(51 551)	
Finance on land costs for total development period @	10.00%	(177 375)	

Basic residual profit appraisal			
Site preparation (£)		(25 000)	
Building costs (£/m² GIA)	969	(1 938 000)	
External costs (£)		(120 000)	
Professional fees (% building costs and external works)	13.00%	(267 540)	
Miscellaneous costs (£)		(80 000)	
Contingency allowance (% construction costs)	3.00%	(72 166)	
Planning fees (£)		(5000)	
Building regulation fees (£)		(20 000)	
Planning obligations (£)		0	
Other fees, e.g. legal, loan, valuation (£)		(95 238)	
Finance on building costs and fees for HALF building period @	10.00%	(194 359)	
Finance on building costs, fees and interest to date for void period:	10.00%	(67 936)	
Letting agent's fee (% ERV)	10.00%	(34 000)	
Letting legal fee (% ERV)	5.00%	(17 000)	
Marketing (£)		(10 000)	
Total development costs (TDC) (£)			(3 968 254)
Developer's profit on completion (£)			793 651
Return on costs			20.00%
Return on NDV			16.67%
Income yield			8.57%

The site price is assumed or is known and can therefore be inserted. Costs associated with site acquisition (typically agent and legal fees) must be added to the costs. Assuming the site was acquired at the very start, interest will accrue on this cost over the whole development period. Here the site costs will incur interest for two years at an annual interest rate of 10%.

The estimated developer's profit can be expressed in several ways to assess the viability of the development and to compare it to other development opportunities. Profit as a percentage of development costs is useful for merchant developers, who need to sell the completed development to raise capital for future projects. Some developers, particularly housebuilders, prefer to express profit as a percentage of value. Development yield is rent expressed as a percentage of development costs and is useful for investor developers who, in contrast to trader developers, retain the development as an investment. Just as the difference between total costs and total capital value represents capital profit, so the difference between the investment yield and development yield represents the developer's annual profit margin over standing investments.

8.3.3   Discounted cash-flow technique

The discounted cash-flow technique requires period-by-period assumptions concerning the breakdown of costs and values during the development period such as the time frame (monthly, quarterly, etc). It thus allows market dynamics through time to be incorporated, such as changes in costs and values, where appropriate. The basic application of a discounted cash flow is to calculate the present residual land value of the estimated costs and revenues over the duration of the development scheme. With all other costs and revenues accounted for, profit is incorporated as a cash sum, usually estimated as a percentage of total costs or revenue.

Whereas a basic residual valuation is often used at an early stage to provide a snapshot of development feasibility, a cash flow provides a more detailed assessment, usually reserved for larger, more-complex proposals. Projecting a cash flow is particularly useful for developments where the initial land acquisition or disposal of the completed development is phased, such as residential or industrial estates, where some units may be sold before others are constructed, or complex central area shopping schemes where parts may be let or sold before the remainder is complete. In short, the advantage of the cash-flow technique is its dynamic capability.

The essential difference between a basic residual valuation and a cash-flow valuation is the way that the *timing* of expenditure and revenue is handled. The basic residual assumes that revenue from the development is received at the end of the development and interest on expenditure is calculated on 50% of all costs over the building period (alternatively, interest is calculated on the total costs over half of the building period). In contrast, a cash flow divides the development project into time periods (usually months or quarters) to allow more refined judgements to be made regarding the flow of income and expenditure. Payments and receipts that were stated as aggregate figures in the residual valuation may now be estimated as to when they are likely to occur. This permits a more accurate calculation of interest payments to be incorporated and allows the valuer to examine how changes in the timing of costs and revenue might affect value or profitability of the development. Throughout the construction phase, adjustments can be made to the cash flow as and when costs and income are realised. This will determine how the project stands at any point in time in terms of potential profit and what began as a valuation becomes an appraisal tool. In a basic residual, it is usually assumed that income is received (and costs incurred) annually in arrears. A cash flow typically assumes that costs and revenue are incurred and received quarterly in arrears. There may be a mixture of timings for incurring expenditure and receiving revenue: construction costs are usually paid in arrears whereas income from property in the form of rent is usually receivable quarterly in advance.

A key advantage of a cash-flow valuation is that it can deal with non-standard patterns of revenue and expenditure. Whereas a basic residual valuation assumes sales must come at the end of the development (albeit after a possible void period), the cash-flow method easily deals with phased schemes, allowing rental income to be accounted for when rent commences before the investment is sold. For example, the leasing of advertising space on hoardings or securing short-term tenancies (for example surface car parking) can help to offset costs before and during the development phase. This is simple to include by incorporating two income lines: one for rent and one for sales. Where phased sales occur, the associated costs, such as agent and legal fees, should also appear in the calculation at the appropriate time. Also, when the opening balance becomes positive, no interest should be charged.

The basic approach of the discounted cash-flow approach can be formalised as follows:

$$LV_0 = R_0 + \Sigma \frac{R_i - p}{(1+f)} + \frac{DV_n}{(1+f)} \qquad (8.2)$$

where R = Recurring periodic net revenue received at the end of each period, f = cost of finance, n = number of periods and other variables are as defined above.

In this simple cash-flow valuation, revenue and expenditure are discounted at the finance rate and developer's profit is included as a cash sum at the end. This is the same as the approach adopted in the basic residual, but a cash flow means that where revenue is expected to be received in phases, receipts can be recognised in the cash flow at the appropriate time. Phasing may be appropriate for residential development where homes are built and sold incrementally or for commercial developments where some existing properties are let on a short-term basis while new properties are built. The timing and extent of phased costs and revenue within the development period can be incorporated into the cash flow. Where income-producing properties are included in the development, the timing of lettings, rent-free periods, capital contributions, etc., can also be incorporated into the cash flow and the development period can be extended as necessary.

Using the example from above, the cash flow below replicates the input costs but allocates them across the development period in a more realistic way. The land price is input at the start, construction costs and fees are spread over the eight quarters and developer's profit is paid out at the end. Discounting the cash flow at the finance rate leaves a small negative net present value (NPV). Whereas the basic residual assumed all costs were incurred at the halfway point, in the cash flow, the spread of costs is weighted a little before the halfway point. This leads to slightly higher finance costs and thus a small negative closing balance of -£46 408. This would eat into the developer's profit. To see how much the developer should reduce the land bid price to preserve the profit margin, the higher finance cost can be fed back into a revised land valuation using iteration ('goal seek' in Excel) to set the closing balance to zero and altering the land price input. The revised land price is £757 077.

Cash-flow residual land valuation

	Total	0	1	2	3	4	5	6	7	8
Development value										
Net development value (NDV) after sale costs (£)	4761905	0	0	0	0	0	0	0	0	4761905
Development costs										
Land price (£)	(793090)	(793090)	0	0	0	0	0	0	0	0
Land purchase costs (% land price)	(51551)	(51551)	0	0	0	0	0	0	0	0
Site preparation (£)	(25000)	(25000)	0	0	0	0	0	0	0	0
Building costs	(1938000)	0	0	(387600)	(581400)	(775200)	(193800)	0	0	0
External costs (£)	(120000)	0	0	(60000)	(60000)	0	0	0	0	0
Professional fees	(267540)	0	0	(160524)	(107016)	0	0	0	0	0
Miscellaneous costs (£)	(80000)	0	0	0	0	0	(80000)	0	0	0
Contingency allowance (% construction costs)	(72166)	0	0	(18244)	(22452)	(23256)	(8214)	0	0	0
Planning fees (£)	(5000)						(5000)			
Building regulation fees (£)	(20000)						(20000)			
Other fees, e.g. legal, loan, valuation (£)	(95238)						(95238)			
Marketing (£)	(10000)	0	0	0	0	0	0	0	0	(10000)
Letting agent's fee (% ERV)	(34000)	0	0	0	0	0	0	0	0	(34000)
Letting legal fee (% ERV)	(17000)	0	0	0	0	0	0	0	0	(17000)
Developer's profit (% NDV)										(793651)
Net cash flow before finance		(869641)	0	(626368)	(770868)	(798456)	(402252)	0	0	3907254
Finance										
Opening balance (£)		0	(869641)	(890611)	(1538455)	(2346421)	(3201458)	(3680909)	(3769669)	(3860570)
Interest (in arrears) (£)	(486077)	0	(20970)	(21476)	(37098)	(56581)	(77199)	(88760)	(90901)	(93093)
Closing balance (£)		(869641)	(890611)	(1538455)	(2346421)	(3201458)	(3680909)	(3769669)	(3860570)	(46408)

Iteration can also be used to determine how the developer's profit might change if the land price had to remain at £793 090. In this case, the profit sum drops to £747 242.

In practice, developers rarely if ever use 100% debt to finance land acquisition and construction. A cash flow can be used to consider various finance arrangements and loan to cost ratios. In doing so, the cash flow is constructed on the equity provided by the developer and the target rate of return is based on the risk of that equity. Depending upon the financial arrangements, that risk would normally be higher than the overall project risk (this assumes lower risk exposure by the lender) and the equity target rate of return would normally be in excess of the project target rate of return.

To summarise, there is a great deal of uncertainty in the residual method. Although there may be a predictable set of costs associated with parts of a development project, there will inevitably be unforeseen costs and delays. Typically, these are handled by a contingency allowance and a suitable risk-adjusted profit margin for the developer. Nevertheless, given very high yet relatively predictable costs (building, fees, finance, etc.) and more volatile revenue (sale prices, rents, yields, letting voids, etc.), developers face high operational gearing. Most projects are lengthy, and costs, values and market activity will change during the development time frame. Little forecasting is undertaken in residual land valuations so this presents additional risk to the developer. Because of inherent uncertainty in the model and the highly geared nature of the residual land value, the method comes with two major health warnings: it is highly site specific and has a very limited shelf-life.

Key points

- The replacement cost method is used to value properties that very rarely trade on the open market and therefore there is little or no evidence of comparable market prices on which to base value estimates.
- As a valuation method, it is generally regarded as a 'method of last resort' because it does not really produce an estimate of market value, at least not of the building component anyway.
- The method does however have wide application in the valuation of public and private sector specialised property assets for accounts purposes and it is also used to estimate building reinstatement costs for insurance purposes.
- Deterioration and obsolescence are the causes; depreciation in value is the effect. Property, although rightly regarded as a long-term investment and factor of production, does depreciate over time. Physical deterioration can be mitigated through an active property management programme. Obsolescence is harder to predict and control, but good design helps.
- The residual method is based on a very simple economic concept, that the value of the land is calculated as a surplus remaining after all estimated development costs have been deducted from the estimated value of the completed development.
- The residual valuation of a site is calculated by first estimating the value of the proposed development and then deducting construction costs, including payments for any money borrowed and expected profit.
- In practice, the residual method is first employed in its simplest form at the evaluation stage and then the complexity level increases as development plans crystallise.

Notes

1. A generic term covering all additions to the land including services, paving, fencing, etc., even though they may have different uses and economic lives.
2. The economic life of a building can be thought of as that period over which it proves to be the most appropriate (least cost) built asset needed to deliver the business function. If kept beyond its economic life, a building may continue to give service, but a replacement will do the job more cost effectively.
3. There is a mathematical relationship between the two measures as follows:

$$x(DC + LV) = DV\left(1 - \frac{1}{1+x}\right)$$

   where x is > 0 and < 1
4. Even if money is not borrowed to fund land purchase or construction, the opportunity cost of funds used should be reflected in the valuation. In the basic residual, the interest rate is regarded as a proxy for the opportunity cost of capital.

References

Marshall, P. and Kennedy, C. (1993). Development valuation techniques. *J. Prop. Valuat. Invest.* 11 (1): 57–66.

RICS (2019). *RICS Professional Standards and Guidance, Global, October 2019: Valuation of Development Property*, 1e. London: Royal Institution of Chartered Surveyors.

Scarrett, D. (1991). *Property Valuation: The 5 Methods*. London: Spon Press.

Whipple, R. (1995). *Property Valuation and Analysis*. Sydney: The Law Book Company.

Questions

Replacement cost method

[1] A local authority owns a recycling centre that must be valued for their register of assets. The property is expected to have a further 25 years of useful life remaining from its original life expectancy of 40 years. The building comprises 5000 square metres gross internal area and construction costs are estimated at £1098 per square metre. A replacement building and compound would take 9 months to construct, fees are estimated at 12% and the local authority could be expected to secure finance at 4.5%. The building is located on a site comprising 18 000 square metres and the price of land per hectare (per 10 000 square metres) for refuse and recycling centre use is estimated at £18 500. Value the freehold of the building.

[2] A purpose-built glass works comprises specialised industrial buildings, a warehouse, canteen and office accommodation, details of which are shown in the schedule below. The property has been developed over several years on a site comprising 2.5 ha. Market evidence shows that current land values are approximately £52 000 per hectare for heavy industrial use.

Building description	Life expectancy	Remaining useful life	Gross replacement cost (including fees and finance per building)
Main glass works	60	20	£6 000 000
Laboratory	60	20	£1 000 000
Canteen and offices	30	5	£800 000
Warehouse	60	40	£9 000 000

a) Assuming a replacement building would take 21 months to reinstate, and that the owner could secure a finance rate of 7% per annum, value the freehold interest.

b) If you were aware that the main glass works building had an inefficient layout for modern manufacturing methods, how might you adjust your valuation?

c) If you were aware that modern manufacturing plants to replace this building would now be located on a site comprising 1.75 ha, how might you adjust your valuation?

Residual method

[1] You have been asked to value an office development site based on the following assumptions:

- Gross internal area of 5000 square metres and an efficiency ratio of 85%.
- The market rent is estimated to be £200 per square metre and the investment yield 6%.
- Building costs are £1200 per square metre, external works £250 000 and ancillary costs £200 000.
- Lead in period: 0.50 years, building period: 1.50 years, void period: 0.75 years
- Professional fees are estimated to be 10% of building costs and external works.
- There is a contingency allowance of 3% of building costs, external works, ancillary costs and professional fees.
- Other costs and fees include site investigations of £10 000, planning application fees of £5000, building regulations of £20 000.
- Finance can be arranged at a rate of 7% per annum.
- The letting agent fee is 10% of the ERV and the letting legal fee is 5% of the ERV. The marketing budget is £150 000.
- Developer's profit is 15% of all development and land costs.

[2] A property development company is proposing to develop an office building on a recently acquired town-centre site. The purchase price was £1.6m and the previous owner had obtained outline planning consent for a four-storey building on the site some two years previously but had not pursued this proposal. A local commercial letting agent has indicated that demand for office space in the town is high and leases have recently been negotiated at rental levels of

£220 per square metre per annum. The agent is not aware of any other new office development proposals in the town. Recent investment transactions show capitalization rates around 8%. The developer has appointed local architects and surveyors and, on the basis of their knowledge, the architect has drawn up a design that provides a building with a gross internal area of 4667 square metres. Analysis of the design shows that the efficiency ratio will be 90%. The developer's agent reports that, among other expressions of interest, they have received firm enquiries from several tenants about leasing space in the development. The quantity surveyor has indicated that building costs for good-quality speculative offices are expected to be £1075 per square metre. External works are expected to cost approximately £250000 and a separate contract has already been let for site clearance and the demolition of some small buildings on the site at a price of £100000. The project manager has drawn up an outline procurement programme that allows six months for design, 15 months for construction and a six-month letting void. The developer intends to borrow the cost of the development from a bank. Current interest rates on project loans of this nature are around 8% per annum. Calculate the developer's profit and report it as a cash sum, as a percentage of costs and as a percentage of value.

[3] Estimate the residual land value from the following net cash flow, assuming a finance rate of 10% per annum.

Quarter	0	1	2	3	4	5	6
Net cash flow	(138 284)	(500 000)	(500 000)	(500 000)	(500 000)	(500 000)	4 000 000

[4] Estimate the residual land value from the following development revenue and expenditure items over a one-year development period, assuming a finance rate of 10% per annum.

Quarter	1	2	3	4
Sales	0	0	300 000	300 000
Construction	(25 000)	(25 000)	(25 000)	(25 000)
Demolition	(20 000)	0	0	0
Prof. fees	(2250)	(1250)	(1250)	(1250)
Sale fees	0	0	(4500)	(4500)
Contingency	(2363)	(1313)	(1538)	(1538)
Profit				(120 000)

[5] Estimate the residual land value of the following development cash flow. The development has a gross internal area of 900 square metres and an efficiency ratio of 80%. Site clearance costs of £100 000 are anticipated in Q1. Building costs are £900 per square metre, with 25% due in Q2, 50% in Q3 and 25% in Q4. Professional fees are 10% of building costs and contingencies are 5% of building costs and professional fees. A simultaneous letting and sale is anticipated in Q7. The MR is estimated to be £220 per square metre, the yield 6%, letting fee 10% of MR and sale fee 2% of NDV. Developer's profit is 17% of GDV and finance can be arranged at a cost of 8% per annum.

[6]   Estimate the developer's profit from the following inputs. Assume that the build costs are spread evenly between Q2 and Q5.

INPUTS	
Land price (£)	850 000
Yield	7.00%
Gross internal area (m²)	2000
Efficiency ratio	90%
Build cost (£/m²)	1000
Estimated MR (£/m²)	175
Land purchase costs (% land price)	5.75%
Investment sale fee (% NDV)	5.75%
Letting fee (% MR)	10%
Lead-in period (quarters)	1
Building period (quarters)	4
Void period (quarters)	2
Contingencies (% build costs and fees)	5%
Professional fees (% build costs)	10%
Finance rate (per quarter)	2.00%

Answers

Replacement cost method

[1]

Building area (m²)	5000		
Building cost (£/m²)	1098		
Cost of modern equivalent building (£)		5 490 000	
Fees (% build cost)	12%	658 800	
Finance on build cost and fees over half build period @	4.50%	102 336	
Gross replacement cost of building (£)		6 251 136	
Depreciation allowance (age\estimated economic life)		38%	
Net replacement cost of building (£)			3 906 960
Site area (ha)	1.80		
Land value (£/ha)	18 500		
Value of land (£)		33 300	
Finance on land over build period @	4.50%	1118	
Cost of land (£)			34 418
Valuation before PCs (£)			3 941 378

[2]

a)   *Main glass works*

Gross replacement cost inc. fees and finance (£)			6 000 000
Depreciation allowance (% remaining life)	20	60	33%

Net replacement cost (£)			2 000 000
Laboratory			
Gross replacement cost inc. fees and finance (£)		1 000 000	
Depreciation allowance (% remaining life)	20	60	33%
Net replacement cost (£)			333 333
Canteen and offices			
Gross replacement cost (inc. fees and finance)		800 000	
Depreciation allowance (% remaining life)	5	30	17%
Net replacement cost (£)			133 333
Warehouse			
Gross replacement cost (inc. fees and finance)		9 000 000	
Depreciation allowance (% remaining life)	40	60	67%
Net replacement cost (£)			6 000 000
Land			
Site area (ha)		2.5	
Estimate value (£/ha)		52 000	
			130 000
Finance on land cost (FV £1 for 1.75 years at 7% p.a.)	1.75	7.00%	1.1257
			146 341
Replacement cost valuation (£)			8 613 007

b) A further percentage deduction might be made in addition to the straight-line depreciation charge made on the main chemical works due to its size.

c) Recalculate the land costs based on 1.75 ha. Also consider if the remaining hectare has an alternative use value and could be valued separately to reflect any development land value.

Residual method

[1]

INPUTS	
Areas	
Gross internal area (GIA) (m²)	5000
Efficiency ratio (net/gross area)	85%
Net internal area (NIA) (m²)	4250
Time	
Lead-in period (years)	0.5
Building period (years)	1.5
Void period (years)	0.75
Values	
Estimated rental value (ERV) (£/m² NIA)	200
Net initial yield	6.00%
Costs	
Site preparation (£)	10 000
Building costs (£/m² GIA)	1200

External works (£)	250000
Professional fees (% construction costs and external works)	10%
Miscellaneous costs (£)	200000
Contingency allowance (% of all construction costs)	5.00%
Planning fees (£)	5000
Building regulation fees (£)	20000
Planning obligations (£)	0
Other fees, e.g. legal, loan, valuation (£)	0
Short-term finance rate (annual)	7.00%
Letting agent's fee (% ERV)	10%
Letting legal fee (% ERV)	5%
Marketing (£)	150000
Sale costs (% NDV)	2.00%
Land purchase costs (% land price)	6.50%
Developer's profit (% development costs)	15.00%

RESIDUAL LAND VALUATION
Development value

Net internal area (NIA) (m²)	4250		
Estimated rental value (ERV) (£/m² NIA)	200		
		850000	
Net initial yield	6.00%	16.6667	
Gross development value (GDV) before sale costs (£)			14166667
Net development value (NDV) after sale costs (£)			13888889

Development costs

Site preparation (£)		(10000)	
Building costs (£/m² GIA)	1200	(6000000)	
External works (£)		(250000)	
Professional fees (% construction costs and external works)	10.00%	(625000)	
Miscellaneous costs (£)		(200000)	
Contingency allowance (% of all construction costs)	5.00%	(353750)	
Planning fees (£)		(5000)	
Building regulation fees (£)		(20000)	
Planning obligations (£)		0	
Other fees (£)		0	
Finance on above costs for *half* building period @	7.00%	(388514)	
Finance on above costs for void period @	7.00%	(408738)	
Letting agent's fee (% ERV)	10.00%	(85000)	
Letting legal fee (% ERV)	5.00%	(42500)	
Marketing (£)		(150000)	
Developer's profit on above costs @	15.00%	(1280775)	
Total development costs (TDC) (£)			(9819278)
NDV – TDC (£)			4069611

Land costs

Developer's profit on land costs @	15.00%	(530819)	3538792	
Finance on land costs over total development period @	7.00%	2.75	0.8302	
Residual land value before purchase costs (£)				2937986
Residual land value after purchase costs (£)	6.50%			2758672

[2]

INPUTS

Areas

Gross internal area (GIA) (m²)	4667
Efficiency ratio (net/gross area)	90%
Net internal area (NIA) (m²)	4200

Time

Lead-in period (years)	0.50
Building period (years)	1.25
Void period (years)	0.50

Values

Estimated rental value (ERV) (£/m² NIA)	£220.00
Net initial yield	8.00%

Costs

Land price (£)	1 600 000
Site preparation (£)	100 000
Building costs (£/m² GIA)	1075
External works (£)	250 000
Professional fees (% construction costs and external works)	13.00%
Miscellaneous costs (£)	0
Contingency allowance (% of all construction costs)	3.00%
Planning fees (£)	0
Building regulation fees (£)	0
Planning obligations (£)	0
Other fees, e.g. legal, loan, valuation (£)	0
Short-term finance rate (annual)	8.00%
Letting agent's fee (% ERV)	10.00%
Letting legal fee (% ERV)	5.00%
Marketing (£)	0
Sale costs (% NDV)	2.00%
Land purchase costs (% land price)	6.50%

RESIDUAL PROFIT VALUATION

Development value

Net internal area (NIA) (m²)	4200		
Estimated rental value (ERV) (£/m² NIA)	220		
		924 066	
Net initial yield	8.00%	12.5000	
Gross development value (GDV) before sale costs (£)			11 550 825
Net development value (NDV) after sale costs (£)			11 324 338

Development costs

Land price (£)		(1 600 000)
Land purchase costs (% land price)	6.50%	(104 000)
Finance on land costs for total development period @	8.00%	(322 157)
Site preparation (£)		(100 000)
Building costs (£/m² GIA)	£1075	(5 017 025)
External works (£)		(250 000)
Professional fees (% construction costs and external works)	13.00%	(684 713)
Miscellaneous costs (£)		0

Contingency allowance (% of all construction costs)	3.00%	(178 552)
Planning fees (£)		0
Building regulation fees (£)		0
Planning obligations (£)		0
Other fees, e.g. legal, loan, valuation (£)		0
Finance on above costs for half building period @	8.00%	(307 005)
Finance on above costs for void period @	8.00%	(256 461)
Letting agent's fee (% ERV)	10.00%	(92 407)
Letting legal fee (% ERV)	5.00%	(46 203)
Marketing (£)		0
Total development costs (TDC) (£)		(8 958 524)

Developer's profit on completion (NDV – TDC) (£)	2 365 814
Profit as a percentage of costs	26.41%
Profit as a percentage of value (NDV)	20.89%
Income yield (ERV as a percentage of costs)	10.31%

[3]   Convert 10% per annum to quarterly equivalent rate: $(1.10)^{0.25} - 1 = 2.41\%$ p.q.

Quarter	0	1	2	3	4	5	6
Net cash flow (£)	(138 284)	(500 000)	(500 000)	(500 000)	(500 000)	(500 000)	4 000 000
PV @ 2.41%	1	0.9765	0.9535	0.9310	0.9091	0.8877	0.8668
PV of cash flow (£)	(138 284)	(488 227)	(476 731)	(465 506)	(454 545)	(443 843)	3 467 137

NPV (residual land value before PCs) (£)	1 000 000
Residual land value after PCs @ 6.5% (£)	938 967

[4]   Convert 10% per annum to quarterly equivalent rate: $(1.10)^{0.25} - 1 = 2.41\%$ p.q.

Quarter	1	2	3	4
Net cash flow (£)	(49 613)	(27 563)	267 712	147 712
PV @ 2.41%	(48 445)	(26 280)	249 243	134 284

NPV (residual land value before PCs) (£)	308 802

[5]

Inputs	
Gross internal area (m²)	900
Efficiency ratio	80%
Build cost (£/m²)	900
Professional fees (% build cost)	10%
Contingencies (% build cost and fees)	5%
Rent (£/m²)	220
Yield	6.00%
Investment sale fee	2.00%
Letting fee (% MR)	10%
Land purchase costs (% land value)	6.50%
Developer's profit (% GDV)	17%
Finance rate (% p.a.)	8%

Calculations

Estimated MR (£/m^2)	158 400
Gross development value (£)	2 640 000
Net development value (£)	2 478 873
Quarterly finance rate (% p.q.)	1.94%

Cash flow

Quarter	1	2	3	4	5	6	7
NDV (£)							2 478 873
Site clearance (£)	(100 000)						
Building (£)		(202 500)	(405 000)	(202 500)			
Professional fees (£)		(20 250)	(40 500)	(20 250)			
Contingencies (£)		(11 138)	(22 275)	(11 138)			
Letting fee (£)							(15 840)
Developer's profit (£)							(448 800)
Net cash flow (£)	(100 000)	(233 888)	(467 775)	(233 888)	0	0	2 014 233
PV £1 @ finance rate	0.9809	0.9623	0.9439	0.9259	0.9083	0.8910	0.8740
PV of cash flow (£)	(98 094)	(225 058)	(441 539)	(216 563)	0	0	1 760 428

NPV (residual land value before PCs) (£)	779 173
Residual land value after PCs @ 6.5%	731 618

Calculations

Build costs (£)	2 000 000
Market rent (£ p.a.)	315 000
Gross development value (GDV) (£)	4 500 000
Net development value (NDV) (£)	4 255 319

Cash flow

Quarter	0	1	2	3	4	5	6	7	TOTALS
Land price (£)	(850 000)								
Land purchase costs (% land price)	(48 875)								
Building costs (£)			(500 000)	(500 000)	(500 000)	(500 000)			
Professional fees (% build costs)			(50 000)	(50 000)	(50 000)	(50 000)			
Contingencies (% build costs and fees)			(27 500)	(27 500)	(27 500)	(27 500)			
Letting fee (% MR)								(31 500)	(3 240 375)
Net development value (NDV)								4 255 319	
Net cash flow (£)	(898 875)		(577 500)	(577 500)	(577 500)	(577 500)		4 223 819	

Finance

	0	1	2	3	4	5	6	7	TOTALS
Opening balance (£)		(898 875)	(916 853)	(1 512 690)	(2 120 443)	(2 740 352)	(3 372 659)	(3 440 112)	
Interest (arrears) (£)		(17 978)	(18 337)	(30 254)	(42 409)	(54 807)	(67 453)	(68 802)	(300 040)
Closing balance (£)	(898 875)	(916 853)	(1 512 690)	(2 120 443)	(2 740 352)	(3 372 659)	(3 440 112)	714 904	

Profit sum (£)	714 904	
Profit on cost	20.19%	Final closing balance divided by all costs including interest
Profit on GDV	15.89%	Final closing balance divided by GDV
Development yield	8.90%	Market rent divided by costs
Profit erosion by rent (years)	2.27	Closing balance divided by market rent

Section B

Section C
Valuation Application

Chapter 9
Valuation of Investment Property

9.1 Introduction

This chapter covers the valuation of *investment property*, that is, property that can produce a rental income that may be capitalised to arrive at a capital value. Chapter 10 covers the valuation of *development property*.

Two internationally recognised bases of value are considered: market value and investment value. Market value is equivalent to value-in-exchange and investment value is equivalent to value-in-use. Investment value has regard to the specific requirements of a purchaser, which might encompass:

- The financial resources available for a property acquisition, including the split between debt and equity finance;
- The timescale for holding a property asset, and
- The tax position, personal tastes and specific requirements of the decision-maker.

Moreover, these specific requirements may relate to the way in which the property is to be managed; a small-scale niche investor may wish to manage the property much more actively than a large institutional investor. Wider considerations relating to an investment portfolio may also need to be considered.

In a perfect market, where buyers and sellers have instant access to market information, their economic requirements are identical and properties are homogeneous, it might be presumed that market participants would arrive at similar decisions and thus individual investment values might converge on a market value. In other words, there would be no difference between exchange prices, market valuations and investment valuations for each homogeneous property. However, the property market is not perfect; the product is heterogeneous, as are buyers and sellers, and there are many typologically and geographically distinct sub-markets.

Property Valuation, Third Edition. Peter Wyatt.
© 2023 John Wiley & Sons Ltd. Published 2023 by John Wiley & Sons Ltd.
Companion website: www.wiley.com/go/wyatt/propertyvaluation3e

9.2   Analysis of rents

When analysing rents that have been agreed on investment properties, it is important to consider the contractual terms contained in the lease. The following order of adjustments is recommended, and each is considered below:

1) Rental lease incentives: Rent-free periods and stepped rents
2) Capital lease incentives: Capital contributions and premiums
3) Value of any surrendered lease
4) Repairs and insurance
5) Rent-review pattern
6) Zoning for retail premises (see Chapter 6)

9.2.1   Rental lease incentives

Landlords sometimes offer incentives to tenants and the financial impact on *rental value* needs to be considered. This is done by converting the incentive into a financially equivalent rental value, which is deducted from the agreed *headline rent* to arrive at an *effective market rent*.

9.2.1.1   Rent-free periods

A rent-free period refers to a predetermined length of time within the term of a lease during which no rent is paid. If a rent-free period is offered to a prospective tenant as an incentive to take occupation of a particular property and such an incentive is not regarded as standard practice for the property type and location in question, then a valuer may wish to calculate the market rent of the property assuming no incentive was granted (the effective rent).

It should be borne in mind that it is common, especially in the case of retail property, for a landlord to grant a short (say three months or so) rent-free period for fitting out the premises. If this is the case, and the tenant is not trading from the premises during the 'fitting-out period', then its financial benefit to the tenant (or financial loss to the landlord) should be ignored when estimating the effective rent.

Landlords offer rent-free periods as a way of maintaining a headline rent so that at rent review a case can be made for a revision to market rent, which includes the financial value of the incentive. Also, declaring a headline rent rather than the effective rent can be beneficial in terms of bank lending ratios if debt has been used to help finance the purchase of the property investment and, in the case of long leases and an occupier of high quality, will assist in raising the valuation of the property (Sayce et al. 2006).

To determine the financial effect of a rent-free period on market rent it is necessary to amortise the capital value of the rent that is actually paid, known as the headline rent, over an appropriate period. The aim is to calculate a set of rent payments that, in the absence of the incentive, would produce an identical present value over that period. For example, a 10-year lease has been agreed at a *headline* rent of £50 000 per annum with a one-year rent-free period and a rent review in year five. Assuming a yield of 8%, what is the effective market rent assuming that the benefit of the incentive is amortised over the period to the rent review?

Headline rent (£ p.a.)	50 000	
YP for amortisation period of 4 years @ 8%	3.3121	
PV £1 for rent-free period of 1 years @ 8%	0.9259	
Capital value of headline rent (£)		153 339
YP for amortisation period of 5 years @ 8%		3.9927
Effective market rent (£ p.a.)		38 405

Using a yield (which implies rental growth) to calculate the effective market rent implies that the gap between the headline rent and the market rent widens over the amortisation period when in fact it should not. However, because the amortisation period is relatively short, the impact of this error is minor.

If there is a fitting-out period of, say, three-months, before the one-year rent-free period, the effective market rent will be slightly lower, reflecting the longer wait for rent. It is calculated as follows:

Headline rent (HR) (£ p.a.)	50 000	
YP for amortisation period (less rent-free and fitting-out period) of 3.75 years @ 8%	3.1336	
PV £1 for rent-free and fitting-out period of 1.25 years @ 8%	0.9083	
Capital value of HR (£)		142 311
YP for write-off period (less fitting-out period) of 4.75 years @ 8%	3.8274	
PV £1 for fitting-out period of 0.25 years @ 8%	0.9809	
Deferred YP		3.7545
Effective MR (CV of HR/deferred YP) (£ p.a.)		37 904

The period over which the headline rent is amortised affects the calculation of the effective market rent. If it was felt that the effective market rent would not grow sufficiently to overtake the headline rent by the rent review, then the headline rent should be capitalised and amortised up to the next review opportunity, in this example, the end of the lease.

Headline rent (HR) (£ p.a.)	50 000	
YP for amortisation period (less rent-free and fitting-out period) of 8.75 years @ 8%	6.1254	
PV £1 for rent-free and fitting-out period of 1.25 years @ 8%	0.9083	
Capital value of HR (£)		278 180
YP for amortisation period (less fitting-out period) of 9.75 years @ 8%	6.5976	
PV £1 for fitting-out period of 0.25 years @ 8%	0.9809	
Deferred YP		6.4719
Effective MR (CV of HR/deferred YP) (£ p.a.)		42 983

Amortising the rent-free period over a shorter duration, such as five years, favours the tenant and over a longer period favours the landlord, particularly if the rent review is upwards only. Keeping the headline rent high protects the landlord's reversionary interest. The key determinant of the length of the amortisation period is when the effective market rent is expected to overtake the headline rent, and this depends on the rental growth rate assumption. In the above example, the growth rate necessary for the effective rent to overtake the headline rent at the rent review in year five is 5.70% per annum, calculated as follows:

$$37904 \times (1+g)^5 = 50000 \; g = 0.0570 \text{ or } 5.70\% \text{ p.a.}$$

whereas the growth rate required for the effective rent to exceed the headline rent in year 10 is 1.52% per annum. This sort of growth rate analysis can be used to help decide the period over which the value of the incentive should be amortised prior to estimating the effective rent being paid by the tenant (Crosby and Murdoch 1994).

9.2.1.2 Stepped rents

Stepped rents are a series of rent changes at intervals more frequent than the standard five-year rent-review pattern usually seen in UK commercial leases. Stepped rents can help the tenant's cash-flow at the start of a lease if the initial rent is less than the market rent. The steps may be reviewed to pre-agreed sums, but this need not necessarily be the case.

In cases where a stepped rent is paid, it may be necessary to determine the effective rent so that the transaction can be used as comparable evidence. This is done by calculating the present value of each rent step and then calculating the annual equivalent of the sum of these present values over the period of the incentive. The discount rate should be the appropriate yield.

For example, a property has just been let on a 15-year lease with 5-year rent reviews but, during the first 5 years the rent payments are stepped as follows: £200 000 in year one, £225 000 in year two, £250 000 in year three, £275 000 in year four and £300 000 in year five. After year five, the rent reverts to the market level. Assuming a yield of 9%, the capital value (sum of the present values) of these stepped rents is:

Year	Rent (£ p.a.)	PV £1	PV (£)
1	200 000	0.9174	183 480
2	225 000	0.8417	189 383
3	250 000	0.7722	193 050
4	275 000	0.7084	194 810
5	300 000	0.6499	194 970
Capital value (£)			955 693

This capital value is then amortised over the period to the first rent review when the stepped rents end and the market rent is payable.

Capital value (£)		955 693
Divided by YP 5 years @ 9%		3.8897
Effective market rent over first five years (£ p.a.)		245 698

As the tenant is paying £300 000 per annum in year five and this is greater than the effective market rent of £245 700 per annum, the tenant should consider whether rental growth over the next five years will mean that the market rent at that time will exceed £300 000 per annum. If it does not, and the lease provides for upward-only rent reviews, the property will become over-rented at this point.

9.2.2   Capital lease incentives

9.2.2.1   Capital contributions

A capital contribution is a financial payment by a landlord to induce a tenant to take occupation. It usually takes the form of a financial payment but may also be for fitting out, taking financial responsibility for an existing lease or some other liability. In lieu of making such a capital contribution, the landlord would expect to receive a rent from the tenant above the market rent.

The calculation of the effective market rent of a property where a capital contribution has been made is conducted by applying the same principles as for rent-free periods: determine the amount of the contribution and the length of the amortisation period. For example, a landlord offers a tenant £100 000 to induce occupation under a new 15-year lease with 5-year rent reviews at a rent of £300 000 per annum. Amortising the capital contribution at a yield of 5% over the period to the first rent review, the effective market rent is calculated as follows:

Headline rent (£ p.a.)		300 000
Capital contribution (£)	(100 000)	
YP for 5 years @ 5%	4.3295	
Annual equivalent of capital contribution (£ p.a.)		(23 097)
Effective MR (£ p.a.)		276 903

The rental growth rate required for the effective MR to overtake the headline rent by the end of this five-year write-off period is 1.62% per annum.

9.2.2.2   Premiums and 'key money'

A *premium* is a capital payment from a tenant to a landlord, often paid in return for the grant or renewal of a lease on favourable terms. These might be a rent reduction, less-frequent rent reviews, a percentage-based rent at review (say 80% of the market rent, known as a geared review), the landlord taking responsibility for repairs or insurance or a wider user-clause. The payment is usually financial but can be non-pecuniary such as the carrying out of repairs or improvements. The benefit of a premium to a landlord is a cash flow where a capital sum is received early and the benefit to the tenant will be an immediate profit rent. A premium may also be paid by the assignee when a lease is assigned and there is a profit rent available because the contract rent is below the market rent. Payments for business goodwill, fixtures, fittings, equipment and stock should be excluded from the amortised sum.

Key money is a payment for a monopoly position for a certain trade. When there is strong demand, tenants may pay *key money* to secure a property – effectively a premium in addition to rent. This key money should be treated as the capital value of additional rent and amortised over the period for which future occupation is assumed (in perpetuity in some cases) and added to the contract rent. It is important that the valuer determines the reason for the payment of any capital sum that has been paid when valuing a property or when analysing comparable evidence. Whereas a premium would be amortised until first rent-review opportunity (because that is when any profit rent would end), key money may be amortised over a longer period.

A premium is simply capitalised rent so, assuming there is a normal situation where a tenant pays market rent, the size of any premium that might be paid will depend on the size of any rent reduction from the market rent. In effect, the landlord is 'selling' part of the market rent, and the tenant is 'buying' it in the form of a profit rent. To calculate a premium, the agreed rent reduction (profit rent) should be capitalised. For example, a property is let on a lease with four years remaining at a rent of £12 500 per annum. The current market rent is estimated to be £15 000 per annum. If the tenant assigns the lease, what premium should be paid by the assignee to compensate for the profit rent? Capitalising the profit rent over the four years and assuming a yield of 10%:

Profit rent (£ p.a.)	2500	
YP 4 years @ 10%	3.1699	
Premium (£)		7925

To calculate the market rent when a premium has already been agreed, amortise the premium over the period of the benefit. For example, at the start of a new lease with five-year rent reviews, the tenant agrees to pay a rent of £10 000 per annum plus a premium of £11 750. What is the effective market rent?

Contract rent (£ p.a.)		10 000
Premium (£)	11 750	
Divided by YP 5 years @ 10%	3.7908	
Annual equivalent of premium (£ p.a.)		3100
Effective market rent (£ p.a.)		13 100

If a premium is to be paid at some point in the future, the amount should be discounted over the relevant deferment period.

Sometimes a lease might specify that, at each rent review, the rent is reviewed to a proportion of market rent. In other words, the tenant receives a discount in the form of a profit rent at each review. A premium might be paid by the tenant to compensate the landlord for offering such an incentive. For example, a tenant pays a premium of £10 000 at the start of a 10-year lease where the rent is reviewed to 70% of market level in year five. If the initial contract rent was £5000 per annum and the yield 8%, the effective market rent can be calculated using either algebra (on paper) or iteration (goal-seek in Excel):

Effective rent for first 5 years (£ p.a.)	x	
Less contract rent for first 5 years (£ p.a.)	(5000)	
Profit rent (£ p.a.)	$x - 5000$	
YP 5 years @ 8%	3.9927	
		$3.9927x - 19964$
Effective rent for second 5 years (£ p.a.)	x	
Less contract rent at review (£ p.a.)	(0.7x)	
Profit rent (£ p.a.)	0.3x	
YP 5 years @ 8%	3.9927	
PV 5 years @ 8%	0.6806	

$$
\text{Capital value of profit rent (£ p.a.)} \qquad\qquad \frac{0.8152x}{4.8078x - 19\,964}
$$

Premium to landlord should compensate for the profit rent to tenant, so

£10 000   = $4.8078x - 19\,964$

x or effective rent (£ p.a.)   =   6232

It may be necessary to consider the value of premiums and associated profit rents from both the landlord and tenant's viewpoints. The values will differ if different yields are used to amortise the rent reduction and the actual amount of premium may therefore require a negotiated settlement in practice.

A *reverse premium* is a capital payment usually made by an assignor of a lease to induce the assignee to take occupation. This situation may arise in a depressed market where the supply of accommodation exceeds demand, and the current rent exceeds the market rent, the property is thus *over-rented*. If the lease contains upward-only rent reviews and the difference between the contract rent and the market rent is significant, the property may remain over-rented for some time. The assignor of a lease on a property that is over-rented may pay a reverse premium to the assignee equivalent to the capital value of the overage rent.

For example, a property was let two years ago for £250 000 per annum on a 10-year lease with an upward-only rent review in the fifth year. The tenant wishes to assign the lease but is aware that the current market rent for the property is £235 000 per annum. What size of reverse premium should the assignor pay the assignee? This is calculated by determining the size of the overage rent (£15 000 per annum in this case) and then deciding over how long this overage rent would be paid for, bearing in mind that the rent review is upwards only, and the future level of market rent will not be known. If we assume that market rental growth for this property will be negligible over the remaining term of the lease, we can capitalise the overage for eight years at a yield based on fixed income investments suitably adjusted for risk. A relatively high yield of 12% has been used here to reflect the over-rented nature of the interest.

Market rent (£ p.a.)	235 000	
Contract rent (£ p.a.)	250 000	
Overage (£ p.a.)	(15 000)	
YP 8 years @ 12%	4.9676	
Reverse premium (£)		(74 514)

If a valuer is seeking to use a property on which a reverse premium has been paid as comparable evidence, the market rent of the property is calculated by deducting the annual equivalent of the reverse premium from the contract rent. Using the example above, assume the tenant assigned the lease and paid a reverse premium of £75 000 to the assignee. Assuming the rent review is upwards only, the market rent is calculated as follows:

Contract rent (£ p.a.)		250 000
Reverse premium (£)	(75 000)	
Divided by YP 8 years @ 12%	4.9676	
		(15 098)
Market rent (£ p.a.)		234 902

9.2.3   'Surrendered' leases

Sometimes, as leases approach their termination dates, tenants may wish to surrender them before they expire and negotiate new ones. This might help to preserve goodwill attached to a particular trading location or remove future uncertainty surrounding the terms of a new lease. Similarly, landlords may be keen to encourage tenants to surrender their old leases and grant new ones if they plan to sell a property investment; new leases will be more attractive to potential purchasers.

If an existing lease is surrendered in return for the grant of a new one, then the capital value of any profit rent that the tenant was entitled to under the existing lease can be reflected as a rent reduction under the proposed lease, usually for the initial term until the first rent review. This means calculating the capital value of the tenant's surrendered profit rent, amortising this sum over the period in the new lease until the initial rent is reviewed (usually the first rent review), and then deducting it from the market rent. Depending on whether the landlord's or tenant's yield is used to capitalise and amortise, the rent will differ and is therefore usually negotiated between the parties to ensure that neither party jeopardise their existing financial positions.

For example, a tenant wishes to surrender the remainder of an existing lease in return for the grant of a new one. The present lease has three years to run with no review and the rent passing is £20 000 per annum. The estimated market rent is £27 000 per annum and comparable evidence suggests that the yield for freehold investments in similar properties is 10%. The landlord is willing to accept a surrender of the current lease and grant a new 15-year lease with upward-only rent reviews every 5 years. The proposed initial term rent can be calculated as follows:

Assuming a freehold yield of 10%:

Market rent (£ p.a.)	27 000	
Current rent (£ p.a.)	(20 000)	
Profit rent (£ p.a.)	7000	
YP for 3 years @ 10%	2.4869	
Valuation of tenant's profit rent (£)		17 408
/YP for 5 years for 5 years @ 10%		3.7908
Annualised value of profit rent over five years (£ p.a.)		4592

Market rent less amortised profit rent (£ p.a.): 27 000 – 4592 = 22 408. Repeating this calculation using a leasehold yield of, say 11%, produces a rent of £22 372 per annum. The landlord and tenant would seek to agree a rent based on these two valuations.

Taking another example, this time involving a long lease and a geared profit rent: a tenant on a long lease at a low rent occupies a property at a passing rent of £250 per annum, without review, and there are 35 years unexpired. The estimated market rent of the property with vacant possession is £400 000 per annum. The landlord has agreed to accept a surrender of the tenant's interest upon grant of a new 125-year lease at a rent of 10% of the market rent. The rent will be reviewed every five years on this basis. Assuming a freehold yield of 6% and using this as a basis for estimates of long lease yields, calculate the premium that should be paid upon commencement of the new lease. The freehold and leasehold interests are values in their present and proposed arrangements, with the difference in value representing the premium.

Tenant's present interest		
MR (£ p.a.)	400000	
Rent paid (£ p.a.)	(250)	
Profit rent (£ p.a.)	399750	
YP 35 years @ 7.5%	12.2725	
Gross valuation (£)		4905936
Valuation (£) net of purchaser's costs @ 6.5%		4606513

Tenant's proposed interest		
MR (£ p.a.)	400000	
Less rent passing @ 10% MRV (£ p.a.)	(40000)	
Profit rent (£ p.a.)	360000	
YP 125 years @ 7%	14.2827	
Gross valuation (£)		5141765
Valuation (£) net of purchaser's costs @ 6.5%		4827948

The difference in value is £4606513 – £4827948 = – £221435.

Landlord's present interest		
Term rent (£ p.a.)	250	
YP 35 years @ 7%	12.9477	
		3237
Reversion to MR (£ p.a.)	400000	
YP perp @ 7%	14.2857	
PV £1 35 years @ 7%	0.0937	
		535217
Gross valuation (£)		538454
Valuation (£) net of purchaser's costs @ 6.5%		505590

Landlord's proposed interest		
Proposed rent (£ p.a.)	40000	
YP perp @ 6%	16.6667	
Gross valuation (£)		666667
Valuation (£) net of purchaser's costs @ 6.5%		625978

The difference in value is £505590 – £625978 = – £120388. Assuming the parties are willing, a figure for the premium would be agreed by negotiation.

9.2.4   Repairs, insurance, and ground rents

Usually, for commercial leases in the United Kingdom, the tenant is responsible for internal and external repairs and insurance costs. This arrangement is known as a *full repairing and insuring* (FRI) lease. However, if the lease requires the landlord to take responsibility for these costs, then it would be rational to expect the rent to be higher (to compensate for paying these costs). Therefore, adjustments should be made to the gross rent to arrive at a net effective rent as follows (it is important to note that the percentage adjustments are not universal and constant; valuers adopt heuristics and they may vary):

- Repairs: Reduction of typically 15% of the gross rent (10% for external repairs, 5% for internal repairs)
- Insurance: Reduction of say 2–2.5% of the gross rent

Since FRI is the standard letting arrangement in the United Kingdom, it is best practice to try and adjust comparable evidence to this basis. For example, if a comparable property was let on IRI terms, then a deduction would be made to its rent to put it on FRI terms, the logic being that the IRI rent is higher than FRI rent because the landlord must pay for external repairs. Another way to think about it is that you are trying to determine the *net* rent in each case. FRI is a net rent (known as a 'triple net' rent in the United States) so no adjustment is needed. If the rent is on IRI terms, then a deduction is needed from this (gross) rent to make it a net rent. This logic is helpful when considering the subject property. If the subject property is let on, say, IRI terms, then the (gross) rent would need to be reduced to get to the net (FRI) rent because the landlord must pay these costs out of the rent received. If the subject property is to be let on non-FRI terms, IRI terms for example, and the comparable evidence is on FRI terms, then the adjustment of the comparable evidence is upwards, so you would make an addition to the FRI rent from the comparable properties to put it on IRI terms.

For example, a first-floor office suite of 1000 square metres has just been let at £150 000 per annum. The landlord is liable for maintaining the structure and common parts and for insuring the building. A service charge covers the cost of heating and lighting. The net rent to the landlord might be calculated as follows.

Annual rent (£ p.a.)	150 000	
Area (m²)	÷ 1000	
Gross rent per square metre (£ p.a./m²)		150.00
Less adjustments for:		
▪ External repairs at, say, 10% of the gross rent	(15.00)	
▪ Internal repairs of common parts, say 2.5%	(3.75)	
▪ Insurance at 2.5%	(3.75)	
▪ Management at 5%	(7.50)	
Making a total deduction of 20% of gross rent		(30.00)
Net rent per square metre (£ p.a./m²)		120.00

In addition to rent, tenants may pay a service charge to the landlord, particularly if the premises are part of a multi-let building, in a managed shopping centre or situated on an industrial estate for example. The service charge typically covers running costs, maintenance, repair and replacement expenditure on repair and maintenance of the building, maintenance of the estate, plant and machinery and the provision of services such as security, reception facilities and so on. If these costs are to be apportioned between tenants this may be done as fixed amounts or percentages, weighted by rent or floor area. Payment is usually at the same time as rent and increases in the charge are negotiated and stated in the lease, typically linked to inflation or pre-determined percentage increases. The extent of land-lord's obligations for provision of works and services is dependent upon the wording of the lease, and it should be noted that the charge is a reimbursement, not a profit stream.

In the case of properties subject to ground leases, the ground rent should be deducted when valuing the head-lease. For example, a property is subject to a head-lease at a fixed ground rent of £10 000 per annum plus 2% of any sub-rent. The head-lessee must also pay the freeholder non-recoverable costs of £500 per annum. The property is sublet at a rent of £400 000 per annum. The head-rent that the freeholder receives is: £10 000 + (£400 000 × 2%) = £18 000 per annum. The net rent that the head-lessee receives is calculated as follows:

£400 000 − £18 000 − £500 = £381 500 per annum

9.2.5   Rent-review pattern

In the United Kingdom, a rent-review clause is usually a feature of leases with terms longer than five years. The clause ensures that rent is periodically revised to 'market' level and are typically every five years and upwards only (or at least the rent cannot move below the current rent passing). The rent-review clause in the lease will set out the basis on which the rent at review must be determined. A dif-ficulty for valuers in this regard is obtaining the necessary information from leases to enable informed value adjustments to be made to rent passing. The availability of information may be constrained by confidentiality clauses, so assumptions are often made. Examples of rent-review assumptions include:

- The premises are vacant and available to let.
- Both landlord and tenant are willing parties to the contract.
- The premises are fit for occupation and use.
- The premises are to be let on the same terms as the actual lease.
- The tenant has complied with lease terms.
- There is a 15-year term to expiry at each review and a prospect of renewal at the end of the lease.
- The value of the tenant's actual occupation and any effect of goodwill are disregarded.
- The value of any tenant's improvement is also disregarded.

A rent review is usually activated by the landlord giving notice to the tenant of the new rent. If the tenant is not happy, then the mechanism for agreeing it is specified in the clause in the lease. The rent-review clause usually specifies the rent as market rent assuming the premises are fitted out and ready for occupation and that the tenant has received a rent-free in respect of fit-out.

Sometimes, it may be useful to calculate the rent that should be paid under non-standard rent review-pattern so that it is equivalent to the rent paid under a standard rent-review pattern. The formula for this calculation was derived by Rose (1979):

$$k = \frac{(1+y)^a - (1+g)^a}{(1+y)^a - 1} \times \frac{(1+y)^n - 1}{(1+y)^n - (1+g)^n} \tag{9.1}$$

where k = constant rent factor, y = yield, g = rental growth rate, a = abnormal rent-review period and n = normal rent-review period. For example, assume the market rent under a five-year rent-review pattern is £10 000 per annum, and this is expected to grow at a constant growth rate of 2% per annum. With a yield of 6%, the equivalent rent factor for a 21-year rent-review pattern is 1.1341. Multiplying this factor by the £10 000 rent gives an equivalent rent of £11 341 per annum.

9.3　Analysis of yields

9.3.1　Equivalent yield

Valuers also analyse comparable transactions to obtain yield evidence. This requires rent and price information to calculate the yield. If the valuer is satisfied that a rent is a market rent (i.e. the investment is *rack rented*), and the recent sale price is also known, the *net initial* yield can be calculated. However, for property investments where the rent passing is lower than the market rent but is likely to revert to market rent in the future, this is referred to as a *reversionary* investment. Here, the initial yield does not provide the full picture because it is a ratio of the rent passing to the price paid for the investment. There will be a reversionary yield too, which is the ratio of the reversionary market rent to the price.

For example, if a property has a rent passing of £250 000 and recently sold for £3 m, the initial yield net of 6.5% purchase costs would be 7.82% (£250 000/ (£3 000 000 × 1.065)). If the market rent is estimated to be £300 000, the net reversionary yield would be 9.39% (£300 000/£3 195 000). There are two yields. If this transaction is to be used as comparable evidence for the yield, it is helpful to reduce these to a single *equivalent* yield.

Continuing the example, assume the next rent review is in four years' time. In terms of cash flow, this investment generates four annual payments of £250 000 and then £300 000 per annum in perpetuity (ignoring rental growth because it is handled in the yield). The equivalent yield is the rate at which this cash flow must be discounted so that it equates to the purchase price of £3 195 000. The calculation can be performed using iteration (the 'goal-seek' function) on a spreadsheet to set the net present value (NPV) to zero by altering the equivalent yield:

Term (contract) rent (£ p.a.)	250 000	
YP for term of 4 years @ 8.94%	3.2443	
		811 069
Reversion to estimated MR (£ p.a.)	300 000	
YP perpetuity @ 8.94%	11.1907	
PV £1 for term of 4 years @ 8.94%	0.7101	
		2 383 931
Valuation including purchase costs (£)		3 195 000
Purchase price including purchase costs (£)		(3 195 000)
Net present value (£)		0

9.3.2   Weighted average unexpired lease term

Understandably, investors are particularly concerned with the risk that vacancies and voids present. These are gaps in income flow that are uncertain in duration and often entail costly outlays while the space remains empty. For single-let property investments, these income gaps are readily apparent, but for multi-let properties or portfolios of property investments, they can occur at various times depending on the lease terms for each occupier. In this case, the calculation of the weighted average unexpired lease term, or WAULT for short, can be helpful when adjusting yield evidence to reflect potential gaps in income.

WAULT is calculated by summing the remaining contractual lease payments between the valuation date and lease expiry for each tenant, then totalling these for all tenants and finally dividing this amount by the total annual contractual rent for the property or portfolio. WAULT is expressed in years and, usually, a longer WAULT is preferable if income stability is required, but if redevelopment or refurbishment is planned, a shorter WAULT might be desirable. WAULT calculates an average unexpired lease term that is weighted by rent. It is possible to use other weights such as floor areas.

As an example, a property investment comprises three tenancies as follows:

Lease	Area (m²)	Lease expiry (years)	Rent (£ p.a.)	Contractual rent remaining (£ p.a.)
1	1000	1	150 000	150 000
2	2000	2	200 000	400 000
3	3000	3	250 000	750 000
Total	6000	6	600 000	1 300 000

A simple average unexpired lease term would be two years. A rent-weighted WAULT would be: £1 300 000/£600 000 = 2.17 years, and an area-weighted WAULT would be: $(1000\,m^2 \times 1/6) + (2000\,m^2 \times 1/3) + (3000\,m^2 \times 2/3) = 2.33$ years.

9.4   Market valuation of investment property

9.4.1   Voids and break options

Short leases and leases with options to break before the end of the lease term mean greater diversity of lease contracts and increased uncertainty for investors. In the case of a short lease, will the tenant renew? If not, will there be a rent void and how long will it be? On re-letting, what will the rent and new lease terms be and the quality of the new tenant? In the case of a break option, will it be exercised? If so, then all the questions relating to re-letting apply here too. The value of a property investment will depend on the answers to these questions. The risks associated with short leases and break options will be accentuated in times of economic downturn, but also may provide opportunity in an active economy, by re-letting at higher rent and securing more favourable lease terms to the investor.

Such diversity in lease terms presents a challenge to valuers. It can be hard to find comparable evidence and justify small but often cumulative adjustments to the yield. Valuers tend to focus on the worse-case scenario and assume there will be a rent void at the end of the (short) lease or that a break option will be exercised.

Consider the following example: a modern office property has just been let on a 15-year FRI lease at a market rent of £50 000 per annum. There is a break option for the tenant in year five, just before the rent review (to prevent the tenant from using it as a bargaining tool). Comparable evidence suggests that rack-rented office investments let on 15-year FRI leases with five-year rent reviews to market rent but without breaks sell at net initial yields of 7%. The break option adds a degree of uncertainty to the income that the investor would receive after year five. Possible outcomes at the break are; the tenant continues in occupation, the break is exercised but there is no void, or the break is exercised and a rent void follows. Faced with such uncertainty, a valuer might increase the yield slightly, here from 7 to 8%.

Market rent (£ p.a.)	50 000	
YP perpetuity @ 8%	12.5	
Valuation before PCs (£)		625 000

If the lease had no break option and was valued using a 7% yield the capital value would be £714 286, so the yield adjustment leads to a 12.5% reduction in value. This simple approach benefits from a direct relationship with comparable evidence, assuming there is enough available, but it hides a lot of assumptions (Havard 2004). Another approach might be a modified term and reversion valuation where the yield is adjusted by a lesser amount and a rent void is incorporated after the break. The valuer needs to be sure (via market evidence) that the void duration is realistic, and an advantage of this approach is that different yields can be used for the existing and new leases (Havard 2004) but, again, only if justified by market evidence. The valuation below incorporates a void period of one year after the break option in year five and the yield has been adjusted upwards to 7.5%, resulting in a slightly higher valuation.

Market rent – first lease (£ p.a.)	50 000	
YP 5 years @ 7.5%	4.0459	
		202 950
Market rent – new lease (£ p.a.)	50 000	
YP perpetuity @ 7.5%	13.3333	
PV 6 years @ 7.5%	0.6480	
		432 000
Valuation before PCs (£)		634 950

Another approach is to produce a range of valuations under different scenarios; the break clause is/is not exercised, the rent void does/does not occur, a void lasts for six months, one year and so on. As a way of summarising the valuation outcomes, probabilities could be assigned and a weighted average 'expected' valuation calculated (French 2001).

Baum (2003) found that the most popular financial adjustment for short leases (say less than five years) and leases with break options in a similar time frame was the inclusion of a rent void, but one that did not reflect the 'true' expected costs of the void. Instead, it was moderated to reflect an estimated probability of the tenant breaking or not renewing. If it was certain that the tenant would exercise the break option or not renew the lease, then a full void allowance was included. For breaks, the notice period and any penalty payment would be factored in (i.e. a long notice period and big rent penalty would neutralise void allowance). When valuing shopping centres in which units are let on short leases, the valuer might incorporate a running void assumption in the cash flow based on the average void rate and expected average void period.

Many break clauses coincide with 5- and 10-year review dates, while some occur sooner, say after three years, known as a short-term break. These tend to be a feature of less valuable properties. The period of notice that a tenant is required to give and the penalty payment (if any) for exercising the break option does vary, typically between 6 and 12 months of rent. Also, there might be more than one break opportunity, the break option may be tenant-only (usual), landlord-only (very rare) or landlord and tenant activated. If a tenant secures a break clause in a lease in a market where they are uncommon, it may require the tenant to pay a rent above market rent. In such a case, if the break option is not exercised, the total cost of the lease to the tenant will be higher than if there was no break clause and a market rent was paid. The likelihood that a tenant might exercise a break or not renew a lease may depend on the amount of financial penalty, the expected cost of dilapidations, the amount spent on fitting out the premises, the availability of alternative premises, estimated relocation costs, growth or contraction of the tenant's business and expected rental growth (Baum 2003).

If a tenant vacates at the end of a short lease or at a break opportunity, the landlord incurs fixed and variable costs. Fixed costs will include fees for finding a new tenant and variable costs will include management and maintenance costs while the property is empty, loss of rent until a new tenant is found and for the duration of any rent-free period offered to a new tenant, and the cost of any other incentives that might need to be offered (McAllister 2001). The magnitude of these variable costs will depend on the length of the void period.

A higher rent to compensate for the break option or short lease might be agreed but the level of the headline rent and the length of time over which it should be amortised will depend on views about rental growth over the lease term because, under a standard lease with upward-only rent reviews, the rent cannot fall whereas with a break the tenant could vacate. It will also depend on the size of any penalty payment. The level of this higher rent can be calculated by valuing the lease with a rent void and perhaps a higher yield too and then valuing the same property assuming standard lease terms. Algebra (shown below) or iteration (goal-seek in Excel) can be used to equate the two valuations by adjusting the rent reserved for the first five years (French 2001). For example, calculate the rent that should be paid for the first five years of a ten-year lease, which has a break option and a rent review in year five. It is assumed that there is a six-month void at the break after which the rent reverts to the market rent of £300000 per annum and there is a one-year void at the end of the lease (to cover marketing and any rent-free period granted) after which the property reverts to a standard lease.

Term 1 rent (£ p.a.)	x	
YP 5 years @ 6%	4.2124	
		4.2124x
Term 2 rent (£ p.a.)	300 000	
YP 4.5 years @ 6%	3.8442	
PV 5.5 years @ 6%	0.7258	
		837 036
Reversion to MR on standard lease (£ p.a.)	300 000	
YP perpetuity @ 6%	16.6667	
PV 11 years @ 6%	0.5268	
		2 634 005
Valuation before PCs (£)		3 471 041 + 4.2124x

Now assume that the standard lease arrangement for this property is a 15-year lease with five-year upward-only rent reviews let at a market rent of £300 000 per annum. The capital valuation would be as follows.

Market rent (£ p.a.)	300 000	
YP perpetuity @ 6%	16.6667	
Valuation before PCs (£)		5 000 000

If we assume that the capital value of the property subject to the break option and the standard lease arrangement should be the same, we can state that:

$$£3\,471\,041 + 4.2124x = £5\,000\,000 \qquad \therefore x = £362\,966$$

So, the initial contract rent must be set £62 966 per annum above the £300 000 per annum market rent to compensate for the estimated voids. Of course, there may be other adjustments to make including voids costs or raising the yield on the short lease, but the valuer must be careful not to double-count the financial implications of flexible terms. Some may argue that the rent at the break point might not drop to £300 000 but the tenant would undoubtedly exercise the break to ensure the rent is the market rent (although this may incur costs).

9.4.2   Statutory considerations

In many countries, there will be legislative context within which valuations take place. These might take the form of rent control, provision of security of tenure for tenants, compensation for improvements undertaken by tenants and so on. The United Kingdom is no exception and a body of legislation and case law, known as landlord and tenant law, governs the legal relations between parties to a lease. Key statutes that regulate business tenancies and affect their valuation are described below.

The *Landlord and Tenant Act 1927* (as amended by *Landlord and Tenant Act 1954 Part III*) requires the landlord to compensate a tenant who leaves at the end of a lease for 'qualifying[1]' improvements made during the lease. Shops, for example, are quite likely to have been subject to tenant's improvements. Landlord's consent is normally required before the improvements can qualify but, under the Landlord and Tenant Act 1988, this consent cannot be unreasonably withheld.

The amount of compensation is calculated as the lesser of the value added due to the improvements or the cost of the improvements at the lease termination date. The value added must relate to the intended use, so no compensation is payable if the property is to be demolished. If the tenant renews the lease, the value of the improvement is disregarded (deducted) from the estimated market rent for a period of 21 years. Assuming the improvements qualify for compensation, the initial valuation problem is determining the extent to which they impact on value.

The *Landlord and Tenant Act 1954 Part II* (as amended by the *Law of Property Act 1969*) provides business tenants with security of tenure by allowing the original lease term to continue but subject to certain grounds that the landlord can establish to regain possession. The occupying tenant is entitled to automatic continuance of the original lease until terminated in accordance with the Act. The tenant's interest is assignable and therefore valuable. In addition to the right of automatic continuance, the landlord or tenant can apply for a new lease. Where a new lease is granted to the existing tenant, the rent payable is normally the market rent but disregarding the effect on rental value of the fact that the tenant or predecessors in title have been in occupation, any goodwill from the existing tenant, qualifying improvements for a period of 21 years and any licences that belong to the tenant in respect of licenced premises.

The tenant may continue to pay the existing rent beyond the end of the lease (known as 'holding over') but, while the terms of the new lease are being agreed, the landlord or the tenant can apply for an *interim rent*. In cases where the lease is not renewed or renewal proceedings are opposed, the interim rent is determined under Sections 34(1) and (2) of the 1954 Act (i.e. a market rent disregarding any qualifying tenant's improvements) but assuming; that the tenancy is from year to year, the rent would be reasonable for a tenant to pay, and regard is had to the passing rent and rent payable under any sub-leases within the property. If the renewal is unopposed, the interim rent is the same as the rent agreed under the new lease (usually a market rent) subject to adjustment to reflect any difference in market conditions or lease terms during the interim period. In such cases, the determination rules under an opposed renewal apply subject to these adjustments. Landlords could try and have an 'upward-only penultimate day review' drafted into the lease to ensure that the interim rent is not less than the rent passing.

If the parties cannot agree the terms of the new tenancy, then the courts are able to grant the tenant a new lease of up to 15 years on expiry of the existing lease at the market rent assuming similar terms as the original lease. The prospective landlord and tenant can agree in writing to 'contract out' of (exclude themselves from) the provisions of the 1954 Act but the lease must be for a fixed term and the landlord cannot contract out of disturbance compensation (see below) liability if the lease is longer than five years.

The landlord is entitled to counter the tenant's application for a new lease by establishing one of seven grounds for possession prescribed by the Act. If the landlord regains possession on the grounds that the rent for the property would be increased if let as a whole, redevelopment is intended or the property is required for own occupation, then the tenant is entitled to 'disturbance compensation' for loss of *goodwill*. The amount of disturbance compensation that is payable is equivalent to the *rateable value* (RV) of the property (or twice the RV if the business has been in continuous occupation for the past 14 years or more).

Two examples illustrate the impact of some of the legislative points described above on the valuation of business property.

9.4.2.1 Example 1

A factory is held on a 15-year lease with five years left at a contract rent of £5000 per annum. The tenant carried out qualifying improvements four years ago, which increased the market rent by 20%. The cost of these improvements today would be £7500. The market rent, including the value of the improvements, is £10000 per annum, the RV of the property is £12000 and yields for investments in this type of property average 8%. Value the landlord's interest in the property assuming:

a) The tenant vacates on at the end of the existing lease
b) A new 10-year lease with a rent review in year five (with a clause that states that the value of improvements is disregarded) is granted to the existing tenant on expiry of the current lease
c) The landlord repossesses the property at the end of the existing lease for own occupation
d) The landlord repossesses the property at the end of the existing lease for redevelopment and the site value is estimated to be £100000

The tenant has the right to two types of compensation if required to vacate the premises at the expiry of the existing lease:

1) Disturbance compensation at twice the current RV of the premises. This equates to £24000.
2) Improvements compensation at the lesser of the cost of the works or value added. The cost (as at the valuation date) is £7500 and the value added is calculated as the capital value of the increase in rent resulting from the improvements:

Increase in market rent (£ p.a.)[a]	1000	
YP perpetuity @ 8%	12.5	
Capital value of improvements (£)		12500

[a] 20% of the £5000 contract rent.

Cost therefore prevails as improvements compensation.

But these are future liabilities of the landlord, and it is important to consider possible changes in the amounts due to a rating revaluation or inflation in building costs for example. Here it is assumed that the RV remains constant and building costs rise at 3% per annum.

a) Valuation assuming the tenant vacates on termination:

Term (contract) rent (£ p.a.)	5000	
YP 5 years @ 8%	3.9927	
		19964
Reversion to market rent (£ p.a.)	10000	
YP in perpetuity @ 8%	12.5000	
PV £1 5 years @ 8%	0.6806	
		85075

Cost of improvements (£ p.a.)	(7500)
inflated over 5 years @ 3% p.a.	1.1593
	(8695)
PV £1 5 years @ 8%	0.6806
	(5918)
Valuation before PCs (£)	99 121

b)   Valuation assuming a new lease is granted at end of lease:
It is helpful to sketch a timeline and mark important dates as in Figure 9.1. The tenant benefits from a rent reduction for 20 years that reflects the value added by the improvements, after which the rent reverts to the market rent *including* the value added by the improvements.

Capital value of first 5 years' rent (as above) (£)		19 964
Subsequent 15 years rent (£ p.a.)[a]	8000	
YP 15 yrs. @ 8%	8.5595	
PV 5 yrs. @ 8%	0.6806	
		46 605
Final reversion market rent (£ p.a.)	10 000	
YP in perpetuity @ 8%	12.5000	
PV £1 20 years @ 8%	0.2145	
		26 813
Valuation before PCs (£)		93 382

[a] This is the market rent of £10 000 less 20% to disregard the value added by tenant's improvements.

c)   Valuation assuming the landlord repossesses at the end of current lease for own occupation:

Value (as [a]) (£)		99 121
Less improvements (as [a]) (£)	(6199)	
Less disturbance; 2 x RV (£)	(24 000)	
	(30 199)	
PV £1 5 years @ 8%	0.6806	
		(20 553)
Valuation before PCs (£)		78 569

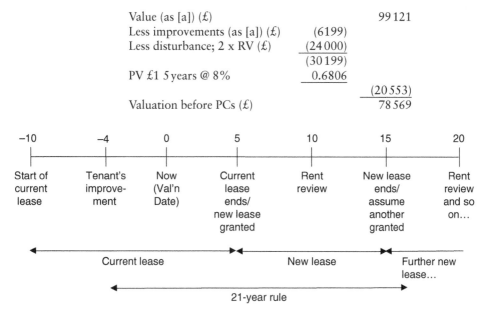

Figure 9.1   Events timeline.

d) Valuation assuming the landlord repossesses at end of existing lease for redevelopment:

The landlord must have owned the property for at least five years to regain the property at the end of the lease. There is no compensation for improvements because their value to the landlord will be zero in the case of redevelopment.

Term (contract) rent (£ p.a.)	5000	
YP 5 years @ 8%	3.9927	
		19 964
Reversion to site value (£ p.a.)	100 000	
Less disturbance: 2 x RV (£)	(24 000)	
	76 000	
PV £1 5 years @ 8%	0.6806	
		51 726
Valuation before PCs (£)		71 690

9.4.2.2   Example 2

The tenant of a shop holds a 15-year internal repairing (IR) lease granted 11 years ago at a current rent of £24 000 per annum. Six years ago, the tenant obtained consent to carry out improvements costing £60 000. The current freehold yield is 6% and the market rent on FRI terms is £50 000 per annum, £5000 of which can be attributed to the improvements made by the tenant. The RV of the premises is £50 000 and building cost inflation is averaging 10% per annum. Value the current interests of the landlord and tenant assuming:

a) The landlord will get permission for his own occupation at the end of the lease
b) The tenant will continue in occupation under a new lease with a typical rent-review pattern

As in the previous example, disturbance compensation is twice the RV, producing a figure of £100 000. Compensation for improvements is estimated as the lesser of the cost of or value added by the improvements:

Value added by improvements (£ p.a.)	5000	
YP perpetuity @ 6%	16.6667	
		83 333
Cost of improvements (£)	60 000	
Inflated at 10% pa over 6 years	1.7716	
		106 296

The value added produced the lower figure in this case.

a) Valuation assuming the landlord gets permission for their own occupation at the end of the lease:

Valuation of the landlord's interest:

Term (contract) rent (£ p.a.)	24 000	
Less external repairs @ 10% of market rent on FRI terms (£)	(5000)	
Less insurance @ 2% of market rent on FRI terms (£)	(1000)	
Net income (£ p.a.)	18 000	
YP 4 years @ 6%	3.4651	
		62 372
Reversion to market rent on FRI terms (£ p.a.)	50 000	
YP perpetuity @ 6%	16.6667	
PV £1 4 years @ 6%	0.7921	
		660 085
Less disturbance compensation (£)	(100 000)	
Less improvements compensation (£)	(83 333)	
	(183 333)	
PV £1 4 years @ 6%	0.7921	
		(145 218)
Valuation before PCs (£)		577 239

Valuation of the tenant's interest:

Market rent on FRI terms (£ p.a.)	50 000	
External repairs (£ p.a.)	5000	
Insurance (£ p.a.)	1000	
Market rent on IR terms (£ p.a.)	56 000	
Rent paid (£ p.a.)	(24 000)	
Profit rent (£ p.a.)	32 000	
YP 4 years @ 8%[a]	3.3121	
		105 987
Compensation (as above) (£)		145 218
Valuation before PCs (£)		251 205

[a] Risky, terminable, non-growth investment.

b) Valuation assuming the tenant will continue in occupation under a new lease with a typical rent-review pattern. Figure 9.2 illustrates the timeline.

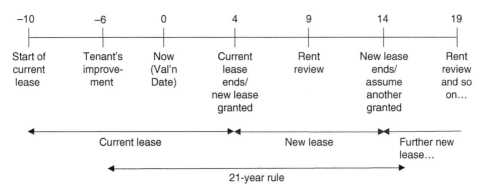

Figure 9.2   Events timeline.

Valuation of the landlord's interest:

Term net income (as above) (£ p.a.)	18 000	
YP 4 years @ 6%	3.4651	
		62 372
Reversion to MR on IR terms, excluding improvements [a]		
MR on FRI terms, excluding improvements (£ p.a.)	45 000	
External repairs (as above)	(5000)	
Insurance (as above)	(1000)	
	39 000	
YP 15 years @ 6%[b]	9.7122	
PV £1 4 years @ 6%	0.7921	
		300 028
Reversion to MR on IR terms, including improvements		
MR on FRI terms, including improvements (£ p.a.)	50 000	
External repairs (as above)	(5000)	
Insurance (as above)	(1000)	
	44 000	
YP perpetuity @ 6%	16.6667	
PV £1 for 19 years @ 6%	0.3305	
		242 367
Valuation before PCs (£)		604 767

[a] Under the 1954 Landlord & Tenant Act, the terms of the new lease will be based on the existing lease.
[b] This yield may be reduced below the freehold yield to reflect security afforded to a tenant occupying on IR terms, but the unattractiveness of an investment returning a non-market rent for 15 years may counter this. Consequently, the yield remains at 6%.

Valuation of the tenant's interest:

Profit rent (as above) (£ p.a.)	32 000	
YP 4 years @ 8%	3.3121	
		105 987
Reversion to profit rent equals to the increase in	5000	
MR made by improvements at lease renewal (£ p.a.)		
YP 15 years @ 7%[a]	9.1079	
		45 540
Valuation before PCs (£)		151 527

[a] Growth potential due to possible rent reviews in sub-lease, so yield is based on freehold yield plus leasehold risk premium.

9.4.3 Over-rented properties

Over-renting occurs when the rent payable under a lease with upward-only rent reviews exceeds the market rent. Some valuers value *over-rented properties* as perpetual cash flows at the passing rent when the lease is long, providing there are upward-only rent reviews and there is no break clause. Because of the higher risk associated with the element of rent that exceeds the market rent, known as *over-age*, other valuers use a modified layer (core and top slice) approach. Here, a yield based on rack-rented freehold comparable evidence is used to capitalise the core rent (the market rent at the time of the valuation) and a usually higher yield that reflects risk associated with the receipt of this tranche of rent (but tempered by the covenant strength of the tenant) to capitalise the overage.

For example, a property was let four years ago at a rent of £220 000 per annum on a 15-year lease with five-year upward-only rent reviews. The current market rent is £200 000 per annum. Comparable properties have recently sold for yields averaging 6%. The valuation is as follows:

Core (market) rent (£ p.a.)	200 000	
YP in perpetuity @ 6%	16.6667	
		3 333 340
Overage (£ p.a.)	20 000	
YP 11 years @ 7%[a]	7.4987	
		149 974
Valuation before PCs (£)		3 483 314

[a] Increased to reflect additional risk.

No attempt has been made to estimate the length of time that the property will remain over-rented for. In the example, the overage has been capitalised for the remainder of the lease, but the market rent grows and the market rent may overtake the contract rent before the end of the lease. As a result, part of the overage is capitalised twice, and the property will be over-valued. This is illustrated in Figure 9.3.

One way to resolve this problem is to project the market rent at a growth rate to determine when it will overtake the contract rent. Continuing the example,

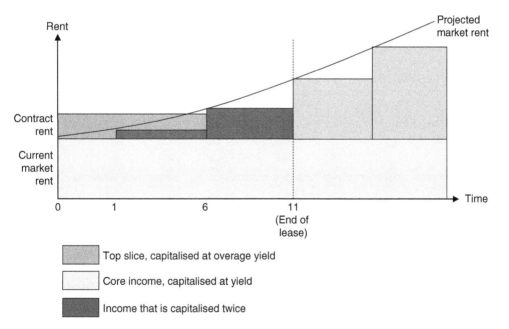

Figure 9.3   Over-rented property.

assume comparable evidence suggests a rental growth rate of 2% per annum. The market rent will grow to the following amounts at the next two rent reviews:

$$\pounds 200\,000 \times (1+0.02)^1 = \pounds 204\,000$$
$$\pounds 200\,000 \times (1+0.02)^6 = \pounds 225\,232$$

The market rent is predicted to overtake the contract rent between the first and second rent reviews. So, the valuation can be reworked with a shorter overage period. The valuation is lower because the double counting has been reduced.

Core (market) rent (£ p.a.)	200 000	
YP in perpetuity @ 6%	16.6667	
		3 333 340
Overage (£ p.a.)	20 000	
YP 6 years @ 7%	4.7665	
		95 330
Valuation before PCs (£)		3 428 670

9.4.4 Turnover leases

Turnover leases allow landlords to participate in the underlying potential profitability of the tenant's business by basing some or maybe all of the rent on the tenant's turnover. In the United Kingdom, these are popular in the case of shop units in shopping centres, airports and other transport termini, and sometimes found in high street retail and petrol stations. Table 9.1 summarises the advantages and disadvantages of turnover leases from the landlord and tenant perspectives.

In a shopping centre, the income generated by a turnover rent structure is dependent upon the performance of the shopping centre and on the success of individual retailers. Tenant mix is important, and the provision of leisure facilities can increase retail trade, as can public areas and food courts. Anchor tenants may be subject to beneficial turnover percentages to reflect their contribution to the success of the centre (Sayce et al. 2006).

Regarding individual retailers, the most common turnover rent arrangement is a base rent plus a turnover rent. The base rent is often a percentage of the market rent of the property, say 75–80%, and either subject to five-yearly rent reviews or indexed to inflation annually. If the base rent is not set as a percentage of market rent, it may be reviewed annually, on an upward-only basis, to a % of the previous year's rent. The turnover rent is based on a percentage of the turnover of the business (usually calculated with reference to annual audited accounts). Turnover rent, which is usually paid monthly in arrears, can be specified in various ways: percentage of gross sales, receipts, turnover above base rent. Until the set turnover level is achieved, the tenant pays base rent only. Alternatively, turnover rent can be calculated as a set percentage of gross receipts/sales/turnover without any minimum base rent. This is considered higher risk to the landlord.

Comparable evidence helps determine the base rent and select the appropriate percentage for turnover. The latter is determined by sales volume and profit

Table 9.1 Advantages and disadvantages of turnover leases.

Advantages for landlord	Disadvantages for landlord	Advantages for tenant	Disadvantages for tenant
Annually reviewable income	Uncertainty over income	Lower risk by tying rent to trade	Contracting out of LTA54 generally appropriate
Knowledge of trading activity of all tenants in a shopping centre and can actively manage the centre to optimise turnover	Difficult to obtain turnover information	Smaller jumps at rent review as only affects base rent	Must disclose turnover information, which may fall into competitors' hands or be used by landlord in lease or rent negotiations for shops in other centres
Rent reviews in base rent could be less contentious	Growing multi-channel shopping complicates turnover leases. How should sales that start online and close in-store be accounted for? How to determine whether a sale was driven by online or instore presence? Also, returned goods problem.	Might benefit from landlord's active management by reducing the number of competitors and in terms of the retail quality of the centre	User clauses are usually absolute and detailed.
Can be employed to attract smaller tenants to a shopping centre, which improves tenant mix	Can be hard to obtain institutional funding as base rent is the only guaranteed income	Need to get the turnover % right, because if profit falls then the proportion paid as turnover rent rises (could make the turnover % reviewable, but this is rare, or make the lease short)	Lease usually includes covenants to trade, for example open during normal hours.
Information on the performance of the centre as a whole may indicate optimum time to refurbish	Valuations may be lower if valuers place risky rates on the turnover rent	Landlord usually controls alienation to ensure tenant mix, so a less-alienable interest for the tenant	Performance clauses are common: minimum turnover, usually linked to inflation and usually linked to provisions that allow the landlord to determine the lease.
For pre-lets, the landlord does not miss out on growth between date of pre-let and opening of scheme, to the extent that growth is reflected in increased turnover	Tenant may exploit trading policies to reduce turnover rent liability	Improvements less easily disregarded from turnover rent although tenant's improvements to shops on turnover leases are rare (inside centres)	
	Management costs likely to be higher than for rack-rented properties	Encourages landlord to be proactive in assisting tenants to increase turnover, especially in the case of a shopping centre	

margins obtainable from different trades. The higher the sales volume or lower the profit margin, the lower the percentage, and vice versa. For example, a jeweller might be 10–11%, shoe shops and ladies' fashion 8–10%, men's fashion 8–9%, sports 9%, electrical and catering 7%, grocer 6%, baker 5% and butcher 4–5%. The percentage can depend on covenant strength as much as trade type. It is also determined by the level of base rent; the lower the base rent, the higher the percentage applied to turnover. In the case of airports, turnover percentages are high, and a base rent may not be paid. Food sales from supermarkets trade on large volumes and narrow profit margins whereas jewellery for example is very much the opposite.

A turnover rent is usually derived from a percentage of turnover net of VAT, staff discounts, returned goods, goods traded in, defective goods, charges made by credit card companies and bad debts. The percentage applied to turnover is usually fixed for the term of the lease, but there may be provisions for variations to take account of changes in use, occupation or longer term changes in retailing practice and profitability. Turnover lease terms can be complex, requiring a minimum trade performance level, notional turnover if closed for several days and restrictions that only allow assignments to similar trades for example. A clause may be inserted into the lease allowing the landlord to terminate the lease contract if a certain level of turnover is not attained during a specified period. The tenant will normally try to cap the turnover rent at say 120% of the market rent, and the ability to reduce this to a lower percentage will depend on the covenant strength of the tenant.

McAllister (1996) found that the most common type of turnover lease in the United Kingdom is where the tenant must pay either a market rent or a turnover rent, whichever is highest. A stepped base rent plus a turnover rent is where the base rent increases annually to levels specified in the lease. When estimating the capital value of a property subject to a turnover rent, the yield used to capitalise the turnover rent may be higher than that used to capitalise the base rent because, it is argued, it will vary annually and perhaps quite markedly. It is difficult to accurately predict turnover so capitalisation is usually of current turnover with an assumption that it will continue. The use of a higher yield on the turnover rent will reduce the capital value in comparison to a rack-rented property. This is one reason why base rents account for 75–80% of the total rent and why pure turnover rents are rare. Investment value could be enhanced by providing for reversion to market rent at some point in the future, perhaps at the first rent review.

An example of a capital valuation of a shop subject to a turnover rent appears below. The base rent is 80% of the market rent for this type of property and the turnover rent is calculated as 5% of net turnover.

Base rent @ 80% MR (£ p.a.)	80 000	
YP perpetuity @ 8%	12.5	
		1 000 000
Turnover rent @ 5% turnover (£ p.a.)	20 000	
YP perpetuity @ 10%	10	
		200 000
Valuation before PCs (£)		1 200 000

9.4.5 Long lease investments

The conventional form of commercial property investments involves renting space on leases of 5–15 years' duration, but space can be let longer leases. Examples of long-lease investments can be found in the office market offering returns that are more like index-linked bonds than equities. In other words, they offer investors less above-inflation growth potential but more security and less volatility. The level of return is usually slightly higher than index-linked gilts or corporate bonds of equivalent credit rating due to the illiquidity of the investment (long lease contracts) and the relatively specialist and opaque nature of the market. Investment value will be influenced by the quality of the occupying tenant, the property and the lease terms. Occupiers may consider a long lease arrangement as a means of raising funds, while potentially benefiting from more predictable rent increases, depending on the negotiated lease terms. Three types of lease arrangement are described below (M&G Investments 2018).

Sale and leaseback: An owner occupier sells their freehold interest to an investor, who then leases the property back on a long lease. The lease might provide for say annual rent increases in line with inflation or some agreed percentage uplift. Depending on the length of the lease, most of the value of the investment is likely to come from the rental cash flow rather than the reversionary value of the property at the end of the lease. It is also worth the investor keeping an eye on alternative use (perhaps redevelopment) value, which can be realised at the end of the lease or sooner in the case of tenant default. The rent is usually lower than on a standard commercial lease, thus lowering occupancy costs during the lease as well as providing a capital sum at the point of sale.

Income strip: For occupiers who prefer to retain the freehold interest in their property, income strips refer to the sale of the cash flow generated by the long lease only. It is, therefore, the present value of the future rental income agreed under a long lease. Often the arrangement transfers ownership of the freehold interest during the lease term and then the occupier has the option to buy it back for a nominal sum at the end of the lease. This gives the investor some security in case of tenant default. As the lease term progresses and the freehold transfer option nears, the occupying tenant is less likely to default.

Ground rent: Here the owner occupier sells their freehold interest to an investor, who then leases the property back on a very long lease at a low rent, often with regular uplifts in line with inflation. According to M&G Investments (2018), ground rents on commercial real estate are typically set at 10–20% of operational earnings (cf. 50–70% in typical 25-year sale and leaseback arrangement). In terms of value, the vast majority is the present value of the rental cash flow, given how far away the unencumbered reversion is going to be. The low rent and long lease term mean the price of ground rent investments is generally low, maybe 50% of vacant possession value. This means the investor has good asset cover, while the occupier has good rent cover.

9.4.6 Synergistic value

Synergistic value is the result of a combination of two or more interests where the combined value is more than the sum of the separate values. Break-up value is the opposite of synergistic value and refers to the division of property interests, leading to the value of the resultant separate interests being greater than the whole. Synergistic value might arise when there is a merger of *physical* interests, perhaps two adjacent land parcels in a location earmarked for development. Another example would be two adjacent shop units, both with narrow frontages. If they were combined, they could form a single standard-sized shop unit. The value of each shop in its existing state is say £200 000 but if combined the merged value would be £500 000, giving a marriage value of £100 000. All other things equal you would expect half of this gain to go to each shop owner, assuming they are in the same negotiating position, and neither can hold the other to 'ransom'.

Alternatively, synergistic value might arise when there is a merger of *legal* interests, perhaps superior and subordinate tenure rights in the same property. For example, the freeholder of commercial development land let it to a head-lessee on a 125-year ground lease, which has 24 years remaining at a ground rent of £5000 per annum with no provision for rent reviews. The head-lessee developed the site as offices and sub-let on a typical FRI occupational lease with five-year rent reviews. The current market rent for the offices is £500 000 per annum and a rent review has just taken place. With just 24 years remaining, the head-tenant is considering purchasing the freehold interest and wishes to know how much should be offered, assuming a freehold yield of 6%, a leasehold yield of 8% and a ground lease yield of 10%. The valuation of the freehold interest is:

Term rent (£ p.a.)	5000	
YP 24 years @ 10%	8.9847	
		44 924
Reversion to MR (£ p.a.)	500 000	
YP perpetuity @ 6%	16.6667	
PV £1 24 years @ 6%	0.2470	
		2 058 337
Valuation before PCs (£)		2 103 261

The valuation of the head-leasehold interest is:

Market rent (£ p.a.)	500 000	
Less ground rent (£ p.a.)	(5000)	
Profit rent (£ p.a.)	495 000	
YP 24 years @ 8%	10.5288	
Valuation before PCs (£)		5 211 756

The sum of the separate freehold and head-leasehold interests is: £2 013 261 + £5 211 756 = £7 315 017. If the two legal interests were combined, the valuation of the freehold in possession would be:

Market rent (£ p.a.)	500 000	
YP perpetuity @ 6%	16.6667	
Valuation before PCs (£)		8 333 350

The synergistic value would be: £8 333 350 – £7 315 017 = £1 018 333.

To purchase the freehold interest, the head-tenant could offer an amount equating to the existing value of freehold interest (£2 103 261) plus some proportion of the synergistic value. A simple 50:50 split is one solution, but it might be more equitable to split it in proportion to the value of the original separate interests. So, the freehold proportion of the synergistic value would be:

£2 103 261/(£7 315 017 + £1 018 333) = £292 798 or 29%. This leaves 71% or £725 535 for the head-tenant.

One further example demonstrates how the merger of interests can become quite complex. 47 years ago, the owner of a property let it on a 60-year ground lease at a fixed rent of £40 000 per annum. The long leaseholder sublets the property, and the current 10-year sub-lease was agreed two years ago at a rent of £155 000 per annum. The market rent for the property is estimated to be £160 000 per annum.

Valuation of the freehold with vacant possession assuming a yield of 5.5% (from comparable evidence):

Market rent (£ p.a.)	160 000	
YP in perpetuity at 5.5%	18.1818	
Valuation before PCs (£)		2 909 091

Valuation of the freehold subject to head lease, assuming a yield of 6% (adjusted upwards from 5.5% to reflect the initial 13 years of low rent without growth):

Term rent (£ p.a.)	40 000	
YP 13 years @ 6%	8.8527	
		354 107
Reversion to market rent (£ p.a.)	160 000	
YP in perpetuity at 6%	16.6667	
PV £1 13 years @ 6%	0.4688	
		1 250 237
Valuation before PCs (£)		1 604 345

Valuation of the long leasehold interest, assuming a yield of 7% (based on freehold yields but adjusted upwards due to the terminable nature of this leasehold investment):

Term rent received (£ p.a.)	155 000	
Rent paid (£ p.a.)	(40 000)	
Profit rent (£ p.a.)	115 000	
YP 3 years @ 7%	2.6243	
		301 796
Reversion to market rent (£ p.a.)	160 000	
YP 10 years @ 7%	7.0236	
PV £1 3 years @ 7%	0.8163	
		688 000
Valuation before PCs (£)		989 797

Valuation of occupational lease, assuming a yield of 8% (based on freehold yields but adjusted upwards to reflect the very short terminable nature of this leasehold investment):

Market rent (£ p.a.)	160 000	
Rent paid (£ p.a.)	(155 000)	
Profit rent (£ p.a.)	5 000	
YP 3 years @ 8%	2.5771	
Valuation before PCs (£)		12 885

Value of all three interests in the property: £1 604 345 + £989 797 + £12 885 = £ 2 607 027.

Therefore, the synergistic value is £2 909 091 – £2 607 027 = £302 064. To incentivise the tenants to vacate, the owner is likely to have to pay each a proportion of the synergistic value in addition to the value of their interests. As in the previous example, an equitable way of doing this is to split it in proportion to the values of their individual interests.

9.5   Investment valuation of investment property

An investment valuation is usually performed by determining the risk and return characteristics associated with holding the property and often includes a valuation, forecasting of key variables and some form of performance analysis. For an investor, the future income stream, quality of the tenant and property are important. For an occupier, the cost of the property as a factor of production or its contribution to profit as well as its future sale price or write-off cost will need to be considered. An investment valuation can be undertaken for different clients for different reasons. For example, a pension fund may need to know how an asset might contribute to portfolio performance whereas a property company might be more interested in a building's redevelopment potential. An occupier will evaluate the business requirements and the cost of debt and equity capital among other things. The aspirations and therefore the appraisal assumptions (such as discount rate, holding period and so on) will undoubtedly vary to some degree. Having said this, groups of similar types of investors and occupiers will behave in a similar way, such as institutional investors, and so certain assumptions can be made.

It is therefore normal for a range of investment valuations to exist for a property but only one exchange price. Differences between price or value of a property and its worth to an individual emerge because of different perceptions about either the utility to a business or potential return to an investor that the property may offer. Perceptions may vary in terms of how utility or return will vary over time (its volatility) and how long that utility or return will last for. So, market value and investment value can be different and provide evidence of mispricing from the perspective of certain decision-makers. This leads on to the debate concerning market efficiency and the fact that the property market offers opportunities for buyers to exploit pricing inefficiencies, mainly due to informational gaps and inaccuracies. Also, market correction is likely to be slower than is the case for other more liquid investment markets such as equities and bonds.

Ball et al. (1998) claim that 'the influence of valuations on price and the focus on price estimation, rather than worth, can lead to systematic mispricing'. What is being suggested is that because property is a thinly traded and heterogeneous investment asset or factor of production, valuers are not only *interpreting* market prices when attempting to estimate the market value of a property but also *influencing* them. Not only that, but valuation methods may also erroneously focus on market pricing rather than investment valuation, looking too much at market price data rather than the fundamental requirements of clients. These criticisms are harsh but not unfounded: conventional valuation techniques are increasingly being supplemented and, in some cases, replaced by contemporary approaches that place more reliance on client fundamentals than market signals. But it should be remembered that valuation must always have interpretation of market activity at its heart and market-generated price signals will always provide a very reliable source of intelligence. It is important to distinguish, then, between market value and investment value, which may coincide if the buyer's decision criteria are typical of other buyers in the market.

A holder of a property would periodically compare its investment value to its market value. This helps to decide whether to hold, refurbish, redevelop or dispose of the property. Investment valuations are also required to help choose between different investment opportunities, to assess the viability of redevelopment or refurbishment projects and as a decision tool for financing arrangements.

9.5.1   Inputs and assumptions

Investment valuation involves making explicit judgements (based on evidence) about depreciation, risk, expenditure, exit value, any rental growth, taxation, financing and all costs. Information needs range from the property-specific to the macro-economic. Table 9.2 is an attempt to classify the information typically sought prior to conducting an appraisal of worth.

Many of these factors can be grouped together and handled by adjusting either the cash flow or the target rate of return. It is important to focus most on those factors considered to affect value to the greatest extent. In an investment valuation these are rent, target rate of return, holding period and exit yield.

9.5.1.1   Rent and rental growth

Forecasts of market rents and rental growth are available and typically relate to prime business space in the locality concerned because the thorny issue of how rents may depreciate as premises age can be avoided. Forecasts of rents are normally undertaken in real terms and then inflation adjusted to give nominal rental value change. These forecasts are produced at a national, regional or local level and are usually based on econometric models of the economy and the property market. Forecasts of rent and rental growth at the town level may be misleading if they are applied at individual property level.

There is a clear need for appraisal to allow for such items as obsolescence and deterioration, but particular care is needed when considering how these phenomena affect value. There have been several empirical studies of the impact of depreciation on rental growth and these are summarised in Crosby et al. (2012). Using a longitudinal approach, Crosby et al. (2012) found that standard retail units depreciated at 0.3% per annum on average over the period 1993–2009, offices at 0.8% per annum and industrial premises at 0.5% per annum. These rates are net of annual capital expenditure rates of 0.3%, 0.5% and 0.2% per annum. The authors note considerable variation in depreciation rates for retail units when they are categorised by type.

It is important to ensure that double counting does not occur. For example, if refurbishment expenditure is included in the cash flow then any enhanced value

Table 9.2 Investment valuation information requirements.

Information	Example
Economic indicators	Economic output and productivity Employment and unemployment statistics Movements in corporate profits (by sector), money supply, public sector borrowing, inflation and interest rates
Market indicators	Current market rents Rental growth and depreciation rates Future redevelopment or refurbishment costs Current yields and forecasts of exit yields Purchase and sale costs Movements in market indices
Portfolio information	Asset returns and correlations (to aid diversification) Sales and purchases Risk indicators
Property information	Physical attributes (areas, building specification, quality, improvements, ancillary space, parking, access to public transport) Financial details (yield, rent passing, rental growth, market rent, capital value) Legal terms (title and lease details, number of tenants, expiry dates, review dates and terms, break clauses, voids, future leases) Outgoings and capital expenditure (vacancies, voids, unrecoverable service and management costs, letting, re-letting and rent review costs, purchase and sale costs) Depreciation, costs and timing of redevelopment and refurbishment, cost inflation Planning Taxation (income and capital gains taxes, business rates, VAT, capital allowances) Occupancy/holding costs (management, review, purchase, sale costs) Dilapidations, service charge and other payments for repairs and insurance if leasehold
Client-specific information	Target rate of return (discount rate) Tax position Holding period Loan facilities Risk profile

should be reflected either in the estimated rental value, the rental growth rate or the exit yield.

It is important to consider the potential impact of gaps or voids on the receipt of rent particularly as lease lengths shorten and break clauses become more prevalent. Of key concern is the likelihood that a tenant operates a break clause or vacates the premises at the end of a lease. Other matters then follow including the costs of holding a vacant property, the length of time to re-let and any works that need to be done to enable a new letting.

9.5.1.2   Target rate of return

Target (or hurdle) rates are widely used in investment decision-making (IPF 2015). The target rate of return from an investment should compensate an investor for the risk taken. It is typically derived by adding a risk premium to a 'benchmark' risk-free rate of return such as income yields on medium/long dated (15–25 year) gilts. The rationale for basing the risk-free rate on this benchmark was because the term coincided with typical lease lengths. As lease lengths shorten, it may be more appropriate to base this risk-free rate on short-dated gilts or 5–10-year swap rates.

A risk premium is added to the risk-free rate to compensate for holding a property asset. This risk premium is difficult to estimate for property as each asset is unique. Investment characteristics that are best handled by adjusting the target rate are generally market-related and include liquidity, rental growth prospects, possible yield movements and depreciation. Property-related risks include the quality of tenant, potential for letting voids, cost of ownership and management and lease structure. The financial impact of these factors can be built into the cash flow. But determining a risk premium for each factor is difficult given paucity of data, complexity of the market and confidentiality of client data. Also, the significant overlapping influence of these risk factors complicates this sort of analysis. Consequently, rather than attempt to derive risk premiums for individual property assets, it may be more helpful to group similar types of property to determine a property risk premium for each group. A market risk premium can then be adjusted up or down to reflect risk associated with the sub-sector being analysed. So, for example:

Risk-free rate
+ market risks (sub-sector risk of market failure, such as illiquidity, poor rent or yield performance, allowance for sub-sector depreciation)
+ property risks (including property-specific risks such as adjustments for tenant quality, and grouped property risks such as adjustments for sub-sector lease structures)
= Risk-adjusted discount rate
Remaining costs (fees, management, dilapidation, etc.) are incorporated in the cash-flow

This 'risk-adjusted discount rate' approach to deriving a target rate of return is frequently used by property analysts and investors but according to Sayce et al. (2006) there are two main limitations. First, only one discount rate is applied to all cash flows, and it therefore fails to distinguish those parts of the cash flow that are risky and those that are not. For example, rental income return might be regarded as secure whereas capital return might be more volatile over the holding period. It is possible to discount different parts of cash flow at different rates using a 'sliced income approach' (Baum and Crosby 1995) or an arbitrage approach (French and Ward 1995), but such methods are not frequently used in property investment appraisal. In property valuations, a core and top-slice approach is used when the risk profile of a rent changes significantly at some future date. The second limitation is that the target rate heavily discounts distant cash flows regardless of whether they are riskier. It is unlikely that the growth in risk is going to be at the same exponential rate as the growth inherent in the risk premium. Furthermore, cash flow after a refurbishment or redevelopment programme is likely to be more uncertain.

9.5.1.3   Holding period

The holding period is normally specified by the investor. As a rule of thumb, large institutional investors might have longer holding periods of say 10–15 years, whereas niche investors and investor–developers may be more interested in the capital growth opportunities afforded by redevelopment potential than long-term income growth. Consequently, their holding periods may be much shorter, say three to five years. The duration of the holding period can also be influenced by lease terms, such as the dates of any break clauses and lease expiry, or by the physical nature of property itself, particularly depreciation factors and redevelopment potential. A longer holding period will mean that it is more difficult to predict the values of key variables in the medium to long term (a problem that is usually hidden by using an exit yield or exit value at the end of a shorter holding period). So, a long holding period is associated with greater risk of fluctuation from predictions of long-term trends and a greater chance of error in selecting exit variables. An additional consideration is whether the market is assumed to be stable over the holding period.

9.5.1.4   Exit value

Exit value refers to the value of the property at the end of the holding period. The usual method of calculating an exit value is to capitalise the forecast rent at the end of the holding period, although the exit value may reflect land values if demolition is anticipated. In selecting an appropriate exit yield at which to capitalise the rent, the important consideration is what yield a purchaser would require for the property at that point of (notional) sale.

The exit yield is usually based on initial and equivalent yields derived by comparison with similar investments. It is important to consider the impact of depreciation, but care should be taken so as not to double-count the effect on value by, say, reducing the forecast rent and raising the exit yield.

The choice of exit yield is central to the investment valuation when the holding period is less than 20 years as the resulting exit value forms a substantial element of the overall value of the investment.

9.5.2   Investment valuation using a discounted cash flow

A discounted cash flow (DCF) is a summation of the present values of all revenue, including rent, premiums and sale price, and expenditure, such as the purchase price and any periodic expenditure. The present value of a future sum, whether it is revenue or expenditure, is dependent on the discount rate and the length of time over which it is discounted: the higher the discount rate and/or the longer the discount period, the lower the present value. A key advantage of a DCF over income capitalisation methods is that the cash flow can be adjusted in each period to account for changes in inflation, rental growth, voids, lease renewals and so on.

The application of DCF to market valuation was described in Chapter 7. Because DCF can be expanded to incorporate explicit assumptions about rental growth, holding period, depreciation, refurbishment, redevelopment, management and transfer costs, and financing costs, it is also used for investment valuation. In fact, DCF techniques are often used by investors to analyse asking prices and market valuations of investments.

A DCF can be used to calculate a net present value (NPV) or an internal rate of return (IRR), and corporate finance textbooks provide detailed explanations of these two approaches to investment appraisal. The focus here is on NPV because that is a capital value which can be compared to market values and prices in the property investment market.

An NPV is a capital sum that is the sum of the present values of a known or projected cash flow over a holding period. The present values are calculated by discounting the cash flow at an appropriate rate, which is usually the target rate of return of the investor. Earlier income is deemed more valuable as the effect of discounting diminishes the value of more distant cash flow. Mathematically, NPV takes the form of a geometric progression:

$$NPV = \frac{1}{(1+r)} + \frac{1}{(1+r)^2} + \frac{1}{(1+r)^3} \cdots \frac{1}{(1+r)^n} = \sum_{i=1}^{n}\frac{1}{(1+r)^i} \qquad (9.2)$$

And, for any other income, A:

$$NPV = \sum_{i=1}^{n}\frac{A}{(1+r)^i} \qquad (9.3)$$

9.5.2.1   Single tenancy property investments

A rack-rented freehold property investment opportunity is on the market for £100000. An investment valuation is required to determine whether this opportunity is one that your client, who has a target rate of 16%, should pursue. The rent is £12000 per annum, rent reviews are every five years, the assumed holding period is 20 years, over which time you expect rent to grow at an average rate of 5% per annum. At the end of the holding period, assume a sale at an exit yield of 11%.

Period	Income (£)	Net cash flow (£)	Growth rate	Real cash flow (£)	YP 5 years @ target rate	PV £1 @ target rate	Discounted income (£)
0–4	12000	12000	1.0000	12000	3.2743	1.0000	39292
5–9	12000	12000	1.2763	15315	3.2743	0.4761	23876
10–14	12000	12000	1.6289	19547	3.2743	0.2267	14508
15–19	12000	12000	2.0789	24947	3.2743	0.1079	8816
20-Perp	12000	12000	2.6533	31840	9.0909[a]	0.0514	14874
NPV (£)							101365

[a] YP perpetuity at exit yield of 11%.

The cash flows from this investment have been concatenated into five-yearly income blocks because the annual rental income between each rent review is identical. The exit yield may well be higher than current initial yields because the property will be 20 years older, so it is important to use comparable evidence of similar but 20-year older properties than the subject property. Also, the rate of rental growth will probably decline, become static or even negative, so a spreadsheet can be used to model various outcomes.

Consider another example but this time a reversionary freehold property investment. An investor is thinking of purchasing the freehold interest in an office refurbishment opportunity. The current lease is on FRI terms, the rent passing is £100 000 per annum and the final review is in two years' time. The asking price is £1 200 000. The investor plans to hold the property until lease expiry, refurbish and then sell the freehold interest. The current cost of refurbishment is £1 000 000 and will take one year to complete. The current market rent of the property in its existing state is £120 000 per annum and £200 000 per annum when refurbished. The freehold yield after refurbishment is 7%. Rental growth for the existing property is estimated to be 4% per annum and for the refurbished property 7% per annum. Building cost inflation is averaging 6% per annum. Assuming the investor's target rate of return is 15%, calculate the investment value of the property. The DCF is year by year so the YP column is dispensed with.

Year	Description	Cash flow (£)	PV @ 15%	DCF (£)
1	Rental income	100 000	0.8696	86 957
2	Rental income	100 000	0.7561	75 614
3	Rental income	129 792[a]	0.6575	85 341
4	Rental income	129 792	0.5718	74 209
5	Rental income	129 792	0.4972	64 530
6	Rental income	129 792	0.4323	56 113
7	Rental income	129 792	0.3759	48 794
8	Sale proceeds	4 909 138[b]	0.3269	1 604 797
	Refurbishment costs	(1 593 800)[c]	0.3269	(521 013)
NPV (£)				1 575 343

[a] MR of £120 000 p.a. compounded over two years at 4% p.a. rental growth rate.
[b] £200 000 p.a. compounded over 8 years at 7% p.a. rental growth rate and capitalised at a yield of 7%.
[c] £1 000 000 building cost compounded over 8 years at 6% build cost inflation rate.

9.5.2.2 Multi-tenancy property investments

An office building consists of three tenancies. The ground floor is let on a lease that expires in 18 months' time, at which point it is estimated that £500 per square metre (net internal area) will need to be spent on refurbishment. Following this expenditure (which, it is assumed, is incurred in a single quarter), there is expected to be a void period. Costs will be incurred during the void period plus, after three months, property tax. Following the void period, a re-letting is assumed subject to a rent-free period. The lease of the first floor expires in four years' time after which, once again, a refurbishment is assumed lasting one quarter and costing £400 per square metre (net internal area). Following this, a void and a rent-free is assumed. The lease of the second to fifth floors runs for another seven years and nine months, subject to a rent review in two years and nine months' time, at which a 5% rent discount is applied to the market rent at the rent review. Assuming a five-year holding period, an exit yield of 5% and target rate of return in 10%, estimate the investment value of the property using a DCF. The letting fee is 10% of the agreed rent and purchase costs are 6.5% of the purchase price.

Tenancy schedule

	Net internal area (m²)	Rent passing (£ p.a.)	Market rent (£ p.a.)	Rent review (years from now)	Lease expiry (years from now)
Ground floor	350	52 500	59 500	–	1.5
First floor	500	65 000	75 000	–	4
Second to fifth floors	2000	260 000	300 000	2.75	7.75

Void assumptions at lease expiry

	Void (quarters)	Rent-free (quarters)	Void costs (£/m² p.a.)	Void taxes (£/m² p.a.)	Refurb cost (£/m²)
Ground floor	4	4	60	90	500
First floor	2	2	60	90	400
Second to fifth floors	2	2	60	90	400

Years		1				2				3				4				5			Exit
Quarters	0	1	2	3	4	5	6	7	8	9	10	11	12	13	14	15	16	17	18	19	20
Growth rates																					
Market Rent	–	0.5%	0.5%	0.5%	0.5%	1.0%	1.0%	1.0%	1.0%	1.0%	1.0%	1.0%	1.0%	2.0%	2.0%	2.0%	2.0%	2.0%	2.0%	2.0%	2.0%
Expenses and refurb costs	–	1.0%	1.0%	1.0%	1.0%	1.0%	1.0%	1.0%	1.0%	1.0%	1.0%	1.0%	1.0%	1.0%	1.0%	1.0%	1.0%	1.0%	1.0%	1.0%	1.0%
Index values																					
Market Rent	1.000	1.000	1.000	1.000	1.000	1.002	1.005	1.007	1.010	1.013	1.015	1.018	1.030	1.035	1.040	1.046	1.051	1.056	1.061	1.067	1.072
Expenses and refurb costs	1.000	1.002	1.005	1.007	1.010	1.013	1.015	1.018	1.020	1.023	1.025	1.028	1.030	1.033	1.035	1.038	1.041	1.043	1.046	1.048	1.051
Cash flow																					
Ground Floor rent	13125	13125	13125	13125	13125	13125	void	void	void	void	r/free	r/free	r/free	r/free	15099	15099	15099	15099	15099	15099	1267707
Refurb cost							–177632														
Void costs and taxes							–5329	–13356	–13389	–13422											
Letting fee											–6039										
First floor rent	16250	16250	16250	16250	16250	16250	16250	16250	16250	16250	16250	16250	16250	16250	16250	16250	void	void	r/free	r/free	1604824
Refurb cost																	–208121				
Void costs and taxes																	–7805	–19560			
Letting fee																			–7959		
Second to fifth floor rent	65000	65000	65000	65000	65000	65000	65000	65000	65000	65000	65000	72502	72502	72502	72502	72502	72502	72502	72502	72502	6352665
	94375	94375	94375	94375	94375	94375	–101710	67894	67861	67828	75211	88752	88752	88752	103850	103850	–128325	68040	79641	87600	9225196
NPV @ 7% p.a.																					7725209
Valuation (after PCs @ 6.50%)																					7253718

Calculation of exit values

Ground floor	Rent passing at exit (£ p.a.)	60 395		(15 099 × 4)
	YP 2.5 years @ 5%	2.2966		
			138 702	
	Reversion to MR (£ p.a.)	63 773		(59 500 × 1.072)
	YP perpetuity @ 5%	20.0000		
	PV £1 2.5 years @ 5%	0.8852		
			1 129 004	
	Exit value (£)		1 267 707	

First floor	Rent passing at exit (£ p.a.)	79 595		(75 000 × 1.061)
	YP 4.5 years @ 5%	3.9425		
			313 801	
	Reversion to MR (£ p.a.)	80 400		(75 000 × 1.170)
	YP perpetuity @ 5%	20.0000		
	PV £1 4.5 years @ 5%	0.8029		
			1 291 024	
	Exit value (£)		1 604 824	

Second to fifth floors	Rent passing at exit (£ p.a.)	290 008		(72 502 × 4)
	YP 2.75 years @ 5%	2.5112		
			728 275	
	Reversion to MR (£ p.a.)	321 600		(300 000 × 1.072)
	YP perpetuity @ 5%	20.0000		
	PV £1 2.75 years @ 5%	0.8744		
			5 624 390	
	Exit value (£)		6 352 665	

At exit:

Total of term rents passing (£ p.a.)	429 997
Total of reversionary market rents (£ p.a.)	465 773
The total exit value is (£)	9 225 196

For the second to fifth floors, lease end is 2.75 years away, so the possible void and rent-free costs are implied in the exit yield.

References

Ball, M., Lizieri, C., and MacGregor, B. (1998). *The economics of commercial property markets*. London, UK: Routledge.

Baum, A. (2003). *Pricing the options inherent in leased commercial property: a UK case study*. Stockholm, Sweden: European Real Estate Society conference.

Baum, A. and Crosby, N. (1995). *Property Investment Appraisal*, 2e. London: Routledge.

Crosby, N. and Murdoch, S. (1994). Capital value implications of rent-free periods. *J. Prop. Valuat. Invest.* 12 (2): 51–64.

Crosby, N., Devaney, S., and Law, V. (2012). Rental depreciation and capital expenditure in the UK commercial real estate market, 1993–2009. *J. Prop. Res.* 29 (3): 227–246.

French, N. (2001). Uncertainty in property valuation: the pricing of flexible leases. *J. Corp. Real Estate* 3 (1): 17–27.

French, N. and Ward, C. (1995). Valuation and arbitrage. *J. Prop. Res.* 2: 1–11.

Grover, R. (2014). Leasehold enfranchisement and graphs of relativity. *J. Prop. Invest. Finance* 32 (6): 642–652.

Havard, T. (2004). *Investment Property Valuation Today*. London, UK: Estates Gazette.

IPF (2015). *An investigation of hurdle rates in the real estate investment process*, Investment Property Forum.

M&G Investments (2018). *A guide to long lease real estate: long-dated secure income for institutional investors*.

McAllister, P. (1996). Turnover rents: comparative valuation issues. *J. Prop. Valuation Invest*. 14 (2): 6–23.

McAllister, P. (2001). Pricing short leases and break clauses using simulation methodology. *J. Prop. Invest. Financ*. 19 (4): 361–374.

Rose, J. (1979). *Tables of the Constant Rent*. The Technical Press.

Sayce, S., Smith, J., Cooper, R., and Venmore-Rowland, P. (2006). *Real Estate Appraisal: From Value to Worth*. Oxford: Blackwell Publishers.

Questions

Analysis of rents

Rental lease incentives

Rent-free periods

A landlord grants a 10-year lease at £300 000 per annum with a five-year upward-only rent review and a two-year rent-free period.

a) Calculate the effective rent using a yield of 8% over five years.
b) Calculate the effective rent using a yield of 8% over 10 years.
c) For (a) and (b), calculate the growth rate that would be required for the effective rent to reach the headline rent.
d) Repeat (a) and (b) but this time assume a three month for fitting-out period before the two-year rent-free incentive.

Stepped rents

[1]   Assuming a discount rate of 8%, which of the following two offers would you prefer? The rent will be payable by a large corporate. Offer 1 is the right to receive £100 000 for five years receivable annually in arrears. Offer 2 is a stepped rent as follows:

Year	Rent (£ p.a. receivable in arrears)
1	60 000
2	80 000
3	100 000
4	120 000
5	140 000

[2]   A shop has just been let on a 15-year lease with five-year rent reviews at a rent of £200 000 in year one, £225 000 in year two, £250 000 in year three, £275 000 in year four and £300 000 in year five. After year five, the rent reverts to the market level. Assuming a yield of 9%, calculate the effective rent of this property over the first five years.

[3]   You are letting a high street shop in a major provincial city and a tenant's agent has offered you three alternative rents for a five-year lease:

- Offer 1: £150 000 at the end of year one, £200 000 in year two, £250 000 in year three, £300 000 in year four and £400 000 in year five.
- Offer 2: £200 000 in year 1 with an annual increase of 10% of the previous year's rent.
- Offer 3: £250 000 fixed for the term of the lease.

Assuming a 10% yield, advise the landlord which offer is the best.

Capital lease incentives

Capital contributions

[1]   Analyse the following transaction to find the effective rent. The landlord grants a 10-year lease at £300 000 per annum with five-year upward-only rent reviews and pays a capital contribution of £450 000 to the tenant at the start of the lease. There is no rent-free period other than three months for fitting-out. Assume a yield of 6% and a write-off over the lease term.

[2]   Two neighbouring properties, 15 and 17 High Road, have recently let.

- Number 15 has a net internal area of 3000 square metres and let at a rent of £600 000 per annum on a 10-year FRI lease with an upward-only rent review in year five. The lease was subject to a rent-free period of 18 months and the landlord paid a capital contribution of £150 000 to the tenant upon signing the lease.
- Number 17 has a net internal area of 2000 square metres and let at a rent of £375 000 per annum on a 10-year FRI lease with an upward-only rent review in year five. The landlord gave the tenant a three-month fitting out period and then stepped the rent up over the first three years of the lease. The rent for the nine months following the fitting-out period was equivalent to £175 000 per annum; for the year after that it was £275 000 per annum, rising to £375 000 per annum the year after. The landlord also paid the tenant's fitting-out costs of £300 000 and contributed an additional capital payment of £100 000.

Assuming a yield of 8%, calculate the effective market rents of 15 and 17 High Road.

Premiums

[1]   A tenant leased a property on a 15-year FRI lease with five-year reviews at a rent of £20 000 per annum. In addition to the rent, the tenant paid a premium of £25 000 to obtain the lease. Assuming a yield of 8% and a write-off period of five years, what is the effective market rent?

[2]   A property with a market rent of £20 000 per annum is to be let for 15 years at £15 000 per annum with rent reviews at five-yearly intervals to the same proportion of market rent as at commencement. Calculate the premium required assuming a freehold yield of 10%.

[3]   A landlord is proposing to let a property but wishes to obtain a premium of £40 000. The current market rent is £20 000 per annum and the proposed lease would be for a term of 20 years with five-yearly rent reviews to an appropriate proportion of market rent at each review. Assuming the rent reduction is spread over the whole term of the lease, calculate the rent to be paid under the proposed lease assuming a freehold yield of 8%.

Surrendered leases

[1] A high street shop is currently let at £10 000 per annum on IR terms. The lease (which has seven-year rent reviews) has only three years left to run. The current market rent of the shop is £25 000 per annum on FRI terms and five-year upward-only rent reviews. The tenant wishes to instal a new shop front so that it matches a new corporate style. The existing shop front is adequate, and the new shop front will not affect the rental value. The tenant has approached the landlord with a view to surrendering the exist-ing lease and agreeing a new one but for a term of 20 years. The landlord is agreeable in principle and would like to take advantage of the opportunity to improve the investment by securing new lease terms modified to be on FRI terms and to include five-year rent reviews. Estimate the rent that should be agreed. Assume a yield of 9% freehold at market rent on FRI terms.

Zoning

[1] A shop was recently let on a 15-year FRI lease with five-year upward-only rent-review provisions. The rent agreed was £288 750 per annum. It has an 11-metre frontage and is 15 metres deep. It is a two-storey property with the first floor being used for storage. The first floor is also 11 metres by 15 metres. A square metre of the first-floor storage has a market rent of 5% of a square metre of zone A. Assuming six-metre zones, what is the market rent per square metre in terms of zone A (ITZA) indicated by the letting?

[2] a. Value Shop 1 using a zone A MR of £1600/m² per annum. It has a first floor (with the same dimensions as the ground floor) accessed from the stairs shown shaded. Use six-metre-deep zones.

b. Value Shop 2, which has a return frontage and no upper floor. The high street is the busiest frontage. Use the same zone A rate.

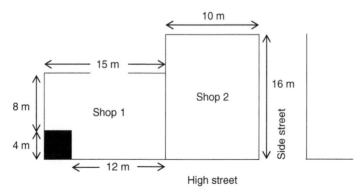

[3] Three adjacent shops are all two-storey and have the same dimensions on the first floors as on their ground floors. Shop B has just let on an FRI lease for ten years with an upward-only rent review after five years at a rent of £200 000 per annum. Estimate the market rent of shops A and C.

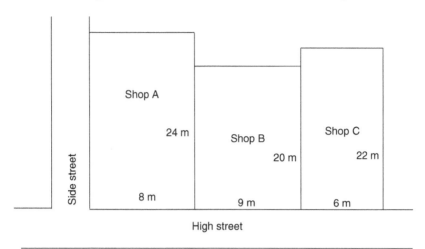

[4] Estimate the market rent of a vacant shop measuring 6.5 metres frontage by 23 metres depth with a first floor of 6.5 metres by 18 metres, half used for storage and half used for sales. A comparable shop recently let on a 10-year FRI lease with an upward-only rent review after five years for £75 000 per annum. The lease includes a one-year rent-free period, the first three months of which is for fitting out. The shop measures 7-metre internal frontage by 28-metre internal depth and has storage accommodation on the first floor, measuring 7 metres by 14 metres. Net initial yields for comparable shops are 7%.

Analysis of yields

Equivalent yield

[1] Property 1 is a 15 000 square metre (net internal area) office building constructed three years ago. It was let three years ago on a 15-year lease on an FRI basis with upward-only rent reviews every five years. The rent passing is £2 500 000 per annum. Property 2 (16 000 square metres NIA) was let a month ago at a rent of £3 200 000. Property 3 (14 500 square metres NIA) was let recently at a rent of £2 900 000 and has just sold for £48 330 000. Property 4 (15 000 square metres NIA) has just sold for £48 050 000. It was let two years ago on a 10-year FRI lease with a five-year upward-only rent review at a rent of £2 850 000 per annum. Assuming all four properties are comparable and purchase costs are 6.5% of purchase price:
a) What is the market rent of property 1?
b) What is the net initial yield (NIY) from the sale of property 3?
c) What is the equivalent NIY of property 4?
d) What is the value of property 1?

[2] Three neighbouring office premises have recently been the subject of transactions:

- 15 Commercial Road has a net internal area of 2100 square metres and recently let on a 10-year FRI lease with five-yearly upward-only rent reviews and with a two-year rent-free period at a headline rent of £500 000 per annum. It is expected that the effective market rent will not exceed the headline rent until the second review.
- 17 Commercial Road has a net internal area of 1000 square metres and also recently let on a 10-year FRI lease with five-yearly upward-only rent reviews. The headline rent is £220 000 per annum. The landlord has paid the tenant's fitting-out costs of £100 000. It is anticipated that effective market rent will exceed the headline rent at the first review in the lease.
- 19 Commercial Road has a net internal area of 1500 square metres and has just been sold for £5 000 000. It is let on a lease for 10 years with five-yearly upward-only rent reviews on FRI terms at a current rent passing of £250 000 per annum. This lease has three years unexpired.
  a) Assuming a yield of 5%, calculate the effective rents for 15 and 17 Commercial Road.
  b) Using the results from (a), estimate the MR and equivalent yield of 19 Commercial Road.

Market valuation of property investments

Voids, break options

A property was let three years ago on a 15-year FRI lease with five-year upward-only rent reviews. The current rent is £210 000 per annum, considered to be close to the market rent. The tenant has the option to break the lease at the next rent review. Transaction evidence indicates that similar properties have been selling at prices that reflect an NIY of 6%. Value the property.

Statutory considerations

[1] Fourteen years ago, a tenant took a 15-year lease of a shop on FRI terms. As a condition of the lease, the tenant agreed to pay for and instal a new shop front. The shop then had a frontage of 6 metres and a depth of 18 metres. The lease contains a user clause 'that the premises can only be used as a florist'. Four years into the lease the tenant, at her own cost but with the permission of the landlord, extended the shop at the side by three metres for the total depth. Comparable shops, but with wide user clauses, have recently let for zone A rents in the region of £100 per square metre. The freehold yield is 10% and the RV is £5500. Current building costs are estimated to be £500 per square metre.
a) What rent should the tenant pay if granted a new lease?
b) What compensation should the tenant receive if the landlord obtains possession of the premises for their own use?

[2] The tenant of a shop holds a 15-year IR lease granted 11 years ago at a current rent of £24000 per annum. Six years ago, the tenant obtained consent for improvements costing £60000. The current market rent on FRI terms is £50000 per annum, £5000 of which can be attributed to the improvements made by the tenant. The lease is protected by the security of tenure provisions of the Landlord and Tenant Act 1954. The RV of the premises is £50000, freehold yields are 6% and building cost inflation is at 5% per annum. Value the freehold and leasehold interests assuming:
   a) The landlord will get permission for own occupation at the end of the lease
   b) The tenant will continue in occupation

Over-rented properties

[1] Value a property that was let four years ago on a 15-year FRI lease with five-year upward-only rent reviews. The rent passing is £120000 per annum. The market rent is £105000 per annum on FRI terms. Assume an NIY of 7% and a rental growth rate of 2% per annum.

[2] A property was let seven years ago on a lease for 15 years with five-year upward-only rent reviews to market rent. The original rent was £850000 per annum on FRI terms, and it remained unchanged at the last rent review because the market rent had fallen since the initial letting. The market rent is currently £800000 per annum FRI. Assuming similar properties let at market rent to strong tenants sell for 6% NIY, and a rental growth rate of 1.5% per annum:
   a) Assess the risk of this property as an investment explaining the factors that influence your assessment.
   b) Value the property, noting any assumptions you make.

[3] A property is let on a lease that has two years left to run at a rent of £500000 per annum. The market rent is estimated to be £300000 per annum. Rack-rented net initial yields for this type of property are estimated to be 5%. Value the freehold interest assuming that at the end of the current lease: (a) the tenant remains in occupation and (b) the tenant vacates.

Turnover leases

[1] A shop is let on a turnover lease. The annual rent passing is 5% of the turnover generated from the property each year but subject to a minimum base rent of £175000 per annum. Last year, the total income from the shop was £4.5m and over the next five years this is expected to grow at 5% per annum. Assuming a yield for the base rent of 6% and a discount rate of 9% for the turnover rent, but making all other assumptions necessary, value the freehold interest in the shop.

[2] A shop was let one year ago on a five-year FRI year lease with provision for a turnover-related payment. The base rent is £375000 per annum and the lease stipulates that the tenant pays 5% of turnover above a threshold of £10m per annum. Turnover last year was £11464000 and is expected to grow at 5% per annum. Comparable evidence indicates that the market rent is £430000 per annum and the net initial yield for investments let on

conventional terms (i.e. no turnover provisions) is 5.25%. A discount rate of 13% should be applied to the turnover-related income. Value the freehold interest in the shop.

Synergistic value

[1] A one-acre site has been identified by the local planning authority for residential development. The local plan states that 15 houses could be accommodated on the site. Residential development land is currently selling for £1m per acre at this plot density. Currently, the site is under three separate ownerships:
- A single detached dwelling plus garden, currently valued at £150 000.
- A small industrial unit totalling 5000 square feet, which is currently vacant. Recent comparable evidence suggests similar units are selling at a price of £40 per square foot.
- A small paddock of 0.5 acres. Paddock land is selling for £50 000 per acre in the locality.

Calculate the synergistic value of the site and allocate this between the current owners.

[2] The freehold interests in the five plots shown above are separately owned. Plots 1, 2 and 3 are each worth £100 000 as development plots. Plots 4 and 5 are worth £25 000 each as part of the grounds of the adjacent properties shown hatched. If the five parcels of land were to be combined into a single site, planning permission would be available for redevelopment and the combined site with planning consent would be worth £475 000. What is the synergistic value and how might it be apportioned between the owners of the plots?

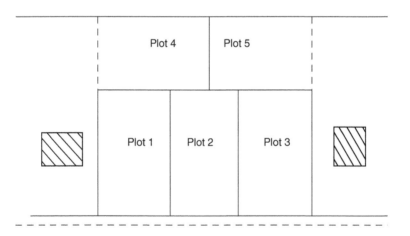

Market valuation of single-let property investments

[1] Value an office building that has just let on a 10-year FRI lease with a 5-year upward-only rent review and a break option after 3 years. The rent passing is

£30 000 per annum for a net internal area of 500 square metres. The following three recent lettings provide comparable rental evidence. The third comparable has just been sold as an investment, showing an NIY of 7%.

	NIA (m²)	Lease length and rent reviews	Rent (£/m²)
1	480	10-year FRI lease, five-year upward-only rent review	£72
2	520	5-year FRI lease, no rent review	£77
3	550	15-year FRI lease, five-year upward-only rent reviews	£70

[2]   Your client is a national high street chain of booksellers and stationers that occupies one of the larger units in a covered shopping centre in a medium-sized market town. The unit is occupied on a 15-year IR lease with five-yearly upward-only rent reviews to market rent and your client wishes to negotiate a new 15-year lease on similar terms. It is situated near the entrance to the shopping centre where the pedestrian flows are best.

Assumptions:
- All units are rectangular in shape.
- The prime pitch in the town is on the High Street either side and opposite the entrance to the shopping centre.
- The pitch deteriorates further into the shopping centre.
- A service charge is payable by tenants of the shopping centre.

Comparable evidence schedule:

Retail unit and location	Tenant details	Internal width (m)	Depth (m)	First-floor sales/storage space (m²)	Rent agreed	Lease terms
Subject unit	Independent stationers	10	20	100-m² first--floor sales space plus 80-m² storage	To be determined	Similar to current lease
Comp 1 Smaller unit in same shopping centre; further from main entrance	Clothing for babies and infants	6	11	30-m² basement storage	Rent recently reviewed to £64 800 p.a.	Internal repairing lease, 10-year term five years ago
Comp 2 Marginally smaller unit on high street adjoining entrance to shopping centre	Regional chain of discount women's fashion	8	14	50-m² first-floor storage	Recent letting at £126 000 p.a. Contribution to fitting out by way of six-month rent free	FRI

Retail unit and location	Tenant details	Internal width (m)	Depth (m)	First-floor sales/storage space (m²)	Rent agreed	Lease terms
Comp 3 Marginally smaller unit towards the back of the same shopping centre	Vacant	9	15	70 m² on first floor	Asking rent £95 000 p.a.	To let on 10-year internal repairing lease with rent review year 5
Comp 4 Small unit on high street opposite entrance to shopping centre	National mobile phone network provider and retailer	5	8	Small amount of storage space at back of ground floor	Recent letting at £78 000 p.a.	Five-year IRI lease

a) Analysing the comparable evidence presented above, advise your client as to the level of rent that should be agreed under the new lease, explaining your reasoning.

b) Make recommendations to your client about improvements to the terms of any new lease that they might agree.

[3] Your client owns an industrial unit on Parkway North, a large industrial estate, measuring 286 square metres, which has been vacant and to let for nine months. Two offers have been received:

- Offer 1 is from a private company that manufactures kitchen units. It is seeking to relocate from a smaller unit elsewhere on the estate. It has offered to take a 10-year lease of the property, with a five-year upward-only rent review, but requires a tenant's break option in year three. The rent offered is £23 000 per annum on an FRI basis.
- Offer 2 is from a printing company, part of a national-listed parent company. It offers to take a nine-year lease with three-yearly upward-only rent reviews. It requires the freeholder to make alterations to the property, costing £18 000 but these are unlikely to increase its rental value. The rent offered is £20 000 per annum on an FRI basis.

Analysis of recent rent and yield evidence:

Unit	Area (m²)	Rent (£/m²)	Lease terms	Comments
27 Parkway North	210	£84.00	Recently let for 9 years with 3-year rent reviews and a 6-month rent-free period	Local company, kitchen design and fitting

Unit	Area (m²)	Rent (£/m²)	Lease terms	Comments
58 Parkway North	188	£79.00	Recently let for 10 years with a 5-year rent review and an option for the tenant to break at year 5	Regional distribution company
71 Parkway North	295	£72.00	Recently let for 5 years.	Food manufacture, local company
77 Parkway North	350	£61.00	Let two years ago on a 10-year lease with a 5-year rent review. The freehold recently sold for £240 000.	National company, document storage

Explaining your assumptions, value the unit subject to each of the two offers.

[4] You have been asked to estimate the market rent and market value of a shop at 22 High Street. It has two storeys, with the first floor used for storage. The property is vacant and available to let on FRI terms, but the client would also contemplate a freehold sale. Its area ITZA is 75 square metres. The following comparable evidence is available:

High street no.	NIA ITZA[a]	Lease terms (Note: all rent reviews are upward only)	Rent (£ p.a.)	Condition/ Construction	Tenant details	Sale price (where relevant)
27	86	Let 5 years ago on a 10-year FRI lease with a rent review in year 5	£172 000 (agreed at recent rent review)	Moderate/ traditional	Independent (clothes) fashion	n/a
12	102	Let 1 year ago on a 15-year IRI lease with 5-year rent reviews and a tenant's break clause in year 5	£250 000	Good/ modern	Sports clothing	Sold six months ago for £2.8 m
17	45	Let 6 months ago on a 5-year FRI lease with 6-month rent-free period at the start of the lease	£118 000	Good/ traditional	Mobile phones	n/a
30	65	Let 1 year ago on a 10-year IRI lease with a rent review in year 5	£140 000	Poor/1960s	Health food	Sold at auction 1 year ago for £1.5 m

[a] All of the units are rectangular in plan and there are no return frontages.

Market valuation of multi-let property investments

[1] Value the freehold interest in Acacia House, a mixed-use commercial property. The building has four storeys, with a shop on the ground floor and offices above. The shop has an internal frontage of 7 metres and an internal depth of 26 metres. There are 18 square metres of storage at the rear. Offices A and B on the first and second floors have net internal areas of 435 square metres each and office unit C on the third floor has an NIA of 400 square metres. The tenancy schedule for the property is shown below.

Floor	Unit	Area (m²)	Current rent (p.a.)	Lease terms
Ground	Shop	182	£85 000	15-year FRI lease with five-year upward-only rent reviews and seven years unexpired
1	Office A	435	£82 000	10-year FRI lease with five-year upward-only rent reviews and five years unexpired (rent review now due)
2	Office B	435	£95 000	5-year FRI lease with four years unexpired. There is a break option in the tenant's favour in one year's time.
3	Office C	400	£100 000	20-year FRI lease with five-year upward-only rent reviews and eight years unexpired

Recent lettings suggest that the MR is £1200 per square metres ZA for retail and for offices it is £230 per square metres on FRI terms. You are aware of a recent investment transaction relating to a similar mixed-use building, Beazer House. The freehold interest is sold for £5 m. There is one shop in the property, which has an internal frontage of 6 metres and a depth of 25 metres plus storage at the rear of 20 square metres. The offices have a total area of 800 square metres. The building was let on a 10-year FRI lease and there are 3 years unexpired. The current rent is £230 000 per annum.

[2] Value the freehold interests in two adjacent properties. Both are three storeys with retail use on the ground floor and offices above. The retail sales space in each property has an internal frontage of 6 metres and an internal depth of 25 metres. In addition, there are 15 square metres of storage space at the rear of each shop. The offices within each property have a net internal area of 300 square metres. The tenancy details are shown below.

Property	Unit	Use	Rent passing (£ p.a.)	Lease terms	Additional information
1	A	Retail	60 000	10-year FRI lease, 8 years unexpired	5-year upward-only rent review
	B	Office	70 000	15-year FRI lease, 5 years unexpired	5-year upward-only rent reviews, last review now due

Property	Unit	Use	Rent passing (£ p.a.)	Lease terms	Additional information
2	A	Retail	Recently let on a 10-year FRI lease with a stepped rent over the first five years (£50 000 in year 1, £60 000 in year 2, £70 000 in year 3, £80 000 in year 4 and £90 000 in year 5) and a rent review to MR thereafter.		
	B	Office	50 000	5-year FRI lease, 3 years unexpired	Tenant undertook 'qualifying' improvements last year costing £40 000 and contributing £4000 p.a. to the MR

A comparable office unit of 400 square metres recently let for £95 224 per annum. on a 10-year lease with a 5-year upward-only rent review. There was a rent-free period of one year. Net initial yields for office investments are currently 6%. You are aware of a recent sale of the freehold interest in a shop at a price of £1.21 m. The shop sales space measured 7 metres internal frontage by 26 metres depth plus storage space of 20 square metres. The whole property is currently let on a 15-year FRI lease with five-year upward-only rent reviews at a rent of £72 550 per annum. The lease has seven years unexpired. The current market rent for comparable shops is £1000 per square metres ZA.

[3] An office property comprises a basement (used for storage) with five floors above (including the ground floor). Externally, notable features include glazed exterior cladding, a high-quality entrance and reception area on the ground floor and a secure barrier to the car park at the rear. The office accommodation is open plan and finished to a reasonable specification. There are two lifts serving all floors. Regarding maintenance of the building, each occupying tenant pays a portion of the annual service charge to the landlord. The floor area that each tenant occupies is used to apportion the service charge between tenants. The service charge pays for the cleaning of common parts, general repairs, services, lighting to common parts, lifts, insurance and management. The tenants pay for their own cleaning and lighting.

Y is the landlord of the site which was let to Z on a 125-year ground lease 17 years ago. The initial rent that was agreed was £10 000 per annum and the landlord has no responsibility for the insurance or repairs of the office building on the site. The rent payable under the ground lease is reviewed every 25 years. At each review, the rent is reviewed to the existing ground rent plus 5% of the estimated market rent of the head-lease in excess of the existing ground rent. The wording of the rent review clause in the ground-lease permits the head-lease to be valued assuming the building is vacant and to let.

All the occupational sub-leases specify that the sub-tenants are responsible for all repairs and insurance (non-internal repairs and insurance payable via the service charge) and are subject to five-year upward-only rent reviews. The schedule below lists the details of the sub-leases.

Floor	Tenant	Use	Business	Covenant[a]	Area (m²)	Current rent (£)	Date lease commenced	Length of lease (years)
Basement	A	Store	Solicitors	Good	305	21 350	8 years ago	15
Ground	A	Office	Solicitors	Good	251[b]	40 160	2 years ago	10
First	B	Office	Insurance	Good	449	76 330	This year	15
Second	C	Office	Travel	Poor	449	49 390	17 years ago	25
Third	D	Office	Surveyors	Average	449	69 595	4 years ago	10
Fourth	E	Office	Publishers	Poor	398	55 720	8 years ago	15
Totals					2301	312 545		

[a] Describes the quality of the tenant in terms of ability to meet the terms of the lease. It is a subjective measure of the security of the income.
[b] Entrance and reception areas are on this floor.

Three properties have recently been the subject of transactions that provide comparable rental evidence for your subject property:

The basement of the office building next door was recently leased to the publishers who occupy the fourth floor of the subject property for additional archiving and general storage. The lease was agreed on standard terms for a period of five years at a rent of £90 per square metre. This provides evidence of the current market rent for storage space in this type of building.

The letting of the first floor of the subject property to the insurance company was recent and agreed on standard terms. It therefore provides good evidence of current market rent for the office space. The rent agreed equates to £170 per square metre.

The fifth (top) floor of the office building next door was recently let on standard terms. The lease was for a term of 15 years at a rent that equates to £150 per square metre. However, on inspection of this building it is noted that the lift only goes up to the fourth floor and clearly a reduction to the 'normal' market rent for office space in this area has been made to take this into account.

a)   Calculate the rent-weighted WAULT and the area-weighted WAULT.
b)   Analyse the comparable evidence.
c)   Assuming a yield of 10%, value the freehold interest.
d)   Justifying the choice of yield, value the head-leasehold interest.

Investment valuations

[1]   A property has just been let on a 10-year FRI lease with an upward-only rent review in year five at a rent of £100 000 per annum. Assuming a target rate of return of 7%, a five-year holding period, rental growth of 2.5% per annum and an exit yield of 5%, estimate the investment value of the property investment.

[2]   Estimate the investment value of the freehold interest in the following property. The property is ready for refurbishment upon expiry of the current lease in seven years' time. The current net rent is £100 000 per annum and the final rent review is in two years' time. The owner plans to hold the

property until lease expiry, refurbish and then sell. Assume the following information:

- The current cost of refurbishment is £1 000 000 and will take one year.
- The current market rent of the property in its existing state is £120 000 per annum and in its refurbished state is £200 000 per annum.
- Rental growth of the existing property is 4% per annum and of the refurbished property is 7% per annum.
- The yield after refurbishment is estimated to be 7%.
- Your client's target rate of return is 15%.
- Building costs are inflating at an average of 6% per annum.

[3] Estimate the investment value of the following property investment. It is an office building with 5000 square metres gross internal area and an efficiency ratio of 80%. The rent passing is £110 000 per annum, which is regarded as the market rent. The investor's holding period is five years, and the target rate of return is 8%. The exit yield is 5%, purchase costs are 5.75% of the purchase price and sale costs are 2.4% of the sale price.

[4] The following freehold property investment opportunity has arisen:

- There are 10 years remaining on the current lease, which is on FRI terms with an upward-only rent review in 5 years' time. The rent passing is £70 000 per annum.
- The market rent of the property in its existing state is £80 000 per annum and, if refurbished, it is £100 000 per annum. Rents are forecast to grow at an average rate of 2% per annum and the depreciation rate for the existing property is 0.5% per annum.
- The investor's target rate of return is 8% and the plan is to hold the property until the end of the lease, refurbish and then sell. The refurbishment is estimated to take one year, cost £250 000, with building costs forecast to inflate at 2% per annum on average.
- Other assumptions are: acquisition costs at 6.5% of acquisition price, sale costs at 2% of sale price, rent review costs at 0.5% of the new rent and management costs at 1% of rental income.

Estimate the investment value of the property assuming:

a) No refurbishment, and an exit yield of 5%
b) Refurbishment, and an exit yield of 4.5%

Answers

Analysis of rents

Rental lease incentives

Rent-free periods
a) £166 011
b) £220 272
c) 12.56% and 3.14%
d) £160 225 and £218 913

Stepped rents
[1]

Year	Offer 1	Offer 2
1	£100 000	£60 000
2	£100 000	£80 000
3	£100 000	£100 000
4	£100 000	£120 000
5	£100 000	£140 000
NPV @ 8%	£399 271	£387 011

These NPVs can then be amortised to find the effective rents, but this is not necessary if the aim is just to compare the two offers. Offer 1 is the best.

[2]

Year	Rent (£ p.a.)	PV £1 @ 9%	PV (£)
1	200 000	0.9174	183 486
2	225 000	0.8417	189 378
3	250 000	0.7722	193 046
4	275 000	0.7084	194 817
5	300 000	0.6499	194 979
Capital value (CV)			955 706
Spread CV over 5 years:			
/YP 5 years @ 9%			3.8897
Effective rent (£)			245 705

[3]

Year	Option 1 (£)	Option 2 (£)	Option 3 (£)
1	150 000	200 000	250 000
2	200 000	220 000	250 000
3	250 000	242 000	250 000
4	300 000	266 200	250 000
5	400 000	292 820	250 000
NPV @ 10%	942 754	909 091	947 697
/YP factor	3.7905	3.7905	3.7905
= Annual equivalent	248 696	239 816	250 000

Option 3 is the best.

Capital lease incentives

Capital contributions
[1] Effective MR = £237 704

[2] 15 High Road: No write-off period was specified. One approach is to assume it is between the first and second reviews (the end of the lease in this case), i.e. 7.5 years. Another approach is to look at the rental growth rates that are

generated by various write-off periods. Doing this reveals a growth rate of 2.30% per annum with a 10-year write-off and this seems the most realistic. The effective market rent is therefore £478 138 per annum. Moving on to the capital contribution, a five-year write-off generates a rental growth rate of 1.65%, which seems realistic. So, the effective market rent after the capital contribution is £440 570 per annum equating to £147 m².

17 High Road: Assume a write-off until the rent review.

Stepped rent for year 1 (three quarters of 175 000 because of the three-month fitting-out period) (£ p.a.)	131 250	
PV £1 1 year @ 8%	0.9259	
	121 524	
Stepped rent for year 2 (£ p.a.)	275 000	
PV £1 2 years @ 8%	0.8573	
	235 758	
Stepped rent for years 3–5 (£ p.a.)	375 000	
YP 3 years @ 8%	2.5771	
PV £1 2 years @ 8%	0.8573	
	828 505	
Capital value of rent over write-off period (£)	1 185 787	
/YP 4.75 years @ 8%	3.8274	
Effective MR over write-off period after stepped rent (£ p.a.)		309 815
Capital contributions (£)	(400 000)	
/YP 5 years @ 8%	3.9927	
		(100 183)
Effective MR over write-off period after stepped rent and capital contribution (£ p.a.)		209 632

Effective market rent = 209 632/2000 = £105/m²

Premiums

[1]   Annual equivalent: 25 000/YP 5 years @ 8% = 3.9927 = £6261
      Rental value: £20 000 + £6261 = £26 261 per annum

[2]

Inputs/Assumptions:		Calculation of capital payment:	
Market rent (£ p.a.)	20 000	Market rent (£ p.a.)	20 000
Reduced rent (£ p.a.)	15 000	Reduced rent (£ p.a.)	15 000
Yield	10%	Profit rent (£ p.a.)	5000
Write-off period (years)	15	x YP @ discount rate for write-off period	7.6061
		Capital value of profit rent/capital payment (£)	38 030

[3]   The period over which the rent reduction is to be spread is a key considera-tion. In this question, the rent reduction will be spread over the whole 20-year term. In another case, it might only be until the next review date. It is vital that this period is correctly identified in any calculation. Reference to the lease terms is usually necessary in practice.

Landlord's point of view:

Market rent (£ p.a.)		20 000
Premium (£)	(40 000)	
Divided by YP 20 years @ 8%	9.8181	
Annual equivalent of premium (£)		(4077)
Rent required (£ p.a.)		15 923

The landlord will require a premium of £40 000 plus a rent of £15 923 per annum. Tenant's point of view:

Market rent (£ p.a.)		20 000
Premium (£)	(40 000)	
Divided by YP 20 years @ 9%	9.1285	
Annual equivalent of premium (£)		(4381)
Rent affordable (£ p.a.)		15 619

The tenant can afford to pay £40 000 premium plus £15 619 per annum in rent. A mid-point compromise in this example would result in a premium £40 000 (as agreed) plus a reduced initial rent of £15 771 per annum (say £15 750 per annum) to the first review. The difference is due to the use of different yields. In the market, there may be other factors that affect the bargaining strengths of the parties; for example for tax or other reasons, one of the parties would prefer a premium rather than more rent. In practice, the parties may compromise based on their bargaining strengths and individual desire to conclude the transaction rather than solely mathematically.

The question indicates that the rent reduction is to be spread over the whole 20-year term rather than just until the first rent review. A common solution is to apportion the rent at review in the same proportion of market rent as at commencement. Thus, at commencement, the rent payable is £15 750, which is 79% of the market rent of £20 000. So rent reviews would be to 79% of market rent at each review. The final answer to this example would therefore be a 20-year lease at a commencing rent of £15 750 per annum with 5-yearly rent reviews to say 80% of market rent plus payment of a premium of £40 000 at commencement.

Surrendered leases

[1] Valuation of landlord's present interest:

Term rent on IR terms (£ p.a.)	10 000	
External repairs @ say 10% MR (£ p.a.)	(2500)	
Insurance @ say 2% MR (£ p.a.)	(500)	
Net rent (£ p.a.)	7000	
YP 3 years @ 9%	2.5313	
		17 719
Reversion to MR (£ p.a.)	25 000	
YP in perp @ 9%	11.1111	
PV £1 3 years @ 9%	0.7722	
		214 500
Valuation before PCs (£)		232 219

Valuation of landlord's proposed interest:

Proposed rent for first 5 years (£ p.a.)	x	
YP 5 years @ 9%	3.8897	
		3.8897x
Reversion to MR (£ p.a.)	25 000	
YP in perp @ 9%	11.1111	
PV £1 5 years @ 9%	0.6499	
		180 528
Valuation before PCs (£)		180 528 + 3.8897x

Section C

Rent required by landlord: Present interest = Proposed interest: £232 219 = £180 528 + 3.8897x, so x = £13 289. At a rent of £13 289 per annum for the first five years, the landlord's proposed interest would be of similar value to his present interest. In addition, the landlord would gain the benefit of modern FRI lease terms and would secure the tenant's covenant for 20 years instead of the remaining three years.

Valuation of tenant's present interest:

MR on FRI terms (£ p.a.)	25 000	
Gross rent paid (£ p.a.)	(10 000)	
Profit rent (£ p.a.)	15 000	
YP 3 years 10%[a]	2.4869	
Valuation before PCs (£)		37 304

[a] Freehold yield adjusted upwards by 1% to reflect relative unattractiveness of short leasehold investment.

Valuation of tenant's proposed interest:

MR on FRI terms (£ p.a.)	25 000	
Rent to be paid (£ p.a.)	(x)	
Profit rent (£ p.a.)	25 000 − x	
YP 5 years @ 10%	3.7908	
Valuation before PCs (£)		94 770 − 3.7908x

Rent required by tenant: £37 304 = £94 770 − 3.7908 x, so x = £15 159
The outcome is likely to depend on the bargaining strengths of the parties and the extent to which either or both may be keen to conclude the transaction. There may be other benefits to either party, which are not reflected in the valuation calculation.

Zoning

[1]   The aim is to convert the actual area of the shop into an area 'in terms of zone A' (ITZA).

Zone	Width (m)	Depth (m)	% Zone A	ITZA (m²)
A	11	6	100%	66
B	11	6	50%	33
C	11	3	25%	8.25
First-floor storage	11	15	5%	8.25
Total				115.5

Dividing the market rent by the total area ITZA will express the rent in terms of ZA.

So, £288 750 p.a./115.5 m² = £2500/m² ZA

[2]

Shop 1

Zone A (m²)	6 m × 15 m	90
Less	3 m × 4 m	(12)
		78
Zone B (m²)	(6 m × 15 m)/2	45
Area ITZA (m²)		123
Zone A rent (£/m²)		1600
MR of ground floor (£ p.a.)		196 800
Assume first floor is valued at zone A/10 (i.e. 10% of ZA value)		
MR of FF: 168 m² × £160/m² (£ p.a.)		26 880
MR of shop 1 (£ p.a.)		223 680

Shop 2

Zone A: 6 m × 10 m (m²)		60
Zone B: (6 m × 10 m)/2 (m²)		30
Zone C: (4 m × 10 m)/4 (m²)		10
Area ITZA (m²)		100
Zone A rent (£/m²)		1600
MR of ground floor (£ p.a.)		160 000
Plus 10% end allowance[a] for return frontage (£ p.a.)		16 000
MR of shop 2 (£ p.a.)		176 000

[a] The ground-floor area ITZA of 100 m² could be adjusted upwards by 10% to 110 m² and this would give the same result. The adjustment is usually only to the ground floor area, not upper floors or basements.

[3]

Shop B: Analysis of comparable evidence

		Space	Depth (m²)	Area (m²)	% Zone A	Area ITZA (m²)
Frontage length (m)	9					
Depth (m)	20	Zone A	6.00	54.00	100.0%	54.00
Zone depth (m)	6	Zone B	6.00	54.00	50.0%	27.00
Other sales space (m²)[a]	120	Zone C	6.00	54.00	25.0%	13.50
Storage space (m²)	60	Remainder	2.00	18.00	12.5%	2.25
		Other sales space		120.00	20.0%	24.00
		Storage area		60.00	10.0%	6.00
						126.75

[a] It has been assumed that the upper floor is split as two third for sales and one third for storage in all properties, but different assumptions could be made.

The ZA rent is £200 000 p.a./126.75 m² = £1578/m².

Shop A: Valuation

Assume shop A has a return frontage to the side street.

		Space	Depth (m²)	Area (m²)	% Zone A	Area ITZA (m²)
GF frontage length (m)	8					
GF depth (m)	24	GF zone A	6.00	48.00	100%	48.00
Zone depth (m)	6	GF zone B	6.00	48.00	50%	24.00
Other sales space (m²)	128	GF zone C	6.00	48.00	25%	12.00
Storage space (m²)	64	GF remainder	6.00	48.00	12.5%	6.00
		End allowance (as % of GF area ITZA)			10%	9.00
		Other sales space		128.00	20%	25.60
		Storage area		64.00	10%	6.40
						131.00

The estimated MR is $131\,m^2 \times £1578 = £206\,718$ p.a.

Shop C: Valuation

		Space	Depth (m²)	Area (m²)	% Zone A	Area ITZA (m²)
GF frontage length (m)	6					
GF depth (m)	22	GF Zone A	6.00	36.00	100%	36.00
Zone depth (m)	6	GF Zone B	6.00	36.00	50%	18.00
Other sales space (m²)	88	GF Zone C	6.00	36.00	25%	9.00
Storage space (m²)	44	GF remainder	4.00	24.00	12.5%	3.00
		Other sales space		88.00	20%	17.60
		Storage area		44.00	10%	4.40
						88.00

The estimated MR is $88\,m^2 \times £1578 = £138\,864$ p.a.

[4]   Analysis of comparable evidence

Zone A:	$7\,m \times 6\,m = 42\,m^2$ @ 100% = $42\,m^2$ (A)
Zone B:	$7\,m \times 6\,m = 42\,m^2$ @ 50% = $21\,m^2$ (A/2)
Zone C:	$7\,m \times 6\,m = 42\,m^2$ @ 25% = $10.5\,m^2$ (A/4)
Remainder:	$7\,m \times 10\,m = 70\,m^2$ @ 10% = $7\,m^2$ (A/10)
First floor:	$7\,m \times 14\,m = 98\,m^2$ @ 10% = $9.8\,m^2$ (A/10)
Total (area ITZA):	$90.3\,m^2$

MR is £75 000 but this is a headline rent so we need to estimate the effective MR. Using a yield of 7%:

- Spread over 5 years until rent review: £61 499, growth rate = 4.05% per annum
- Spread over 10 years until lease end: £67 317, growth rate = 1.09% per annum

Looking at the growth rates, the 10-year write-off period looks more realistic, suggesting an effective MR of £67 317 per annum. The ITZA was $90.3\,m^2$ so that is a zone A MR/m² of £745/m².

Application to subject property

Zone A:	6.5 m × 6 m = 39 m² @ 100% = 39 m² (A)
Zone B:	6.5 m × 6 m = 39 m² @ 50% = 19.5 m² (A/2)
Zone C:	6.5 m × 6 m = 39 m² @ 25% = 9.75 m² (A/4)
Remainder:	6.5 m × 5 m = 32.5 m² @ 10% = 3.25 m² (A/10)
First-floor sales:	6.5 m × 9 m = 58.5 m² @ 20% = 11.7 m² (A/5)
First-floor store:	6.5 m × 9 m = 58.5 m² @ 10% = 5.85 m² (A/10)

The MR of the subject property is its area ITZA ($89.05\,\text{m}^2$) × the MR/m² ZA ($£745/\text{m}^2$) = £66 342 p.a.

Analysis of yields

Equivalent yield

[1]

  a) Analysing property 2:

$£3\,200\,000/16\,000\,\text{m}^2 = £200/\text{m}^2$

Applying this to property 1:

$£200/\text{m}^2 \times 15\,000\,\text{m}^2 = £3\,000\,000$ per annum

  b) $£2\,900\,000 / (£48\,330\,000 \times 1.065) = 5.63\%$

  c) Equivalent yield is 5.82%

  d)

Term (contract) rent (£ p.a.)		2 500 000	
YP for term @	5.82%	1.8380	
			4 595 069
Reversion to estimated MR (£ p.a.)		3 000 000	
YP perpetuity @	5.82%	17.1821	
PV £1 for term @	5.82%	0.8930	
			46 032 309
Valuation before PCs (£)			50 627 378
Valuation after PCs (£)			47 537 444

[2]

  a) 15 Commercial Road: As MR will exceed the headline rent at the second review, the write-off period should be 10 years. The effective MR is £379 599.

17 Commercial Road: As MR exceeds headline rent at the first review, the write-off period should be five years. The effective MR is £196 903.

  b) 19 Commercial Road: Analysing the effective rents of 15 and 17 Commercial Road above:

$$15 \text{ Commercial Road} = \frac{£379\,587}{2100} = £181/\text{m}^2$$

$$17 \text{ Commercial Road} = \frac{£196\,903}{1000} = £197/\text{m}^2$$

Say the market rent of 19 Commercial Road is £188/m². So, $1500\,\text{m}^2 \times £188 = £282\,000$ p.a. and the equivalent yield is 5.21%.

Market valuation of property investments

Voids, break options

One approach is to capitalise the MR at 6% and say that breaks like this are typical for the market:

Market rent (£ p.a.)		210000	
YP in perpetuity @	6.00%	16.6667	
Valuation before PCs (£)			3 500 000

Another is to adjust the yield up slightly to reflect the risk of a break in rent should the option be exercised:

Market rent (£ p.a.)		210000	
YP in perpetuity @	6.50%	15.3846	
Valuation before PCs (£)			3 230 769

Alternatively, include a break in the rent to reflect the loss of rent if the option is exercised:

Market rent – current lease (£ p.a.)		210000	
YP till break/lease end @	6.00%	1.8334	
			385 012
Market rent – new lease (£ p.a.)		210000	
YP perpetuity @	6.00%	16.6667	
PV £1 for current lease + void period[a]	3 years	0.8396	
			2 938 667
Valuation before PCs (£)			3 323 680

[a] This valuation could include void costs.

The no-break and break valuations could incorporate some probability analysis:

No-break valuation	60%	3 500 000
Break valuation	40%	3 323 680
Weighted average valuation		3 429 472

Statutory considerations

[1]

a) Rent to be paid under the new lease: Section 34 of Landlord and Tenant Act 1954 states that improvements are to be disregarded if carried out other than as a lease obligation, provided they were carried out during current tenancy or less than 21 years before renewal. Therefore, the improvements carried out at the start of the lease can be considered, but those carried out four years into the lease must be disregarded.

Area of shop excluding the extension:

Zone	Width (m)	Depth (m)	Zone	Area ITZA (m²)
A	6	6	100%	36
B	6	6	50%	18
C	6	6	25%	9
			Total	63

Adjusting £100/m² ZA to £90/m² ZA to reflect restrictive user clause. So, MR = £90/m² ZA × 63 m² area ITZA = £5670 p.a.

b)   Compensation to be received if the landlord obtains possession for own use.

Compensation for disturbance – S37 LTA54 (as amended)
Tenant for more than 14 years, so 2 x RV: 2 x £5500 = £11 000
Compensation for improvements – S1 LTA27
Lesser of net addition in value or cost of improvements at the end of the lease less the cost of any repairs. Area of shop including the extension:

Zone	Width (m)	Depth (m)	Zone	Area ITZA (m²)
A	9	6	100%	54
B	9	6	50%	27
C	9	6	25%	13.5
			Total	94.5

MR = £90/m² × 94.5 m² = £8505 p.a.

Addition in rental value (£8505 – £5670) (£ p.a.)		2835	
YP in perp	@ 10%	10	
Value of improvements (£)			28 350
Cost of improvements (£) 54 m² @ £500 m²			27 000

Compensation payable:

Disturbance (£)	11 000
Improvements (£)	27 000
TOTAL (£)	38 000

[2]

a)   Landlord gets possession
Compensation for improvements:

Lower of:	Value added by improvements (£ p.a.)	5000	
	YP in perp @ 6%	16.6667	
			83 333
Or:	Cost (£)	60 000	
	Inflated @ 5% p.a. over 6 years	1.3401	
			80 406

Compensation for disturbance:

2 x RV (£)   100 000

Valuation of landlord's interest:

Term rent passing (£ p.a.)	24 000
External repairs (£ p.a.)	(5000)
Insurance (£ p.a.)	(1000)
Net income FRI (£ p.a.)	18 000
YP 4 years @ 6%	3.4561
	62 372

Reversion to market rent (£ p.a.)		50000	
YP in perp @ 6%		16.6667	
PV £1 4 years @ 6%		0.7921	
			660132
Compensation for improvements (£)	(80406)		
FV £1 4 years @ 5% p.a.	1.2155		
		(97733)	
Compensation for disturbance (£)		(100000)	
		(197733)	
PV £1 4 years @ 10%[a]		0.6830	
			(135052)
Valuation before PCs (£)			587452

Section C

[a] Due to risk of increase in improvement compensation and disturbance compensation may increase if there is a rating revaluation.

Valuation of tenant's interest:

Market rent (FRI) (£ p.a.)	50000	
External repairs (£ p.a.)	5000	
Insurance (£ p.a.)	1000	
Market rent (IR) (£ p.a.)	56000	
Rent paid (£ p.a.)	(24000)	
Profit rent (£ p.a.)	32000	
YP 4 years @ 10%[a]	3.1699	
		101437
Compensation (as above) (£)		135052
Valuation before PCs (£)		236489

[a] Risky, terminable, non-growth investment.

b)   Tenant continues in occupation

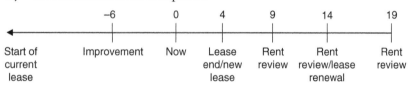

–6	0	4	9	14	19
Start of current lease	Improvement	Now	Lease end/new lease	Rent review	Rent review/lease renewal	Rent review

Valuation of landlord's interest:

First term:			
Net income, as above (£ p.a.)		18000	
YP 4 years @ 6%		3.4561	
			62372
Second term (excluding improvements):			
Gross MR, IR terms, reduced by value of improvements (£ p.a.) [a]		51000	
Cost of external repairs and insurance (£ p.a.)		(6000)	
Net MR, IR terms, reduced by value of improvements (£ p.a.)		£45000	
YP 15 years @ 6%		9.7122	
PV £1 4 years @ 6%		0.7921	
			346187
Reversion:			
Gross MR including improvements (£ p.a.)		56000	

Less cost of external repairs and insurance			(6000)
Market rent, FRI terms (£ p.a.)			50 000
YP in perp @ 6%			16.6667
PV £1 19 years @ 6%			0.3305
			275 483
Valuation before PCs (£)			684 042

[a] MR (FRI) including improvements (£ p.a.)			50 000
Rental value due to improvements (£ p.a.)			(5000)
Cost of external repairs (£ p.a.)	@ say	10% of MR	5000
Cost of insurance (£ p.a.)	@ say	2% of MR	1000
MR of new 15-year IR lease (£ p.a.)			51 000

Valuation of tenant's interest:

Profit rent from existing lease (as above) (£ p.a.)	32 000	
YP 4 years @ 10%	3.1699	
		101 437
Profit rent from new lease due to improvements (£ p.a.)	5000	
YP 15 years @ 9%[a]	8.0607	
PV £1 4 years @ 9%	0.7084	
		28 551
Valuation before PCs (£)		129 988

[a] Growth potential due to possible rent reviews in sub-lease- so yield based on FH yield plus LH risk premium.

Over-rented properties

[1] Because there are rent reviews before the end of the lease, we need to decide the write-off period for the overage. The decision depends on the rental growth rate assumption. $(1+g)^6 \times 105\,000 = 120\,000$, so $g = 2.25\%$ per annum. This is similar to the rental growth rate given in the question so assume a write-off period for the overage of six years.

Core market rent (£ p.a.)	105 000	
YP in perpetuity @ 7%	14.2857	
		1 500 000
Overage (£ p.a.)	15 000	
YP 6 years @ 9%	4.4859	
		67 289
Valuation before PCs (£)		1 567 289

[2]

a) Identify that the property is over-rented and describe risk to overage rent. The tenant's covenant strength is a crucial risk factor and suggests only a small extra risk from over-renting. Consider unexpired term and degree of over-renting and their implications for risk to overage.

b) Identify and justify realistic overage yield and decide how long to value overage for.

Core market rent (£ p.a.)	800 000	
YP in perpetuity @ 6%	16.6667	
		13 333 333
Overage (£ p.a.)	50 000	
YP 8 years @ 8.25%	5.6925	
		284 625
Valuation before PCs (£)		13 617 958

[3]

a)   Re-let to existing tenant:

Core (market) rent (£ p.a.)	300 000	
YP in perpetuity @ 5%	20.0000	
		6 000 000
Overage (difference between rent passing and MR) (£ p.a.)	200 000	
YP for 2 years @ 7%	1.8080	
		361 604
Valuation (£) before PCs		6 361 604
Valuation (£) after PCs @ 6.50%		5 973 337

NB: if there are no reviews before the end of the lease, then there is no double counting.

b)   Tenant leaves, assume a one-year void period:

Overage (£ p.a.)	200 000	
YP 2 years @ 7%	1.8080	
		361 604
MR until end of lease (£ p.a.)	300 000	
YP 2 years @ 5%	1.8594	
		557 823
Reversion to MR on new lease (£ p.a.)	300 000	
YP perpetuity @ 5%	20.0000	
PV 3 years [a] @ 5%	0.8638	
		5 183 026
Valuation before PCs (£)		6 102 453

[a] PV three years to account for two years remaining term plus one-year void.

Turnover rents

[1]

Assume that the turnover rent arrangement continues. This means the base rent can be capitalised into perpetuity and the turnover rent calculation will also need an 'exit' to perpetuity.

Valuation of base rent:

Base rent (£ p.a.)	175 000	
YP perp @ 6%	16.6667	
		2 916 673

Valuation turnover rent:

Year	Turnover (£ p.a.)	5% of turnover	Less base rent of £175 000 (£)	PV @ 9%	PV (£)
1	4 725 000	236 250	61 250	0.9174	56 166
2	4 961 250	248 063	73 063	0.8417	61 519
3	5 209 313	260 466	85 466	0.7722	65 979
4	5 469 778	273 489	98 489	0.7084	69 730
5	5 743 267	287 163	112 163/0.07[a]	0.6499	1 041 353
					1 294 747

[a] The assumption here (in the absence of anything to the contrary in the question) is that the turnover rent continues in perpetuity and is capitalised at 7% (risk-adjusted base rent yield).

Valuation before purchase costs: £2 916 673 + £1 294 747 = £4 211 420.

[2]   The approach outlined below assumes: no void in the property at lease expiry, no yield adjustment to reflect increased risk, a turnover rent received for only four years and growth in turnover will continue at previous levels for the next four years.

Valuation of base rent:

Base rent over term (£ p.a.)	375 000	
YP 4 years @ 5.25%	3.5255	
		1 322 045
Reversion to MR (£ p.a.)	430 000	
YP in perpetuity @ 5.25%	19.0476	
PV £1 4 years @ 5.25%	0.8149	
		6 674 531
Value of base rent and reversion (£)		7 996 576

Valuation of turnover rent:

	Projected turnover (£)	Surplus (£)	Payable (£)	PV @ 13%
This year	12 037 200	2 037 200	101 860	90 142
Next year	12 639 060	2 639 060	131 953	103 339
Year 3	13 271 013	3 271 013	163 551	113 349
Year 4	13 934 564	3 934 564	196 728	120 657
Value of turnover rent (£)				427 486

Valuation before PCs: £7 996 576 + £427 486 = 8 424 062

Synergistic value

[1]

Valuation of existing interests:	
Dwelling (£)	150 000
Industrial unit: 5000 ft² @ £40/ft² (£)	200 000
Paddock land: 0.5 acres @ £50 000/acre (£)	25 000
Total (£)	375 000
Valuation of combined interests:	
Development site of one acre (£)	1 000 000
Synergistic value: 1 000 000 – 375 000 (£)	675 000

Apportionment of synergistic value between existing interests might be done in several ways; a simple three-way split, allocations weighted by existing use value or by area of land occupied.

[2]   Value of combined site with planning permission is £475 000.

Value of existing separate plots:

1	£100 000
2	£100 000
3	£100 000
4	£25 000
5	£25 000
Total	£350 000
Synergistic value	£125 000

The value of each existing individual plot would be affected by the element of marriage value. The sums achieved are likely to be dependent on the bargaining strengths of the parties, but a common way to apportion the marriage value is in the ratio of the individual plot values to the value of the whole.

Plot	Separate value (£)	Portion of marriage value (£)	Total (£)
1	100 000	(100 000 × £125 000)/350 000 = £35 714	£135 714
2	100 000	(100 000 × £125 000)/350 000 = £35 714	£135 714
3	100 000	(100 000 × £125 000)/350 000 = £35 714	£135 714
4	25 000	(25 000 × £125 000)/350 000 = £8929	£33 929
5	25 000	(25 000 × £125 000)/350 000 = £8929	£33 929
Check		£125 000	£475 000

Market valuation of single-let property investments

[1]

- Identify that lease length appears to be the factor explaining differences in rents. Perhaps anchor on first comparable (£72/m²) in the absence of any other details.
- Note that the comparables do not have break clauses so use this as justification for valuing the subject property with an adjustment to value due to its break option.

Decide on valuation approach; some are more rigorous than others. Also need to decide whether to (i) assume the break-option lease should be valued in perpetuity or (ii) value the break-option lease for its term and then revert to standard terms, as indicated by the comparable evidence. (ii) is probably more defensible, so need to assume an MR, say £72/m² × 500 m² = £36 000 p.a.

Could adjust the yield of 7% from the comparable to reflect uncertainty surrounding the exercise of the break option in the subject property.

Term (contract) rent (£ p.a.)	30 000	
YP for 10 years @ 7.5%	6.8641	
		205 923
Reversion to estimated MR (£ p.a.)	36 000	
YP in perpetuity @ 7%	14.2857	
PV £1 for 10 years @ 7%	0.5083	
		261 437
Valuation before purchase costs		467 360
Valuation net of purchase costs @ 6.5%		438 836

A more robust approach would be to make reasonable assumptions about exercise of break clause and re-letting either after break or at lease end and carry out a valuation accordingly. Void costs are assumed to be 50% of MR and PV'd from half-way through the void period.

Rent passing (£ p.a.)	30 000	
YP for 3 years @ 7%	2.6243	
		78 729
Market rent – new lease (£ p.a.)	36 000	
YP perpetuity @ 7%	14.2857	
PV £1 for current lease + void: 4 years @ 7%	0.7629	
		392 346
Void costs (£)	(18 000)	
PV £1 from mid-way through void period: 3.5 years @ 8%	0.7639	
		(13 750)
Valuation gross of purchaser's costs (£)		457 325
Valuation net of purchase costs @ 6.5%		429 413

A probability-weighted approach would be even more rigorous:
Value assuming void (£429413 above)
Value assuming no void (as in 1 but no yield adjustment)

Term (contract) rent passing (£ p.a.)	30 000	
YP for 10 years @ 7%	7.0236	
		210 708
Reversion to estimated MR (£ p.a.)	36 000	
YP perpetuity @ 7%	14.2857	
PV £1 for 10 years @ 7%	0.5083	
		261 437
Valuation before purchase costs		472 145
Valuation net of purchase costs @ 6.5%		443 329

Weight them 70:30 in favour of no break, referring to lease events review as evidence for the weightings.

[2]

Comp 1: IR lease

Zone	Width (m)	Depth (m)	% Zone A	ITZA (m²)
A	6	6	100%	36
B	6	5	50%	15
Storage	30m²		10%	3
			Total	54 m²

£64 800/54 = £1200/m² ZA p.a. Deduct 12% to allow for landlord's external repairs and insurance. So, £1056/m² ZA.

Comp 2:

Zone	Width (m)	Depth (m)	% Zone A	ITZA (m²)
A	8	6	100%	48
B	8	6	50%	24
C	8	2	25%	4
Storage	50 m²		10%	5
			Total	81 m²

£126 000/81 = £1400/m² ZA p.a. Deduct 10% of rent to allow for six months rent-free, so £1260/m² ZA.

Comp 3: IR lease but assume equivalent to FRI due to service charge structure

Zone	Width (m)	Depth (m)	% Zone A	ITZA (m²)
A	9	6	100%	54
B	9	6	50%	27
C	9	3	25%	6.75
Storage	70 m²		10%	7
			Total	94.75 m²

£95 000/94.75 = £1000/m² ZA.

Comp 4: IRI lease

Zone	Width (m)	Depth (m)	% Zone A	ITZA (m²)
A	5	6	100%	30
B	5	2	50%	5
			Total	35m²

£78 000/35 = £2000/m² ZA. IRI lease so deduct 10% for landlord's external repairs. So, £1800/m² ZA.

Comp 4 is a very small unit let on a short lease, hence the high rent. Comp 3 is an asking rent so may be overvalued, which may explain why still vacant. Comps 1 and 2 are better. Comp 2 is most similar in terms of size but on High Street and has a lease incentive. Comp 1 is in an inferior location and smaller, let to an independent retailer. Suggest rent somewhere between the two, say £1100/m² ZA on FRI terms.

The subject unit is let on an IR lease, so the FRI rent from the comparable evidence needs to be adjusted upwards to take this into account, say by 12% to £1232/m².

Zone	Width (m)	Depth (m)	% Zone A	ITZA (m²)
A	10	6	100%	60
B	10	6	50%	30
C	10	6	25%	15
Remainder	10	2	10%	2
First-floor sales	100 m²	–	20%	20
Storage	80 m²	–	10%	8
			Total	135 m²

$135\,m^2 \times £1232/m^2 = £166\,320$ p.a.

[3]   This is slightly different to the normal 'value the following property. . .' It asks for two offers to be considered. They are both for light industrial uses. Offer 1 is from a local private company whereas Offer 2 is backed by a larger national company. This is relevant to the covenant strength of the two tenants. A small local company potentially may have little asset backing but loyal to area whereas a larger national company may benefit from asset backing from the parent company but may have little commitment to the area.

Turning to the lease terms, Offer 1 is a conventional 10-year lease with an upward-only rent review in year five, whereas Offer 2 is a 9-year lease with upward-only rent reviews in years three and six. So, despite the slightly shorter lease length, the rent reviews are more frequent. There is probably not much difference between the two offers in value terms. More importantly is that Offer 1 includes a break clause in year three, and the value implications of that should be considered. Offer 2 requires the landlord to make a capital contribution of £18 000. It is stated that this investment is unlikely to increase the rental value of the property.

The calculations below assume all leases are on FRI terms and purchase costs are 6.5% of price.

Analysis of comparable evidence

Analysis of #77 for the equivalent yield:
The term rent is given ($350\,m^2 \times £61/m^2 = £21\,350$ per annum), but the market rent will need to be estimated. This will need to be done using #58 and #71 because #27 includes a rent-free period and this cannot be analysed until we have a yield. So, compared to #77:

- #71: £72/m², similar size, local firm, shorter lease
- #58: £79/m², smaller size, similar type of firm, similar lease terms but with a break

Estimate market rent of #77 to be £75/m², giving a rent of 350 m² × £75/m² = £26 250 per annum.

Net term initial yield	8.35%
Net reversion yield	10.27%
Equivalent yield (approx.)	9.87%

Assume that this yield can be applied without adjustment to all of the industrial units since our adjustments for lease terms and covenant strength are being handled by adjusting the rent.

Analysis of #27 for the effective rent:
We need to assume a write-off period. The calculation below assumes a nine-year write-off, i.e. to the end of the lease, but it could also be six years (second rent review), which implies a rental growth rate of 1.9% per annum. A three-year write-off would be too short (implied rental growth of 7.14% per annum).

Headline rent (HR) (£ p.a.)					84
YP for write-off period (less rent-free period)	8.50	years @	9.87%	5.5796	
PV £1 for rent-free period	0.50	years @	9.87%	0.9540	
Capital value of HR (£)					447
YP for write-off period	9.00	years @	9.87%		5.7889
Effective MR/m² (CV of HR / YP)					77

Estimation of MR for subject property:
This MR will be the reversion rent when valuing the subject property subject to the two offers.

Value factor	#71	#58	#27	Subject
Break option	None	Year 5	None	Assume none
Lease length	5 years	10 years	9 years	Assume 10 years
Rent review(s)	None	Year 5	Years 3 and 6	Assume year 5
Covenant	Okay	Good	Okay	Assume okay
Size	295 m²	188 m²	210 m²	286 m²
Comment	Good comparable Similar size	Smaller unit. Break clause.	Good comparable Little smaller	Estimating the rent to be between #71 and #27, but closer to #71.
Rent	£72/m²	£79/m²	£77/m²	£74/m²

Valuation of subject property based on each offer

Offer 1:
The analysis of comparable evidence above has estimated an MR for the subject property of £74 per square metre assuming no break option. So, when considering offer 1, the valuation should allow for the proposed break and a possible void. There are several ways of doing this (adjust yield, assume a 'shortened' void for example), but the most rigorous approach is a weighted value between no-break and break scenarios. For this property the proposed rent is higher than the market rent (to sweeten the break option deal) but it is assumed, given how close they are, that the MR will overtake by the lease end.

No-break scenario:

Term (contract) rent passing (£ p.a.)	23 000	
YP for 10 years @ 9.87%	6.1790	
		142 118
Reversion to estimated MR (£ p.a.)	21 164	
YP perpetuity @ 9.87%	10.1317	
PV £1 for 10 years @ 9.87%	0.3901	
		83 655
Valuation before PCs (£)		225 772
Valuation after PCs @ 6.5% (£)		211 993

Break scenario (assumes a one-year void):

Rent passing (£ p.a.)	23 000	
YP for 3 years @ 9.87%	2.4926	
		57 329
Market rent – new lease (£ p.a.)	21 164	
YP perpetuity @ 9.87%	10.1317	
PV £1 for current lease + void: 4 years @ 9.87%	0.6863	
		147 151
Void costs (£)	(10 582)	
PV £1 from mid-way through void period: 3.5 years @ 8%	0.7639	
		(8083)
Valuation before PCs (£)		204 480
Valuation after PCs @ 6.5% (£)		192 000

Weighted average value assuming 70% weight on no-break scenario and 30% weight on break scenario: $(0.7 \times 211\,993) + (0.3 \times 192\,000) = £205\,995$.

Offer 2

The valuation should allow for the cost of the proposed modifications (and possibly build in a slight delay to the lease start, but that has not been included below).

Term (contract) rent passing (£ p.a.)	20 000	
YP for 3 years @ 9.87%	2.4926	
		49 851
Reversion to estimated MR (£ p.a.)	21 164	
YP perpetuity @ 9.87%	10.1317	
PV £1 for 3 years @ 9.87%	0.7540	
		161 675
Less cost of modifications		(18 000)
Valuation before PCs (£)		193 526
Valuation after PCs (£)		181 715

The two offers are very similar in value, with offer 1 slightly higher. As it does not involve any upfront capital expenditure, offer 1 is probably the one to recommend, despite the tenant being a local firm. It is perhaps worth indicating that all elements of either offer can be negotiated, including rent, lease length, payment for modifications, etc.

[4] Estimate MR of property: Consider unit numbers to see which way the rents might be increasing or decreasing along the street, calculate the areas ITZA, look at dates, type of deals (might want to check if new leases were to previous tenants), any unusual terms (break for unit 12, rent-free for unit 17, IRI terms for units 12 and 30), construction, condition and tenant covenants. Then it is a case of weighing up each, deciding on the adjustments and determining the MR.

Unit	Type of deal	Date	Adjustment	Adjusted zone A rent
27	Rent review	Recent	None	£2000/m²
12	New lease	1 year ago	IRI – deduct say 10%	£2206/m²
			Break – no adjustment	
17	New lease	6 months ago	Deduct 6 months rent-free (1/10th)	£2360/m²
30	New lease	1 year ago	IRI – deduct say 10%	£1938/m²

Most similar is no 27, although evidence is a rent review rather than a new letting. Evidence would suggest an MR in the region of £2300/m² ZA. So, market rent would be £2300/m² × 75 m² = £172 500 per annum.

Estimate the MV of the property: Consider dates of transactions (both were a while ago) and the fact that unit 30 was sold at auction. Analysis of comparable evidence:

- No 12 adjusted rent £225 012/(£2.8 m × 1.065) = 7.55% NIY
- No 30 adjusted rent £125 970/(£1.5 m × 1.065) = 7.89% NIY
- No 12 issue of tenant's break clause may cause some increase in yield.
- No 30 one year out of date and inferior pitch, etc.
- Suggest valuing at a yield of say 7.75%

Market rent (£ p.a.)	172 500	
YP perp @ 7.75%	12.9032	
Valuation before PCs (£)		2 225 802

Market valuation of multi-let property investments

[1]

Analysis of comparable evidence

The market rents for office and retail space are provided. The investment transaction for Beazer House needs analysing to estimate an equivalent yield. The retail space needs to be zoned beforehand:

Zone A:	6 m × 6 m = 36.00 m²
Zone B:	(6 m × 6 m)/2 = 18.00 m²
Zone C:	(6 m × 6 m)/4 = 9.00 m²
Remainder:	(6 m × 7 m)/8 = 5.25 m²
Store:	20 m²/10 = 2.00 m²
Total area ITZA:	70.25 m²

The MR of retail space in Beazer House is $70.25\,m^2 \times £1200/m^2 = £84\,300$ p.a. The MR of the office space is $800\,m^2 \times £230/m^2 = £184\,000$ p.a.

The total MR for Beazer House is £268 300 per annum. Together with the current rent of £230 000 per annum and the sale price of £5 m (assume transaction costs at 6.5% of this price), this information can be used to calculate the equivalent yield of 4.95% (approx.)

Valuations: unit by unit
Shop

Zone A:	$7\,m \times 6\,m = 42.0\,m^2$	
Zone B:	$(7\,m \times 6\,m)/2 = 21.0\,m^2$	
Zone C:	$(7\,m \times 6\,m)/4 = 10.5\,m^2$	
Remainder:	$(7\,m \times 8\,m)/8 = 7.0\,m^2$	
Store:	$18\,m^2/10 = 1.8\,m^2$	
Total area ITZA:	$82.3\,m^2$	

$MR = £1200/m^2 \times 82.3\,m^2 = £98\,760$ p.a.

Term (contract) rent passing (£ p.a.)	85 000	
YP for 2 years @ 4.95%	1.8607	
		158 162
Reversion to estimated MR (£ p.a.)	98 760	
YP perpetuity @ 4.95%	20.2020	
PV £1 for 2 years @ 4.95%	0.9079	
		1 811 386
Valuation before PCs (£)		1 969 548

Office A

Market rent (£230/m² × 435 m²) (£ p.a.)	100 050	
YP perp @ 4.95%	20.2020	
Valuation before purchase costs (£)		2 021 212

Office B

Term (contract) rent passing (£ p.a.)	95 000	
YP for 4 years @ 4.95%	3.5501	
		337 257
Reversion to estimated MR	100 050	
YP perpetuity @ 4.95%	20.2020	
PV £1 for 4 years 4.95%	0.8243	
		1 666 027
Valuation before PCs (£)		2 003 284

There are several ways to handle the break option: do nothing, which assumes breaks are endemic in the sector; adjust the term yield upwards, which is a bit simplistic; incorporate a void; or assign a probability to the break and non-break options and then calculate a weighted average value.

Office C
The MR for office C is $£230/m^2 \times 400\,m^2 = £92\,000$ p.a., so it is over-rented. Use a modified core and top-slice technique. The valuer must decide what

yield to capitalise the overage at. Here it is increased by 1% – it does not make much difference because the overage amount is very small relative to the core MR. The valuer must also decide how long the overage might last for. One approach would be to try the next rent review (in three years) and see what the implied growth rate is. If it is considered to be too high, then try the lease end (eight years). This should reveal that eight years is appropriate. A short-cut approach is to value the overage until halfway between the next rent review and the lease end, i.e. 5.5 years.

Rent passing (£ p.a.)	100 000	
Less MR (£ p.a.)	(92 000)	
Overage (£ p.a.)	8000	
YP 8 years @ 5.95%	6.2221	
		49 777
MR (£ p.a.)	92 000	
YP perpetuity @ 4.95%	20.2020	
		1 858 586
Valuation before PCs (£)		1 908 382

Total valuation before purchase costs: £1 969 548 + £2 021 212 + £2 003 284 + £1 909 382 = £7 903 426

[2]
Analysis of comparable evidence
> *Office lease*
> To estimate the effective MR, the valuer needs to decide on a write-off period. Could be the rent review (5 years) or the lease end (10 years), or both and then calculate the rental growth rates and decide which one looks the most sensible. Alternatively, use a write-off period between the rent review and the end of the lease.

Using a yield of 6% and writing off mid-way between rent review and lease end:		
Headline rent (£ p.a.)	95 224	
YP @ 6% for 6.5 years	5.2547	
PV £1 @ 6% for 1 year	0.9434	
Capital value of headline rent (£)		472 053
/YP @ 6% for 7.5 years		5.9007
Effective MR (£ p.a.)		80 000

The office is 400 square metres so the MR is £200 per square metre.

> *Shop sale*
> Sale price is £1 288 650 with 6.5% acquisition costs
> Area ITZA of retail space is 82.5 m²
> Rent passing is £72 550 p.a.
> MR is 82.5 m² × £1000/m² = £82 500 p.a.
> Initial yield = £72 550/1 288 650 = 5.63%
> Reversionary yield = 82 500/1 288 650 = 6.40%
> Equivalent yield = 6.31%

Valuations

Unit 1A
Area ITZA = 69.75 m² so MR is £69 750 p.a.
Reversionary investment valuation:

Term (contract) rent (£ p.a.)	60 000	
YP for initial term @ 6.31%	2.6578	
		159 465
Reversion to estimated MR (£ p.a.)	69 750	
YP perpetuity @ 6.31%	15.8479	
PV £1 for term @ 6.31%	0.8323	
		920 010
Valuation before PCs (£)		1 079 475

Unit 1B
MR is 300 m² × £200/m² = £60 000 p.a. so the unit is over-rented.
Using modified core and top-slice, consider the period over which overage
will be received and whether to increase the yield to reflect risk of this part
of the income.

Core (MR) (£ p.a.)	60 000	
YP perpetuity @ 6%	16.6667	
		1 000 000
Top-slice (overage) (£ p.a.)	10 000	
YP 5 years @ 8%^{a}	3.9927	
		39 927
Valuation before PCs (£)		1 039 927

^{a} Increased yield due to risky top-slice.

Unit 2A
The valuer needs to calculate the PV of each of the five years of stepped rents
(using an assumed 8% discount rate) and then add them up.

Value of stepped rents:		
Year 1	50 000	
PV £1 for 1 year @ 8%	0.9259	
		46 296
Year 2	60 000	
PV £1 for 2 years @ 8%	0.8573	
		51 440
Year 3	70 000	
PV £1 for 3 years @ 8%	0.7938	
		55 568
Year 4	80 000	
PV £1 for 4 years @ 8%	0.7350	
		58 802
Year 5	90 000	
PV £1 for 5 years @ 8%	0.6806	
		61 252
		273 360

Value of reversion:

Rent at review (MR)	69 750
YP perp @ 6.31%	14.8148
PV £1 5 years @ 6.31%	0.7634
	788 846
Valuation before PCs (£)	1 062 206

Unit 2B

The MR is £60 000 per annum so this is a reversionary investment valuation, but on reversion, which is the end of the lease, various things could happen. The tenant could renew, vacate or the landlord could regain possession. If the tenant renews lease, then £4000 per annum should be deducted from the MR for 21 years from the date of the improvements. The latter two options would be a simple term and reversion valuation plus an assessment of compensation for the improvements (and disturbance compensation could be noted in the case of the landlord regaining possession).

[3]

a)   Rent-based WAULT and area-based WAULT

Tenant	Area (m²)	Current rent (£ p.a.)	Length of lease (years)	Remaining term (years)	Rent (£ p.a.) WAULT	Area WAULT
A	305	21 350	15	7	149 450	1
A	251	40 160	10	8	321 280	1
B	449	76 330	15	15	1 144 950	3
C	449	49 390	25	8	395 120	2
D	449	69 595	10	6	417 570	1
E	398	55 720	15	7	390 040	1
TOTAL	2301	312 545			2 818 410	
WAULTS					9.02	9

b)   Analysis of comparable evidence

It is decided that the comparable evidence in (c) will be classed as secondary due to the poor lift access. Thus, the current market rent for office space in this locality is estimated to be £170 per square metre. The comparable evidence of market rents for storage and office space is used to calculate the current market rents for each floor of the subject property, shown below.

Floor	Tenant	Contract rent (£ p.a.)	Next rent review	Current market rent (£ p.a.)
Basement	A	21 350	2 years' time	27 450
Ground	A	40 160	3 years' time	42 670
First	B	76 330	5 years' time	76 330
Second	C	49 390	3 years' time	76 330
Third	D	69 595	Next year	76 330
Fourth	E	55 720	2 years' time	67 660
Totals		312 545		366 770

c)   Valuation of the freehold interest

Term:

Current (contract) head rent (£ p.a.)	10 000	
YP 8 years @ 8%[a]	5.7466	
		57 466

Reversion:

Market rent of occupation leases (£ p.a.)	366 770		
Current head rent (£ p.a.)	(10 000)		
	356 770		
5% share	0.05		
		17 839	
Rent passing (£ p.a.)		10 000	
		27 839	
YP in perpetuity @ 10%		10.000	
PV £1 for 8 years @ 10%		0.4665	
			129 871
Valuation before PCs (£)			187 337

[a] Very secure income.

d)   Valuation of the head-leasehold interest

Given the long length of the ground-lease (125 years) and the relatively low ground rent (currently £10 000) this interest will be valued as though it were a freehold. The difference is negligible, the YP for the remainder of the ground lease (108 years) at 11% is 9.0906 whereas the YP in perpetuity at 11% is 9.0909. The main decision that a valuer must make is the choice of yield. Although this long leasehold interest is, in many ways, like a freehold interest, it is ultimately a wasting asset and usually not as desirable. The yield should reflect such market perception as well as opportunity cost of capital, potential for growth and a return for risk taken. Yield choice is always difficult and particularly so with interests such as this where comparable evidence is hard to obtain. In practice, different yields may be applied to the capitalisation of the various rental income streams. For example, a higher yield may be adopted for the capitalisation of the reduced profit rent receivable after the review of the ground rent. Similarly, different yields may be chosen depending on which sub-tenant the rental income originates from. This may help to reflect the security value of each portion of the rental income. It is assumed here, for simplicity, that the yield is 10%.

Tenant A – basement:

Current rent (£ p.a.)	21 350	
YP 2 years @ 10%	1.7355	
		37 053
Reversion to market rent (£ p.a.)	27 450	
YP perp @ 10%	10	
PV £1 2 years @ 10%	0.8264	
		226 847
		263 900

Tenant A – ground floor:

Current rent (£ p.a.)	40 160	
YP 3 years @ 10%	2.4869	
		99 874

Reversion to market rent (£ p.a.)	42670	
YP perp @ 10%	10	
PV £1 3 years @ 10%	0.7513	
		320580
		420454

Tenant B – first floor:

Current rent (market rent) (£ p.a.)	76330	
YP perp @ 10%	10	
		763300

Tenant C – second floor:

Current rent (£ p.a.)	49390	
YP 3 years @ 10%	2.4869	
		122828
Reversion to market rent (£ p.a.)	76330	
YP perp @ 10%	10	
PV £1 3 years @ 10%	0.7513	
		573467
		696296

Tenant D – third floor:

Current rent (£ p.a.)	69595	
YP 1 years @ 10%	0.9091	
		63269
Reversion to market rent	76330	
YP perp @ 10%	10	
PV £1 1 year @ 10%	0.9091	
		693916
		757185

Tenant E – fourth floor:

Current rent (£ p.a.)	55720	
YP 2 years @ 10%	1.7355	
		96702
Reversion to market rent (£ p.a.)	67660	
YP perp @ 10%	10	
PV £1 2 years @ 10%	0.8264	
		559142
		655844

Capital value of sub-leases (£)		3556979

Head-lease:

Current ground rent (£ p.a.)	10000	
YP 8 years @ 10%	5.3349	
		53349
Reversion to market ground rent (£ p.a.)	27839	
YP perp @ 10%	10	
PV £1 8 years @ 10%	0.4665	
		129869
Value of head-lease (to be deducted) (£)		183218

Investment valuation of property investments

[1]

Year	Rent (£)	Exit value (£)	Net cash flow (£)	PV @ 7%	PV (£)
1	100000		100000	0.9346	93458
2	100000		100000	0.8734	87344
3	100000		100000	0.8163	81630
4	100000		100000	0.7629	76290
5	100000	2262816[a]	2362816	0.7130	1684655
Total PV (valuation before PCs) (£)					2023377

[a] This is the current MR projected over five years at 2.5% p.a. and capitalised at the exit yield of 5%: [100000 × (1.025)^5]/0.05 = 2262816.

[2]

Year	Description	Cash flow (£)	PV @ 15%	PV (£)
1	Rent	100000	0.8696	86957
2	Rent	100000	0.7561	75614
3	Rent	129792[a]	0.6575	85341
4	Rent	129792	0.5718	74209
5	Rent	129792	0.4972	64530
6	Rent	129792	0.4323	56113
7	Rent	129792	0.3759	48794
8	Sale proceeds	4909138[b]	0.3269	1604797
	Refurb costs	(1593800)[c]	0.3269	(521013)
Total PV (valuation before PCs) (£)				1575343

[a] MR of £120000 compounded over two years at 4% p.a. rental growth rate.
[b] £200000 p.a. compounded over eight years at 7% p.a. rental growth rate and capitalised at 7%.
[c] £1000000 build cost compounded over eight years at 6% p.a. build cost inflation rate.

[3]

Year	1	2	3	4	5
Rent (£ p.a.)	110000	110000	110000	110000	110000
Exit value (£)[a]					2200000
Sale costs (£)[b]					(49929)
Cash flow (£)	110000	110000	110000	110000	2260071
PV £1 @ 8%	0.9259	0.8573	0.7938	0.7350	0.6806
PV (£)	101849	94303	87318	80850	1538204

[a] 110000/0.05.
[b] 2080378×0.024.

Total PV (valuation before PCs) = £101849 + £94303 + £87318 + £80850 + £1538204 = £1902524

[4]

a)   No refurbishment

Rental growth rate net of depreciation:

$[(1+0.02)/(1+0.005)] - 1 = 1.49\%$ p.a.

MR at rent review $= (1+0.0149)^5 \times 80\,000 = 86\,151$

MR at lease end/new lease $= (1+0.0149)^{10} \times 80\,000 = 92\,775$

Years	1–5	6–10	11+
Gross rent (£ p.a.)	70 000	86 151	92 775
Management costs (£)	(700)	(862)	Implied in exit yield
Net rent (£ p.a.)	69 300	85 290	92 775
YP 5 years @ 8%	3.9927	3.9927	
YP perp @ 5%			20.0000
Rent review cost (£)		(431)	Implied in exit yield
PV £1 @ 8% for 0, 5 and 10 years	1	0.6806	0.4632
Exit value before sale costs (£)			859 456
Sale cost @ 2% sale price (£)			(16 852)
PV (£)	276 695	231 470	842 604

Total PV = £276 695 + £231 470 + £842 604 = £1 350 769

b)   Refurbishment

First 10 years as above (£):			508 165
Then refurbishment:			
Cost now (£)	(250 000)		
Inflation at 2% p.a. over 10 years	1.2190		
		(304 749)	
PV £1 at 8% over 10 years		0.4632	
			(141 160)
Then sale:			
MR after refurbishment (£ p.a.)	100 000		
Inflated at 2% p.a. over 11 years	1.2434		
		124 337	
YP perp @ 4.5%		22.2222	
PV £1 11 years @ 8%		0.4289	
			1 185 074
Valuation before PCs (£)			1 552 079

9.A   Appendix: valuation of residential dwellings let on long leases

In England, residential dwellings are let on either short or long leases. Since the 1980s, most short leases are Assured Shorthold Tenancies (ASTs). These tenancies provide little security of tenure; they are for a minimum fixed term of six months with a two-month statutory notice period, and they are not subject to rent control.

Tenants must be individuals, i.e. no corporate lets and the maximum rent is £100 000 per annum. Estimating market rents for ASTs is relatively straightforward given the abundance of comparable rental evidence. Sales of the landlords' investment interests will be less prevalent so two valuation methods are frequently used. First, a comparison method based on vacant possession sales, perhaps adopting a percentage of vacant possession value and maybe incorporating an allowance for time to vacate. Second an income capitalisation method based on market rents and yields.

Tenants of residential dwellings let on long leases usually pay a relatively large capital sum at the start of the lease and then a relatively low ground rent during the lease. They have more security of tenure that those on short leases, and the legislation that provides this security is different for long leases of flats from long leases of houses.

9.A.1 Flats let on long leases

For flats let on long leases, tenants who meet certain criteria have an individual right to extend their leases by 90 years, and a collective right (i.e. two or more flats in a block) to enfranchise their leasehold interests.

When extending a lease, the tenant must compensate the landlord for any reduction in value of the landlord's interest and for half of the marriage value (where it applies). For example, the lease of a flat with 68 years unexpired has a market value of £150 000 but this would be £165 000 if the lease term was extended by 90 years. The ground rent is £50 per annum.

Landlord's present interest:

Ground rent (£ p.a.)	50	
YP 68 years @ 8%	12.4333	
Value of term (£)		622
Reversion to MV (£)	165 000	
PV 68 years @ 8%	0.0053	
Value of reversion (£)		875
		1 497

Landlord's proposed interest: after lease extension, reversion is 158 years away, so nil value.

Marriage Value: Value of proposed merged interests – (value of L's present interest + value of T's present interest) = £165 000 – (£150 000 + £1 497) = £13 503. Half of this equates to £6 752. Therefore, the tenant must pay the landlord £6 752 + £1 497 = £8 249.

The rules governing enfranchisement require the tenant to compensate the landlord for:

a) The value of ground rent until the end of the lease (the term)
b) The reversionary value of the freehold on lease expiry (the reversion), or, if the tenant is entitled to lease renewal at lease expiry, the right to receive a modern ground rent
c) A share of synergistic (or marriage) value that may arise from enfranchising the lease (since it converts an encumbered freehold interest and a leasehold interest into an unencumbered freehold interest). It requires estimation of the

value of the existing leasehold interest and the value of the unencumbered freehold interest (vacant possession value), and the difference is split equally between the freeholder and leaseholder. The legislation requires the existing leasehold to be valued as if it were in a 'no-enfranchisement' world and this is difficult because comparable evidence will relate to transactions that are subject to the enfranchisement legislation. Case law shows a lack of consistency in the way the value of these leasehold interests is adjusted to take account of the legislative environment. Following the decision in Sloane Stanley Estate v Mundy (2016), two approaches appear available: adjustments based on market evidence, and adjustment based on published 'graphs of relativity' (Grover, 2014). These graphs show the percentage of freehold value of a dwelling that a lease of a given unexpired term comprises. The percentage drops at an increasing rate as the unexpired term decreases.

d) Compensation (injurious affection) for any diminution in value of another property because of the forced sale. For example, loss of access to an adjoining site or the removal of a future opportunity for development of a joint site. It may also include compensation for the loss of possible development value of the property subject to the action, for example the opportunity to build on the roof of a block of flats.

The legislation requires that the value of the interest to be acquired should be determined in accordance with market values, i.e. assuming a willing buyer and willing seller. Freehold dwellings subject to long leases are normally valued using an income approach. For example, a block of 10 flats is let on long leases with 68 years unexpired. The annual ground rent for each flat is £50 per annum, so the total ground rent for the block is £500 per annum. The current leasehold value of each flat is £150 000, so for 10 flats, this is £1 500 000. Leasehold values decline as the lease expiry approaches, but enfranchising leaseholders can extend their leases, and this will increase the leasehold values. The amount of this increase will largely depend on the length of the unexpired term before extension, and the valuer will have to estimate this, based on comparable evidence. Also, where a leaseholder has made improvements, their impact on value must be disregarded. To calculate the reversion, a value must be ascribed to the flats representing what they could be sold for when the current term expires. For this purpose, it must be assumed that the most favourable leases will then be granted to maximise value, e.g. a 999-year term.[2] In this example it is assumed that the acquisition of the freehold would increase the market value of each flat by 10% to £165 000, so the improved value is £165 000 × 10 = £1 650 000.

Value of the term:		
Ground rent (£ p.a.)	500	
YP for 68 years @ 5% [a]	19.2753	
		9638
Value of the reversion:		
Reversionary capital value of flats (£)	1 650 000	
PV £1 for 68 years @ 5%	0.0362	
		59 730
Valuation before PCs (£)		69 368

[a] Based on case law.

The synergistic value is the uplift from the current value of the freehold and leasehold interests to the enfranchised value:

$$£1\,650\,000 - (£1\,500\,000 + £69\,368) = £80\,632$$

The law requires that this is shared equally between the parties. When lease expiry is close, synergistic value can be very high relative to the value of the freehold interest. The legislation stipulates that for any flat held by a participating member where the unexpired term of the lease is more than 80 years, any synergistic value is to be ignored.

9.A.2   Houses on long leases

Certain tenants of houses[3] have the right to enfranchise the freehold (and any intermediate leasehold interest). The qualifying conditions are set out in the legislation, and they open up two routes to the estimation of the enfranchisement price. Either way, the freehold interest must be valued assuming an open market sale by a willing seller to a willing purchaser.

9.A.2.1   The site value basis

If the lease satisfies the original low rent test and the house satisfies the value limits (set out in the legislation), then the purchase price is the value of the site. The principle is to assess today's site value in terms of the ground rent the landlord could receive. The law provides an approach to this based on the tenant's right to a new 50-year lease on expiry of the current lease. The modern ground rent for the new lease would be assessed on the first day of the lease as a proportion of the site value. The valuation assumes that the tenant takes up the 50-year extension at the modern ground rent. Therefore, the freeholder's interest becomes the right to receive the rent and any increased rent on review during the term of the lease, followed by the right to receive a modern ground rent for 50 years, followed by the right to vacant possession of the house at the end of the lease.

For example, assume a house is let on a long lease at a fixed ground rent of £10 per annum, with 30 years remaining on the lease.

Term 1 value:		
Ground rent (£ p.a.)	10	
YP for 30 years @ 5%	15.3725	
		154

The next step is to value the modern ground rent. As there is rarely any evidence of sales of plots of land, the site value is assumed to be a percentage (say between 30% and 50%) of the freehold vacant possession value of the house developed to its full potential (assumed here to be £1 600 000). This site value is then decapitalised at a yield of 5% to arrive at the modern ground rent.

Term 2 value:

Site value assumed to be 40% of £1 600 000 (£)	640 000
Modern ground rent taken as 5% of site value (£ p.a.)	32 000
YP for 50 years @ 5%	18.2559
PV of £1 in 30 years @ 5%	0.2314
Ground rent (£ p.a.)	135 181

The last step is to value the final reversion. This is based on the value of the house as it stands. If this is assumed to be £1 140 000, the reversionary value is:

Standing house value (£)	1 140 000
Present value of £1 in 80 years @ 5%	0.0202
	23 028

The purchase price for the freehold is therefore the sum of the three values:

$$£154 + £135 181 + £23 028 = £158 363$$

9.A.2.2 The market value basis

In all other cases, including cases where the original lease had been extended, the purchase price is the value of the house, including a share of synergistic value. This method assumes that the freeholder gains possession of a house and not a cleared site. The method also recognises the leaseholder's special interest in the purchase, and this is applied as synergistic value, half of which is added to the market value of the landlord's interest when the long lease expires. No synergistic value is payable in cases where the unexpired period of the lease is more than 80 years.

For example, a house with an unexpired term of 65 years and a rent of £50 per annum has a vacant possession value estimated at £1 200 000. On expiry of the lease, the freeholder will receive the house with vacant possession. It is assumed that the acquisition of the freehold will produce an increase in value of 10%, to £1 320 000.[4]

Term value:		
Ground rent (£ p.a.)	50	
YP for 65 years at 5%	19.1611	
		958
Reversion value:		
Value of the reversion (£)	1 320 000	
PV £1 for 65 years @ 5%	0.0419	
		55 308
Valuation before PCs (£)		56 266

The synergistic value is £63 734 (£1 320 000 minus the leaseholder's interest of £1 200 000 and the freeholders interest £56 266). The tenant would have to pay half this figure in addition to the value of the freeholder's interest. Any additional value in the property arising from improvements by the tenant is ignored[5].

Notes

1. To qualify for compensation, the improvements must have been made after the 25 March 1928 and not in pursuance of a statutory or contractual obligation (except that after 1954, those in pursuance of a statutory obligation will qualify).

2. Most long leaseholders have statutory protection to revert to assured tenancies on expiry of their existing leases. In the valuation of leasehold interests subject to protected occupancy, the improved value is sometimes discounted by a percentage to reflect that the landlord will not receive a vacant flat on expiry but a tenant paying a full weekly rent. This is ignored here.

3. The qualifying conditions are set out in the legislation.

4. A valuer may also account for the tenant's right to occupy the property after the lease has ended. The law provides that when a long lease expires the leaseholder can continue to occupy the property, under an assured tenancy, paying a market rent. This right can in some cases reduce the value of the freehold interest. This is ignored for the purpose of this example.

5. Leasehold Advisory Service (https://www.lease-advice.org).

Chapter 10
Valuation of Development Property

10.1 Introduction

A development property is defined in the International Valuation Standards (IVS 410) as a property 'where redevelopment is required to achieve the optimum use, or where improvements are either being contemplated or are in progress at the valuation date'. This includes undeveloped land, the redevelopment of previously developed land for the same or alternative uses and the improvement and alteration of existing buildings.

A development property is, therefore, a hypothetical construct because it does not exist at the time of the valuation. It cannot be inspected, measured, surveyed; it can only be imagined. The definition states that the development should be to the optimum use, so that provides a steer, but valuers are likely to vary in their opinions of the optimum use when it is precisely defined in the valuation. Therefore, sensitivity analysis and scenario modelling may form part of the valuation of development property.

As with the valuation of investment property, valuations of development property can be undertaken using various valuation bases, primarily market value and investment value.

10.2 Market valuation of development property

The market value of a property will reflect its highest and best (or optimum) use, and this includes alternative uses as well as the existing use. These alternative uses must be possible, legally permissible, and financially feasible. As such, they can

---

Property Valuation, Third Edition. Peter Wyatt.
© 2023 John Wiley & Sons Ltd. Published 2023 by John Wiley & Sons Ltd.
Companion website: www.wiley.com/go/wyatt/propertyvaluation3e

include development or redevelopment potential, having regard to current and prospective economic and market circumstances and planning conditions. So, the market value of a development property refers to its optimum value assuming (re) development.

This means the valuation is likely to be the subject of special assumptions concerning the proposed or anticipated development in addition to the usual ones associated with a market valuation of an existing use. Examples include:

- The valuation assumes that the site has vacant possession.
- If the site owned by more than one owner, arrangements for allocating development costs and value have been agreed.
- No claims for ransom value for the provision of access and/or services.
- No abnormal costs associated with the construction of infrastructure.

It may be that a market valuation assuming existing use and market valuation assuming development to an optimum alternative use (or an improved existing use) are both required.

The extent of information necessary for a valuation is determined by a range of factors including the stage at which the valuation is being prepared, the purpose and the individual characteristics of the property being valued, and any assumptions or special assumptions made. As well as the factors listed in Chapter 5, the planning regime is an important factor in the value of development property and particular consideration should be paid to (RICS 2019, pp. 36–7):

- Permitted use if buildings are to be retained, or the possibility of identifying an established use;
- Any extant permissions to undertake development, and if close to expiry, whether a similar permission would be granted again;
- The existence of a development plan and land allocations for specific uses;
- The possibility, nature and extent of development that might be possible without the need for a planning permission and
- The existence of regulations relating to, for example, heritage protection, conservation, environmental features, view corridors, sight lines or buffer zones.

When assessing development potential, it is important to specify the proposed development and make assumptions clear in the valuation report. It may be necessary to consider what planning consent might be forthcoming, together with any conditions and obligations.

An assessment of the form and extent of the physical development that can be accommodated on the site is essential, having regard to site features, characteristics of the surrounding area and the likelihood of obtaining permission. Matters that may need to be considered include:

- Building-related issues such as the period estimated to complete the new buildings, achieving optimum occupational efficiency ratios, car parking standards and/or restrictions, regulations concerning energy efficiency, biodiversity enhancement and carbon reduction.

- Any conditions attached to the planning consent, such as the provision of infrastructure, community facilities and low-cost housing. This may be particularly important if the magnitude of provision is linked to the value of the development property.
- Potential for realising synergistic development value acquiring adjacent property or interests in land.

Market valuations of development property are normally undertaken in two ways, the comparison method and the residual method, and one can be used to cross-check the other.

10.2.1 Comparison method

The comparison method is normally preferred, but it does require information on similar assets in a similar location. In the case of development property, valuation by comparison is potentially reliable if evidence of sales can be found and analysed on a common unit basis, such as price per hectare, site price as a proportion of the value of the completed development or price per unit or habitable room. Care is required because simple unit metrics can hide factors that may influence value in individual cases.

The comparison method may be most appropriate where there is an active market and/or a relatively straightforward low-density form of development is proposed. Examples might include greenfield land in rural areas where infrastructure costs are consistent and not excessive, or small residential developments or small commercial and industrial estates where it is likely that the density, form and unit cost of the development will be similar. Less frequently, it may be possible to compare larger sites for housing or other developments on this basis.

In comparing sites, the condition of each site, any remediation costs, construction costs, infrastructure and service requirements will need careful consideration. Site prices can differ considerably within a small geographical area.

For assets where work on the development has commenced but not completed, the market approach is unlikely to be the most appropriate choice because comparable transactions of partly completed developments are unlikely to be found. The residual method is more likely to take account of the individuality that will exist.

Where development has begun, the assumption of optimum development is expected to apply. Where the actual development taking place is not the optimum development, the cost of removing the existing works must be included unless improving them forms part of the optimum development.

When valuing a partly completed development property, all the inputs, including the cost of completing the development, must be assessed as at the valuation date. The valuation should reflect the risks remaining at the valuation date, which may be different from the commencement of the scheme and a re-assessment of the rate of return is required. This may be affected by the stage the project has reached, whether building contracts remain in place or whether any agreements to purchase/let the whole or part of the completed development are in place. A project that is nearing completion will normally be viewed as being less risky than one at an early stage.

10.2.2 Residual method

As explained in Chapter 8, the residual method is based on the simple concept that the value of a property with development potential is the value of the completed development minus the cost of undertaking that development. Because it is derived from what remains after costs have been deducted from the value of the completed development, it is widely referred to as residual land value. The method can be used to determine other residual sums such as developer's profit, but the focus here is on land value as the residual amount.

The simple residual valuation concept is complicated by the fact that development takes time while the valuation is snapshot. Two techniques have evolved to handle this, a basic residual valuation technique and a more detailed discounted cash-flow (DCF) technique. Choice of technique will depend on the purpose of the valuation, the stage in the development process that the valuation is performed and the type of development property that is being valued. A basic residual valuation might be used for less-complex assets or early in the development process to consider various proposals in the search for the optimum development. A DCF valuation may be used for more complex assets with phased construction or disposal, where the timing of events needs to be fully accounted for in the valuation.

Revenues and costs that are typically considered in residual valuations are:

- Value of completed development, usually based on sale prices or capitalised rental income
- Site clearance, remediation and preparation costs
- Costs of construction, including professional fees and a contingency allowance
- Costs and fees relating to planning, including infrastructure levies, development fees or planning obligations
- Finance costs of debt used to fund construction and site costs
- Developer's profit
- Any other costs or inflows related to the development

In Chapter 8, the residual method of valuation was described, covering two techniques, the basic residual and the DCF. These techniques are discussed further below.

10.2.2.1 Basic residual technique

This method was described in Chapter 8. Here, an example is provided as a reminder of how the technique proceeds.

A site has planning permission for an office building comprising 1875 square metres gross internal area (GIA; 80% efficiency ratio). The market rent for new offices in this location is £390 per square metre and new, recently let office buildings in this location have recently sold to investors for a price reflecting an initial yield of 7.50%. Construction costs are expected to be £1775 per square metre and site preparation work and external works are estimated at £30 000 and £100 000 respectively. Professional fees are estimated at 13% of build costs and external works, and there is a contingency allowance set at 5% of build costs, external works and professional fees. The developer has a target profit margin of 20% of development costs and has secured bank lending at 8% per annum. The construction is expected to take 15 months and a void period of 3 months is anticipated. Using the basic residual technique, the valuation might proceed as follows.

Development value			
Net internal area (NIA) (m²)	1500		
Estimated rental value (ERV) (£/m²)	390		
		585000	
Net initial yield	7.50%	13.3333	
Gross development value (GDV) before sale costs (£)			7800000
Net development value (NDV) after sale costs (£)			7647059
Development costs			
Site preparation (£)		(30000)	
Building costs (£/m² GIA)	1775	(3328125)	
External costs (£)		(100000)	
Professional fees (% building costs and external works)	13.00%	(445656)	
Contingency allowance (% construction costs)	5.00%	(193689)	
Finance on costs and fees for half building period @	8.00%	(201908)	
Finance on costs and finance for void period @	8.00%	(83522)	
Letting agent's Fee (% ERV)	10.00%	(58500)	
Letting legal fee (% ERV)	5.00%	(29250)	
Developer's profit on total development costs (%):	20.00%	(894130)	
Total development costs (TDC) (£)			(5364780)
NDV – TDC (£)			2282278
Land costs (£)			
Developer's profit on land costs (%)	20.00%	(380380)	1901899
Finance on land costs over total development period	8.00%	1.50	0.8910
Residual land value before purchase costs (£)			1694540
Residual land value after purchase costs (£)			1591117

The basic residual technique is a simple means of estimating the value of development property. However, its simplicity comes at a price and the technique is often criticised for several reasons.

First, the handling of finance. By calculating interest on half of the building costs over the construction period (or on all the building costs over half of the building period), it is assumed that these costs are incurred evenly. Often, the initial build-up of costs tends to be gradual, peaks at 60% and then tails off, taking the form of an s-curve. Typically, only 40% building costs are incurred halfway through the construction period whereas the residual method assumes 50%. Consequently, the estimated finance cost can be higher than that which is actually incurred. As a counter to this over-estimate, the residual method assumes that interest accumulates annually rather than quarterly.

Second, development projects take time, sometimes many years, and a return that is expressed as a cash margin does not reflect the timing of cost outlays or the receipt of profit and therefore might not compensate developers appropriately for the risks taken. To reflect uncertainty caused by time, a void period might be incorporated, and this would increase the development costs. However, developer's profit is calculated as a percentage of those costs so there is no obvious penalty (or risk) associated with an extended void period for the developer. Instead, land value reduces, and the landowner is penalised. The internal rate of return (IRR) of the development project also falls owing to the delay in receiving revenue that is associated with the void, but this is not apparent to the developer using a cash margin metric.

This point raises a wider issue. When developer's profit is expressed as a cash margin, it could represent very different IRRs depending on the length of the development period and the scale of construction costs relative to revenues. Research by Crosby et al. (2020) revealed that the two main drivers of variation between a simple profit on cost and the project IRR are the length of time a development takes and the ratio of costs to value, while finance rates have very little impact. If the same cash margin is spread over more years, then the associated IRR will naturally fall as the development period lengthens. Interestingly, the IRR and the cash margin are roughly the same, around 15–20%, for development periods of around two to three years. There is a counter-effect in that the longer the period, the higher the interest charges in the residual model and the lower the residual land value. A reduced land value will then reduce the initial costs of the scheme, but this is more than offset by the normal discounting of future revenue flows, so as the development period gets longer, the IRRs reduce. This pattern contrasts with what literature from corporate finance suggests regarding the behaviour of required rates of return, that these should, all else equal, be higher with longer duration projects (see Cornell 1999).

Also, as the ratio of construction costs to development value increases (i.e. the initial outlay on land reduces), the IRR increases. This is because the land cost becomes a smaller share of the development cost, and, proportionately, more costs are incurred later in the development period, thus having a beneficial impact on the IRR.

Finally, the basic residual technique cannot deal with revenue that may be received and expenditure that may be due at various times during the development period. Revenue might be received in phases when part of the development is sold or let during the development period. For residential developments, sales of dwellings often occur in stages. This requires a cash-flow format to reflect finance costs when debt can be paid down during the scheme.

10.2.2.2   Discounted cash-flow technique

As with investment property, valuations of development property may be undertaken using a DCF technique, as introduced in Chapter 8.

For example, your client owns the freehold interest in a development site that has planning permission for a three-storey mixed-use retail and office property. Each floor will measure 10 metres by 30 metres GIA. The ground floor will be developed as a retail unit having sales space of seven metres internal frontage by 20 metres depth, with 20 square metres of storage space at the rear. The two upper floors will be used for offices and have lettable floor space of 250 square metres on each floor. Site preparation is estimated to cost £100000. Market rent for offices is £200 per square metre and overall market rent for retail is £400 per square metre for ground floor sales and £150 per square metre for storage space. Yields are currently 6% for this kind of development. Costs of construction are estimated to be around £900 per square metre of GIA, professional fees are 10% of building costs and contingencies are 5% of building costs and professional fees. Development finance is available at 8% per annum. A lead-in period of six months is anticipated, the construction period is estimated to be nine months and a void period of six months is also anticipated. The developer's required profit margin is 17% of gross development value (GDV). Calculate the market value of the site using the DCF technique.

Assumptions:

- Site preparation costs are incurred in the first quarter.
- Building costs are spread over quarters two to four as follows: 25%, 50%, 25%.
- Land purchase costs are 6.50% of the land price.
- Investment sale costs are 2% of the NDV.
- Letting agent and legal fee are 10% of the annual rent.

Calculations:

- Estimated rental income: retail $(140\,m^2 \times £400/m^2) + (20\,m^2 \times £150/m^2)$; offices $500\,m^2 \times £200/m^2 = £159\,000$ p.a.
- Gross development value: $£159\,000/0.06 = £2\,650\,000$ and net development value: $£2\,650\,000/(1 + 0.02) = £2\,598\,039$
- Total building cost: $900\,m^2 \times £900/m^2 = £810\,000$
- Finance at 8% per annum = 1.94% per quarter

Cash flow:

	0	1	2	3	4	5	6	7
Net development value (£)								2598039
Site preparation (£)		(100000)						
Building costs (£)			(202500)	(405000)	(202500)			
Professional fees (£)			(20250)	(40500)	(20250)			
Contingencies (£)			(11138)	(22275)	(11138)			
Letting and legal fee (£)								(15900)
Developer's profit (£)								(450500)
Net cash flow (£)	0	(100000)	(233888)	(467775)	(233888)	0	0	2131639
PV £1 @ 1.94%	1.0000	0.9809	0.9623	0.9439	0.9259	0.9083	0.8911	0.8742
PV of cash flow (£)	0	(98094)	(225058)	(441539)	(216563)	0	0	1863479
NPV (residual land value before PCs) (£)	882225							
Residual land value after PCs @ 6.5% (£)	828380							

Understanding the nature of development risk is crucial to the identification of the appropriate developer's profit for undertaking the development. Many of these risks will affect the revenue and costs of the development and small changes in these inputs can lead to large shifts in the output residual land value. For example, site A is a prime city-centre location with high land cost relative to other costs and site B is out-of-town, in a greenfield location with low land cost relative to other costs. Residual valuations have produced the following estimates of development value, development cost and site value:

Site A – Development on a prime site:

Development value (£)	10 000 000
Development cost, including finance (£)	(7 000 000)
Residual land value (£)	3 000 000

Site B – Development on a cheap site:

Development value (£)	10 000 000
Development cost, including finance (£)	(9 000 000)
Residual land value (£)	1 000 000

Three scenarios may be constructed based on changes to development cost and value over the period of the development:

i) Development value and cost increase by the same percentage:
   If this happens, site value at both locations will increase by the same percentage amount.

ii) Development value increases by 25% and cost by 5%:

Site A	Development value (£)	12 500 000	
	Development cost, including finance (£)	7 350 000	
	Residual land value (£)	5 150 000	+72%
Site B	Development value (£)	12 500 000	
	Development cost, including finance (£)	(9 450 000)	
	Residual land value (£)	3 050 000	+205%

iii) Development value increases by 5% and cost by 25%:

Site A	Development value (£)	10 050 000	
	Development cost, including finance (£)	(8 750 000)	
	Residual land value (£)	1 300 000	–57%
Site B	Development value (£)	10 050 000	
	Development cost, including finance (£)	(11 250 000)	
	Residual land value (£)	(1 200 000)	–220%

If the residual land value is small relative to other costs, changes in development value and development cost will magnify changes in the residual so much so that it can easily disappear. This volatility will inform the level of profit that developers aim for.

It is possible to incorporate the developer's profit sum as a percentage return on development value or on development costs. Then the cash flow is usually discounted at a finance rate that assumes both land and construction costs are wholly debt financed to calculate a residual land value. This approach, which is closest to the basic residual valuation, is criticised for incorporating a profit measure that does not reflect the timescale of the investment. For instance, all else equal, the profit level (if expressed as a ratio of development costs or value) would be the same for a 1- or 10-year scheme, whereas the internal rates of return would be different. The application of an absolute (in cash terms) profit margin invariant with the time frame of a development implies an assumption that developers are indifferent to whether £1 is received next year or in 10 years. The approach has also been criticised for incorporating finance in a project appraisal, and at a debt level that is unrealistic. The approach does not conform to mainstream project appraisal in other asset classes where the investment decision is usually separate from the financing decision. There is little direct connection between the rate at which a developer can borrow and the appropriate discount rate to be applied to a particular development project. This is particularly so when the cash flows are subject to a high degree of project risk as in many property developments.

A DCF residual land valuation can be adapted so that profit is not input as a lump sum receivable at the end of the scheme. Instead, it can be handled by using a developer's *target rate of return* as the discount rate. This means that financing is handled separately. This approach is best illustrated using an example. A site has planning permission for a six-storey mixed-use retail and office building. The dimensions of each floor are 20 metres by 30 metres (gross). The five upper storeys will be used for offices with a net internal area of 500 square metres on each floor. The ground floor will be developed as two large retail units, each with 8 metres of frontage, 20 metres of depth sales space and 7 metres depth of storage space at the rear of each unit. The site has been cleared, serviced and is ready for development. It is estimated that the development will take 18 months. Annual rental values are estimated to be £220 per square metre for offices and £440 per square metre for ground floor sales on an overall basis (not zoned) for retail space. Storage space is currently letting at around and £150 per square metre. Investment sales for this type of mixed-use scheme are yielding 7%. Building costs are estimated to be £1200 per square metre of GIA, professional fees are 10% of building cost and contingencies are 5% of building costs and professional fees. The building costs are assumed to be spread as follows over the one-and-a-half-year development period:

	0	1	2	3	4	5	6
Building costs (£)		10%	15%	50%	15%	10%	

Development finance is available at 7% per annum. Assuming a developer's target rate of return of 12% per annum and using a quarterly cash flow, the residual land value is calculated as follows:

Inputs and Assumptions:

GIA (m²)	3600
Office NIA (m²)	2500
Retail sales (m²)	320
Retail storage (m²)	112
Building costs (£/m²)	1200
Office rent (£/m²)	220
Retail rent (£/m²)	440
Storage rent (£/m²)	150
Finance (% p.a.)	7.00%
Target rate (%)	12.00%
Yield (%)	7.00%
Purchase costs (% NDV)	6.50%
Professional fees (% building costs)	10%
Contingency (% building costs and professional fees)	5%
Letting fee (% estimated MR)	15%
Disposal costs (% NDV)	2%

Calculations:

Construction cost (£)	4 320 000
Office ERV (£ p.a.)	550 000
Retail ERV (£ p.a.)	140 800
Storage ERV (£ p.a.)	16 800
Total ERV (£ p.a.)	707 600
YP perp % yield	14.2857
GDV (£)	10 108 571
NDV (£)	9 910 364
Quarterly finance rate	1.71%
Quarterly target rate	2.87%

Quarter:	0	1	2	3	4	5	6	Totals
Building costs (£)	0	(432 000)	(648 000)	(2 160 000)	(648 000)	(432 000)	0	(4 320 000)
Professional fees (£)	0	(43 200)	(64 800)	(216 000)	(64 800)	(43 200)	0	(432 000)
Contingencies (£)	0	(23 760)	(35 640)	(118 800)	(35 640)	(23 760)	0	(237 600)
Letting fee (£)	0	0	0	0	0	0	(106 140)	(106 140)
NDV (£)	0	0	0	0	0	0	9 910 364	9 910 364
Net cash flow before finance (£)	0	(498 960)	(748 440)	(2 494 800)	(748 440)	(498 960)	9 804 224	
Finance:								
Opening balance (£)	0	0	(498 960)	(1 255 912)	(3 772 136)	(4 584 923)	(5 162 095)	
Interest on loan (£)	0	0	(8 512)	(21 424)	(64 347)	(78 212)	(88 058)	(260 552)
Closing balance (£)	0	(498 960)	(1 255 912)	(3 772 136)	(4 584 923)	(5 162 095)	(5 356 292)	
Residual land value (£)								4 554 072
PV of residual land value before PCs (£)	3 734 808							
Residual land value after PCs (£)	3 506 862							

As well as adjusting the developer's profit requirement, risks associated with a development can be factored into the valuation by either adjusting valuation inputs perhaps using sensitivity analysis or scenario modelling. These risk-analysis techniques may be used to identify variation in valuations of development property and the source of that variation (see Chapter 14). The presence of future possible outcomes or options inherent in the development process suggest some additional analysis of the valuation would help contextualise the spot estimate. With larger sites that take longer to develop, options within the development process become more likely, including phasing the development, thereby using the experience from the first stage to manage the development of later phases, opting to wait until the optimum timing of development can be identified or even the option to abandon the current scheme. These options have been the subject of several studies using option pricing techniques from financial markets to quantify them (Geltner et al. 2018). Chapter 14 shows how risk-analysis techniques can address some of these issues and can help indicate likely variation around the valuation.

10.3   Investment valuation of development property

The investment value basis of valuation may be appropriate where assumptions made within a financial appraisal or valuation relate to a specific entity and/or client. Here, investment value is taken to refer to the estimation of value to the developer, in other words, the developer's profit or return. Both the basic residual and the DCF technique can be used to estimate the investment value of development property, and these are considered below. After that, the creation of more complex financial appraisals is discussed.

10.3.1   Estimating the investment value of development property

10.3.1.1   Basic residual technique

For example, a property development company is considering the acquisition of a development site, which is on the market at an asking price of £3.5m. The site has outline planning consent for an office building with a GIA of 5000 square metres. A local commercial letting agent has indicated that demand for office space in the town is steady and rents have recently been agreed at £220 per square metre per annum. Recent office investment transactions show capitalization rates around 6%. Building costs are estimated to be £1100 per square metre. Site clearance and external works are expected to cost £400 000. After a lead-in period of 6 months, construction is expected to take 15 months. Current interest rates on development loans of this nature are around 8% per annum. Making assumptions where necessary, calculate developer's profit and report it as a cash sum, a return on costs and a return on value.

Development value

Net internal area (NIA) (m²)		4250	
Estimated rental value (ERV) (£/m²)		220	
			935 000
YP in perp @	6.00%	16.67	
Gross development value before sale costs (£)			15 583 333
Net development value (NDV) after sale costs (£)	2.00%		15 277 777

Development costs

Land price (£)		(3 500 000)	
Land purchase costs (% land price)	6.50%	(227 500)	
Building costs (£/m² GIA)		(5 500 000)	
External works (£)		(400 000)	
Professional fees (% construction costs and external works)	10.00%	(590 000)	
Contingency allowance (% construction costs)	3.00%	(194 700)	
Finance on site price and purchase costs for development period	8.00%	(704 718)	
Finance on construction costs and fees for half building period	8.00%	(329 397)	
Finance on construction costs, fees and interest for void period	8.00%	(275 166)	
Letting agent's fee (% ERV)	15.00%	(140 250)	
Total development costs (TDC) (£)			(11 861 731)
Developer's profit on completion (NDV − TDC) (£)			3 416 046
Return on costs			28.80%
Return on value			22.65%

10.3.1.2   Discounted cash-flow technique

In a DCF, the land price can be inserted in the cash flow and the discount rate represents the developer's profit expressed as a period rate of return. The cash flow can be used to calculate the NPV given a target rate or it can be used to calculate the IRR of the project. It is important to note that the target rate and the IRR are before finance. If the valuer wants to estimate profit as a lump sum at the end of the development, the land value is again inserted at the beginning of the cash flow. Then, interest on this land cost and all other development costs is compounded to the end of the cash-flow (assuming 100% borrowing). The residual amount at the end of the cash flow is the developer's profit.

Take the following example:

Inputs and Assumptions:

Land price (£)	(1 000 000)
Yield	6.75%
Gross internal area (GIA) (m²)	2500
Build cost (£/m²)	(1250)

Estimated rental value (ERV) (£/m²)	210
Build period (years)	1.25
Efficiency ratio	85%
Finance rate (% p.a.)	7.50%
Target rate (% p.a.)	15.00%
Land purchase costs (% land purchase price)	5.75%
Development sale costs (% NDV)	2.00%
Letting fee (% ERV)	15%
Lead-in period (years)	0.25
Void period (years)	0.50
Contingencies (% build costs and professional fees)	5.00%
Professional fees (% build costs)	10.00%

Calculations:

Net internal area (NIA) (m²)	2125
Build costs (£)	(3 125 000)
Estimated MR (£ p.a.)	446 250
Gross development value (GDV) (£)	6 611 111
Net development value (NDV) (£)	6 481 481
Finance rate (% p.q.)	1.82%
Target rate (% p.q.)	3.56%

CASH FLOW

	0	1	2	3	4	5	6	7	8	9
NDV (£)										6 481 481
Land price (£)	(1 000 000)									
Land purchase costs (£)	(57 500)									
Building costs (£)			(625 000)	(625 000)	(625 000)	(625 000)	(625 000)			
Professional fees (£)			(62 500)	(62 500)	(62 500)	(62 500)	(62 500)			
Contingencies (£)			(34 375)	(34 375)	(34 375)	(34 375)	(34 375)			
Letting fee (£)										(66 938)
Net cash flow (£)	(1 057 500)	0	(721 875)	(721 875)	(721 875)	(721 875)	(721 875)	0	0	6 414 544
PV £1 @ target rate	1.0000	0.9657	0.9325	0.9005	0.8696	0.8397	0.8109	0.7830	0.7561	0.7302
PV (£)	(1 057 500)	0	(673 152)	(650 038)	(627 717)	(606 163)	(585 349)	0	0	4 683 771
NPV (£) (before finance)	483 851									
Finance										
Opening balance (£)	0	(1 057 500)	(1 076 794)	(1 818 314)	(2 573 364)	(3 342 189)	(4 125 041)	(4 922 175)	(5 011 978)	(5 103 420)
Interest (arrears) (£)	0	(19 294)	(19 646)	(33 174)	(46 950)	(60 977)	(75 260)	(89 803)	(91 442)	(93 110)
Closing balance (£)	(1 057 500)	(1 076 794)	(1 818 314)	(2 573 364)	(3 342 189)	(4 125 041)	(4 922 175)	(5 011 978)	(5 103 420)	1 218 014
After finance profit (£)	1 218 014									
As a % cost	(4 733 813)	25.73%								

Developer's target profit metrics, be they cash margins such as profit-on-cost or profit-on-value, or a rate of return, can vary significantly between projects. Many risks relate to the volatility of profit relative to uncertainty regarding the major inflows and outflows over the development period. A target rate of return has an advantage over the profit on cost or value approaches because it accounts for project duration, and it can be compared with rates of return from other types of projects. The rate is subject to scheme-specific risk, market risk and often high levels of operational gearing. Rates of return that may be observable from other developments will need careful consideration if they are to be used as comparable evidence. If comparable evidence is lacking, another approach is to build a target rate from a risk-free rate using components of market and asset-specific risk premia. Consideration should be given to scheme specific as well as market risks, and this should be set out by reference where possible to market data and scheme-specific inputs.

What is an acceptable risk-adjusted target rate of return for development activity? It depends on the type of developer, type and location of the development and the state of the market. According to the RTPI (2018), developers usually base required returns on experience rather than on sophisticated modelling. Geltner et al. (2007, chapter 29) stress that, although difficult, estimating a required rate of return is an unavoidable element of all project evaluations and inherent to the process. They suggest several possible approaches, contingent upon the stage in the development process, which draw upon real option pricing, the use of a 'reinterpreted' weighted average cost of capital (WACC) or historic return data from 'pure play' real-estate development companies. Brown and Matysiak (2000) discuss risk grouping, risk ratios, capital asset pricing model, arbitrage pricing theory and WACC. Estimating a required rate of return for development opportunities requires data that typically do not exist or assumptions that are difficult to verify but, while problematic, it is important to acknowledge that required rates of return are implicit in all conventional development appraisal techniques when applying simple profit-on-value and profit-on-cost ratios.

As it is a risk-adjusted return, it is important to ensure that risk is not double counted by allowing for it in, say, a contingency fund in the appraisal and in the risk-adjusted discount rate. All else equal, a higher contingency allowance should be compensated by a relatively lower target rate of return. Also, the treatment of finance within the residual appraisal is linked to the formulation or target rates of return and appropriate metric or proxy. Development risk is compounded when a combination of debt and equity funds are used to finance real-estate development.

10.3.2 Financial appraisals of development property

It is possible to use the DCF technique to conduct appraisals which take account of the level of borrowing and different costs of borrowing on different forms of debt. When developing these models, the role and purpose of the valuation must be fully recognised. Market valuations require market-based inputs and assumptions as to highest and best use. Specific funding arrangements and rates of return required by individual developers are not necessarily based on market indicators.

As the purpose of the development appraisal shifts from the estimation of market value to investment value, the 100% debt financing assumption in the basic residual and DCF techniques can be altered. Instead, it is useful to consider a combination of debt and equity funding. Appraisals of real-estate development projects (as opposed to standing investments) often involve multiple sources of finance. Therefore, a post-finance cash flow is an important component of the appraisal and is used to estimate a return on equity. Perusal of various development appraisals will reveal that there is no standard form that these post-finance cash flows take but, in general, they follow a pattern. The following example is an attempt to illustrate the rudiments of a typical financial appraisal of a development project.

A simple one-year quarterly cash flow is shown below.

	0	1	2	3	4
Costs	(10000)	(10000)	(10000)	(10000)	
Revenue					45000
Net cash flow	(10000)	(10000)	(10000)	(10000)	45000

The IRR of the net cash flow is 4.77% per quarter (20.48% per annum). If it is assumed that, rather than using 100% debt finance to fund all the costs, 75% is financed by 'senior' debt at a cost of 2.41% per quarter (10% per annum equivalent), accrued in arrears. The remaining 25% of costs are funded by the developer's own funds (equity). The debt and the equity funding are provided on a side-by-side basis and because the revenue from the project is received at the end, no debt is repaid until this point and interest on the debt is rolled up until the end of the scheme. The senior debt finance is shown below.

Opening balance	0	(7500)	(15181)	(23047)	(31102)
Interest	0	(181)	(366)	(555)	(750)
New costs	(7500)	(7500)	(7500)	(7500)	0
Closing balance	(7500)	(15181)	(23047)	(31102)	(31852)

The equity costs and receipts are shown below. The total equity costs are 10000 and the total revenue (after the finance costs have been paid) is 45000 – 31852 = 13148. This means the cash profit to the developer is 13148 – 10000 = 3148.

Equity costs	(2500)	(2500)	(2500)	(2500)	0
Equity revenue	0	0	0	0	13148
Equity cash flow	(2500)	(2500)	(2500)	(2500)	13148

Three common performance metrics for equity return are:

- Equity multiple: equity revenue/equity costs = 13148/10000 = 1.31
- Return on equity: equity profit/equity costs = 3148/10000 = 31.48%
- Equity IRR: 53.20% per annum

Gearing the funding in this way has increased the developer's IRR from 20% to 53%.

Further 'mezzanine' debt finance is now introduced so that 15% of the costs are funded on a side-by-side basis at an annual cost of 18% per annum (4.22% per quarter). The mezzanine debt finance is shown below.

Opening balance	0	(1500)	(3063)	(4693)	(6391)
Interest	0	(63)	(129)	(198)	(270)
New costs	(1500)	(1500)	(1500)	(1500)	0
Closing balance	(1500)	(3063)	(4693)	(6391)	(6660)

And the equity cash flow is as follows.

Equity costs	(1000)	(1000)	(1000)	(1000)	0
Equity revenue	0	0	0	0	6488
Equity cash flow	(1000)	(1000)	(1000)	(1000)	6488

Total equity costs are now 4000 and revenue is 45 000–31 852–6660 = 6488. Cash profit is therefore 2488.

- Equity multiple: 6488/4000 = 1.62
- Return on equity: 2488/4000 = 62.20%
- Equity IRR: 109.56% per annum

Additional gearing from the mezzanine finance increases the equity IRR further.

Some lenders might be keen to participate in some of this equity return, in the form of a profit-sharing arrangement perhaps. Continuing the example above, assume the developer sets a threshold equity return of say 25%, and the mezzanine lender takes a 50% share of any profit earned over this rate. By participating in the equity in this way, the mezzanine lender offers to reduce the interest rate on their debt to say 14%. The senior debt is as before, but with mezzanine finance at 14% per annum (3.33% per quarter), the mezzanine debt table is shown below.

Opening balance	0	(1500)	(3050)	(4652)	(6306)
Interest	0	(50)	(102)	(155)	(210)
New costs	(1500)	(1500)	(1500)	(1500)	0
Closing balance	(1500)	(3050)	(4652)	(6306)	(6516)

The developer's threshold return at 25% per annum (5.74% per quarter) is calculated as follows.

Opening balance	0	(1000)	(2057)	(3175)	(4358)
Interest	0	(57)	(118)	(182)	(250)
New costs	(1000)	(1000)	(1000)	(1000)	0
Closing balance	(1000)	(2057)	(3175)	(4358)	(4608)

The profit after all these costs is: 45 000 – 31 852 – 6516 – 4608 = 2024. 50% of this (1012) is paid to the mezzanine funder and 1012 is paid to the developer, whose equity cash flow is shown below.

Equity input	(1000)	(1000)	(1000)	(1000)	0
Receipts	0	0	0	0	5620
Equity cash flow	(1000)	(1000)	(1000)	(1000)	5620

Total equity input is 4000. Total revenue is the closing balance from the threshold return calculation plus the profit share: $4608 + 1012 = 5620$. Cash profit is $5620 - 4000 = 1620$ and the performance metrics are:

- Equity multiple: $5620/4000 = 1.40$
- Return on equity: $1620/4000 = 40.05\%$
- Equity IRR = 69.34% per annum

Sharing profit reduces the IRR significantly.

The next example shows how revenue might be received in phases rather than all at the end of the scheme and pays back the debt in a specified order of priority. A developer has recently acquired a residential development site for £10m. To finance the project, two options are being considered:

a) Senior debt finance of 65% of any negative cash flow per quarter at an interest rate of 2% per quarter. Interest and principal are rolled up until the cash flow turns positive and then the loan is paid off before any return on equity.

b) As (a) plus mezzanine finance for 15% of any negative cash flow per quarter (excluding interest) at an interest rate of 4% per quarter. The mezzanine finance is second priority loan but receives a 50% share of any surplus above a 5% per quarter return on equity.

The table below begins with the pre-finance cash flow. The IRR is 22% and when the net cash flow is in credit, there are funds available for distribution to lenders as and when necessary. The second stage of the appraisal calculates the return to equity following the debt finance arrangement described in (a) above. 65% of the costs are debt funded and interest is calculated on the balance carried forward from the previous quarter. The remaining 35% is funded by equity. Distribution funds are used to pay down the debt from the fourth quarter until the sixth quarter. In the sixth quarter, the debt is paid off and there is a net cash flow to equity which continues until the eighth and final quarter. The IRR on equity is 39%; the gearing has had a positive impact compared to the project IRR.

The third stage incorporates the mezzanine finance that is used to meet 15% of the costs in each quarter. The calculations work in the same way as the senior debt but because the senior debt is serviced first, there are no funds available to pay down the mezzanine debt until the sixth quarter. The debt is paid off in full by the seventh quarter and the IRR on equity is 60%. The fourth stage introduces the profit share arrangement. A surplus occurs in the last two quarters and must be distributed evenly between the equity provider and mezzanine finance provider. This is done by taking the flow to equity in [C], discounting it at the equity provider's target rate of return (5% per quarter). This takes care of the 'normal' return to the equity provider. Deducting the discounted equity cash flow from the net equity cash flow leaves the surplus, half of which goes to the equity provider and is therefore added to the DCF. Having done this in quarters seven and eight, the resulting IRR on equity is 46%.

Section C

Quarter	0	1	2	3	4	5	6	7	8	IRR
A. Project cash flow: pre-finance										
Total revenue					5 000 000	5 000 000	5 000 000	5 000 000	5 000 000	
Cash outflows (dev. costs)	(10 000 000)		(1 000 000)	(2 000 000)	(2 000 000)	(1 000 000)	(1 000 000)	(2 000 000)	(2 000 000)	
Net cash flow (pre-finance)	(10 000 000)		(1 000 000)	(2 000 000)	3 000 000	4 000 000	4 000 000	3 000 000	3 000 000	22.38%
Available for distribution					3 000 000	4 000 000	4 000 000	3 000 000	3 000 000	
B. Project cash flow: post senior debt contribution proportion @ 65% LTC, interest @ 2% per qtr										
Opening balance			(6 500 000)	(6 630 000)	(7 412 600)	(8 860 852)	(6 038 069)	(2 158 830)		
Quarterly interest (in arrears)			(130 000)	(132 600)	(148 252)	(177 217)	(120 761)	(43 177)		
Contribution to current quarter's costs		(6 500 000)		(650 000)	(1 300 000)					
Repayment						3 000 000	4 000 000	2 202 007		
Closing balance		(6 500 000)	(6 630 000)	(7 412 600)	(8 860 852)	(6 038 069)	(2 158 830)			
Net cash flow to equity	(3 500 000)		(350 000)	(700 000)			1 797 993	3 000 000	3 000 000	38.85%
C. Project cash flow: as above plus mezz debt contribution @ 15% LTC, interest @ 4% per qtr										
Opening balance			(1 500 000)	(1 560 000)	(1 622 400)	(1 687 296)	(1 754 788)	(1 824 979)		
Quarterly interest (in arrears)			(60 000)	(62 400)	(64 896)	(67 492)	(70 192)	(72 999)		
Contribution to current quarter's costs		(1 500 000)								
Repayment							1 797 993	103 985		
Closing balance		(1 500 000)	(1 560 000)	(1 622 400)	(1 687 296)	(1 754 788)	(1 824 979)	(99 986)		
Net cash flow to equity	(2 000 000)		(200 000)	(400 000)				2 896 015	3 000 000	60.18%
D. Project cash flow: as above plus mezz debt provider receives 50% profit share above developer's 5% TRR										
Net cash flow to equity (from above)	(2 000 000)		(200 000)	(400 000)				2 896 015	3 000 000	
Discounted at 5% p.q. (developer's TRR)	1.0000	0.9524	0.9070	0.8638	0.8227	0.7835	0.7462	0.7107	0.6768	
Discounted cash flow	(2 000 000)		(181 406)	(345 535)				2 058 144	2 030 518	
Less 50% surplus to mezzanine debt provider								(418 936)	(484 741)	
Net cash flow to equity	(2 000 000)		(200 000)	(400 000)				2 477 079	2 515 259	45.68%

Profit-sharing arrangements, like the ones illustrated in the examples above, are used to provide a developer or operating partner (referred to as the sponsor) in a joint venture (JV) with an extra share of profit, known as a promote. Hence, they are often referred to as promote structures. They can be structured in a variety of ways. One relatively common approach is a waterfall model (imagine cascading pools of water that fill up with cash flow and then spill over to the next pool once full). In a waterfall model, the sponsor receives an increased share of profit if the project IRR is greater than expected and a reduced share if it is less than expected. Common waterfall model components include:

- Return hurdle: This is the rate that must be achieved to trigger model stages (profit splits for example). IRRs and equity multiples are common return hurdles. It is necessary to state from which perspective the return hurdle is to be measured – project, sponsor equity or investor equity.
- Preferred return or 'pref': This is the first claim on profit until first return hurdle is achieved. Key parameters are:
  - Who gets pref? Some or all the equity investors?
  - Is it cumulative? This is relevant if there is not sufficient cash flow to pay the pref in any one period. If the pref is cumulative, then it will be added to the investment balance for the next period and accumulate until it is paid out.
  - Is it compounded? If the pref is cumulative, is unpaid cash flow compounded at the pref as it accumulates? If so, what is the compounding frequency?
- The Lookback Provision: This allows sponsor to look back at the end of the project and, if the investor does not achieve a pre-determined rate of return, the sponsor will be required to give up a portion of its already distributed profits so that the investor achieves a pre-determined return. This provision is typically favoured by the sponsor.
- The Catchup Provision: This provides the investor with all the profit until a pre-determined rate of return is achieved; then all the profit goes to the sponsor until caught up. This provision is typically preferred by the investor.

Below is an example of a promote structure. A promoter (developer or sponsor) invests 10% of required equity and receives 10% of shares in the Special Purpose Vehicle (SPV) formed to undertake a development project. An investor invests 90% of the equity required and receives 90% of shares in the SPV. Shareholders receive a pref of 10% per annum (2.41% per quarter) IRR generated by the levered project. The promoter receives a promote payment of 25% of any cash surplus above the 10% pref. All remaining funds will be distributed to shareholders in accordance with shareholdings.

The table below shows the cash flow for this project. The IRR of the project (SPV) cash flow is 4.07% per quarter (17.28% per annum). The IRR of the promoter's net cash flow is 7.26% per quarter (32.36% per annum) and the IRR of the investor's net cash flow is 3.67% per quarter (15.50% per annum).

Quarter		0	1	2	3	4	5	6	7	8
Contributions										
Project (SPV) cash flow		(8 000 000)	(1 000 000)	(3 000 000)	(4 000 000)	(2 000 000)	1 000 000	10 000 000	10 000 000	1 000 000
Promoter equity contribution	10%	(800 000)	(100 000)	(300 000)	(400 000)	(200 000)	0	0	0	0
Investor equity contribution	90%	(7 200 000)	(900 000)	(2 700 000)	(3 600 000)	(1 800 000)	0	0	0	0
Total equity contribution		(8 000 000)	(1 000 000)	(3 000 000)	(4 000 000)	(2 000 000)	0	0	0	0
Cash flow available for distribution		0	0	0	0	0	1 000 000	10 000 000	10 000 000	1 000 000
Distributions										
Tier 1 Balance (bop)		0	(8 000 000)	(9 192 800)	(12 414 346)	(16 713 532)	(19 116 328)	(18 577 032)	(9 024 738)	0
Equity contribution		(8 000 000)	(1 000 000)	(3 000 000)	(4 000 000)	(2 000 000)	0	0	0	0
Accrual at hurdle rate	2.41%	0	(192 800)	(221 546)	(299 186)	(402 796)	(460 704)	(447 706)	(217 496)	0
Accrual distribution		0	0	0	0	0	1 000 000	10 000 000	9 242 235	0
Balance (eop)		(8 000 000)	(9 192 800)	(12 414 346)	(16 713 532)	(19 116 328)	(18 577 032)	(9 024 738)	0	
Promoter tier 1 equity distribution	10%	0	0	0	0	0	100 000	1 000 000	924 223	0
Investor tier 1 equity distribution	90%	0	0	0	0	0	900 000	9 000 000	8 318 011	0
Remaining cash flow for tier 2		0	0	0	0	0	0	0	757 765	1 000 000
Tier 2 Promoter tier 2 bonus	25%	0	0	0	0	0	0	0	189 441	250 000
Remaining cash flow for tier 3		0	0	0	0	0	0	0	568 324	750 000
Tier 3 Promoter share of tier 3	10%	0	0	0	0	0	0	0	56 832	75 000
Investor share of tier 3	90%	0	0	0	0	0	0	0	511 492	675 000
Returns										
Promoter net cash flow		(800 000)	(100 000)	(300 000)	(400 000)	(200 000)	100 000	1 000 000	1 170 497	325 000
Investor net cash flow		(7 200 000)	(900 000)	(2 700 000)	(3 600 000)	(1 800 000)	900 000	9 000 000	8 829 503	675 000

The next example illustrates a slightly more complex three-tier waterfall promote structure. The sponsor invests 10% of the project costs and the investor invests 90% of the project costs. Profits are split pari passu up to 10% IRR and then profits are split disproportionately, i.e. both sponsor and investor receive 10% per annum as their pref on invested capital (pari passu). If distributions fall below 10% in any year, the deficiency is carried over to following years and compounded annually at the pref (the pref is cumulative and compounded). After the 10% pref hurdle is achieved, all profit up to 15% IRR is allocated at 20% to sponsor and 80% to investor. So, the sponsor gets an additional 10% profit above the original 10% pro rata share. This additional 10% is the promote. Above 15% IRR, the split is 60% to the investor and 40% to the sponsor. So, the sponsor gets a 30% promote after the final 15% IRR is achieved. The structure can be summarised as follows.

		IRRs[a]		Profit split	
		From	Up to	Sponsor	Investor
Tier 1	Hurdle 1	0%	10%	10%	90%
Tier 2	Hurdle 2	10%	15%	20%	80%
Tier 3	Hurdle 3	15%		40%	60%

[a] IRR hurdle calculations are at the project level.

		0	1	2	3	4	5
Project	Project cash flow	(1 000 000)	(50 000)	(100 000)	(100 000)	500 000	3 000 000
	Promoter equity Contribution	(100 000)	(5 000)	(10 000)	(10 000)	0	0
	Investor equity contribution	(900 000)	(45 000)	(90 000)	(90 000)	0	0
	Total equity contribution	(1 000 000)	(50 000)	(100 000)	(100 000)	0	0
	Cash available for distribution	0	0	0	0	500 000	3 000 000
Tier 1	Balance (bop)	0	(1 000 000)	(1 150 000)	(1 365 000)	(1 601 500)	(1 261 650)
	Total equity contribution	(1 000 000)	(50 000)	(100 000)	(100 000)	0	0
	Tier 1 accrual	0	(100 000)	(115 000)	(136 500)	(160 150)	(126 165)
	Tier 1 accrual distribution	0	0	0	0	500 000	1 387 815
	Balance (eop)	(1 000 000)	(1 150 000)	(1 365 000)	(1 601 500)	(1 261 650)	0
	Investor equity distribution	0	0	0	0	450 000	1 249 034
	Promoter equity distribution	0	0	0	0	50 000	138 782
	Promoter promote cash flow						

(Continued)

Section C

		0	1	2	3	4	5
	Remaining cash to distribute to tier 2	0	0	0	0	0	1612185
Tier 2	Balance (bop)	0	(1000000)	(1200000)	(1480000)	(1802000)	(1572300)
	Equity contributions	(1000000)	(50000)	(100000)	(100000)	0	0
	Tier 2 accrual	0	(150000)	(180000)	(222000)	(270300)	(235845)
	Tier 2 accrual distribution	0	0	0	0	500000	1808145
	Balance (eop)	(1000000)	(1200000)	(1480000)	(1802000)	(1572300)	0
	Investor equity cash flow	0	0	0	0	0	336264
	Promoter equity cash flow	0	0	0	0	0	42033
	Promoter promote cash flow	0	0	0	0	0	42033
	Remaining cash to distribute to tier 3	0	0	0	0	0	1191855
Tier 3	Investor cash flow	0	0	0	0	0	715113
	Promoter equity cash flow	0	0	0	0	0	119186
	Promoter promote cash flow	0	0	0	0	0	357557
Returns	Investor equity contributions	(900000)	(45000)	(90000)	(90000)	0	0
	Investor distributions	0	0	0	0	450000	2300411
	Investor net cash flows	(900000)	(45000)	(90000)	(90000)	450000	2300411
	Investor IRR	22.19%					
	Investor equity multiple	2.44					
	Promoter equity contributions	(100000)	(5000)	(10000)	(10000)	0	0
	Promoter distributions	0	0	0	0	50000	699590
	Promoter net cash flows	(100000)	(5000)	(10000)	(10000)	50000	699590
	Promoter IRR	47.41%					
	Promoter equity multiple	6.00					

Key points

- The basic residual technique assumes all the development costs are debt financed. Developer's profit is usually specified as a cash margin on either the total cost or total value of the proposed scheme. The effect of time on the valuation is handled by discounting the residual value to present value using the finance rate.
- Some DCF techniques discount the cash flow at a finance rate and include developers profit as a cash sum, others include profit as a target rate per period.

- The residual valuation method is often regarded as inflexible and sensitive to small but compounded changes in the increasing number of variables that are input as a development progresses.
- The cash-flow method enables the valuer to be explicit about the breakdown of costs and revenue. It provides a more granular breakdown of monetary flows over a specified period.
- Providing debt capital often involves receipt by the lender of a fixed return in the form of interest. Availability, type and terms of debt vary with borrower, scheme and market conditions. Bank loans are a common form of project finance for property development.
- Debt changes the risks and returns of a project for the developer. Finance structures may mitigate risks but at the expense of rewards. Use of debt (gearing) can increase potential return on equity but it can also increase the volatility of returns.
- Providing equity capital involves direct participation by the investor in the profitability of the project or company.
- For the developer, profit sharing means giving away profit but taking less risk.

References

Brown, G. and Matysiak, G. (2000). *Real Estate Investment: A Capital Market Approach.* Harlow, UK: FT Prentice Hall.

Cornell, B. (1999). Risk, duration, and capital budgeting: New evidence on some old questions. *Journal of Business* 72 (2): 183–200.

Crosby, N., Devaney, S., and Wyatt, P. (2020). Performance metrics and required returns for UK real estate development schemes. *J. Prop. Res.* 37 (2).

Geltner, D., Miller, N., Clayton, J., and Eicholtz, P. (2007). *Commercial Real Estate Analysis and Investments*, 2e. US: Cengage Learning.

Geltner, D., Kumar, A., and Van de Minne, A. (2018). Riskiness of real estate development: a perspective from urban economics and option value theory. *Real Estate Econ.* 48 (2): 406–445.

RICS (2019). *Valuation of Development Property*, Guidance Note, 1e. Royal Institution of Chartered Surveyors.

RTPI (2018) *Planning Risk and Development: how greater planning certainty would affect residential development*, RTPI Research Paper, April 2018, Royal Town Planning Institute.

Taylor, L., Phaneuf, D., and Lui, X. (2016). Disentangling property value impacts of environmental contamination from locally undesirable land uses: Implications for measuring post-cleanup stigma. *J. Urban Econ.* 93: 85–98.

Questions

[1]   A development site has permission for three self-contained two-storey office buildings, one with GIA 3000 square metres and the other two having 2500 square metres each. Costs of construction are £1500 per square metre of GIA, the efficiency ratio is 80% and it is estimated that construction will take 18 months. The rental value is estimated to be £300 per square metre of net internal area and the yield is 6.5%. Making any further assumptions regarding other costs of construction, development periods and phasing of lettings and

sales, construct a quarterly cash flow to estimate the land value. The developer's target rate of return is 15% per annum, and this includes the cost of finance. Value the site.

[2] You have been asked to estimate the investment value of an office development opportunity. The site has been acquired for £1.5m (plus acquisition costs) and the proposed scheme has a GIA of 2000 square metres with an efficiency ratio of 90%. Site preparation costs are £250 000, building costs are £1000 per square metre and professional fees of 11% of building costs and contingencies are 5% of site preparation costs, construction costs and professional fees combined. The market rent is estimated to be £220 per square metre, the yield is 6.50%, disposal costs are 2% of NDV and a letting fee of 10% of market rent. The scheme will take 1.5 years to complete.

a) Construct quarterly cash flows assuming:
    i) An even spread of construction costs over Q2–Q6 and site prep costs in Q1
    ii) An s-curve of construction costs over Q2–Q6 and site prep costs in Q1
    iii) 5% per annum growth in costs and values
    iv) A nine-month void after completion
b) Calculate the NPV of (i) to (iv) using a target rate of 5% per quarter.
c) Calculate the IRR for (i) to (iv) and state them as annual rates.
d) Add 100% finance to the cash flow from (iv) using a quarterly rate of 4% per quarter.

Answers

[1]

Inputs and assumptions	
Land price (£)	8 000 000
Yield	6.50%
GIA (m²)	8000
NIA (m²)	6400
Build cost (£/m²)	1500
Rent (£/m²)	300
Land purchase costs (% land price)	6.50%
Sale fee	2.00%
Letting fee (% ERV)	10%
Lead-in and void periods	6 months each
Contingency (% building cost and professional fees)	5%
Professional fees (% building cost)	10%
Calculations	
Build costs	12 000 000
Estimated MR (£ p.a.)	1 920 000
GDV (£)	29 538 462
NDV (£)	28 959 276
Quarterly target rate of return	3.56%

Using a spreadsheet, it is possible to estimate the land value by iteration (the Goal Seek function in Excel). The NPV is set to zero by adjusting the land price input, which, here, is £8 653 760.

	0	1	2	3	4	5	6	7	8	9
Land price (£)	(8 653 760)									
Land purchase costs (£)	(497 591)									
Building costs (£)			(600 000)	(1 800 000)	(3 600 000)	(3 600 000)	(1 800 000)	(600 000)		
Professional fees (£)			(60 000)	(180 000)	(360 000)	(360 000)	(180 000)	(60 000)		
Contingencies (£)			(33 000)	(99 000)	(198 000)	(198 000)	(99 000)	(33 000)		
Letting fee (£)										(192 000)
NDV (£)										28 959 276
NET CASH FLOW (£)	(9 151 351)	0	(693 000)	(2 079 000)	(4 158 000)	(4 158 000)	(2 079 000)	(693 000)	0	28 767 276

NPV @ 15% p.a.   0
(3.56% p.q.)

[2]

Inputs

Contingencies (% site prep, construction and fees)	5%
Professional fees (% construction costs)	11%
Building costs (£/m²)	1000
Site preparation costs (£)	250 000
GIA (m²)	2000
Efficiency ratio	90%
Yield	6.50%
Estimated MR (£/m²)	220
Disposal fees (% NDV)	2.00%
Land purchase costs (% land price)	6.50%
Land price (£)	1 500 000
Development period (years)	1.5
Letting fee (% MR)	10%
Target rate of return	5.00%

Calculations

NIA (m²)	1800
Estimated MR (£ p.a.)	396 000
GDV (£)	6 092 308
NDV (£)	5 972 851
Building cost (£)	2 000 000

a. i. Even spread of construction costs over Q2–Q6 and site prep costs in Q1

	0	1	2	3	4	5	6
Land price (£)	(1 500 000)						
Land purchase costs (£)	(97 500)						
Site preparation costs (£)		(250 000)					
Building cost (£)			(400 000)	(400 000)	(400 000)	(400 000)	(400 000)
Professional fees (£)			(44 000)	(44 000)	(44 000)	(44 000)	(44 000)
Contingency (£)		(12 500)	(22 200)	(22 200)	(22 200)	(22 200)	(22 200)
Letting fee (£)							(39 600)
NDV (£)							5 972 851
Net cash flow (£)	(1 597 500)	(262 500)	(466 200)	(466 200)	(466 200)	(466 200)	5 467 051
b. NPV (£)	657 695						
c. IRR	42.56%						

Section C

a. ii. S-curve spread of construction costs over Q2–Q6 and site prep costs in Q1

	0	1	2	3	4	5	6
Land price (£)	(1 500 000)						
Land purchase costs (£)	(97 500)						
Site preparation costs (£)		(250 000)					
Building cost (£)			(200 000)	(400 000)	(800 000)	(400 000)	(200 000)
Professional fees (£)			(22 000)	(44 000)	(88 000)	(44 000)	(22 000)
Contingency (£)		(12 500)	(11 100)	(22 200)	(44 400)	(22 200)	(11 100)
Letting fee (£)							(39 600)
NDV (£)							5 972 851
Net cash flow (£)	(1 597 500)	(262 500)	(233 100)	(466 200)	(932 400)	(466 200)	5 700 151
b. NPV (£)	659 523						
c. IRR	42.77%						

a. iii. 5% per annum growth in costs and values equates to 1.23% p.q.

	0	1	2	3	4	5	6
Quarterly growth rate	1.0000	1.0123	1.0247	1.0373	1.0500	1.0629	1.0759
Land price (£)	(1 500 000)						
Land purchase costs (£)	(97 500)						
Site preparation costs (£)		(253 068)					
Building cost (£)			(204 939)	(414 908)	(840 000)	(425 154)	(215 186)
Professional fees (£)			(22 543)	(45 640)	(92 400)	(46 767)	(23 670)
Contingency (£)		(12 653)	(11 374)	(23 027)	(46 620)	(23 596)	(11 943)
Letting fee (£)							(42 607)
NDV (£)							6 426 368
Net cash flow (£)	(1 597 500)	(265 721)	(238 856)	(483 575)	(979 020)	(495 517)	6 132 96
b. NPV (£)	897 869						
c. IRR	49.91%						

a. iv. 9 months void after completion

	0	1	2	3	4	5	6	7	8	9
Quarterly growth rate	1.0000	1.0123	1.0247	1.0373	1.0500	1.0629	1.0759	1.0891	1.1025	1.1160
Land price (£)	(1 500 000)									
Land purchase costs (£)	(97 500)									
Site preparation costs (£)		(253 068)								
Building cost (£)			(204 939)	(414 908)	(840 000)	(425 154)	(215 186)			
Professional fees (£)			(22 543)	(45 640)	(92 400)	(46 767)	(23 670)			
Contingency (£)		(12 653)	(11 374)	(23 027)	(46 620)	(23 596)	(11 943)			
Letting fee (£)										(44 195)
NDV (£)										6 665 881
Net cash flow (£)	(1 597 500)	(265 721)	(238 856)	(483 575)	(979 020)	(495 517)	(250 799)	0	0	6 621 687
b. NPV (£)	402 606									
c. IRR	28.73%									
d. Finance accounting at 4.00% p.q.										
Opening balance (£)	0	(1 597 500)	(1 927 121)	(2 243 063)	(2 816 361)	(3 908 035)	(4 559 874)	(4 993 068)	(5 192 791)	(5 400 502)
Interest (£)	0	(63 900)	(77 085)	(89 723)	(112 654)	(156 321)	(182 395)	(199 723)	(207 712)	(216 020)
Closing balance (£)	(1 597 500)	(1 927 121)	(2 243 063)	(2 816 361)	(3 908 035)	(4 559 874)	(4 993 068)	(5 192 791)	(5 400 502)	1 005 164
Total interest payable (£)	1 305 533									

10.A   Appendix – example development cash flow

Sheet 1: Inputs

Revenues	£/m²	NIA (m²)		Totals
Residential – market dwellings	2000	5000		£9 803 922
Residential – affordable dwellings	250	1000		£245 098
Commercial space (rent, yield)	200	4000	5.00%	£15 686 275
Other revenue				£3 000 000
Costs				
Land				
Land acquisition price, including transaction costs		£5 281 642	2.00%	£5 387 274
Site preparation, infrastructure, utilities				£1 000 000
Construction	£/m²	Gross: net		
Residential – market dwellings	1000	80%		£6 250 000
Residential – affordable dwellings	1000	80%		£1 250 000
Commercial space	800	85%		£3 764 706
Abnormal costs				£200 000
Fees and other costs				
Professional fees (% total construction costs)			10.00%	£1 146 471
Contingency (% total construction costs)			3.00%	£343 941
Planning fees				£5000
Building control, NHBC, etc.				£20 000
Planning obligations				£50 000
Infrastructure levy				£100 000
Other fees (e.g. legal, loan, valuation)				£200 000
Marketing costs				£100 000
Other assumptions				
Sale transaction costs (% sale price)	2.00%			
Letting transaction costs (% annual rent)	15.00%			

Sheet 2: Cash flow

	0	1	2	3	4	5	6	7	8
Revenues									
Residential – market	9 803 922	–	–	–	–	–	4 901 961	–	4 901 961
Residential – affordable	245 098	–	–	–	–	–	122 549	–	122 549
Commercial space	15 686 275	–	–	–	–	–	–	–	15 686 275
Other revenue	3 000 000	–	–	–	–	–	–	–	3 000 000
TOTAL REVENUE	28 735 294	–	–	–	–	–	5 024 510	–	23 710 784
Costs									
Land costs	5 387 274	(5 387 274)	–	–	–	–	–	–	–
Site preparation, infrastructure, utilities	1 000 000	(1 000 000)	–	–	–	–	–	–	–
Residential – market dwellings	6 250 000	(312 500)	(312 500)	(625 000)	(1 250 000)	(1 875 000)	(1 250 000)	(625 000)	–
Residential – affordable dwellings	1 250 000	(62 500)	(62 500)	(125 000)	(250 000)	(375 000)	(250 000)	(125 000)	–
Commercial space	3 764 706	(188 235)	(188 235)	(376 471)	(752 941)	(1 129 412)	(752 941)	(376 471)	–
Abnormal costs	200 000	(10 000)	(10 000)	(20 000)	(40 000)	(60 000)	(40 000)	(20 000)	–
Professional fees	1 146 471	(57 324)	(57 324)	(114 647)	(229 294)	(343 941)	(229 294)	(114 647)	–
Contingency	343 941	(17 197)	(17 197)	(34 394)	(68 788)	(103 182)	(68 788)	(34 394)	–
Planning fees	5 000	(250)	(250)	(500)	(1 000)	(1 500)	(1 000)	(500)	–
Building control, NHBC, etc.	20 000	(1 000)	(1 000)	(2 000)	(4 000)	(6 000)	(4 000)	(2 000)	–
Planning obligations	50 000	(2 500)	(2 500)	(5 000)	(10 000)	(15 000)	(10 000)	(5 000)	–
Infrastructure levy	100 000	(5 000)	(5 000)	(10 000)	(20 000)	(30 000)	(20 000)	(10 000)	–
Other fees	50 000	(2 500)	(2 500)	(5 000)	(10 000)	(15 000)	(10 000)	(5 000)	–
Marketing costs	100 000	(5 000)	(5 000)	(10 000)	(20 000)	(30 000)	(20 000)	(10 000)	–
TOTAL COSTS	19 667 392	(6 387 274)	(664 006)	(1 328 012)	(2 656 024)	(3 984 035)	(2 656 024)	(1 328 012)	–
Net cash flow		(6 387 274)	(664 006)	(1 328 012)	(2 656 024)	(3 984 035)	2 368 486	(1 328 012)	23 710 784

Sheet 3: Residual land valuation

A lump sum developer's profit of 20% of development value has been added to the development costs and the cash flow has been discounted at the quarterly equivalent of a 6% per annum finance rate.

		0	1	2	3	4	5	6	7	8
TOTAL REVENUE								5 024 510	–	23 710 784
TOTAL COSTS		(6 387 274)	(664 006)	(664 006)	(1 328 012)	(2 656 024)	(3 984 035)	(2 656 024)	(1 328 012)	–
Net cash flow		(6 387 274)	(664 006)	(664 006)	(1 328 012)	(2 656 024)	(3 984 035)	2 368 486	(1 328 012)	23 710 784
Developer's profit (% development value)	20%									(5 747 059)
Net cash flow including Developer's profit		(6 387 274)	(664 006)	(664 006)	(1 328 012)	(2 656 024)	(3 984 035)	2 368 486	(1 328 012)	17 963 725
NPV (£) (discounted at finance rate)	6.00%	1 790 950								

It is then possible to use the cash flow to iterate the residual land value by setting the NPV to 0, as shown below.

		0	1	2	3	4	5	6	7	8
TOTAL REVENUE								5 024 510	–	23 710 784
TOTAL COSTS		(8 178 224)	(664 006)	(664 006)	(1 328 012)	(2 656 024)	(3 984 035)	(2 656 024)	(1 328 012)	–
Net cash flow		(8 178 224)	(664 006)	(664 006)	(1 328 012)	(2 656 024)	(3 984 035)	2 368 486	(1 328 012)	23 710 784
Developer's profit (% development value)	20%									(5 747 059)
Net cash flow including developer's profit		(8 178 224)	(664 006)	(664 006)	(1 328 012)	(2 656 024)	(3 984 035)	2 368 486	(1 328 012)	17 963 725
NPV (£) (discounted at finance rate)	6.00%	–								

Sheet 4: Ungeared (unlevered) cash flow

Reverting to the original cash flow (but keeping the revised land price), it is possible to calculate performance metrics. Profitability has increased compared to the residual land valuation because finance costs are not included in this project cash flow.

	0	1	2	3	4	5	6	7	8
TOTAL REVENUE	–	–	–	–	–	–	5 024 510	–	23 710 784
TOTAL COSTS	(8 178 224)	(664 006)	(664 006)	(1 328 012)	(2 656 024)	(3 984 035)	(2 656 024)	(1 328 012)	–
Net cash flow	(8 178 224)	(664 006)	(664 006)	(1 328 012)	(2 656 024)	(3 984 035)	2 368 486	(1 328 012)	23 710 784
Project (ungeared) IRR	26.63%								
Total costs (£)	18 802 318								
Total revenue (£)	26 079 271								
Profit (£)	7 276 952								
Equity multiple	1.39								
Profit on cost	38.70%								

Sheet 5: Geared (levered) cash flow – senior debt – side-by-side arrangement

	0	1	2	3	4	5	6	7	8
TOTAL REVENUE	–	–	–	–	–	–	5 024 510	–	23 710 784
TOTAL COSTS	(8 178 224)	(664 006)	(664 006)	(1 328 012)	(2 656 024)	(3 984 035)	(2 656 024)	(1 328 012)	–
Net cash flow	(8 178 224)	(664 006)	(664 006)	(1 328 012)	(2 656 024)	(3 984 035)	2 368 486	(1 328 012)	23 710 784
Senior debt									
Drawdown	4 906 934	398 404	398 404	796 807	1 593 614	2 390 421	–	796 807	–
Repayment	–	–	–	–	–	–	(2 368 486)	–	(9 651 797)
Interest paid	–	–	–	–	–	–	–	–	–
Total senior debt Cash flow	4 906 934	398 404	398 404	796 807	1 593 614	2 390 421	(2 368 486)	796 807	(9 651 797)
Levered equity cash flow	(3 271 290)	(265 602)	(265 602)	(531 205)	(1 062 409)	(1 593 614)	–	(531 205)	14 058 988
IRR	53.19%								
Equity cost (£)	7 520 927								
Equity revenue (£)	14 058 988								
Profit (£)	6 538 061								

(Continued)

	0	1	2	3	4	5	6	7	8
Equity multiple	1.87								
Return on equity	86.93%								
Balance available for repayment	–	–	–	–	–	–	2 368 486	–	23 710 784
Senior debt table									
Balance (bop)	–	4 906 934	5 365 557	5 829 808	6 698 160	8 373 975	10 867 164	8 632 042	9 534 783
Interest due @ annual rate of	–	60 219	65 847	71 545	82 201	102 767	133 364	105 934	117 013
interest paid current (1 = yes, 0 = no)	–	–	–	–	–	–	–	–	–
Drawdown @ loan-to-cost ratio of	4 906 934	398 404	398 404	796 807	1 593 614	2 390 421	–	796 807	–
repayment	–	–	–	–	–	–	(2 368 486)	–	(9 651 797)
Balance (eop)	4 906 934	5 365 557	5 829 808	6 698 160	8 373 975	10 867 164	8 632 042	9 534 783	–

Assumes senior lender pays share of ALL costs, including land. If developer is to pay all land cost, then set drawdown in period 0 to zero.

Sheet 6: Geared (levered) cash flow – senior and mezzanine debt – side-by-side arrangement

	0	1	2	3	4	5	6	7	8
TOTAL REVENUE	–	–	–	–	–	–	5 024 510	–	23 710 784
TOTAL COSTS	(8 178 224)	(664 006)	(664 006)	(1 328 012)	(2 656 024)	(3 984 035)	(2 656 024)	(1 328 012)	–
Net cash flow	(8 178 224)	(664 006)	(664 006)	(1 328 012)	(2 656 024)	(3 984 035)	2 368 486	(1 328 012)	23 710 784
Mezzanine debt									
Drawdown	1 226 734	99 601	99 601	199 202	398 404	597 605	–	199 202	–
Repayment	–	–	–	–	–	–	(2 368 486)	–	(680 988)
Interest paid	–	–	–	–	–	–	–	–	–
Total mezz debt Cash flow	1 226 734	99 601	99 601	199 202	398 404	597 605	(2 368 486)	199 202	(680 988)
Senior debt									
Drawdown	4 906 934	398 404	398 404	796 807	1 593 614	2 390 421	–	796 807	–
Repayment	–	–	–	–	–	–	–	–	(12 078 773)
Interest paid	–	–	–	–	–	–	–	–	–
Total senior debt Cash flow	4 906 934	398 404	398 404	796 807	1 593 614	2 390 421	–	796 807	(12 078 773)
Levered equity cash flow	(2 044 556)	(166 001)	(166 001)	(332 003)	(664 006)	(996 009)	–	(332 003)	10 951 024

Metric	Value
IRR	76.38%
Equity cost (£)	4 700 580
Equity revenue (£)	10 951 024
Profit (£)	6 250 444
Equity multiple	2.33
Return on equity	132.97%

		1	2	3	4	5	6	7	8	9
Balance available for repayment		–	–	–	–	–	–	2 368 486	–	23 710 784
Mezzanine debt table										
Balance (bop)		–	1 226 734	1 350 166	1 475 996	1 703 871	2 135 375	2 774 463	459 875	668 011
Interest due @ annual rate of	8.00%	–	23 831	26 229	28 673	33 100	41 483	53 898	8 934	12 977
interest paid current (1 = yes, 0 = no)	0	–	–	–	–	–	–	–	–	–
Drawdown @ loan-to-cost ratio of	15%	1 226 734	99 601	99 601	199 202	398 404	597 605	–	199 202	–
repayment		–	–	–	–	–	–	(2 368 486)	–	(680 988)
Balance (eop)		1 226 734	1 350 166	1 475 996	1 703 871	2 135 375	2 774 463	459 875	668 011	–
Balance available for repayment		–	–	–	–	–	–	–	–	23 029 797
Senior debt table										
Balance (bop)		–	4 906 934	5 365 557	5 829 808	6 698 160	8 373 975	10 867 164	11 000 528	11 932 336
Interest due @ annual rate of	5.00%	–	60 219	65 847	71 545	82 201	102 767	133 364	135 001	146 436
interest paid current (1 = yes, 0 = no)	0	–	–	–	–	–	–	–	–	–
Drawdown @ loan-to-cost ratio of	60%	4 906 934	398 404	398 404	796 807	1 593 614	2 390 421	–	796 807	–
repayment		–	–	–	–	–	–	–	–	(12 078 773)
Balance (eop)		4 906 934	5 365 557	5 829 808	6 698 160	8 373 975	10 867 164	11 000 528	11 932 336	–

Section C

Sheet 7: Geared (levered) cash flow – senior debt – equity first arrangement

	Param	0	1	2	3	4	5	6	7	8
TOTAL REVENUE		–	–	–	–	–	–	5 024 510	–	23 710 784
TOTAL COSTS		(8 178 224)	(664 006)	(664 006)	(1 328 012)	(2 656 024)	(3 984 035)	(2 656 024)	(1 328 012)	–
Net cash flow		(8 178 224)	(664 006)	(664 006)	(1 328 012)	(2 656 024)	(3 984 035)	2 368 486	1 328 012	23 710 784
Senior debt										
Drawdown		–	258 893	664 006	1 328 012	2 656 024	3 984 035	2 656 024	1 328 012	–
Repayment		–	–	–	–	–	–	(2 368 486)	–	(10 968 212)
Interest paid		–	–	–	–	–	–	–	–	–
Total senior debt Cash flow		–	258 893	664 006	1 328 012	2 656 024	3 984 035	287 537	1 328 012	(10 968 212)
Levered equity Cash flow		(8 178 224)	(405 113)	–	–	–	–	2 656 024	–	12 742 572

IRR	36.09%
Equity cost (£)	8 583 337
Equity revenue (£)	15 398 596
Profit (£)	6 815 259
Equity multiple	1.79
Profit on cost	79.40%

	Param	0	1	2	3	4	5	6	7	8
Accumulated Expenditure	8 583 337	8 178 224	8 842 230	9 506 236	10 834 248	13 490 271	17 474 306	20 130 330	21 458 342	21 458 342
Equity cash flow		8 178 224	405 113	–	–	–	–	–	–	–
Balance available for repayment		–	–	–	–	–	–	2 368 486	–	23 710 784
Senior debt table										
Balance (bop)		–	–	258 893	926 076	2 265 453	4 949 279	8 994 053	9 391 967	10 835 239
Interest due @ annual rate of	5.00%	–	–	3 177	11 365	27 802	60 739	110 377	115 260	132 973
interest paid current (1 = yes, 0 = no) of	0	–	–	–	–	–	–	–	–	–
Drawdown @ loan-to-cost ratio of	60%	–	258 893	664 006	1 328 012	2 656 024	3 984 035	2 656 024	1 328 012	–
repayment		–	–	–	–	–	–	(2 368 486)	–	(10 968 212)
Balance (eop)		–	258 893	926 076	2 265 453	4 949 279	8 994 053	9 391 967	10 835 239	–
Senior debt remaining balance		12 875 005	12 616 112	11 952 106	10 624 094	7 968 071	3 984 035	1 328 012	–	–

Sheet 8: Geared (levered) cash flow – senior and mezzanine debt – equity first arrangement

	0	1	2	3	4	5	6	7	8
TOTAL REVENUE	–	–	–	–	–	–	5 024 510	–	23 710 784
TOTAL COSTS	(8 178 224)	(664 006)	(664 006)	(1 328 012)	(2 656 024)	(3 984 035)	(2 656 024)	(1 328 012)	–
Net cash flow	(8 178 224)	(664 006)	(664 006)	(1 328 012)	(2 656 024)	(3 984 035)	2 368 486	(1 328 012)	23 710 784
Mezzanine debt									
Drawdown									
Repayment									
Interest paid									
Total mezz debt cash flow									
Senior debt									
Drawdown	–	–	–	105 077	2 656 024	3 984 035	2 656 024	1 328 012	–
Repayment	–	–	–	–	–	–	(2 368 486)	–	(8 671 978)
Interest paid	–	–	–	–	–	–	287 537	–	–
Total senior debt cash flow	–	–	–	105 077	2 656 024	3 984 035	287 537	1 328 012	(8 671 978)
Levered equity cash flow	(8 178 224)	(664 006)	(664 006)	(1 222 935)	–	–	2 656 024	–	15 038 806

IRR	32.10%
Equity cost (£)	10 729 171
Equity revenue (£)	17 694 829
Profit (£)	6 965 659
Equity multiple	1.65
Profit on cost	64.92%

	0	1	2	3	4	5	6	7	8
Accumulated Expenditure	8 178 224	8 842 230	9 506 236	10 834 248	13 490 271	17 474 306	20 130 330	21 458 342	21 458 342
Equity cash flow	4 291 668	–	–	–	–	–	–	–	–
Balance available for repayment									
Mezzanine debt table									

(Continued)

		0	1	2	3	4	5	6	7	8
Balance (bop)		–	3 886 556	4 697 481	5 539 061	6 971 384	7 234 916	7 508 410	7 792 243	8 086 805
Interest due @ annual rate of	16.00%	–	146 920	177 574	209 388	263 532	273 494	283 833	294 562	305 697
interest paid current (1 = yes, 0 = no)	0	–	–	–	–	–	–	–	–	–
Drawdown @ loan-to-cost ratio of	30%	3 886 556	664 006	664 006	1 222 935	–	–	–	–	–
repayment		–	–	–	–	–	–	–	–	(8 392 502)
Balance (eop)		3 886 556	4 697 481	5 539 061	6 971 384	7 234 916	7 508 410	7 792 243	8 086 805	–
Mezzanine debt remaining balance	6 437 503	2 550 947	1 886 941	1 222 935	–	–	–	–	–	–
Balance available for repayment		–	–	–	–	–	–	2 368 486	–	237 710 784
Senior debt table										
Balance (bop)		–	–	–	–	105 077	2 762 390	6 780 326	7 151 073	8 566 844
Interest due @ annual rate of	5.00%	–	–	–	–	1 290	33 901	83 210	87 760	105 134
interest paid current (1 = yes, 0 = no)	0	–	–	–	–	–	–	–	–	–
Drawdown @ loan-to-cost ratio of	50%	–	–	–	105 077	2 656 024	3 984 035	2 656 024	1 328 012	–
Repayment		–	–	–	–	–	–	(2 368 486)	–	(8 671 978)
Balance (eop)		–	–	–	105 077	2 762 390	6 780 326	7 151 073	8 566 844	(0)
Senior debt remaining balance	10 729 171	10 729 171	10 729 171	10 729 171	10 624 094	7 968 071	3 984 035	1 328 012	(0)	(0)

Sheet 9: Geared cash flow – senior debt side-by-side arrangement with a simple profit share arrangement

	0	1	2	3	4	5	6	7	8
Levered equity Cash flow[a]	(3271290)	(265602)	(265602)	(531205)	(1062409)	(1593614)	–	(531205)	14058988
Promote structure[b]									
Inputs:									
Equity TRR (promote agreement) 20%	4.66%								
Proportion of surplus to investor 50%									
Proportion of surplus to developer 50%									
Promote arrangement									
Balance (bop)	–	(3271290)	(3689449)	(4127109)	(4850782)	(6139409)	(8019335)	(8393318)	(9315946)
Compounded at developer's IRR	–	(152557)	(172058)	(192468)	(226217)	(286312)	(373983)	(391424)	(434450)
Levered cash flow (from above)	(3271290)	(265602)	(265602)	(531205)	(1062409)	(1593614)	–	(531205)	14058988
Balance (eop)	(3271290)	(3689449)	(4127109)	(4850782)	(6139409)	(8019335)	(8393318)	(9315946)	4308591
Surplus	–	–	–	–	–	–	–	–	4308591
Proportion of surplus to lender	–	–	–	–	–	–	–	–	2154296
Net cash flow to equity	(3271290)	(265602)	(265602)	(531205)	(1062409)	(1593614)	–	(531205)	11904692

IRR	37.33%
Equity cost (£)	7520927
Equity revenue (£)	11904692
Profit (£)	4383765
Equity multiple	1.58
Return on equity	58.29%

[a] The net cash-flow line could be swapped for any other on which the promote structure could be applied.

[b] Tier 1 = 20% return to developer; tier 2 = split of surplus between developer and investor.

Sheet 10: Geared cash flow - senior debt side-by-side arrangement with a waterfall promote structure

		0	1	2	3	4	5	6	7	8
Levered Equity Cash Flow		(3 271 290)	(265 602)	(265 602)	(531 205)	(1 062 409)	(1 593 614)	–	(531 205)	14 058 988

Promote structure

Promoter Equity Contribution (% Equity Proportion)	10%
Investor Equity Contribution (% Equity Proportion)	90%

	Preferred Return:		Profit Split:	
	From	Up to	Promoter	Investor
Tier 1	0.00%	20.00%	10.00%	90.00%
Tier 2	20.00%	25.00%	25.00%	75.00%
Tier 3	25.00%		40.00%	60.00%

Contributions

		0	1	2	3	4	5	6	7	8
Net cash flow (from above)		(3 271 290)	(265 602)	(265 602)	(531 205)	(1 062 409)	(1 593 614)	–	(531 205)	14 058 988
Promoter Equity Contribution	10%	(327 129)	(26 560)	(26 560)	(53 120)	(106 241)	(159 361)	–	(53 120)	–
Investor Equity Contribution	90%	(2 944 161)	(239 042)	(239 042)	(478 084)	(956 168)	(1 434 253)	–	(478 084)	–
Total Equity Contribution		(3 271 290)	(265 602)	(265 602)	(531 205)	(1 062 409)	(1 593 614)	–	(531 205)	–
Equity Cash Flow Available for Distribution		–	–	–	–	–	–	–	–	14 058 988

Tier 1

		0	1	2	3	4	5	6	7	8
Balance (bop)		–	(3 271 290)	(3 689 449)	(4 127 109)	(4 850 782)	(6 139 409)	(8 019 335)	(8 393 318)	(9 315 946)
Total Equity Contribution		(3 271 290)	(265 602)	(265 602)	(531 205)	(1 062 409)	(1 593 614)	–	(531 205)	–
Balance (bop) compounded at	4.66%	–	(152 557)	(172 058)	(192 468)	(226 217)	(286 312)	(373 983)	(391 424)	(434 450)
Tier 1 IRR										
Tier 1 Accrual Distribution		–	–	–	–	–	–	–	–	9 750 396
Balance (eop)		(3 271 290)	(3 689 449)	(4 127 109)	(4 850 782)	(6 139 409)	(8 019 335)	(8 393 318)	(9 315 946)	–
Promoter Tier 1 Equity Cash Flow										975 040
Investor Tier 1 Equity Cash Flow										8 775 357
Remaining Cash to Distribute to Tier 2										4 308 591

Tier 2

Balance (bop)	–	(3 271 290)	(3 724 570)	(4 203 856)	(4 976 241)	(6 324 143)	(8 280 582)	(8 755 649)	(9 789 176)
Total Equity Contribution	(3 271 290)	(265 602)	(265 602)	(531 205)	(1 062 409)	(1 593 614)	–	(531 205)	–
Compounded at Tier 2 IRR 5.74%	–	(187 678)	(213 683)	(241 181)	(285 493)	(362 824)	(475 067)	(502 323)	(561 617)
Tier 2 Accrual Distribution	–	–	–	–	–	–	–	–	–
Balance (eop)	(3 271 290)	(3 724 570)	(4 203 856)	(4 976 241)	(6 324 143)	(8 280 582)	(8 755 649)	(9 789 176)	10 350 794
Promoter Tier 2 Promote Cash Flow			–	–	–	–	–	–	90 060
Promoter Tier 2 Equity Cash Flow			–	–	–	–	–	–	127 584
Investor Tier 2 Equity Cash Flow			–	–	–	–	–	–	382 753
Remaining Cash to Distribute to Tier 3			–	–	–	–	–	–	3 708 194

Tier 3

Promoter Tier 3 Promote Cash Flow			–	–	–	–	–	–	1 112 458
Promoter Tier 3 Equity Cash Flow			–	–	–	–	–	–	1 038 294
Investor Tier 3 Equity Cash Flow			–	–	–	–	–	–	1 557 441

Equity Cash Flows

Promoter Equity Contribution	(327 129)	(26 560)	(26 560)	(53 120)	(106 241)	(159 361)	–	(53 120)	–
Promoter Distribution	–	–	–	–	–	–	–	–	3 343 436
Promoter Net Cash Flow	(327 129)	(26 560)	(26 560)	(53 120)	(106 241)	(159 361)	–	(53 120)	3 343 436
Investor Equity Contribution	(2 944 161)	(239 042)	(239 042)	(478 084)	(956 168)	(1 434 253)	–	(478 084)	–
Investor Distribution	–	–	–	–	–	–	–	–	10 715 552
Investor Net Cash Flow	(2 944 161)	(239 042)	(239 042)	(478 084)	(956 168)	(1 434 253)	–	(478 084)	10 715 552

Promoter's performance metrics

IRR	161.57%
Equity cost (£)	752 093
Equity revenue (£)	3 343 436
Profit (£)	2 591 343
Equity multiple	4.45
Return on equity	344.55%

Investor's performance metrics

IRR	37.34%
Equity cost (£)	6 768 835
Equity revenue (£)	10 715 552
Profit (£)	3 946 717
Equity multiple	1.58
Return on equity	58.31%

10.B   Appendix: contaminated land

Some sites may be contaminated because of their previous use and valuers have a responsibility to investigate, consider and report on its impact where appropriate. Taylor et al. (2016) investigated the impact of contamination on house prices and found that contamination more than doubles the negative influence commercial properties have on neighbouring dwelling values. However, they found little evidence of stigma effects once a contaminated site is remedied. The negative spill-over effects associated with remediated contaminated sites are largely indistinguishable from spill-over effects from commercial properties with no known contamination.

Valuers should investigate previous land uses of the subject property and neighbours and should report possible or actual contamination if spotted. Types of environmental matters/contamination that valuers should look out for include building materials that are known to cause problems (such as asbestos), disused mines and quarries, flood risk, coastal erosion and other abnormal ground conditions, waste and high-voltage equipment. Information might be obtained from local authority sources such as building control, planning and environmental health. Also, environmental protection authorities, utility companies, historic maps and aerial photography may provide valuable insight.

It is often difficult to find comparable evidence to help value a contaminated site because the variability of location-specific contaminants and resultant severity and extent of contamination will often lead to different estimates of impaired value. The accepted approach, in the likely absence of comparable evidence, seems to be the 'cost to correct' approach, where the valuer values the site assuming the no contamination and then deduct the cost of remediation (where feasible) and any adjustment for the effect of stigma (see below). Heavily contaminated sites may require remediation that costs more than the site is worth, resulting in a negative value (a liability rather than an asset).

Stigma refers to the value impact of potential risk and uncertainty surrounding the future use of a contaminated site, even though the contamination may have been removed. The degree of stigma may depend on future regulations and liability regimes in relation to the contamination. Developers might seek discounts to reflect stigma or may decline development altogether. Initial perception of problem may induce a substantial drop in value (dread factors) but as understanding improves, value may increase to point where it relates to logical factors such as clean-up costs, control measures, delay and contingent liabilities. Not all purchasers will be equally risk sensitive; local developers may be prepared to outbid institutional investors. The attitude of lenders is also important; if the proposed use is residential, lenders may not be prepared to offer mortgages on the dwellings. In practice the valuation impact of stigma is difficult to quantify; it may be accounted for by either adjusting the yield or making an end allowance.

For example, a valuation is required of a freehold factory situated on contaminated land. The freeholder has legal responsibility for the contamination. The current rent is £800 000 per annum and the 15-year lease has two years remaining. The current tenant does not intend to renew the lease and remediation is deemed necessary. An environmental impact assessment suggests a £2 000 000 remediation cost and a period of one year in which to complete the work. The yield for uncontaminated comparable property investments is 9.5%. The current market rent is £850 000 per annum.

Term rent (£ p.a.)	800 000	
YP 2 years @ 9.5%[a]	1.7473	
		1 397 840
Reversion to MR (£ p.a.)	850 000	
YP perpetuity @ 10.5%[b]	9.5238	
PV £1 for 3 years @ 10.5%[c]	0.7412	
		6 000 184
		7 398 024
Remediation costs:		
Clean-up (£)	(2 000 000)	
Finance @ 8% for 6 months[d] (£)	(78 461)	
Cost of environmental impact assessment (£)	(14 000)	
Total (£)	(2 092 461)	
PV £1 for 2 years @ 8%[e]	0.8573	
		(1 793 867)
Valuation before PCs (£)		5 604 157

[a] Although the security of a term rent below market rent would normally attract a reduction from the yield, in this case, because of the contaminated state of the site, the yield has not been reduced.
[b] The yield has been increased by 1% to reflect stigma.
[c] Discounting the reversionary value over three years builds in the one-year clean-up period.
[d] It is assumed the clean-up costs are debt-financed at 8% per annum, but the costs are spread evenly over the year (i.e. interest only paid on total cost over six months).
[e] Costs are deferred until the end of the current lease at the finance rate of 8% (it is assumed money can be invested at the same rate that it can be borrowed).

The adjustment to the yield to account for uncertainty at re-letting due to possible residual contamination and stigma is subjective and it might be argued that an explicit end allowance would be more appropriate. An adjustment to the yield will have a greater effect on property investments that are valued at lower yields than those valued at higher yields. For example, take two investment opportunities, a factory in the north of England and a shop in the West End of London, both valued at £500 000 and both requiring the same expenditure on remediation:

	Factory		Shop	
Unimpaired valuation:				
Income (£ p.a.)	500 000		250 000	
YP perpetuity @ 10% (factory)/5% (shop)	10.0000		20.0000	
Valuation (£)		5 000 000		5 000 000
Impaired valuation:				

(Continued)

Income (£ p.a.)		500 000		250 000
YP perpetuity @ 11% (factory)/6% (shop)		9.0909		16.6667
		4 545 455		4 166 667
Less remediation costs, say (£)		(1 000 000)		(1 000 000)
valuations before PCs (£)		3 545 455		3 166 667
Reduction in value		29%		37%

Ceteris paribus the shop suffers a greater depreciation in value. One solution is to adjust the yield proportionately, say an increase of 10% would mean an impaired yield for the factory of 11% and 5.5% for the shop, thus producing the same diminution in value for the shop and factory.

Chapter 11
Valuations for Financial Statements and for Secured Lending

11.1 Valuing property for financial statements

Commonly referred to as *asset valuations*, these relate to the valuation of property assets for inclusion in financial statements such as company accounts, stock exchange prospectuses and documents for takeovers and mergers. Many types of organisations either are required to or prefer to report the current value (known as *fair value* in accounting) of their property assets rather than report the historic acquisition cost, including business owners and occupiers, investors and public bodies.

Property assets are valued so that their current value is reported, rather than their historic acquisition cost. This helps an entity make more informed decisions about what to do with them and is useful to management, owners and other stakeholders in making economic decisions. Regular revaluations of property assets may be required by statute.

There are implications for the organisation's financial statement when reporting current value rather than historic cost. The income statement will incorporate realised profits (losses) and unrealised valuation gains (losses). Also, because the income statement records net proceeds from disposals, realised profit will not look as high as it would if the carrying amount was historic cost. The income statement may report more volatile earning figures for property investment companies as valuation surpluses and shortfalls will appear, to the extent that they exceed any previous shortfall or surplus. Regarding the balance sheet, the higher fair value of property assets for business occupiers will mean increased shareholders' funds and net asset value, leading to a stronger capital-raising base (but a fall in return on shareholders' equity). It will also increase the figure for capital employed, leading to reduced return on capital employed. Balance sheet strength will rise and fall according to movements in property market as well as trading performance.

Property Valuation, Third Edition. Peter Wyatt.
© 2023 John Wiley & Sons Ltd. Published 2023 by John Wiley & Sons Ltd.
Companion website: www.wiley.com/go/wyatt/propertyvaluation3e

Asset valuations often end up in the public domain and may relate to very large amounts of money. Consequently, there is a need for tight control and accounting standards to regulate this process. Valuations for financial statements are regarded by the Royal Institution of Chartered Surveyors (RICS) as *Regulated Purpose Valuations* and special rules ensure these valuations, which are often undertaken on a regular basis, remain objective and independent.

There are several circumstances where property valuations are required for financial statements, the main ones being (RICS 2019a: UK VPGA 1):

- Reporting the carrying amount of property assets on the balance sheet,
- Measurement of lease assets and liabilities and
- Calculation of depreciation charges and impairment losses.

Appendix 11A summarises other circumstances where property valuations are required for financial statements in the United Kingdom.

The chief role of the valuer is the assessment of fair value, but the overall process of including these valuations in financial statements is also very important because it helps underpin the objectivity and independence of the valuation. It can be summarised as follows:

- Establish basis of reporting measurement
- Categorise property assets
- Determine basis of value
- Undertake the valuation(s)
- Consider any other issues such as leaseholds, depreciation accounting and impairment losses

These stages are discussed below.

11.1.1   Basis of reporting measurement

Valuations of property for inclusion in financial statements must comply with the financial reporting standards adopted by the entity. In this chapter, two sets of reporting standards are considered:

[1]   International Financial Reporting Standards (IFRS): The International Accounting Standards Board (IASB) publishes IFRS but has also adopted the body of standards issued by its predecessor, the International Accounting Standards Committee (IASC), and their standards continue to be designated as International Accounting Standards (IAS).

[2]   UK financial reporting standards (FRS): These are published under the umbrella of UK Generally Agreed Accounting Procedures (GAAP).

In the United Kingdom, company law recognises these two financial reporting frameworks. Publicly listed companies must use IFRS for group accounts but can use either for their individual parent accounts. Other entities can choose between IFRS and FRS. Consequently, depending on the reporting requirements of the entity, valuers will need to refer to international and national financial reporting standards, together with international and national valuation standards and guidance. In addition, other statutes, regulations and guidance might apply. For example, authorised UK bodies can issue Statements of Recommended Practice

Table 11.1   International accounting standards applicable to property or property-related assets.

- IAS 2: Inventories
- *IAS 16: Property, Plant and Equipment*
- *IFRS 16: Leases*
- *IAS 40: Investment Property*
- IFRS 5: Non-current Assets Held for Sale and Discounted Operations
- IFRS 6: Exploration for and Evaluation of Mineral Resources
- IFRS 13: Fair Value Measurement
- IAS 36: Impairment of assets

Section C

to supplement accounting standards and legal requirements. All of this should be agreed with the client and stated in the valuation report. The basis of reporting measurement must be applied consistently to an entire class of property.

Various IFRSs provide guidance on how to treat different categories of property asset in financial statements, including recognition of assets, determination of their carrying amounts, and the depreciation charges and impairment losses that should be recognised. The standards that are pertinent to property assets are listed in Table 11.1.

IAS 16, IFRS 16 and IAS 40 relate to the different categories of property assets that are likely to be encountered, namely owner-occupied property (IAS 16), investment property (IAS 40) and leases (IFRS 16). The other standards listed in the table relate to the reasons why properties may be held by organisations including inventories (IAS 2), non-current assets held for sale or discontinued operations (IFRS 5), and exploration and evaluation of mineral resources (IFRS 6). The remaining two (IFRS 13 and IAS 36) describe how properties should be treated in accounts. Regarding IFRS 13 Fair Value measurement, it is worth noting that what accountants refer to as *measurement,* valuers refer to as *valuation.* The terms can be regarded as synonymous.

The two UK financial reporting standards that are relevant to the valuation of property assets are:

- FRS 102: The Financial Reporting Standard applicable in the United Kingdom and the Republic of Ireland. This standard has been derived from IFRS but is less detailed.
- SSAP 19: Accounting for Investment Properties

Both international and UK reporting standards differentiate property assets according to the purpose for which they are held. This is because their treatment in financial statements differs. The next step, therefore, is to categorise property assets.

11.1.2   Property categorisation

Each property asset (both land and buildings) should be classified as:

- Property, plant and equipment;
- Investment property;
- Inventories and
- Other.

The unit of account is usually the individual property.

Property, plant and equipment are tangible assets held for the purposes of the business. They include owner-occupied properties, own-use plant and equipment and properties fully equipped as an operational entity and used by the owner. They can be sub-categorised as follows:

- Specialised operational
- Non-specialised operational
- Non-operational surplus (surplus to the requirements of the business)
- Under construction

Property, plant and equipment may be valued using either the cost model or revaluation model. Investment property is valued using the fair value model if it can be done so without undue cost and effort. Inventories are valued using the cost model and consequently there is not a significant role for the valuer. For 'other' properties, the valuation model is dependent upon the nature of the asset.

For accounting purposes, investment properties are defined as interests in land and/or buildings held for their investment potential (to earn rent or for capital appreciation or both) rather than for consumption in business operations. They include land and buildings and may be in the course of development. In the case of investment properties, the current value and changes in current value are of prime importance.

In the United Kingdom, SSAP 19 Accounting for Investment Properties requires investment properties to be included in the balance sheet at their open market value (equivalent to market value), but without an allowance for depreciation.

Internationally, if properties are held as investments, then IAS 40: Investment Property prescribes the appropriate accounting treatment. As with operational property, investment properties are initially recognised at cost but subsequently they can be reported either at cost (less-accumulated depreciation and any accumulated impairment losses, as prescribed by IAS 16) or at fair value (but this time without any deduction for subsequent accumulated depreciation and any accumulated impairment losses). Once selected, the measurement model (cost or fair value) must be applied consistently to all investment property. Valuations of investment property under IAS 40 should be conducted on a market value basis, regardless of whether the entity chooses the cost or fair value model, and the report should indicate whether the value was supported by market evidence or was heavily based on other factors because of the nature of the property and lack of comparable market data (Cherry 2006). Investment properties reported at fair values will have their revaluation gains and losses transferred directly to the profit and loss account (unless they reverse previous losses that have been shown against equity). Under present UK GAAP, these are shown in the Statement of Total Recognised Gains and Losses (STRGL) but not in the profit and loss account. Property assets held as investments but still under development should be valued by estimating their end values (indicating whether this is market value on practical completion with the sales income deferred or not) and deducting development costs, including fees and finance.

Inventories are non-current assets that are: held for sale in the ordinary course of business, in the process of production for such sale, or in the form of materials or supplies to be consumed in the production process or in the rendering of services.

Other types of property asset may be categorised according to facts and circumstances, for example specialised properties, mineral resources, options and other contractual rights that may be saleable and of value. Specialised properties include heritage assets, biological assets relating to agricultural activities, mineral rights and reserves.[1]

Sometimes, a property asset may be treated differently. For example, under IASs:

- Assets held for sale: IFRS 5 applies
- Exploration and evaluation assets: IFRS 6 applies
- Mineral rights and mineral reserves such as oil, natural gas and similar non-regenerative resources

Regardless of its category, a property asset should initially be recorded at acquisition or construction cost, but, subsequently, it may be recorded using one of three accounting models:

[1] *Cost model.* The cost model is the cost of the property asset less subsequent depreciation and impairment losses (both accumulated). The cost model is an accounting concept, not to be confused with the cost approach to valuation.

[2] *Revaluation model.* This is the *fair value* (see below) of the property asset less subsequent depreciation and impairment losses (both accumulated). With this model, it must be used for all property assets in the same class (having a similar nature, function or use in the business). Along with the revaluation, the amount that would have been reported under the cost model must also be stated. Revaluations should be undertaken sufficiently regularly, so that the carrying amount of an asset does not differ materially from its fair value at the balance sheet date. This does not necessarily mean every year.

[3] *Fair value model.* The fair value model is used to value investment property when this can be done without undue cost and effort. Fair value valuations of investment property are required at each reporting date, usually annually.

11.1.3   Basis of value

Fundamentally, property assets should be recognised in financial statements when it is likely that their future economic benefits will flow to the entity, and the cost of the asset can be measured reliably. These costs should be recognised at the time they are incurred and can include acquisition or construction costs as well as maintenance and repair costs incurred subsequently.

As explained above, there are different accounting models. Both the revaluation model and the fair value model refer to the valuation basis known as fair value, and a valuer is usually asked to advise on fair value in accordance with these two models.

Fair value is defined in IFRS as:

'the price that would be received to sell an asset or paid to transfer a liability in an orderly transaction between market participants at the measurement date'. (IFRS 13, para 9 and RICS 2019b, p. 56).

IFRS 13 explains how to measure fair value. The fair value basis requires an entity to determine:

- The asset that is the subject of the valuation (consistently with its unit of account)
- For a non-financial asset, the valuation premise that is appropriate for the valuation (consistently with its highest and best use)
- The principal (or most advantageous) market for the asset or liability
- The appropriate valuation technique(s), considering the availability of data with which to develop inputs that represent the assumptions that market participants would use when pricing the asset and the level of the *fair value* hierarchy within which the inputs are categorised. IFRS set out the following hierarchy of inputs (IFRS 13, para 76, 81, 86):
  - Level 1 Observable – quoted prices (unadjusted) in active markets for identical assets or liabilities that the entity can access at the measurement date
  - Level 2 Observable but not quoted – inputs, other than quoted prices, that are observable for the asset or liability, either directly or indirectly
  - Level 3 Unobservable – unobservable inputs for the asset or liability

IFRS and UK GAAP fair value definitions differ in detail, but reported figures will be the same in most cases and usually will be equivalent to the IVSC definition of market value. However, there may be cases, particularly involving properties with future development potential or hope value, where the two values are not the same. The definition of fair value, when applied to a non-financial asset such as a property, assumes that it will be used for its *highest and best use*. Insofar as such factors would be considered by market participants when pricing the asset, this highest and best use must be physically possible, legally permissible and financially feasible. An entity's current use of the property is presumed to be its highest and best use unless market or other factors suggest that a different use by market participants would maximise the value of the property.

If a special purchaser can be identified (other than the reporting entity), special value should be separately stated when reporting fair value, as would be the case when reporting market value. Consideration should be given to the extent that fair value can reflect an increase in value because of specialist facilities, which may detract from value if appraised on a market value basis. Special assumptions are generally not applicable for financial reporting as they do not (by definition) reflect reality.

An asset valuation is usually reported at a specific date, and this can affect the value attributed to the property. For example, if the valuation relates to a large plot of land and the highest and best use would be individual plot sales, then optimum value might be achieved by selling these over a period of time rather than all at once. The valuer may decide to reduce the value of the land to reflect simultaneous sales of all plots at the same time or may decide to use a cash flow to incorporate phased disposal. The valuation technique should be explained in the report.

The fair value of an asset that is being used, and will continue to be used by the entity, will be assessed on the assumption that it is sold in combination with other assets as part of the going concern. On the other hand, an asset that has been declared surplus to requirements, or which has been impaired, will be assessed on the assumption that it is sold separately from the other assets employed in the business.

Assets whose values are likely to change frequently should be valued more regularly, for example investment properties, surplus assets and assets held for sale. Any operations that require statutory consents, permits, licences, etc., might require an assumption that these continue. This should be stated in the report. Where a business is closed and the property has been stripped of furniture, fixtures and fittings, it will normally be available for redevelopment, refurbishment or change of use and should be valued accordingly. Land and buildings in the course of development should be valued on the basis of fair value. The entity's directly attributable acquisition or disposal costs should not be included in the valuation of a property asset on a fair value basis.

11.1.4   Valuation

Property assets held for business purposes can be classified as either non-specialised or specialised. This classification refers to their use. Shops, offices, factories, warehouses, for example, would be classified as non-specialised, and market evidence can be used to estimate fair value so either the income approach or the market approach is appropriate. Hotels, restaurants, casinos and cinemas are examples of trade-related properties and may be valued using the profits method. Schools, chemical works, fire stations and hospitals are examples of specialised properties for which there is no market. For these properties, the replacement cost method or, more specifically, the *depreciated replacement cost* (DRC) method is used to estimate fair value, because there is usually no market evidence. Use of this method must be disclosed in the valuation report.

In undertaking replacement cost valuations, the 'going concern' assumption is key, and it is essential to ensure the valuation can be supported by the potential profitability of the company. If the valuation is for a public body, the assumption is that it is subject to the prospect and viability of the current occupation and use.

The valuation date is the reporting date. Where a valuation is materially affected by events after the reporting date, these must be referred to in the report. Furthermore, a distinction must be made between 'adjusting' events, which provide evidence of conditions that existed at the reporting date, and 'non-adjusting' events, which are indicative of conditions that arose after the reporting date.

The methods used to value non-specialised commercial and industrial properties are covered in Chapter 9 and development valuation methods are covered in Chapter 10. Here the focus is on the application of the replacement cost method to the valuation of specialised properties for financial reporting purposes.

The DRC method is an application of the replacement cost method of valuation used to assess the market value of specialised property assets for financial reporting purposes where market evidence is limited. The method is described in international and UK valuation standards. It is the current cost of modern equivalent building(s) less an allowance for depreciation plus the market value of the land in its existing use considering the constraints, if any, on use imposed by the existing buildings and other improvements made to the land. These components of the DRC method are considered below.

11.1.4.1   Land value

Land value should be assessed by reference to the cost of purchasing a notional replacement site that would be equally suited to the existing use. Sites of similar size and location to the actual site would be preferred as comparable evidence but, because of the specialised nature of the property, usually these are not readily available. Consequently, the search for comparable land prices might require an expanded search that encompasses a wider group of land uses. For example, general industrial land prices might be appropriate comparable evidence for specialised industrial premises. In such cases, planning permission for the existing use or relevant range of uses prevailing in the locality should be assumed if the existing use is highly specialised.

Sometimes the use is so specialised, a power station for example, that the search might extend to land prices agreed as part of compulsory purchase negotiations. If land is held leasehold, the valuation should reflect lease terms in the existing lease.

It is also important to consider how the site requirement for the specific use may have changed over time, perhaps due to technological advances in production or a shift in location from the centre of an urban area to the edge. The extent of land to be included in the valuation should be agreed in advance as some might be vacant but retained for future expansion or it might be surplus to requirements. The aim is to identify the least-cost replacement for the use.

Usually, alternative-use value will be a simple statement that the value may be significantly higher than the value of the whole asset estimated using the DRC method, rather than a detailed valuation. Because of the specialised nature of the buildings, alternative-use value is likely to relate to the land only. The extent to which the value of alternative uses should be considered under the highest and best use assumption depends on several factors:

- How easy the alternative use would be achieved, including planning potential.
- The additional costs of preparing a detailed valuation of the alternative use.
- The number of assumptions and special assumptions that would need to be made.
- Whether it would be used in the financial statement, considering that it might not be higher than existing-use value, particularly if the latter was valued using the replacement cost method.
- Whether it is a public sector asset that may be valued on a current (existing use) value approach only.

11.1.4.2   Replacement cost of buildings and other site improvements

An initial step is to distinguish property from plant and equipment. For some properties, an oil refinery for example, this is not straightforward. Each item of plant, and indeed each building on the site, may differ in terms of use, age and economic life.

The next step is to estimate the replacement cost of a modern equivalent building. It must offer an equivalent service potential, in terms of performance and output. As with the site, a modern equivalent building or buildings might differ in size and specification from existing building(s).

In terms of building costs, these should include everything necessary to complete the construction fit for existing use as at the valuation date. Cost information is available from published sources, from the client or from professional cost estimators such as quantity surveyors and engineers. Developer's profit is not included as the purchaser is assumed to be procuring the building for their own occupation. If the costs are based on the actual building, they may need some adjustment to fit with the specification of a modern equivalent building. As a minimum, the actual (and, by definition, historic) costs would need adjusting for inflation and construction cost indices can be used for this. Component-level costs may be required by the client for depreciation accounting. It is worth bearing in mind that the actual costs and the modern equivalent replacement costs may differ in certain respects:

- Site preparation costs may have been included for the actual site but would not need to be included for the modern equivalent if the comparable evidence relates to prepared, levelled and serviced land.
- The actual site may have been developed in stages whereas the replacement would be constructed at same time.
- There may be other differences such as working conditions and practices, planning requirements and building regulations.

In addition to the building costs, other construction-related costs may be included such as professional fees, contingency allowance and finance costs. Estimation of finance costs requires an assumption regarding the time it would take to construct the building as well as an appropriate finance rate. Some DRC valuations have an 'instant build' assumption, which removes the finance costs and contingency allowance from the calculation. This assumption is, however, only applicable to certain organisations and depends on accounting regulations and guidelines.

If the buildings are of architectural or historic interest and protected by legislation, the modern equivalent property approach still applies but the cost of actual reinstatement or reproduction should be included. If the buildings are not legally protected, then the valuer must decide where the property falls along a spectrum between simple modern alternative and reinstatement of existing. In some cases, the Houses of Parliament for example, reproduction is the only way to provide equivalent service potential. At the extreme, some historic assets may be irreplaceable.

If there is a material difference between the DRC of a property asset and its alternative use value, then the valuer must report the alternative use value if it is clearly identifiable and likely to produce a higher value. Where a potentially more valuable alternative use is uncertain, or is speculative, the valuer should indicate that the value may be higher without necessarily providing a figure.

Where an entity has significantly adapted an asset, it may elect to treat the cost of adaptations separately in the accounts, in which case the valuer would need to add a special assumption to disregard these adaptations from the valuation.

11.1.4.3   Depreciation

If the subject building is not new, then its replacement cost is usually based on the cost of a replacement new building but with a reduction for depreciation. Depreciation in value can result from physical deterioration of the building and the onset of obsolescence.

Physical deterioration is the result of wear and tear over time. One way to quantify the effect of deterioration on value is to compare the value of older properties with the value of comparable new assets. However, such comparable evidence is not normally available in the case of specialised properties. Therefore, the effect of deterioration is usually measured by reference to the estimated physical life of the building. This is done by applying a depreciation factor to the estimated cost of a replacement new building. The depreciation factor is the ratio of the estimated remaining economic life (or physical life if it is shorter) of the existing building to the full economic life of a new equivalent building:

$$\left(\frac{\text{Remaining economic life}}{\text{Full economic life}}\right) \times \text{Replacement cost} = \text{Depreciated amount}$$

Economic life is the number of years a building returns more value than it costs to own, operate and maintain. On multi-building sites, there may be different economic lives, but they can be grouped if there is a strong functional inter-dependency, such as an industrial process. For financial reporting purposes, many buildings are assumed to have an economic life of 50 years, but a valuer may regard the depreciation factor to be higher or lower (and hence the lifespan of a building to be shorter or longer) after considering its type and construction, its use, specification, degree of specialisation and whether any capital investment has extended the life of the building. The usual assumption in this regard is that routine maintenance and repair is undertaken but not replacement and refurbishment. Depending on the nature of the building, the valuer may have regard to differing rates at which components wear out, and this may require a weighted average remaining life to be estimated. For leasehold interests, the remaining economic life should be the lower of the unexpired term of the lease and the remaining economic life of the asset.

The impact of obsolescence on property value is much harder to quantify because it refers to the effect on value caused by buildings becoming outdated or outmoded rather than simply wearing out. The DRC method is used when obsolescence is partial rather than absolute because, in the latter case, the value would be zero. Obsolescence means that the design and/or specification of the building no longer allows it to fulfil its original function, perhaps due to changes in legislation or regulations or advances in technology. A depreciation adjustment might therefore reflect the cost of upgrading to a current specification. If this is not possible, then the adjustment should reflect the financial consequences of reduced efficiency compared to a modern equivalent property. It is important to note that the estimated cost of the modern equivalent replacement property may already reflect the 'optimised' building, and so adjustment for functional obsolescence would not be necessary.

11.1.5   Other issues

11.1.5.1   Leaseholds

Under IFRS16, leases must be reported on balance sheets. Rent that is contractually due over the entire lease term must be shown on the tenant's company balance sheet. The right to use the property during the lease term is shown as an asset and future rent payments are shown as a depreciable liability. Companies that rely heavily on leased properties, such as retailers, leisure operators, service companies and banks, will have higher reporting amounts than other organisations. Also, in the early years of a lease, there will be a higher charge to the income statement because of the amortisation and interest charges. Under UK GAAP, if an entity elects to revalue assets rather than carry at historic cost, long leasehold property can be accounted for under FRS 15 or SSAP 19 if classified as investment property. The carrying amount will be market value. The only requirement is to separately classify leasehold and freehold property assets.

11.1.5.2   Depreciation accounting and impairment of assets

A valuer may be asked to assist with the calculation of provision for depreciation in financial statements. Depreciation in this sense means something different to the depreciation described in the previous section. In valuation, depreciation means reducing or writing down the cost of a modern equivalent asset to reflect actual physical condition and utility together with obsolescence and relative disabilities. In accounting, it is a charge against income to reflect consumption of an asset over an accounting period. This section refers to the accountants' view of depreciation.

For financial reporting purposes, all companies (except property investment companies, which value their fixed assets annually) must depreciate the value of fixed assets that have a limited economic life (wasting assets). Buildings on freehold land and leasehold interests that are used for the purposes of the entity are wasting assets and liable to depreciation. Freehold land is not a wasting asset and is therefore not depreciated, except when it has a limited useful life, such as mineral-bearing land or land subject to a time-limited planning permission.

This means that the valuer may be asked to apportion value between wasting and non-wasting elements so that the wasting element can be depreciated. Depreciation may be applied on a component basis: each part of a property with a cost that is significant in relation to the total cost of the item is depreciated separately. Surplus or investment property is not depreciated. Apportionment is for accounting purposes and does not mean that the freehold and leasehold components must be separately valued, which would involve a hypothetical value separation that would be hard if not impossible to support with any evidence. Instead, it can be done by deducting the land value in its existing use from the valuation of the property as a whole, thus leaving the building value component. Alternatively, it can be done by estimating the current replacement of the building net of any depreciation. Usually, the latter approach is preferable as it would have already been done as part of the DRC valuation. If the property is fully equipped as an operational entity, the valuation may need apportioning between land, buildings, and fixtures and fittings.

Depreciation is a reduction in an asset's stated value in a financial statement over its *useful life*. It is charged against income in the profit and loss account to reflect the use of the asset. Thus, the annual profit and loss account contains a *depreciation charge* in respect of the amount of depreciation suffered in an accounting year. The figure on which this depreciation charge is based is known as the *depreciable amount*. So, the useful life and the depreciable amount are required to establish a depreciation charge. To estimate depreciable amount:

- Apportion the reported value of the property between wasting and non-wasting elements, i.e. between building(s) and land by either:
  - Deducting land value (in existing use) from total cost/value of asset, or
  - Estimating the replacement cost of the buildings
- Estimate the future useful life of the building. Because of the difficulty in doing this, it is common to adopt bandings of say 10–30 years and 30–50 years. The useful life of an asset is not necessarily the same as its economic life. Useful life is determined by the entity, which may have a policy to dispose of a property asset after a certain period. A valuer may be asked to assist in the determination of the useful life of an asset, by considering physical deterioration, obsolescence, environmental factors and information from the entity on its disposals policy. If consulted on the useful life of a leasehold property asset, lease terms will also need to be considered. Components of a building may be depreciated separately.
- Deduct any residual value of the property. This is value of the asset, net of disposal costs, assuming it was already of the age and in the condition expected at the end of its useful life. If the useful life is the same as the physical or economic life of the asset, the residual value might relate to bare land value, net of costs. Or it might have some continuing value after, say, a refurbishment.
- The residual value is deducted from the fair value or the cost of the wasting element of the property, depending upon the basis adopted by the entity. What is left is the depreciable amount.
- The annual depreciation charge is the depreciable amount divided by number of years of remaining economic life, i.e. spread, usually on a straight-line basis, over the useful life of the asset.

For example, an entity estimates the remaining useful life of its property asset to be 20 years while a valuer estimates its remaining economic life to be 40 years. The property has a market value of £500 000, £200 000 of which is regarded as land value. The depreciable amount is estimated as follows:

Market (fair) value (reported value of the property asset) (£)		500 000
Less: Value of land (the non-wasting element) (£)		200 000
Value of building[a] (the wasting element) (£)		300 000
Residual value:		
– Useful life of building (as determined by entity)	20 years	
– Future economic life of building (estimated by valuer)	40 years	
Valuer's estimate of current value of building assuming it was 20 years older with a remaining economic life of 20 years (£)		100 000
Depreciable amount of building (£)		200 000

[a] Could also be estimated using a DRC approach.

The entity can then allocate the £200 000 over 20 years and the resulting annual amount is charged to the profit and loss statement in the annual accounts.

11.1.5.3   Valuing properties individually or collectively

A question arises when valuing a group of properties such as the estate of a business or the portfolio of an investor; should the properties be valued individually or collectively? The market values may be different in each case. For example, adjacent land parcels might be worth more when assembled as part of a development programme than they would be individually. Also, if a group of properties were to be sold at the same time, this could 'flood' the market and the increase in supply might lead to a decrease in the prices obtained for each property. Conversely, an opportunity to purchase the group of properties might persuade a bidder to pay a premium and therefore increase the collective price paid.

The RICS (2019b) advises that the properties should be valued as though they were part of a group and in the way that they would most likely be offered for sale. If the purpose of the valuation is one that would ordinarily assume that a group of properties will remain in existing ownership and occupation (the valuation is for a set of company accounts for example), then it is not appropriate to reduce the value as a result of all properties flooding the market at the same time. But if the group of properties is being valued for, say, loan security then the flooding effect should not be ignored. In such a case, the assumption would normally be that the properties are marketed in an orderly way. Hayward (2009) adds that purchasing a group of functionally or geographically related properties can mean reduced acquisition fees and a shorter transaction time on the part of the purchaser, and this may lead to the payment of a 'lotting' premium. It may also allow the purchaser to obtain valuable personal property such as a brand name or design right. Whatever approach is adopted, all assumptions should be reported with the valuation and both group and individual valuations should be stated if they are different.

11.1.6   Example valuations

11.1.6.1   The valuation of a non-specialised owner-occupied property asset

A single-storey factory with a gross internal area (GIA) of 1000 square metres is owned and occupied for industrial use. The premises were built 17 years ago when it was estimated that their economic life would be 50 years. The market rent of the factory is estimated to be £25 000 per annum on FRI terms. Planning permission has been granted to redevelop the whole site as 2000 square metres GIA of new industrial floor space. It is estimated that the works, which could commence immediately, would be completed within one year and that the finished scheme would let at approximately £40 per square metre on FRI terms. Costs, including building, financing and fees, are estimated to be £220 per square metre. Analysis of recent freehold investment transactions suggests a 9.5% initial yield. Value these premises for inclusion in the occupier's company accounts.

Estimated market rent (£)	25 000	
YP perpetuity @ 9.5%	10.5263	
Existing-use value (£)		263 158

This is the figure that will appear in the balance sheet.

11.1.6.1.1 The calculation of the depreciable amount in respect of the wasting element of the asset is undertaken as follows.

Assume no residual value (i.e. asset wastes away to zero value) and an economic life of 50 years:

Depreciable amount: Gross replacement cost (1000 m² @ £220/m²) = £220 000
Depreciation charge: £220 000 divided by 50 = £4400

The market value, which will include alternative-use value, is estimated by looking at the figures relating to the redevelopment of the site. It was suggested that a pre-let could be obtained at £40 per square metre on a building twice as large as the current one. This is likely to mean that the market value of the property is considerably different to its existing-use value, so it needs to be reported. A simple residual valuation would suffice. Because the property is industrial, both building costs and rental value are estimated on a GIA basis. Assuming a one-year building period, the valuation might be as follows.

Estimated market rent on 2000 m² @ £40/m² (£)	80 000	
YP perpetuity @ 8.5%	11.7647	
Gross development value (GDV) (£)		941 176
Estimated demolition costs (£)		(10 000)
Building costs on 2000 m² @ £220/m² (£)		(440 000)
Agent and legal fees @ 1.25% GDV (£)		(11 750)
Developer's profit @ 10% of GDV (£)		(94 118)
Residual balance (£)		352 969
PV £1 at 7% for 1 year		0.9346
Valuation before purchase costs (£)		329 878

This figure would be included in the valuer's report since it is significantly different from existing-use value.

11.1.6.2   The valuation of a specialised owner-occupied property asset

The property is a fully utilised sports centre, held by the current occupier on lease with 32 years remaining. Most of the buildings that comprise the sports centre were constructed 27 years ago but a swimming pool was added to the centre 16 years ago. Because of the age of the premises and its piecemeal expansion, configuration is poor, and it is expensive to maintain. The flat roofs on the older buildings need renewing at an estimated cost of £169 000. The 1.2-ha site is surrounded by good-quality owner-occupied residential property and current residential land values are estimated to be in the order of £1 200 000 per hectare but for the existing use they are estimated to be in the region of £250 000 per hectare. Demolition and site clearance costs are currently estimated to be £900 000.

Because of the specialised nature of the premises, a DRC valuation is appropriate. The depreciation factor is, once again, estimated using the straight-line method.

Description	Date built	Life expectancy of a modern equivalent (years)	Life expectancy of existing building on valuation date (years)	Estimated gross replacement cost (£)	Depreciation factor	Net replacement cost (£)
Main sports centre building	27 years ago	50	23	8 000 000	23/50	3 680 000
Replacement cost of flat roofs (£)						(169 000)
Swimming pool extension	16 years ago	40	24	2 000 000	24/40	1 200 000
DRC of buildings (£)						4 711 000
Value of land: 1.2 ha @ £250 000/ha						300 000
Valuation before PCs (£)						5 011 000

The alternative-use value of £540 000 should also be brought to the attention of the finance director. This value is based on a residential land value of £1 200 000 per hectare less demolition and site clearance costs of £900 000.

11.2   Valuing property for secured lending purposes

Valuations for secured lending purposes is a substantial area of work for valuers. As part of their risk assessment and due diligence processes, lenders rely on valuations of properties that are to be used as security or collateral for a loan. Due diligence is the process of factual and legal investigation of borrower, property, sponsor and other principal parties typically undertaken by a prospective buyer, lender or investor before a transaction (CREFC 2013). It is undertaken to check if the property is a sound investment or loan security, check veracity of information, identify risks and mitigation strategies, comply with statutory or regulatory requirements and help decide whether to proceed with a transaction. It typically involves a valuation of the property because this underpins the investment or lending criteria, mitigates potential for overpaying, offers a summary of features and crystallises their impact on value.

Valuations, along with associated comments in relation to the property, therefore, are a key part of the lending decision.

'Real property accounts for a significant portion of bank lending. Primarily it is used as security for loans, and valuations of that security are an important part of the secured lending process, both at the commencement of the loan and – particularly for commercial property – during the life of the loan'. (RICS 2018, p. 1).

Lending that requires property as collateral, which includes the commercial and residential mortgage lending markets, is a substantial area of business for banks and other lenders. Moreover, each loan usually involves a large sum of money. Things can go wrong and borrowers may breach loan terms and even default on repayments. This can, and often does, happen in market downturns and therefore lenders can find themselves saddled with properties that are worth less than the loan. In such situations, valuers may become exposed to potential litigation. Litigation cases usually take the form of professional negligence claims where the claimant attempts to prove that a valuer has fallen short of the standard expected from a reasonably competent valuer. It is essential, therefore, that valuers adhere to professional standards and guidance since departure from these will be challenging to defend in court.

11.2.1   Professional standards and guidance

Loan security valuations might be required for property that is owner-occupied, held as an investment or going to be redeveloped or refurbished. In the United Kingdom, such valuations are regulated by the RICS. The relevant global standard is VPGA 2: Valuation of interests for secured lending. It provides examples of matters that might be appropriate when valuing different types of property for secured lending purposes, including owner-occupied, investments, trading entities and developments. There are also UK-specific valuation standards that cover valuations for loan security, including:

- UK VPGA 10: Valuation for commercial secured lending purposes,
- UK VPGA 11: Valuation for residential mortgage purposes and
- UK VPGA 12-14, which deal with valuations for various lending purposes and loan-related products specific to the UK residential property market.

As well as the standard level of professional practice required of all valuers in terms of integrity, independence and objectivity, those engaged with secured lending valuations may be required to take additional measures. These might be the result of client requirements or jurisdictional obligations and often come under the heading of conflicts of interest. Examples of conflicts of interest include situations where the valuer or the valuer's firm (RICS 2019b, p. 69):

- Has a long-standing professional relationship with the borrower or the owner of the property or asset,
- Is introducing the transaction to the lender or the borrower, for which a fee is payable to the valuer or firm,
- Has a financial interest in the asset or in the borrower,
- Is acting for the owner of the property or asset in a related transaction,
- Is acting (or has acted) for the borrower on the purchase of the property or asset,
- Is retained to act in the disposal or letting of a completed development on the subject property or asset,
- Has recently acted in a market transaction involving the property or asset,
- Has provided fee-earning professional advice on the property or asset to current or previous owners or their lenders and/or is providing development consultancy for the current or previous owners.

Ideally, the lender would wish to see a declaration from the valuer that there is no current or recent (within the last two years, say) fee-earning involvement with the property to be valued, with a borrower or prospective borrower, or with any party connected to the transaction for which the loan is required. However, where a conflict of interest is suspected, it must be disclosed, and if the valuer or the potential client considers that the disclosed involvement does create a conflict, then the instruction should be declined. If the valuer or potential client can agree arrangements to avoid the conflict, these should be recorded in writing, set out in the terms of engagement and referred to in the report.

If the instruction is received from a party other than the lender, the borrower or a broker for example, and the identity of the lender is not known, then the valuer must state in the terms of engagement that the valuation may not be acceptable to the lender. The valuer must enquire as to whether the property was recently sold or if a price has been agreed pending a sale. If it has, then the valuer must investigate the price paid (or agreed), the extent of marketing and the nature of any incentives.

One of the underlying principles of valuation for financial reporting is the assumption of continuation of the business; such an assumption does not apply to valuations of properties that are going to be used as security for a loan. This means that valuations for secured lending are usually undertaken on a market value basis, in some cases, with special assumptions. Examples of special assumptions include (RICS 2019b, pp. 72–74):

- Owner-occupied properties:
  - Planning consent has been granted for development of the property, including a change of use.
  - A building or other proposed development has been completed in accordance with a defined plan and specification.
  - All necessary licences and consents are in place.
  - There has been a physical change to the property, such as new construction, refurbishment or removal of equipment or fixtures.
  - The property is vacant when in reality at the valuation date it is occupied.
- Investment properties:
  - A different rent has been agreed or determined, e.g. after a rent review.
  - Any existing leases have been determined, and the property is vacant and to let. Often, this is requested to assess downside risk.
  - A proposed lease on specified terms has been completed.
- Trade-related properties:
  - Assumptions relating to trading performance, including projections that materially differ from current market expectations.
  - Because closure of the business is likely to significantly affect market value, one or more of the following special assumptions may be used to report the impact: the business has been closed and the property is vacant; the trade inventory has been depleted or removed; licences, consents, certificates and/or permits have been lost or are in jeopardy; accounts and records of trade are not available to a prospective purchaser.
- Development properties:
  - The works described had been completed in a good and workmanlike manner, in accordance with all appropriate statutory requirements.
  - The completed development had been let, or sold, on defined terms or
  - A prior agreed sale or letting has failed to complete.
- Special assumptions that may apply to any of the above include:
  - The existence of a special purchaser, which may include the borrower.

- A constraint that could prevent the property being either brought or adequately exposed to the market is to be ignored.
- A new economic or environmental designation has taken effect.
- The property suffers from natural, non-natural or existing use environmental constraints.
- Any unusual volatility in the market as at the valuation date is to be discounted and
- Any lease or leases between connected parties has been disregarded.

The valuation should also include comment on potential demand for alternative uses.

Owner-occupied properties should be valued based on vacant possession. This does not preclude the owner as part of the market but does require that any special advantage of the owner's occupancy, which may be reflected in the value of the business, be separated from the value of the property. This is done because, in the event of default on the financial arrangements, security for the loan can be realised only by a change in occupancy (IVSC 2021). Partly because of this, specialised properties, which have limited marketability and derive value from being part of a business, may not be suitable as separate security for loans. If they are offered as security individually or collectively, they should be valued assuming vacant possession (IVSC 2021). Because trade-related properties are valued with regard to the maintainable profit of an operational business, when valuing them for lending purposes the valuer should notify the lender of any significant difference in value that may result if the business was to close; the inventory removed; licences/certificates, franchises or permits were removed or placed in jeopardy; the property vandalised; or there were other circumstances that may impair future operating performance (IVSC 2021). The valuer may also wish to note any specific circumstances that might put the business's profitability at risk, given that the profits method relies on an assumption of adequate profitability.

Although market value (or market rent where appropriate) is the most widely adopted basis for valuations for secured lending, in some jurisdictions alternative bases may be recognised or expressly required, for example, as a result of statute or regulation. Mortgage lending value is one example, a definition of which is set out in the EU Capital Requirements Regulation (CRR) Article 4(74): 'The value of immovable property as determined by a prudent assessment of the future marketability of the property taking into account long-term sustainable aspects of the property, the normal and local market conditions, the current use and alternative appropriate uses of the property'. Countries that make reference to mortgage lending value within regulations include: Austria, Czech Republic, Germany, Hungary, Luxembourg, Poland, Slovenia and Spain.

In some cases, projected market value is requested by a lender. This is an estimate of market value but with a special assumption that the property will exchange at a future date specified by the valuer. The date assumes marketing begins on the valuation preparation date and reflects the period the valuer considers adequate for marketing and completion of negotiations. It also assumes simultaneous exchange and completion.

As well as the usual matters that must be included in a valuation report (see chapter five), other matters relevant to a loan-security valuation include (RICS 2019b, pp. 71–73):

- A statement regarding any conflict of interest.
- The valuation method adopted, details of comparable transactions relied upon and their relevance to the valuation. The DRC method is not suitable for secured lending valuations but may be a useful cross-check to other methods.
- An explanation of any factor that potentially conflicts with the basis of value or its underlying assumptions.
- A statement regarding the use of information relating to a recent transaction or provisionally agreed transaction on the property being valued.
- A comment on the suitability of the property as security for mortgage purposes and an explanation of any circumstances of which the valuer is aware that could affect the price, including the potential and demand for alternative uses, or any foreseeable changes in the current mode or category of occupation. If the property is, or is intended to be, the subject of development or refurbishment for residential purposes, the impact of giving incentives to purchasers should be commented upon.
- Commentary on the potential occupational demand for the property, any disrepair, deleterious or harmful materials, any environmental or economic designation, environmental issues such as flood risk, historic contamination or non-conforming uses.
- A comment on market trends, marketability over the life of the loan and any other matter revealed during normal enquiries that might have a material effect on value.

For investment properties, there are additional matters, including: a summary of occupational leases, indicating whether leases have been read and the source of any information relied on; a comment on covenant strength, current rental income and a comparison with current market rental value; a comment on sustainability of income over the life of the loan (and any risks to the maintainability of income), with reference to lease breaks or determinations and anticipated market trends; and a comment on any potential for redevelopment or refurbishment at the end of the occupational lease(s).

Similarly, for development properties: a comment on the viability of the proposed project, costs and contract procurement; the implications on value of cost over-runs or contract delays; if the valuation is based on a residual method, an illustration of the sensitivity of the valuation to assumptions; and a comment on the anticipated length of time the redevelopment or refurbishment will take.

If the valuation is for residential mortgage lending purposes, the lender may ask for advice on the extent of the use of the property for residential purposes (RICS 2019a, p. 86). Normally, the effect on value of any reasonable prospect of obtaining planning consent for future development is disregarded by means of a special assumption. UK VPGA 11 (RICS 2019a, pp. 90–91) lists the building components that are to be inspected. Essentially, it should include the interior and exterior of the property that is visible from ground level with the site boundaries and adjacent public and communal areas, and when standing at various floor levels. The valuer does not need to make detailed enquiries into legal, planning and environmental matters, but obvious breaches should be reported. The valuer should also report obvious, recent and significant extensions and alterations. The guidance sets out assumptions and special assumptions that may be made, including for leaseholds and dwellings that are part of a building. It also sets out reporting matters in addition to the standard ones.

As well as the valuation figure, the report is likely to refer to issues considered relevant to the purpose of the valuation, namely loan security. If a loan-to-value ratio has been given, it is possible to calculate the maximum amount of loan that could be secured against this property. The valuer may comment on the suitability of the property as loan security given the terms of the loan, the location, condition and size of the property, nature of the tenants and lease terms, and so on. It is also possible to calculate simple loan risk ratios such as rent cover and interest cover (based on current rent level). Depending on the magnitude of these ratios, it may be possible to offer comment on the risk profile of the property. The marketability of and demand for the property are important should the lender find itself saddled with it. In determining demand, consideration would be given to existing and alternative uses, security of income over the duration of the loan (regarding lease lengths, breaks, market activity) and reversionary value (especially if in loan period) including potential for improvement or redevelopment.

11.2.2   Valuation methods for loan security valuations

For residential and non-specialised commercial properties, the comparison and income approaches are appropriate. Trade-related properties are likely to be valued using the profits method. Valuations that are based on replacement cost are used for specialised properties, which are not bought or sold and are not often used for secured lending purposes, but use is made of this basis to calculate the cost of physical reinstatement for insurance purposes, which is a requirement of commercial mortgages.

Development properties can be valued using the residual method under the assumption that the construction work is complete, but it is important to consider market movements between the valuation date and estimated completion date. Cherry (2006) notes that the following additional matters should be reported:

- A comment on costs and contract procurement
- A comment on viability of the proposed project
- An illustration of sensitivity to assumptions made
- Implications on value of any cost overruns or delays

The valuer should also indicate whether plans and costs have been provided by an architect and quantity surveyor respectively.

11.2.3   Example valuation

This is a valuation, for loan-security purposes, of the freehold interest in a property located in the centre of a city. The six-storey property is situated in a mixed office/retail location and is in a conservation area. It is situated in the middle of a terrace of similar properties and is in relatively good order; the stone-faced walls and slate roof are in good condition. A loan is being sought at an interest rate of 6% per annum over eight years with no repayment of capital before the expiry of the loan. A loan-to-value ratio of 65% is required. Details of the tenants currently occupying the property are shown in Table 11.2.

Table 11.2 Tenancy schedule.

Unit	Floor	Use	NIA (m²)	Current rent (£p.a.)	Lease	Tenant
1	Basement	Storage	75	£105 000	Let 7 years ago on 15-year lease (IRI terms) with 5-year reviews	Telcom Ltd. (assigned from PhonesRus Ltd)
	Ground	Retail	80 (ITZA)			
2	Ground	Retail (bank)	105 (ITZA)	£120 000	Let 6 years ago on a 20-year lease (FRI terms) with 5-year reviews, with a break option at the second review	RNS Bank plc
	Basement	Vault + storage	38			
3	First	Office	150	£63 000	Recently let on a 10-year lease (IRI terms) with 5-year reviews	Paperweight Ltd.
4	Second	Office	145	£59 000	Let 7 years ago on 10-year lease (FRI terms) with 5-year review	Bean Traders Ltd
5	Third	Residential	187	£39 000	Assured shorthold tenancy for one year. Fully furnished, no repairing obligations	Mrs Flint
	Fourth	Two-bed apartment with small balcony				

The appropriate valuation method is a unit-by-unit income capitalisation, with term and reversion valuations for those units that are occupied and where the rent is below market levels. The basis of valuation is market value.

Comparable market evidence:

a) A retail unit of 66 square metres (ITZA) recently let on a 20-year lease with five-year reviews at a rent of £93 000 per annum (FRI terms) and subsequently sold for £1 860 000.
b) The freehold interest in office premises recently sold for £1 500 000. The premises were recently let at a rent of £92 500 per annum on FRI terms.
c) The freehold interest in a similar mixed-use property, but in relatively poorer condition and in a slightly less desirable location, recently sold for £4 550 000. The premises comprise three retail units on the ground floor plus three floors of offices and two self-contained flats above. The gross total rental income is £345 000 per annum. One of the retail units is vacant.
d) The long leasehold interest in a one-bed apartment flat recently sold for £450 000. The lease has 89 years unexpired, and the ground rent is £125 per annum.

Collection and analysis of comparable evidence is crucially important to any valuation but is a process that is difficult to replicate in a textbook. Suffice it to say that good company files, a network of market contacts and experience all help to produce a reliable body of evidence that includes explanation and justification of calculations and assumptions. An expurgated version is set out below:

a) Estimated rent value for retail units: £93 000 p.a./66 m² = £1409/m² ITZA (FRI terms). The net initial yield achieved on the investment sale was 4.69%.
b) Based on the analysis of this investment transaction, the net initial yield for office space is 5.79%.
c) This comparable provides evidence of a yield from the sale of a mixed-use investment property. It is 7.12%. It should be noted that one of the retail units is empty (thus reducing current rental income and increasing the initial yield)
d) The (vacant possession) capital value of the long leasehold interest in a comparable (but one-bed) residential apartment is £450 000.

Note that comparable evidence can also be found in the subject property itself, Unit 3 was recently let at a rent of £63 000 per annum, equating to £420/m². The lease was on IRI terms.

Before proceeding with the valuation, some assumptions can be made. Based on the sales evidence, the net initial yield is assumed to be 5% for retail space and 5.75% for office space. A lease on IRI terms is assumed to produce a net income that is equivalent to the rent on FRI terms less 10% for both retail and office space. At lease renewal, it is assumed that lease terms continue unaltered. The rent applied to basement storage space is estimated as a percentage of retail rent (ITZA), in this case 10%. No void period has been inserted to reflect the break clause in Unit 2's lease but the yield has been adjusted slightly instead. Purchase costs are assumed to be 6.5% of purchase price.

Unit 1

Term rent (£ p.a.)		105 000	
YP 3 years @ 5%		2.7232	
			285 936
Reversion to MR (IRI) (£ p.a.)			
Retail space	(80 m² × £1409/m²)	112 720	
Storage	(75 m² × £141/m²)	10 575	
Less external repairs		(12 330)	
at 10% MR			
		110 695	
YP perp @ 5%		20.0000	
PV £1 3 years @ 5%		0.8638	
			1 912 367
			2 198 303

Unit 2

Term rent (£ p.a.)		120 000	
YP 4 years @ 5%		3.5052	
			420 624
Reversion to MR (FRI) (£ p.a.)			
Retail space	(105 m² × £1409/m²)	147 945	
Storage	(38 m² × £141/m²)	5 358	
		153 303	
YP perp @ 5.5%		18.1818	
PV £1 4 years @ 5.5%		0.8072	
			2 249 928
			2 670 552

Unit 3

MR (IRI) (£ p.a.)	63 000	
Less external repairs at 10% (£ p.a.)	(6 300)	
Net rent (£ p.a.)	56 700	
YP perpetuity @ 5%	20.0000	
		1 134 000

Unit 4

Term rent (£ p.a.)	59 000	
YP 3 years @ 5.75%	2.6854	
		158 439
Reversion to MR (FRI) (£ p.a.)		
145 m² × £420/m²	60 900	
Plus external repairs @ 10% (£ p.a.)	6 090	
	66 990	
YP perpetuity @ 5.75%	17.3913	
PV £1 3 years @ 5.75%	0.8456	
		985 161
		1 143 600

Unit 5

Capital value, say (£)	475 000
Valuation before PCs (£)	7 621 455

11.2.4 *Reinstatement cost assessment*

A lender may require a reinstatement cost assessment as part of a report on the suitability of the property as security for a loan. This allows the lender to check that the property being used as loan security is adequately insured. This assessment consists of an estimation of rebuilding costs, so the valuer must have knowledge of buildings and construction techniques, constraints and costs alongside appropriate valuation skills.

The cost approach is used to assess either the new replacement cost or the DRC. For the latter, allowance should only be made for depreciation arising from physical deterioration, not obsolescence, as the aim is to replace what may be physically lost. The value of the land is not included unless it is subject to an identified risk covered by the insurance policy (for example flooding, contamination or a mudslide). In some countries, such damages are covered separately.

Where a property comprises more than one unit of occupation, it is usual for all units to be insured within one policy, including common areas and ancillary accommodation.

Key points

- As far as international accounting standards are concerned, the IVSC advises that, in all cases, when valuing a property asset, market value is the appropriate basis. But the devil is in the detail.
- Slowly but surely, there will be parity between United Kingdom and international financial reporting standards. In fact, UK standards have provided companies with a choice between reporting property assets at cost or value for some time so the merger will not be too onerous.
- It is important that the valuer discusses the future use of a property asset with the entity before preparing a valuation for financial reporting purposes to ensure that appropriate valuation assumptions are made.
- Lenders are concerned with the financial security of their loans. If a property is being considered as security for a loan, then it is standard practice for a valuation to be commissioned to determine whether it represents adequate collateral. Market value is the basis but there may be special assumptions in particular circumstances, especially regarding loans made in respect of development activity.

Note

1   The IFRS do not refer to heritage assets, and for biological assets, bearer plants are included but the produce is not.

References

Cherry, A. (2006). *A Valuer's Guide to the Red Book*. RICS Books.

CREFC Europe (2013). *Guidelines for Due Diligence on Real Estate in the UK*. Commercial Real Estate Finance Council Europe.

Hayward, R. (2009). *Valuation: Principles into Practice*, 6e. London: Estates Gazette.

IVSC (2021). *International Valuation Standards*. London: International Valuation Standards Council.

RICS (2018). *Negotiating Options and Leases for Renewable Energy Schemes*, RICS Guidance Notes, 2e. Royal Institution of Chartered Surveyors, June 2018.

RICS (2019a). *RICS Valuation – Global Standards 2017: UK Supplement*. London, UK: Royal Institution of Chartered Surveyors.

RICS (2019b). *RICS Valuation – Global Standards 2017*. London, UK: Royal Institution of Chartered Surveyors.

Questions

[1]   An old four-storey warehouse is owner-occupied and comprises 2400 square metres GIA. Market evidence suggests a market rent of £82 per square metre and yields for these older, multi-storey warehouses average 11%. Also, it is understood that the warehouse site could be redeveloped to include a large single storey distribution warehouse of 3000 square metres GIA. The development, which would take one year, would cost £500 per square metre (including demolition, building, contingencies, finance, fees). The market rent and yield of a new warehouse are estimated to be £100 per square metre and 7%, respectively. Value the warehouse for inclusion in the owner-occupier's annual accounts and make recommendations to the directors of the company for any additional information you feel should be reported in the accounts.

[2]   A property comprises two attached brick-built single-storey buildings, originally designed as warehouses, on a 0.79-ha site. Unit 1 is let as it was surplus to the company's occupation requirements and was converted by the current tenants as a condition of the lease to good-quality open plan office accommodation. Unit 2 is used as a warehouse by the company. The table below provides further details:

Unit	Use	Area (m²)	Lease details	Current rent (£ p.a.)
1	Office	GIA = 608 NIA = 517	Let 3 years ago for 15 years with 5-yearly rent reviews on FRI terms	49 870
2	Warehouse	GIA = 1689	Owner-occupied. Two years ago, the company installed bespoke external loading facilities, the current replacement of which is estimated to be £160 000, with an estimated life of 20 years.	
Electricity substation		Site = 50	Let 32 years ago on a 50-year lease with a review at year 25	390

Last year planning permission was granted for conversion of both buildings to 28 two-bed residential apartments, the site being valued at £2.3m at that time.

The tenant of Unit 1 has offered to surrender the current lease and take a new 15-year lease of both the office and warehouse at a rent of £100 per square metre (30% below market levels for the duration of the lease) to reflect their conversion costs of the warehouse to office space. Office yields are 6.5%.

In terms of comparable evidence, first, a warehouse unit with a GIA of 1450 square metres was recently let at a rent of £84 000 per annum. The freehold interest recently sold for £1 028 000. Second, a former public house site of 230 square metres, located in a more prime residential area than the subject property, with planning consent to erect six two-bedroom flats, has just sold for £525 000.

Making appropriate assumptions, value the property for accounts purposes.

[3] A wind farm of 15 ha is situated in a larger site totalling 120 ha that is used for pasture. The wind farm was let 4 years ago on a 30-year lease to an energy company. There are 10 wind turbines with an output capacity of 2 MW each and a capacity factor of 25%. Output revenue is £50 per MWh. The tenant pays a base rent plus turnover component as follows:

- An index-linked base rent of £5000 per annum plus RPI for the first 15 years of the lease, rising to £10 000 per annum plus RPI thereafter.
- The turnover element is 5% of gross income for the first 15 years and then 10% thereafter.

The current market rental value of grazing land in the area is £3000 per hectare. Assume a yield of 9% for wind farm investments and a 3% yield for grazing land. Value the wind farm for the operating company's financial statement.

Answers

[1]

Basis of value: existing-use value. Applying the current market rent to the subject property, we get £82/m² × 2400 m² = £196 800 per annum.

Valuation (existing-use value):

Market rent (£ p.a.)			196 800	
YP in perp	@	11%	9.0909	
Valuation before PCs (£)				1 789 091

Alternative-use value must be considered even though the existing premises have been valued at existing-use value.

Valuation (alternative-use value):

Rent (3000 m² × £100/m²) (£ p.a.)	300 000	
YP in perp @ 7%	14.2857	
Gross development value (£)		£4 285 170
Building costs (3000 m² × £500/m²) (£)		(£1 500 000)
Developer's profit @ 15% of costs		(£225 000)
Residual value (£)		2 560 170
PV £1 @ finance rate for one year		0.9091
@ 10% p.a.		
Valuation before PCs (£)		2 327 451

The alternative-use value of the land is £2 327 451 and the gross development value of the completed development is £4 285 170. Both figures should be included in the directors' report.

[2]

Determine bases of value:

- If Unit 1 is to remain surplus to requirements, it should be valued on a market value basis and if Unit 2 is to remain owner occupied then it should be valued on a market value basis assuming existing use.
- If the company intends to accept the tenant's offer, Units 1 and 2 should be valued on an MV basis.
- If the company intends to take back Unit 1, then the basis of value for the whole property will be market value assuming existing use but subject to the lease of Unit 1 in the interim.

Analysis of comparable evidence:

Office: MR is £100/0.7 = £142/m²

Warehouse: MR is £58/m² (£84 000/1450 m²)

                   Yield of 7.67% (£84 000/£1 094 820)

Redevelopment site value: £87 500 per two-bed flat (£525 000/6) or

                   £2283/m² (£525 000/230 m²) for residential use

Valuations:

Unit 1:

Term (contract) rent (£ p.a.)				49 870	
YP for	2.00	years @	6.50%	1.8206	
					90 795
Reversion to estimated MR (£ p.a.)				73 414	
(£142/m² x 517 m²)					
YP perpetuity @			6.50%	15.3846	
PV £1 for	2.00	years @	6.50%	0.8817	
					995 787
Valuation before PCs (£)					1 086 581

Unit 2:

Market (rack) rent (1689 m² x £58/m²) (£ p.a.)	97 962	
YP in perpetuity @ 7.67%	13.0378	
Valuation before purchase costs (£)		1 277 210
Improvements: 18/20 years = 0.9 x £160 000 (£)		144 000
Valuation before PCs (£)		1 421 210

Electricity substation: £390 × YP perp @ 5% = £7800

Alternative-use value 1: tenant surrender and renewal valuation:

Term rent: (£100/m² × 517 m²) + (£100/m² × 0.85 × 1689 m²) = £195 265

Market rent: (£142/m² × 517 m²) + (£142/m² × 0.85 × 1689 m²) = £277 276

Term (contract) rent passing (£ p.a.)				195 265	
YP for	15.00	years @	6.50%	9.4027	
					1 836 012
Reversion to estimated MR (£ p.a.)				277 276	
YP perpetuity @			6.50%	15.3846	
PV £1 for	15.00	years @	6.50%	0.3888	
					1 658 650
Valuation before PCs (£)					3 494 662

Alternative-use value 2: Redevelopment potential

28 × £87 500 = £2 450 000. Could cross-check against the site value of £2.3m from last year, maybe indexed to PV.

Comments:

- Report alternative-use value for development based on comparable and past site value.
- Comment also on value based on surrender and renewal offer. The valuer should consider how the rent goes from £100/m² to £142/m² and when, e.g. make it a condition of lease that the tenant pays for conversion so include improvement in rent at lease renewal.

[3]

Base rent		
Term rent passing (£ p.a.)	5000	
YP 11 years @ 9%	6.8052	
		34 026
Reversion to (£ p.a.)	10 000	
YP 15 years @ 9%	8.0607	
PV 11 years @ 9%	0.2745	
		22 127
		56 153
Turnover rent		
Turnover: Turbines	10	
Installed capacity (MW)	2	
Capacity factor (i.e. 365 days × 24 hours × 0.25)	2190	
£/MWh	50	
Gross income (£ p.a.)	2 190 000	
1st-term rent (5% gross income for 11 years) (£ p.a.)	109 500	
YP 11 years @ 13%	5.6869	
		622 716
2nd-term rent (10% of gross income for 15 years) (£ p.a.)	219 000	
YP 15 years @ 13%	6.4624	
PV 11 years @ 13%	0.1599	
		226 301
		849 017
Reversion of turbine site to farmland:		
15 ha @ £3000/ha (£ p.a.)	45 000	
PV 26 years @ 3%	0.4637	
		20 867

Remainder of site
105 ha @ £3000/ha (£ p.a.)   315 000
YP perp @ 3%   33.3333
                                                                              10 500 000
                                                                              10 520 867

Total valuation: £56 153 + £849 017 + £7 020 860 = £11 405 170

11.A   Other circumstances where valuations are required for financial statements in the United Kingdom

11.A.1   Valuations for other regulated purposes (UK VPGA 2)

11.A.1.1   Valuations for listings and prospectuses

Property companies seeking Financial Conduct Authority (FCA) approval must include a valuation by an expert valuer in the prospectus. When a UK-listed company proposes acquisition or disposal of a property and it is 25% or more of the value of the company (a Class 1 transaction), the company must seek shareholder approval and include a valuation by an expert valuer in the circular to shareholders. A UK-listed company must also include a valuation where it makes significant reference to the value of a Class 1 property in a circular to shareholders. Valuation reports can be in a condensed form, the framework for which is included in UK VPGA 2.1.

11.A.1.2   Valuations for takeovers and mergers

The first and most important point to note is that the obligations relating to the declaration of interests that are imposed by the Takeover Code apply to all advisors including valuers. Also, potential conflicts of interest may be harder to manage than normal. Valuations must be undertaken by named, independent valuers (members of the RICS who are external valuers). If a valuation relates to land being developed or with immediate development potential, the valuation should include: the value after completion, the value after completion and let, the estimated development cost, the development period and leasing/occupation period, a statement on whether planning permission is in place and, if so, the date and details of any value-significant conditions.

11.A.1.3   Collective investment schemes

Valuations must be in accordance with the FCA Collective Investment Schemes Sourcebook and the Investment Managers Association has issued a Statement of Recommended Practice on the implementation of accounting standards.

11.A.1.4   Unregulated property unit trusts

There is no regulatory requirement for independent valuations, but most trust deeds require them. In all cases, the appropriate basis is market value.

11.A.2 Valuations for assessing the adequacy of financial resources (UK VPGA 3)

Valuations for inclusion in the assessment of the adequacy of financial resources of insurance companies are the same as those adopted by the entity for its accounting purposes. For financial institutions (banks, building societies, investment firms), the value of properties must be monitored on a frequent basis, particularly when market conditions are changing significantly. Statistical methods can be used. Valuations must be reviewed by an independent valuer if values fall materially relative to the market. Loans of more than 5% of a firm's capital resources must have their valuations reviewed at least every three years by an independent valuer. The appropriate basis is market value (mortgage lending value is not normally used in the United Kingdom).

11.A.3 Valuations of local authority assets for accounting purposes (UK VPGA 4)

The Code of Practice on Local Authority Accounting in the United Kingdom adapts IAS 16 so that property, plant, and equipment that is operational (and therefore provides *service potential*) is valued at *existing use value* (EUV) rather than fair value. Where EUV is very different from market value, both are reported. Some assets such as infrastructure, assets under construction and some community assets (parks, cemeteries, allotments for example) are measured at historic cost. Fair value is the appropriate basis for properties categorised as investments, surplus or held for sale. Leases of land and buildings are separated into land and building elements and classified and accounted for separately. Heritage assets can be valued or reported at cost, as appropriate.

11.A.4 Valuation of central government assets for accounting purposes (UK VPGA 5)

As with local authority assets, international accounting standards are adopted but adapted slightly. For operational property, plant and equipment held for their service potential, EUV is the appropriate basis. For assets surplus to requirements, fair value is the appropriate basis, so long as the market has access to them.

The EUV basis as applied to UK local authority and central government accounting is defined as:

'The estimated amount for which a property should exchange on the *valuation date* between a willing buyer and a willing seller in an arm's length transaction after proper marketing and where the parties had acted knowledgeably, prudently and without compulsion, assuming that the buyer is granted vacant possession of all parts of the asset required by the business, and disregarding potential alternative uses and any other characteristics of the asset that would cause its *market value* to differ from that needed to replace the remaining service potential at least cost'. (Source: UK VPGA 6)

11.A.5   Valuation of registered social housing providers' assets for financial statements (UK VPGA 7)

For dwellings held for social housing purposes, the appropriate basis is *existing-use value for social housing* (EUV-SH). For surplus properties, the appropriate basis is fair value.

EUV-SH is an opinion of the best price at which the sale of an interest in a property would have been completed unconditionally for a cash consideration on the *valuation date*, assuming:

- A willing seller
- Prior to the *valuation date*, there had been a reasonable period (having regard to the nature of the property and the state of the market) for the proper marketing of the interest for the agreement of the price and terms and for the completion of the sale
- The state of the market, level of values and other circumstances were on any earlier assumed date of exchange of contracts, the same as on the *date of valuation*.
- No account is taken of any additional bid by a prospective purchaser with a special interest.
- Both parties to the transaction had acted knowledgeably, prudently and without compulsion.
- The property will continue to be let by a body pursuant to delivery of a service for the existing use
- The vendor would only be able to dispose of the property to organisations intending to manage their housing stock in accordance with the regulatory body's requirements
- That properties temporarily vacant pending re-letting should be valued, if there is a letting demand, on the basis that the prospective purchaser intends to re-let them, rather than with vacant possession.
- Any subsequent sale would be subject to all the same *assumptions* above. (Source: UK VPGA 7)

11.A.6   Valuation of charity assets (UK VPGA 8)

The Charities Commission requires charity trustees to obtain a 'Section 36 Valuation' when seeking to buy or sell a property interest with a remaining term greater than seven years. The report must be by a qualified surveyor who must confirm that his or her professional opinion conforms to the relevant legislation and to the RICS Red Book. For acquisitions, the report should consider specific requirements of the charity such as whether any repairs or alterations are required and the estimated costs of doing them. It may also advise on whether it is economically sensible for the charity to acquire given prevailing market conditions, the state of the property, any lease terms and the asking price. For disposals, the valuer should consider whether the property has been adequately marketed and whether it might benefit from alterations or adaptation prior to marketing. If the property is to be auctioned the report must consider the reserve price, scrutinise the auctioneer's reasoning, be conversant with the planning position and aware offers received before the auction. Charities may face tax implications if disposal is subject to an overage clause (regularly used when planning is uncertain).

11.A.7 Local authority disposals of land for less than best consideration

Local authorities in England and Wales are required to seek consent from central government if they sell land at a price below market value but only where the discount is more than £2m. A valuer may be asked to advice on whether an application for consent is necessary or to support a request for consent. The RICS Red Book provides guidance in relation to the valuations that may be required. There are three bases of value that are relevant:

[1] Unrestricted value: This is market value but considering any additional value that might be due to a purchaser with a special interest and ignoring any reduction in value caused by the local authority imposing any conditions on the sale. The valuer should assume the land is offered for sale on terms that maximise its value.

[2] Restricted value: Again, market value but this time having regard to the terms of the transaction. In this way, it should take account of any conditions imposed by the local authority. If tenders are invited for purchase of the land, restricted value will normally be the bid from the preferred bidder. Otherwise, it is the proposed purchase price.

[3] Value of local authority conditions: These are the conditions that may be imposed by local authorities. They must be capable of being quantified in monetary terms and might include operational savings and income.

   The discount is: Unrestricted value – (restricted value* + value of conditions)
   *or value of consideration if different from restricted value

When reporting the valuation, the valuer should include a description of the land and buildings, location and surroundings; summarise the proposed transaction; and provide details of the tenure, attaching a copy of the lease (or at least the heads of terms) if the transfer is leasehold. If the land is a development site, then reasonable assumptions can be made regarding the proposed scheme, including planning assumptions. The valuer should note existing uses, current planning consents and likely permitted uses in line with the development plan. The date of the valuation must be within six months of the application submission date.

11.B Valuation of a hotel refurbishment for secured lending purposes

11.B.1 Overview

The location has a generous supply of good-quality hotels. There are now more than thirty three-star hotels in the town, seven four-star but no five-star establishments. The number and focus of the three-star hotels within the town adds to its attraction as a business and conference venue, although inevitably there is fierce competition that tends to drive down prices, especially outside the summer season. The subject property trades towards the upper end of the three-star range. The operator's improvements of facilities, to the point where it is now ready for assessment to obtain a four-star rating, will enable it to compete at the higher end of the market in the town.

11.B.2   The property

11.B.2.1   Accommodation

11.B.2.1.1   Public Areas

Reception: The entrance lobby has an automatic entry door. Located off this is a cloaks/reception area with seating for four guests and a porter's lodge. The lobby leads through to the main hotel reception area with reception desk and office beyond. There is seating for two customers and access to the passenger lift.

Restaurants: The main restaurant has been fully refurbished. There is a dance floor area, covered in the daytime with a removable carpet, and a small bar servery. This room has space for approximately 200 restaurant covers and up to 260 for a function. At one end, a permanent self-service buffet has been created.

Lounge and Le Café: This is located off the reception area. There is a glazed atrium to the front elevation providing access to the quiet lounge and swimming pool/fitness centre. Le Café to the rear has its own separate bar servery together with a further food servery presented in a carvery style format. Overall, there is capacity for around 140 covers within this area.

Quiet Lounge: A conservatory extension acts as a link between the Lounge and the Fitness Centre with seating for 28 on rattan-style chairs. It extends to approximately 81 square metres.

Fitness Centre: A former outdoor pool has been re-lined and now ranges in depth from 0.5 to 1.5 metres. It is positioned under a conservatory style roof and, including a reception desk to one corner, extends to 216 square metres, or thereabouts. The air conditioning system is this area maintains a constant air temperature at 24 oC and maintains a constant humidity. Located off is the gymnasium, containing thee cycling machines, two rowing machines, three running and three step aerobics machines, together with a range of five weight machines. It extends in total to 80 square metres. Serving these two areas are ladies' and gent's changing rooms and a disabled persons' shower/w.c.

Function/Meeting Rooms: The Sandy Suite is the main function room and is in the main building. It has a timber dance floor, also carpet covered, and small raised stage with multi-lighting. It has its own bar servery and has now been fitted with two extra portable dance floors. The Coastal Suite is also located within the main building and has two syndicate rooms off. The Coral, Clavelle and Bloom Suites are accessed via Mortimer Road. The Coastal and Bloom Suites have a Civil Wedding Licence. Each suite, except the Coral Suite, has a syndicate room off, together with a further syndicate room, known as the Conference Room. A bar servery has recently been provided to the Bloom Suite. Each room is equipped with visual aids and public address systems. Conference equipment includes projectors, video screens and monitors, photocopying and fax facilities, flip charts and wipe boards.

Additional specialist audio visual equipment and secretarial services are available upon request. The accommodation for conferences is as follows:

Room name	Floor area (m²)	Dinner (number)	Reception (number)
Sandy	280	180	200
Coastal	107	70	100
Coral	39	35	40
Cavelle	53	35	60
Bloom	125	70	100

The Health Spa: Stairs from the reception lead to the lower ground floor area. The Health Spa has its own reception and changing facilities. This area is fully air-conditioned. Facilities include a heated indoor air spa pool, jacuzzi and main high-temperature pool. Within this area there are two underwater jets operating in the main pool and the air spa pool alternately. Also within the Health Spa area is a sauna for up to 10 persons, with automatically dispensed aromatherapy oils and an ice zone outside, to include an ice dispenser and two cold showers. The new steam room can accommodate 10 persons and again has automatically dispensed aromatherapy oils. In addition, there is an aromatherapy cave, which is halfway in humidity and heat between a steam room and sauna. This area can accommodate up to 20 and incorporates low-level changing lights and a heated central stool, together with music. There is also a relaxation lounge with eight steamer-style recliners, together with low lights and magazines. There are three beauty therapy treatment rooms, together with a hydro-bathroom, the latter also due to become a treatment room.

Toilets: There are public toilet facilities located off all the suites, functions rooms, restaurants and public areas.

11.B.2.1.2 Letting Accommodation

Layout: The letting accommodation is principally located within the main building, arranged over four floors, and interlinked via various staircases and a shaft lift. The Canford wing has accommodation at both first and second floors and the Meredith wing at first-floor level. Access to these latter areas is by stairwell only.

Floor	Single rooms	Twin/double rooms	Multiple rooms		En suite facilities	No. of bedrooms	Total beds per floor
			Rooms	Beds			
Fourth	0	4	0	0	4	4	8
Third	1	17	3	9	21	21	44
Second	1	21	3	10	25	25	53
First	1	19	7	24	27	27	63
TOTAL	3	61	13	43	77	77	168

Room facilities: All the bedrooms have direct dial telephone, hair dryer and satellite television. Tea and coffee facilities are provided, and the rooms also have a trouser press and iron. The letting bedrooms have been refurbished to four-star standard and there is a rolling programme of refurbishment. Most bathrooms comprise a bath with shower plus toilet and wash hand basin. Four of the rooms are designated as executive suites and have additional items, such as a jacuzzi bath or electrically operated curtains. One is designated as a junior suite. The above table reflects the maximum capacity of the hotel. The multiple-bedded rooms are, however, generally utilised as double/twin rooms; hence, the hotel is usually operated to a bed occupancy capacity of 148 within the 77 rooms.

11.B.2.2   Ancillary areas

Cellar/Storage: The main cellar is thermostatically regulated and lined with wine racks. There is a centrally positioned empty crate store with rear-access doorway and a lockable red wine cellar off in addition. Portable bars and dance floors are utilised for other suites. A small beer cellar serves the restaurant service bar. This unit is not thermostatically regulated.

   Kitchen: Of substantial size. Quarry tiled floor, tiled walls. Stainless steel surfaces throughout. Good range of commercial catering equipment. Extraction system. Washing-up room located off. Walk-in fridge located off. Lockable dry goods cupboard. Still room.

   Laundry: Contains two commercial washing machines, two commercial drying machines and a large rotary iron. All laundry is undertaken on the premises.

   Staff: Staff changing room and locker corridor. Staff dining room off the kitchen. Two accounts offices and sales office. Manager's office off reception foyer. Further manager's office in former syndicate room to Coastal Suite. Two bedrooms and a communal bathroom within the second-floor level of the Meredith hotel building are used as overnight stay rooms for management.

11.B.2.2.1   Residential Accommodation
None.

11.B.2.3   External amenities

The site has an area of over approximately 1.67 acres (0.68 ha). There are three principal car parks providing 120 spaces. There is a sun terrace and adjacent gardens.

11.B.2.4   Statutory and regulatory considerations

11.B.2.4.1   Planning
Established C1 user. The property lies just outside the designated hotel area in which there is a presumption against change of use from hotel usage. No outstanding planning consents.

11.B.2.4.2   Licencing
Suitable Premises Licence is held.

11.B.2.4.3   Environmental Health
No matters remain outstanding from the last visit.

11.B.2.4.4   Fire
A Fire Risk Assessment has been prepared and approved.

11.B.2.4.5   Access

The subject property has direct access from adopted highways.

11.B.2.4.6   Rating

Rateable value (£ p.a.)	171 500
Business rates (£ p.a.)	72 373

11.B.2.4.7   Services

We are advised that the property is connected to all mains services. Water is metered. Three-phase electricity is available. Heating is by means of two gas-fired boilers for the main building, while there are other boilers for the ancillary accommodation. Air conditioning is provided to the Fitness Spa, restaurant, Coastal Suite, Bloom Suite and the Fitness Centre.

11.B.2.5   Environmental issues

11.B.2.5.1   Contaminated Land

This is a period property in an established commercial setting. As such, we are not aware of any contaminated land issues directly affecting it. We have, therefore, valued on the assumption that the property is not subject to land contamination. We do not consider that a full environmental risk assessment is necessary. Should it be established subsequently that contamination exists at the property or on any neighbouring land, or that the premises have been or are being put to any contaminative use, this might reduce the values now reported.

11.B.2.5.2   Asbestos

Under the Control of Asbestos at Work Regulations 2002, it is necessary for all property owners/occupiers to undertake an inspection for asbestos in all non-domestic properties (to include common parts of shared residential dwellings). Following this exercise there are several requirements, to include the need for an asbestos management plan to be drawn up and implemented. For valuation purposes, we have assumed that no significant costs of a capital nature will be identified as part of this investigation process. Should this not be the case then we may wish to review the valuation opinion provided, upon receipt of a costed schedule of the required works.

11.B.2.5.3   Disability Discrimination Act

For valuation purposes, we have assumed that no significant costs of a capital nature are required to achieve compliance with this legislation. Should this not be the case, then we may wish to review the valuation opinion provided, upon receipt of a costed schedule of the required works.

11.B.2.6   Tenure

Freehold, with the assumption of full vacant possession being available. We understand that the Title is subject to various restrictive covenants, and we assume that full indemnity cover is in place in the event of enforcement.

11.B.2.7   Agreements

11.B.2.7.1   Fixtures, Fittings and Equipment

We understand that all items of trade fixtures, fittings and equipment are held unencumbered. We have assumed for the purposes of our valuation that the property would be sold with the benefit of an unencumbered and full trade inventory. Any current or proposed hire purchase, leasing or rental costs have, however, been included in our financial assessments.

11.B.3   The business

11.B.3.1   Background and history

The operator has decided to pursue a policy of upgrading the facilities at the subject concern, ultimately to achieve four-star status. They have already largely completed a programme of room refurbishment. At the time of our inspection, we were provided with a schedule of improvements that have been undertaken to the property in the last three years. This schedule largely details the bedroom conversions that have been completed, together with the work carried out in the former leisure club, now the Health Spa.

Within the last three years, some £470 393 has been spent. While the Fitness Centre, the newly developed area to the front of the property, was completed two years ago, the Health Spa was not completed and opened until last year. This, together with the other improvement works that have been undertaken over the year, has meant that the premises have presented, on some occasions, rather like a building site rather than a high-quality hotel. This has impacted upon the marketing of the subject concern, which has been held up until the completion of the Health Spa.

Accordingly, therefore, the last three years saw significant capital investment in the property, while at the same time the promotion of the business was restricted, owing to the consequences of that capital investment.

11.B.3.2   Customer profile

A significant number of corporate and public sector clients. Functions trade derived from local residents, companies and organisations. Additional hotel trade at weekends derived from tourists and local organisations. Restaurant: weekdays, hotel guests and at weekends non-residents. Fitness Centre and Health Spa: 270 external members and growing.

11.B.3.3   Opening hours

The coffee lounge serves between 8.00 a.m. and 5.00 p.m. Food is available throughout the day and night. The health club operates from 7.00 a.m. to 10.00 p.m., seven days per week. The hotel trades throughout the year. Peak trading periods are December and the six weeks of the summer holidays.

11.B.3.4   Staffing

The operator is involved in the administration, while his son oversees marketing and hotel refurbishment on a full-time basis. In addition, a Finance Director also works full time in the business.

There is a general manager, deputy manager and three managerial staff together with three sales staff, seven receptionists and two part-time administrative staff. Otherwise, the hotel operates on a traditional pyramid style structure with heads of department reporting to the general manager. Departmental heads include the chef, sales manager, restaurant manager, housekeeper, leisure club manager and the reception manager.

11.B.3.5   External staff costs

Year	% of net sales
$n-11$	37.9
$n-10$	36.2
$n-9$	38.1
$n-8$	36.7
$n-7$	35.6
$n-6$	35.0
$n-5$	37.9
$n-4$	39.6
$n-3$	42.2
$n-2$	42.3
$n-1$	50.5
Last year (n)	58.5

It is normal within a corporate establishment of this size and type, effectively run under management, to expect wage costs of around 37% of the turnover to be incurred (this could increase to around 40% for a four-star establishment), although there will be a degree of fluctuation depending upon the trading mix and physical attributes of the building. The significant rises in percentages in years $n-1$ and n are the result of reduced turnover affected by the disruption caused by the extension works and refurbishment, now recently completed.

Wages shown within the management profit and loss accounts for the 337 days ended 31 January year n total £1 154 600, including national insurance contributions and pension costs. On an annualised basis adjusted for seasonality, the current wages liability is believed to be in the region of £1 250 000.

The wage costs are, to a large extent, self-imposed by the operator, who considers this to be necessary to attract and retain high-quality staff. They believe that this has contributed to the success and development of the business. In addition, staffing levels have been purposely increased and extensive training undertaken in preparation for a four-star grading. This is, nevertheless, a very high wage bill in relation to the assessed current level of trade and the industry norm. We assume that the wage costs shown in the management accounts do not include any element of directors' remuneration.

11.B.3.6   Income streams

11.B.3.6.1   Beverages
11.B.3.6.1.1   Tariff

Product	£
Draught bitter	2.40–2.60
Draught lager	2.60–2.80
Mainstream spirits	2.00
Wine (glass)	2.30
Fruit juice	1.10

The above prices are inclusive of VAT.

Prices have been maintained, and in some cases have been reduced, in the last year. No discounting is offered. There is a wine list of 33 bins with prices ranging between £11.00 to £55.00 per bottle. House wines are generally £11.00 per bottle.

11.B.3.6.2   Food
11.B.3.6.2.1   Tariff

Menu	£
Two-course meal	15.00
Three-course meal	21.00
Bar lunches	5.50

The above prices are inclusive of VAT.

Again, prices have been held over the last year. There is, in addition, a detailed banqueting menu with a considerable price range per head available, subject to choice.

11.B.3.6.3   Accommodation
11.B.3.6.3.1   Tariff

Room/Service	£ per room per night
Bed and breakfast	65.00
Dinner, bed and breakfast	75.00

The above prices are inclusive of VAT.

The above tariffs are effective up to 9 June n. There is a £25 supplement for single occupancy of a double room. Short break discounts are also offered. The hotel is still effectively charging a three-star level of tariff. The conference delegate rates range from £76 to £98 for a 24-hour delegate and the day delegate rate is from £29 to £38 per person. Health and Beauty breaks are also now offered from Sunday to Thursday inclusive, with packages from £150 per night to £350 for three nights.

Section C

11.B.3.6.4   Fitness Centre and Health Spa
11.B.3.6.4.1   Tariff

Membership type	£ per person
Gold	540.00
Joint gold	825.00
Bronze	310.00
Silver	365.00
Corporation (minimum 10)	200.00

This represents a significant recent increase. The prices are for new members. Gold membership allows access to all facilities, silver to the Health Spa only and bronze to the Fitness Centre. Price increases for existing members have now been phased in. The hotel now has its own speedboat with two-hour cruises at £27.50/person and four-hour cruises at £37.50/person. This is available both to leisure club members and hotel guests. It is also available for private charter.

11.B.3.7   Purchasing arrangements

Wet stock: Two national suppliers
Food stock: Mainly local suppliers

11.B.3.8   Promotion

The hotel has a three-star rating. It has a website, with a virtual tour and online reservations. The hotel has an in-house team of three sales executives for corporate and function business.

11.B.3.9   Functions/special events

Numerous weddings are catered for each year and there are also regular in-house conferences.

11.B.3.10   Entertainment

Live music is provided on a regular but much reduced basis. A pianist is employed on Saturday evenings. During December, entertainment is provided to each function room nightly. During the six weeks of summer, daily seasonal entertainment is provided.

11.B.3.11   Methods of payment

Cash and all major credit and debit cards. We understand that an average commission of 2.0% is charged in respect of the major credit cards. There are several account customers, given 30 days credit.

11.B.4   Financial information

11.B.4.1   Accounts

11.B.4.1.1   Information Provided

We have been provided with copies of audited accounts for the year ended 25 February $n - 3$. We have also been provided with copies of abbreviated accounts for the year ended 25 February $n - 2$ from the same source. Unfortunately, however, these no longer include the provision of a full trading and profit and loss account. Instead, we were provided with copies of management accounts for the same period, upon which the audited figures are said to be based. We have cross-referenced these two sets of accounts where possible. In addition, we have been provided with management accounts for the year ended 25 February $n - 1$, which we assume can be supported by an Accountant's Certificate.

11.B.4.1.2   Extract of Accounting Information

We provide below a reconstituted extract from the accounts provided:

Year end:	25/02/$n - 3$	25/02/$n - 2$	25/02/$n - 1$
Net sales	2 404 146	2 378 022	2 294 500
Weekly average	46 107	45 606	44 004
Gross profit	1 938 828	1 917 571	1 833 900
Gross yield (%)	80.6	80.6	79.9
Machine/Sundry Income	N/A	N/A	N/A
Adjusted net profit	420 123	380 429	194 900
Adjusted net yield (%)	17.5	16.0	8.5
Rental income	N/A	N/A	N/A

Adjusted net profit has been calculated prior to taxation, depreciation, directors' remuneration and finance costs. We have also excluded hire of equipment.

11.B.4.2   Analysis of accounts

The business property has experienced a decline over the three years, with a marginal decrease in the gross yield during the year ended 25 February $n - 1$. We understand that this reflects a combination of factors, including a decrease in corporate business trade. Furthermore, we understand that two of the most experienced members of the sales office left, taking some customers with them and that the period of re-adjustment and re-organisation that followed caused a significant level of disturbance to the business.

In the last financial year, the business has been affected by the building works programme, most especially in respect of the front of the property. This has significantly reduced passing trade and was further considered a deterrent to regular clientele.

While we have already referred to staffing levels and significantly higher expenditure in respect of staffing costs than we would expect, no staff bonus was paid during the year ended 25 February $n - 1$. The bonus scheme yielded £16 315 in the year ended 25 February $n - 2$.

Motor running expenditure appears high at £13 064 for the year ended 25 February $n - 3$ and £12 800 and £10 700 in the subsequent two years. We understand that the directors own their own cars and recharge business mileage to the subject concern.

In addition, travel expenses are also charged to the business, these totalling £13100 in the year ended 25 February $n - 1$.

During the year ended 25 February $n - 3$, a privately owned motorboat was transferred to the business. It is available for hire by residents and leisure club members. While the income derived is apparently modest, the expenditure for the operation of this motorboat is carried by the business.

Legal and professional fees appear high at £15077 for the year ended 25 February $n - 2$, compared to £11300 in the year ended 25 February $n - 1$. However, during the previous year, the costs of obtaining planning permission were incurred and additional professional fees were incurred in the subsequent construction period.

Television rental, effectively a nominal amount, was apparently included within 'guests' entertainment', the latter totalling £21682 in the year ended 25 February $n - 3$ and £36057 and £17900 in the year ended 25 February $n - 1$. We understand that entertainment expenditure is to further reduce in the forthcoming financial year.

The reduction in sales and gross yield, together with a general increase in levels of expenditure and in particular the wages liability, has led to a significant reduction in the adjusted profit/yield.

We have also now been made aware that income and expenditure concerning the hotel are also included within this accounting information.

11.B.4.3   Trading records

11.B.4.3.1   VAT Returns
No VAT records were made available.

11.B.4.4   Management accounting information

We were provided with a management report in the format of a profit and loss account, which includes figures for the 337 days period ended 31 January n. This revealed the following information:

	337 days to 31/01/n
Net sales	1 973 000
Weekly average	40 982
Gross profit	1 576 200
Gross yield (%)	79.9
Adjusted net profit /(Loss)	*(46 600)*
Adjusted net yield (%)	N/A

The adjusted net loss has been calculated prior to taxation, depreciation, directors' remuneration and finance costs. We have also added back a charge of £25400 relating to Hire of Equipment and a cost of £6900 in respect of Private Health. The gross yield of 79.9% is in line with the historic annual accounts we have been provided with and is assumed to include barrelage discounts. We note that the management accounts include charges of £41200 in respect of Repairs and Maintenance and £23500 in respect of Renewals.

The management reports provided forecast that for the year ended 25 February n, sales were likely to amount to £2087600, compared to £2294500 in the previous financial year. This would represent a reduction of some 9%. In addition, the management accounting information provided to us allowed a monthly comparative analysis for the years ended 29 February $n - 1$ and 28 February n. These figures are as follows:

Month	Year ended February $n - 1$	Year ended February n
March	166200	140800
April	182600	170400
May	160900	146100
June	191200	155800
July	267500	206700
August	222300	210400
September	192800	176100
October	223600	192700
November	157600	127100
December	347000	292600
January	110100	122400
February	137300	150000
Total	2359100	2091100

In addition, we were provided with income figures for March $n + 1$, which reflect sales of £195100 compared to £140800 in March n and £166200 in March $n - 1$.

11.B.4.5   Additional information

At the time of our inspection, we were provided with an analysis of the membership for the Fitness Centre and Health Spa. As of 6 February n, there were a total of 390 members, 69 of whom were on the old tariff. As of 1 May n, there were 421 members with only 31 paying the old tariff. It is apparent that the membership of the Fitness Centre and Health Spa is increasing rapidly and that the quality of the tariffs being achieved is also increasing, as the old membership fees fall out of the system. The management accounting information provided to us reflects the following apportionment of trade for the 337-day period to 31 January n:

	%
Wet	16.6
Food	26.2
Function food	10.6
Accommodation	44.7
Other	1.9
Total	100.0

11.B.4.6   Analysis of trading records

Approximately £21000 per annum of sales are said to relate to turnover generated at the Hotel.

It is apparent that the property has experienced a particularly poor trading period in $n - 1$. Month by month sales achieved during $n - 1$ were considerably lower than those in the equivalent trading period in $n - 2$. July $n - 1$ saw sales 22.7% lower than the equivalent period in $n - 2$. In November $n - 1$ sales were 19.4% lower, while in December sales were 15.7% lower.

However, since the completion of the Health Spa, the business appears to have begun to increase sales significantly. The business no longer appears to be restricted by the impact of the building works either physically or in terms of the company's marketing effort. Sales for January n were 11.2% higher than in the equivalent period in $n - 1$, while sales in February were 9.2% higher and sales in March were 38.6% higher.

Overall, net sales for the quarter ended 31 March n were 20.4% higher than the level achieved in the equivalent period in the previous year.

The subject concern has clearly experienced an extremely difficult two-year period of trade during the refurbishment and extension works. Once released of these difficulties, in the last quarter, business has rebounded considerably.

11.B.4.7   Current trade assessment

Taking the above factors into account, we have prepared the following indicative assessment of the annualised current trading position:

	£
Net sales	2 500 000
Weekly average	47 945
Gross profit	2 000 000
Gross yield (%)	80.0
Machine/Sundry income	N/A
Adjusted net profit	*210 000*
Adjusted net yield (%)	8.4
Rental income	N/A

Adjusted net profit has been calculated prior to taxation, depreciation, directors' remuneration and finance costs.

11.B.4.8   Comparison with accounts

The level of turnover and profitability shown in the current trade assessment reflects our analysis of the latest management accounts and the improvement in trade seen particularly in the first quarter of n. It reflects an apparent recent increase in trade following completion of most of the refurbishment works, and the opening of the Health Spa. This has been extrapolated for seasonality, evidenced by the monthly breakdowns.

The annual wages liability is now assessed at £1 250 000, which is inclusive of staff pension costs. At 50% of net sales, this is considered extremely high in relation to industry norms but reflects your applicants' deliberate policy to attract and retain suitably efficient high-quality staff, to enable a smooth transition to a high grading and increased levels of trade. This policy has been maintained throughout the period of trade affected by the building works.

Other expenses have been assessed having regard to the trading information provided and our knowledge of industry norms.

11.B.4.9   Fair maintainable profit

With two owners fully involved, we estimate that the current Fair Maintainable Profit of the business is £460 000. This assumes that the wages liability could be contained at around 40% of net sales, without the current gearing up for a higher rating/level of sales. We would stress that the Fair Maintainable Profit is assessed for information and valuation purposes only. We have allowed for the indicated level of current owner's involvement in the Current Trade Assessment.

11.B.5   Proposals

11.B.5.1   Property

In the forthcoming year the only work, other than routine maintenance, will be the refurbishment of the last nine bedrooms, at a cost of approximately £3000 each. Two of these rooms will be combined to create an executive suite.

11.B.5.2   Business

11.B.5.2.1   Operational Changes

The business will be operated to its current format. Essentially, the property/business has physically reached the standard now required of a four-star hotel and, given the current staffing levels and your applicants' experience, this should be true in respect of the style of the operation. Your applicants intend to achieve four-star status by the end of 2006, allowing them to slowly increase tariffs, especially for corporate guests. However, your applicants are anxious not to increase tariffs too quickly, to ensure that their existing customer base is not alienated, with consequential disruption to cash flow. Accordingly, therefore, it is likely to be a further full financial year before the full effects of your applicants' recent investment will be realised in terms of sales and profitability.

The completion of the Health Spa will allow a full resumption of marketing at the subject concern, with a beneficial impact on sales and profitability.

11.B.5.2.2   Staffing

All three directors will continue their existing roles. The operator believes that the level of additional staffing required at the subject concern to achieve a four-star rating will be modest. The restaurant will be required to extend last orders by 45 minutes, but this could be achieved by adjustment in rotas. Additional bar staff will be provided to offer table service.

11.B.5.3   Projection

11.B.5.3.1   Borrower's Projection

We have been provided with a projection for the year ending 25 February $n+1$, which has been revised in May n following the significantly better sales generated from January n onwards. This projection shows total income of £2 885 419, generating a gross profit of £2 396 061 (83.0%) and an adjusted net profit of £426 877 (14.8%). This is, however, calculated after allowing for wage costs of £1 440 698 (49.9%).

11.B.5.3.2   Basis of Report Projection

Our own projection is slightly less optimistic than that of the operator and reflects a 10% increase in net turnover over and above the current level of trade generated. We have excluded estimated income and expenditure relating to the hotel. Our projection covers the next 12 months of trading prior to a four-star status being achieved. Thus, our projection excludes potential increases in tariff due to this change in status. We have allowed for a level of wages liability approaching that shown in your applicant company's projection. We have, however, made adjustment in our assessment of the enhanced Fair Maintainable Profit, to reflect what we consider to be a more normal level of wage liability for a corporate establishment of this size and type, effectively run under management.

11.B.5.4   Projection

	£	£	%
Net sales		2 750 000	100.0
Weekly average		52 740	
Less operating costs			
Wages and salaries	1 375 000		50.0
Rates, water, environmental charges	90 000		3.3
Heating and lighting	65 000		2.4
Repairs and maintenance	100 000		3.6
Insurance	55 000		2.0
Telephone	15 000		0.5
Printing, postage, stationery	30 000		1.1
Promotion	55 000		2.0
Accountancy and professional fees	15 000		0.5
Transport	25 000		0.9
Laundry, cleaning, linen hire	45 000		1.6
Entertainment	20 000		0.7
Credit and charge card commissions	18 000		0.7
Sundries (including licence fees)	22 000		0.8
		1 930 000	
Adjusted net profit		270 000	9.8
Rental income		N/A	

Adjusted net profit has been calculated prior to taxation, depreciation, directors' remuneration and finance costs.

11.B.6   Enhanced fair maintainable profit

With two owners fully involved, we are of the opinion that the wages liability can be contained at around 40% of net sales. Upon this basis, the fair maintainable profit is believed to be in the region of £545 000.

11.B.6.1   *Valuation*

The Market Value of the Freehold Interest in the Hotel is £5m as a fully equipped operational entity, valued having regard to trading potential.

Market value is the estimated amount for which an asset should exchange on the date of valuation between a willing buyer and a willing seller in an arm's length transaction after proper marketing wherein the parties had each acted knowledgeably, prudently and without compulsion. Market value is understood as the value of an asset estimated without regard to costs of sale or purchase and without offset for any associated taxes.

Market value could be significantly affected if the business has been closed, the inventory has been depleted or removed, licences have been lost or breached and/ or accounts or records of trade would not be available to a prospective purchaser. Where requested, we have provided valuations on the basis of such or other special assumptions.

Section C

Chapter 12
Valuations for Land and Property Taxation

12.1 Introduction

Tax is required to produce revenue for government, and it can be nationally or locally set and raised. The expenditure is not always aligned to the raising powers; in other words, it can be raised nationally and spent in certain areas. Fairness is a guiding principle for any tax, and this normally means it should fall on those who can afford to pay it, it should not be regressive, it should be hard to avoid, and it should (preferably) be easy and cheap to collect. Land and property are an attractive source (or base) for taxation for several reasons:

- Because there is a relationship between land ownership and wealth (and therefore ability to pay), it can be marginally progressive.
- Occupiers benefit from government services, and it can be an important local tax.
- Market transactions generate price information, which is helpful when establishing the tax base.
- It has a broad base and yet is difficult to avoid or evade.
- It can be relatively cheap to administer, transparent and understandable.
- It is predictable and buoyant (via regular revaluations or reviews of the tax rate).
- It can help optimise economic use of land

Valuations are required for capital and revenue taxation purposes.

Land and property taxation (LPT) policy and legislation are shaped by political culture, socioeconomic beliefs and aspirations, so the way in which land and property are taxed is very jurisdiction specific. For example, if people's wealth depends to a great extent on the value of their residential properties, taxing them is likely to be very unpopular.

Property Valuation, Third Edition. Peter Wyatt.
© 2023 John Wiley & Sons Ltd. Published 2023 by John Wiley & Sons Ltd.
Companion website: www.wiley.com/go/wyatt/propertyvaluation3e

This chapter considers, first, a debate in the field of LPT on whether to tax land only or land and property. It then discusses the various types of land and property taxes that exist, and the role of valuation, before examining specific examples of these taxes that have been implemented in England.

12.2   A land tax or a land and property tax?

The classical and neo-classical economists demonstrated that the economic rent (and its capitalised equivalent, value) land can earn over and above the return generated after optimally employing labour and capital is determined by its scarcity and its location, neither of which is derived from any productive activity on the part of the landowner. Land value is, therefore, the price of monopoly: It is the relative, not the absolute, capability of land that determines its value. The value of a parcel of land always measures the difference between it and the best land that may be had for free. The more scarce and less substitutable a parcel of land is, and the more attractive the location in relation to the market (consumers) and factors of production (labour, raw materials), the more valuable the land. Figure 12.1 illustrates this point. Rent to land is 100% economic (scarcity) rent because land is not a produced input. At the extreme, the supply of *all* land is inelastic and therefore return to landowner is *all* economic rent.

Land-use planning and regulation, which also are not the result of landowner action, create further scarcity and generate demand, thus increasing the value of land in specific locations. At the land parcel level, the grant of permission to develop land (including changing its use) can generate substantial increases in land value. In societies where governments provide infrastructure, services and amenities, landowners may benefit from value uplift as a direct result of this publicly funded investment. Land value is, then, the creation of the community and expresses, in financial terms, the right a community has in land held by an individual.

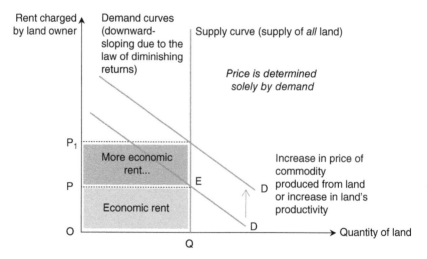

Figure 12.1   Monopoly or economic rent.

One means of recovering this unearned land value is via a tax. Adam Smith argued that a tax on land value would not harm economic activity and would not increase land rents: its supply is fixed and cannot be affected by a tax. With the same amount of land available, people would not be willing to pay any more for it than before, so the present value of an LVT would be reflected in a one-for-one reduction in land price – an example of 'tax capitalization'. The effect of the tax would be a windfall loss incurred by landowners. There would be no effect on buying, selling, developing or using land. The idea of a recurrent tax on land value has been propounded ever since, with Henry George (1879) making the most well-known case for a single tax on land value, arguing that it would:

- Capture unearned land value and help redistribute land-related wealth,
- Pay for the provision of infrastructure services and amenities,
- Help mitigate negative externalities and motivate productive use of land and
- Reduce land prices[1] (and land speculation) by discouraging the holding of vacant land/land-banking/rent-seeking behaviour while having no detrimental economic impact on buying, selling, developing or using land.

George's central argument was nationalisation of land ownership with leases granted to the highest bidders under conditions that guaranteed rights to improvements. But, for pragmatic reasons, he advocated a single 100% tax on 'location' (land) instead. The land tax would replace taxes on wages and goods and would not be levied on improvements made to land unless they had 'blended in' with the land over time. There are no examples of a single land tax as envisaged by George; the political barriers outweigh the ideological case. This might be due to a combination of political reluctance and practical difficulties. Politically, the arguments against a land-value tax centre on the windfall loss incurred by landowners, the difficulty in dismantling centuries of land ownership rights and the impact on other taxes. Practically, the challenges are compiling and maintaining a parcel-level register of land ownership and land use, revising the planning system so that it specifies a permitted (and optimum) land use for each parcel and confers this development right to the landowner, and valuing each parcel of land. Focusing on the valuation challenges:

- A lack of market transaction evidence, particularly in urban areas, means land valuations are hypothetical and therefore contestable.
- Would land be valued assuming its existing use or its optimum? If the latter, how would that be determined? Would that be the role of government (planners) or the market (valuers)? Should the value assume land is developed to its optimum use, but the surroundings are 'as is'?
- For improved land, how would the value of the land be separated from the value of improvements?[2] If the land is to be valued assuming an optimum use that is different to the existing use, how would this development value be apportioned between land and improvements?

It would be necessary to specify development density, open-space provision and so on. There would be difficulties in dealing with restrictions on development such as protected buildings. Land valuations would have to be regularly updated and the impact of value influences (changing transport infrastructure, new development, etc.) would have to be apportioned between the unimproved land and the improvements. Finally, the capitalised price that is paid for a site will depend not only on the level of economic rent but also on finance and discount rates, market sentiment and conditions, hope value of changing permitted use, purchaser's potential to purchase, vendor's potential to sell, taxation, subsidies, etc. For these reasons, instead of a single land tax, many jurisdictions have implemented a mixture of land and property taxes that capture a small fraction of value – small vestiges of George's idea.

12.3   Types of land and property taxes

Land and property taxes can be categorised as:

- Annual or recurrent occupation taxes. These are usually assessed with reference to value (market capital value, market rental value or cadastral value based on parameters such as land use, location or size) of land and/or buildings.
- Transfer and wealth taxes. These include sales or transfer tax, assessed as a percentage (sometimes stepped with thresholds) of price agreed on transfer of ownership; capital gains tax (CGT), which is charged against property asset(s) whose value has appreciated over time; and inheritance tax (IHT), which is charged on the value of property owned at death.
- Betterment taxes. These are levied on any increase in value attributable to the granting of different land-use rights.

A categorisation of land and property taxes is shown in Table 12.1, and Figure 12.2 illustrates the way in which these taxes can be grouped together.

Table 12.1   A categorisation of land and property taxes.

Type of tax	Description	Occurrence	Liability	Incidence
Occupation tax	A tax ostensibly for local services	Annual	Occupiers	
Sales or transfer tax	% price agreed on transfer of ownership	On transfer	Owners	Transfer
Capital gains tax	Accruing to property asset(s) whose value has appreciated over time	On realisation of gain	Owners	Wealth, transfer
Inheritance tax	On the value of property owned at death	On death	Owners	Wealth, transfer
Betterment tax	On increase in value attributable to granting of land-use rights	On grant of planning permission	Owners	General, scheme specific

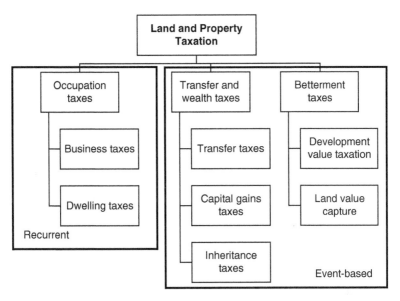

Figure 12.2   Land and property taxes.

12.3.1   Occupation taxes

Occupation taxes are relatively straightforward and cost effective to administer. Although technical expertise is required to assess values, valuations are comprehensible, so long as they are made publicly available. Benefiting from a broad base, they are difficult to avoid or evade since the taxable objects are real, and there is a correlation between assessed value and ability to pay and so they are progressive. With regular revaluations and reviews of the tax rate, tax revenue should remain predictable and buoyant.

In terms of the taxable entity, occupiers are easier to identify but owners are more politically acceptable because the tax is, in large part, a wealth tax. Sometimes, there is confusion over whether an occupation tax is a tax on occupants to help pay for the provision of local infrastructure, services and amenities or whether it is a betterment tax designed to capture uplift in value resulting from the provision of local infrastructure, services and amenities. The confusion stems from the argument that it is an occupation tax that is assessed by reference to values. Is the tax based on values to capture greater taxes from those with higher value properties or is it based on values because those living in higher value properties will use infrastructure, services and amenities more?

A decision needs to be made as to whether to base an occupation tax on unimproved or improved (i.e. land that has been developed in some way) land. In urban areas, very few sales of unimproved land take place so it can be difficult to defend the assessed land values. Sales of improved land on the other hand are more commonplace and therefore provide a more substantial evidence base. Yet a land tax may give better incentives for development of vacant or under-utilised land in urban areas, but the value of both land and buildings might be a better base for distributing the cost of local government. In rural areas, administration of a land tax is probably easier because it does not have to include information about farm buildings and plants. The decision is, therefore, not straightforward: charging an annual tax based on improved capital values can act as a disincentive to improving land, so many annual land taxes rely on assumed or calculated unimproved land values.

To base a tax on either unimproved or improved land, it is necessary to assess differences in utility. This is usually measured by assessing how much someone would be willing to pay for (the value of) property rights but could also be soil quality, size, number of windows or storeys, or some other quantum measure. The value might be based on rental value or capital value, the decision likely to depend on the availability of market transaction evidence. The advantages of using market value (MV) as a basis for assessment is that it encourages people to recognise the value of their assets, it relates revenue to economic performance, and market transactions provide an evidence base.

Assessed values can be based on current or optimum (highest and best) use values. Basing assessments on optimum values can encourage development but requires clear and strong land-use planning. The value of a parcel of land must have regard to its surroundings (but not matters such as site assembly, hope value[3] or marriage value). Optimum land uses must be ones that are permissible.

Administrative responsibilities are usually distributed between the different levels and parts of government. Tasks include:

- Collect and analyse market transaction data (prices, rents, dates, other details)
- Identify, classify (use, size, type, etc.) and inspect taxable objects (land and property)
- Conduct valuations
- Identify taxable entities (owners, occupiers)
- Administer the tax, including appeals

It is regarded as good practice to separate valuations of individual land and property interests from the setting of the tax *rate*. The former would be undertaken by valuers, made transparent and open to appeal and the latter by elected politicians possibly within limits set by central government. The rate would then change annually (perhaps indexed to general inflation or to changes in real prices of property). This preserves buoyancy and avoids the high political costs of periodic revaluations and potentially sharp tax increases. Banding values can simplify administration, especially when evidence is scarce.

Widespread use is made of automated valuation methods to speed up and lower cost of valuations for tax purposes. These *mass appraisal* techniques tend to use the hedonic pricing method if sufficient transactions and adequate records of key attributes are available, including:

- Land area and building area,
- Location,
- Building quality, materials and age,
- Improvements and
- Planning status.

12.3.2   Transfer and wealth taxes

Taxes that are levied on the transfer of land and property rights are usually calculated as a percentage of the transaction price, consequently there is little need for a valuation. Similarly, wealth taxes are usually calculated as a percentage of asset price, but there may be a requirement for valuation input if the assets are not subject to a market transaction at the time of the tax assessment. For example, an IHT payment might be due at the point where an heir inherits land and property assets but there is no market sale at this point.

12.3.3   Betterment taxation

Betterment refers to an increase in land value, and betterment taxation, also known as land-value capture (LVC), is an event-based tax on some or all[4] of the uplift in land value that occurs when investment by the state enhances land values or when development rights are granted. The state can increase land values through planning and regulation, infrastructure provision, environmental improvement, improving the service delivery or image of a locality. Both kinds of planning (development regulation and forward planning) affect land values. Development regulation (land-use controls, building controls) restricts the amount of land for each use and that will mean differential land values (Evans 1983). Forward planning (deciding where a new train route should go, a new bypass or a new town for example) will provide location advantages to affected land. In societies where governments fund the provision of infrastructure, services and amenities, the case for a betterment tax is strong since, it is argued, the landowner benefits from betterment as a direct result of this publicly funded investment. It should be noted that increases in land value can be recovered via occupation taxes too, although regular revaluations are required.

There are many different types of LVC, illustrated in Figure 12.3.

The advantages of LVC are that it:

- Captures 'some' unearned increment
- Meets the costs of new development which general taxation may not be able to cover, such as a new school
- Is payable at the point at which the gain is realised
- Is suited to hypothecation

The disadvantages of LVC are that it:

- Can be avoided by landowners not bringing land forward
- Could deter development
- Needs strong development regulation
- Does not capture wider unearned increment

Valuations are required to determine the size of the uplift in land value. This is usually the increase from the existing-use value (EUV) of the land to the MV of the land with planning consent for the development. Improvements must be disregarded (which is difficult for reasons explained in the section on occupation taxes).

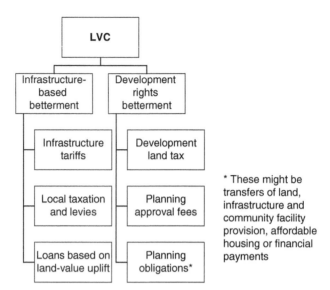

Figure 12.3   Types of LVC.

In some cases, estimation of EUV is straightforward – agricultural land for example. In other cases, it is not so easy, for example urban land that is capable of being used in different ways, subject to planning. Estimation of MV is usually undertaken in one of two ways: by comparison with sale prices of similar sites, where they exist, or by estimating the 'residual' value of the land having estimated the development cost and value of the completed development. Clearly, the first method is easier; however, development sites are often different in many respects and comparison is not possible. This means that the residual method is often relied upon.

12.4   Land and property taxation in England and Wales

In England and Wales, valuations for CGT, IHT and Stamp Duty Land Tax purposes are based on statutory definitions of MV. A definition for the basis of valuation for CGT can be found in Section 272 of the Taxation of Chargeable Gains Act 1992, for IHT it is in Section 160 of the Inheritance Act 1984 and for Stamp Duty Land Tax it is in Section 118 of the Finance Act 2003. These current statutory definitions are similar to those used in earlier tax legislation and, over the years, case law has established that, in arriving at MV, the following assumptions must be made:

- The sale is hypothetical.
- The vendor and purchaser are hypothetical, prudent and willing parties to the transaction (unless the latter is considered a 'special purchaser').
- For the purposes of the hypothetical sale, the vendor would divide the property to be valued into whatever natural lots would achieve the best overall price, known as 'prudent lotting'.

- All preliminary arrangements necessary for the sale to take place have been carried out prior to the valuation date.
- The property is offered for sale on the open market by whichever method of sale will achieve the best price.
- Adequate marketing has taken place before the sale.
- The valuation reflects the bid of any 'special purchaser' in the market (provided they are willing and able to purchase).

Further clarification on detailed aspects of the statutory definitions of MV, as established by case law, can be found in the UK Supplement of the RICS valuation standards (RICS 2019: UK VPGA 15).

12.4.1   Occupation taxes

12.4.1.1   Business rates

Occupiers of commercial premises in England and Wales must pay a property tax, known as *business rates*, or, more formally, national non-domestic rates to the government. The tax liability is calculated by assessing the *rateable value* of the premises and multiplying this amount by a rate known as the uniform business rate or UBR. The rateable value of a property is very similar to its annual rental value but with some simplifying assumptions. Valuers are employed by the Valuation Office Agency (VOA) (an executive agency of the government's Revenue and Customs Department) to assess the rateable value of every business property in the country. Valuers are also employed by occupiers who wish to ensure the rateable value has been correctly assessed.

Business rates are levied annually on individually occupied non-domestic premises including shops, offices, factories, warehouses, workshops, schools, hospitals, universities, places of entertainment, hotels, pubs, town halls, sewage farms, swimming pools, etc. These separately occupied units of business accommodation are legally defined as *hereditaments* and the amount of tax due from each occupier is based on an assessment of the annual 'rateable' value of the hereditament occupied. Some properties are used for both domestic and non-domestic purposes and these 'composite' hereditaments require an apportionment of tax liability between business rates and council tax: each element is valued having regard to the benefit of the other. Certain premises are exempt from business rates, including agricultural premises, fish farms and fisheries, places of religious worship, parks and property used for the disabled. There are also some reliefs, for small businesses for example). The tax is paid by occupiers (or owners if the hereditament is empty).

The tax is collected by Billing Authorities who, in general, are the local authorities. The amount of tax payable is calculated by multiplying the *rateable value* of the hereditament by the UBR, which changes each year in line with inflation. In order to redistribute the tax base in line with shifts in market values, rateable values are revalued every five years. Each hereditament is valued on an antecedent valuation date, two years before each rating list comes into force.

> The revaluation process involves several stages:
> - Gather market rental transaction information.
> - Adjust and analyse market rents.
> - Prepare valuation schemes based on the market rental evidence.
> - Value hereditaments.
> - Publish draft rating lists and supporting valuations.
> - Amend as appropriate and publish final rating lists.

The VOA is the government agency charged with producing and maintaining the rating list. A rateable value is assigned to each rateable hereditament. Market transactions provide evidence of rents, and these are analysed and adjusted into line with the definition of rateable value so that they provide an indication of the level of values as at the valuation date. Then every hereditament is assessed based on the adjusted rental evidence and put into the rating list. There is also a central list that deals with hereditaments that are, generally, in the form of a network throughout the country such as property owned by the water companies and other utilities.

Rating legislation comprises the *Local Government Finance Act 1988* and *Local Government Act 1989* (as amended). The definition of Rateable Value under Paragraph 2(1) Schedule 6 of the Local Government Finance Act 1988 (as amended by the Rating [Valuation] Act, 1999) is '. . .an amount equal to the rent at which it is estimated the hereditament might reasonably be expected to let from year to year on these three assumptions', subject to three assumptions:

[1]   The tenancy begins on the date prescribed
[2]   The hereditament is in a state of reasonable repair, excluding any repairs that a reasonable landlord would consider uneconomic
[3]   The tenant undertakes to pay 'all usual tenant's rates and taxes and to bear the costs of the repairs and insurance and the other expenses (if any) necessary to maintain the hereditament in a state to command the rent'.

The definition is based on the concept of a hypothetical property that is vacant and available to rent on an annual tenancy with a reasonable prospect of continuance. The valuation must account for all possible bids, and it is assumed that the landlord and tenant are commercially prudent yet reasonably minded and the premises are in a reasonable state of repair for the type of property, location and tenant. Because of this latter assumption, the impact on value of any disrepair is normally ignored because the tenant is assumed to maintain the hereditament. Having established these assumptions, the premises themselves are valued as they exist at the valuation date. The Latin phrase *rebus* sic *stantibus* (things standing thus) is often used to explain how the valuation must regard the premises in terms of its extant use, physical attributes and extent and its location.

The extent of the hereditament is determined by four rules: it must be:

[1]   Capable of separate occupation,
[2]   A single geographical unit that is contiguous or otherwise functionally essential,
[3]   In single use and
[4]   In a single definable position.

The rateable occupier is not defined by statute; instead, the meaning is deduced from case law but there are four essential ingredients. Occupation must:

[1]   Be *actual*: Someone there (although empty rates now charged on person with *right* to occupy). This includes an intention to occupy at some time in the future.
[2]   Be *exclusive*: The occupier must be in control of the premises. Difficulties arise where several parties have rights of occupation, such as kiosks on a railway station.
[3]   Be *beneficial*: Occupation must be of value or benefit. It is not the actual occupier who is assumed to pay a rent but the 'hypothetical tenant'. The rental value should reflect the actual use or an alternative use if, under the same mode of occupation, planning permission could be obtained, and no structural alterations were required.
[4]   Have a *degree of permanence*: Not transient. Builders' huts or caravans are not sufficiently permanent. Hereditaments that exist for at least a year are usually regarded as sufficiently permanent under the '12 month' rule. Properties occupied seasonally will be liable.

The VOA has the legal right to inspect premises and gather occupation details. It can issue notices to business occupiers requiring information on accommodation, details of any rent paid, whether outgoings are included in the rent, whether the rent includes other items such as fixtures, fittings and services, details of rent review provisions and so on. The information received is analysed, the rent adjusted to correspond with the definition of rateable value and used together with all other information obtained to keep the current rating list up to date and assist in the compilation of the next one.

The billing authorities help in the preparation of the list to the extent that they have a responsibility to inform the VOA of any changes occurring in their areas, which require an amendment of the rating list. Such things would involve the construction or alteration of property, or the change of use of property, in fact those changes that would normally be apparent from the granting of planning permissions or the approval of work by Building Control Officers. A new or altered property becomes rateable after the local authority serves a 'completion notice', and this can be served up to three months before the expected completion date. When new properties are constructed or when properties are amalgamated or split, this causes reassessments, and these will be based on values as at the antecedent valuation date. This continues to be the case for all new assessments until a new list comes into force. The problem, of course, is that circumstances change through the years so, although we might be attempting to value, say, a shop at the antecedent date, when the location was a peaceful village, at the time of valuation the shop may be in an expensive suburb. Conversely, a particular shopping street might now be worth less in value due to physical changes in the neighbourhood, the building of a shopping centre for instance. Such problems are dealt with in Local Government Finance Act, 1988, Schedule 6, paragraphs 5, 6 and 7, whereby certain matters must be taken into account when a property is valued as at a particular date for entry in the list. These are:

- Matters affecting the physical state or physical enjoyment of the hereditament,
- The mode or category of the occupation of the hereditament,
- The quantity of minerals or other substances in or extracted from the hereditament,
- Matters affecting the physical state of the locality in which the hereditament is situated or which, though not affecting the physical state of the locality, are nonetheless physically manifest there and
- The use or occupation of other premises situated in the locality of the hereditament.

Creating a level of values at a particular historical date is known as 'having regard to the tone of the list'. The publication of a new list triggers the ratepayers' right of appeal, should they disagree with the entry of the hereditament in the list, the extent of the hereditament assessed, the description of the hereditament or, more particularly, the value ascribed to it.

Occupation of part of a hereditament is deemed to be the same as occupation of the whole but if the occupier can establish that there is no intention to reoccupy the vacant part, it may be separately assessed. Separate provision is made to apportion the rateable value where part of the hereditament is vacant for only a short period.

Although there is a statutory definition of rateable value, there is no statutory method. The methods that valuers use to value rateable hereditaments are comparison[5], profits and replacement cost (usually referred to as the contractors' test in the context of business rates) with the addition of either statutory or nationally agreed formulae for the valuation of certain specialised properties such as hospitals, petrol stations and so on.

12.4.1.1.1   Rental comparison

The most widely used valuation method for business rate purposes is rental comparison, where schedules of market rents are prepared, based on rents devalued on a zoned basis for shops or in terms of a main space for office and industrial space. Evidence of market rents can be obtained from several sources, the best being market rental transaction evidence close to the antecedent valuation date. Evidence may also be derived from rent reviews and lease renewals, but such evidence is considered secondary to market transactions. The comparison method can be difficult to apply when there is a lack of market evidence or where the transaction involved specific arrangements including rent-free periods, stepped or turnover rents, premiums, break options, capital contributions, non-standard repairing and insuring obligations or other incentives. The rent may also include the use of other facilities or may be below normal market levels if it is for a unit in a new development such as an anchor tenant. A typical rental valuation for a retail property that, in this case, is zoned on the basis of normal 6.1 metre (20 feet) zones is as follows:

Description	Area (m²)	Fraction of zone A	Unit value (£/m²)	Value (£)
Ground floor:				
Zone A	26.5	A	900	23 850
Zone B	25.4	A/2	450	11 430
Zone C	15.6	A/4	225	3510
Storage	10.3	A/10	90	927
First floor:				
Storage	60.7	A/20	45	2732
Site:				
Parking spaces	3 no.		£400/space	1200
Rateable value (£p.a.)				43 649

It is possible for a range of factors to add value (in the form of an end allowance) to the basic assessment including return frontages to shops, air conditioning or car-parking spaces. The valuer may also consider that a quantum allowance should be applied where a property is exceptionally large compared to comparable evidence and a prospective tenant bidding on the property would reduce the rental bid based on the large amount of space being taken. A reverse quantum allowance is the opposite, where a particularly small property has an addition in value added to its basic assessment. This is known as the 'kiosk' effect. Disability allowances can apply where, for example, a property suffers from some form of geographical or functional factor that would reduce the likely rental bid from a prospective tenant. Temporary allowances are applied, as their name suggests, to situations where a temporary change in the property or its physical location warrants (usually) a reduction in the rateable value. Once the factor causing the temporary allowance has been removed, then the assessment will be reinstated to its full value (but not necessarily the same value!)

Most properties have a rental market upon which evidence can be drawn and therefore it is possible to use this information to arrive at an assessment of rateable value. However, there are hereditaments that do not have an active rental market and these need to be valued using alternative methods.

12.4.1.1.2 Profits (receipts and expenditure) method

There are various types of specialised trading property discussed earlier, such as public houses, bowling alleys, night clubs, cinemas and hotels, for which comparable evidence of rents can be hard to find. In such cases, trading information will give some idea as to how much profit the hereditament makes and hence what it can afford to pay in rent (and thus rates). The method requires the valuer to adjust the accounts to bring them into line with the definition of rateable value. The turnover is adjusted to reflect the cost of purchases and working expenses. The resulting net profit, which is referred to as the *divisible balance* for rating purposes, is apportioned between the occupier or operator (as profit) and the owner (as rent). The rent provides evidence of the rateable value. The actual rates paid by the occupier should be deducted as part of the working expenses so that the bottom line is equivalent to the rent only (see *Thomason v Rowland (VO) (1995) RA 255).*

The split between profit and rent, which can be hypothetical in the case of sole trader, will vary according to the perceived risk of the business; the higher the risk, the higher the operator's required return (Marshall and Williamson 1996). An example of a profits method valuation of a hotel for rating purposes is given below. The trading figures are those that pertain to the hereditament at the antecedent valuation date and are exclusive of value-added tax (VAT).

Turnover (gross receipts from rooms, bar, restaurant) (£p.a.)	1 600 000
Less purchases (£p.a.)	(300 000)
Gross profit (£p.a.)	1 300 000
Less working expenses (wages, utilities, stationery, marketing, insurance, vehicles, rates, etc., repair and maintenance of property, repair and renewal of furniture, fixtures, fittings and equipment) (£p.a.)	(700 000)
Net (trading or operating) profit	600 000
Less interest on tenant's capital (fixtures, fittings, furniture and equipment or FFFE, stock, cash) (£p.a.)	(40 000)
Divisible balance (£)	560 000
Split between:	
Return (normal or residual profit) to operator/tenant (£p.a.)[a]	(260 000)
Rent to landlord (on which the rateable value is based) (£p.a.)	300 000

[a] A percentage based on comparable evidence, in this case 46%.

Care is needed when deriving information from the accounts. Usually, valuers will consider the accounts figures drawn from the three previous years to the valuation date and arrive at a fair maintainable trade (the expected trade that the reasonably minded operator would derive from the property and business). Therefore, abnormally large amounts of expenditure in any one year may be written down over several years to arrive at a fair maintainable trade.

In place of a full profits method valuation as noted above, it is possible to draw up relationships between other figures in the accounts, between gross receipts (turnover) and rent (as a proxy for rateable value) for example. Thus, a reasonably run hotel might expect the percentage of gross receipts paid as rent to be in the region 20%. This short-cut technique is known as the 'shortened profits method'. A formula based on some other method may also be used to value certain hereditaments, such as hotel rent per bedroom or bed space. When the profits method is used to value a hereditament, the valuation will include all plant and machinery used in the business operation. For instance, if a fuel storage depot is valued on a profits basis any rateable plant in the hereditament simply goes to make up the profit – tanks, security fencing, fire-protection equipment and the like all help to produce the profit – if they were not there, the enterprise would either not operate or not be so profitable.

A more detailed example involving a 60-room hotel is shown below. The tenant's share equates to 22% of tenant's capital, 53% of the divisible balance and 11% of fair maintainable trade (FMT). Deducting the tenant's share from the divisible balance leaves £90 598, which is taken to be the rateable value, and equates to approximately 10% of FMT.

Example: Rating valuation of a 60-room hotel

Accounts for year end 31 March	2 years ago	% total income	1 year ago	% total income	Last year	% total income	Adjusted	% total income
Income (£p.a.):								
Accommodation	305 000	38.9%	317 500	35.0%	320 000	37.2%	330 000	36.7%
Food	350 000	44.6%	425 000	46.8%	385 000	44.7%	410 000	45.6%
Drinks	95 000	12.1%	122 500	13.5%	117 500	13.6%	120 000	13.3%
Other	35 000	4.5%	42 500	4.7%	38 500	4.5%	40 000	4.4%
Total income	785 000	100.0%	907 500	100.0%	861 000	100.0%	900 000	100.0%
Cost of sales	160 000		200 000		195 000		200 000	
Gross profit	625 000	79.6%	707 500	78.0%	666 000	77.4%	700 000	77.8%
Working expenses (£p.a.):								
Wages	(225000)	28.7%	(252500)	27.8%	(242500)	28.2%	(250000)	27.8%
Utilities	(40000)	5.1%	(45000)	5.0%	(46500)	5.4%	(47500)	5.3%
Repairs and maintenance	(5000)	0.6%	(22500)	2.5%	(14000)	1.6%	(20000)	2.2%
Marketing	(15000)	1.9%	(15000)	1.7%	(15500)	1.8%	(16000)	1.8%
Stationery	(4500)	0.6%	(4750)	0.5%	(4500)	0.5%	(4750)	0.5%
Cleaning	(25 000)	3.2%	(31 000)	3.4%	(27 500)	3.2%	(30 000)	3.3%
Commissions	(5500)	0.7%	(4500)	0.5%	(6750)	0.8%	(6000)	0.7%
Replacements	(1250)	0.2%	(2000)	0.2%	(1500)	0.2%	(1750)	0.2%
Transport	(5250)	0.7%	(5500)	0.6%	(6500)	0.8%	(4000)	0.4%
Professional fees	(18 500)	2.4%	(2500)	0.3%	(3000)	0.3%	(3500)	0.4%
Rates	(13 250)	1.7%	(15 000)	1.7%	(16 250)	1.9%	(17 000)	1.9%
Insurance	(10 250)	1.3%	(10 000)	1.1%	(12 000)	1.4%	(12 500)	1.4%
Other	(27 500)	3.5%	(31 500)	3.5%	(25 500)	3.0%	(27 500)	3.1%
Bank charges	(2000)	0.3%	(2750)	0.3%	(2500)	0.3%	(2500)	0.3%
Equipment hire	(24 000)	3.1%	(21 250)	2.3%	(24 500)	2.8%	(25 000)	2.8%
Total working expenses	(422 000)	53.8%	(465 750)	51.3%	(449 000)	52.1%	(468 000)	52.0%
Net profit (£p.a.)	203 000	25.9%	241 750	26.6%	217 000	25.2%	232 000	25.8%
Depreciation (£p.a.)	(42 500)	5.4%	(41 000)	4.5%	(18 500)	2.1%		
Loan interest (£p.a.)	(145 000)	18.5%	(140000)	15.4%	(100 000)	11.6%		
Director's remuneration (£p.a.)	(46 000)	5.9%	(50 000)	5.5%	(50 000)	5.8%		
Profit/loss on sale of assets (£p.a.)					(1500)			
Net profit in accounts (£p.a.)	(30 500)	−3.9%	10 750	1.2%	47 000	5.5%		

Fair maintainable trade (adjusted total income) (£p.a.)		900 000	
Adjusted net profit (£p.a.)		232 000	(25.8% of FMT)
Head office expenses (£p.a.)	n/a		
Renewal fund for replacement of non-rateable assets:			
Present value of non-rateable assets (60 rooms × 7000) (£)	420 000		
Residual value (£)	0		
Balance (£)	420 000		
Either sinking fund for 10 years @ 3% (£p.a.)	0.08723	36 637	
Or straight-line depreciation over 10 years (£p.a.)		42 000	
Renewal fund adopted (£p.a.)		40 000	(4.4% of FMT)
Divisible balance (adjusted net profit – renewal fund) (£p.a.)		192 000	
Tenant's share:			
Tenant's working capital (estimated as either a % of FMT or as a % of expenses)			
Assume 2 weeks' FMT	34 615		
Present value of non-rateable assets (60 rooms × 7000)	420 000		
Total tenant's capital (£)	454 615		
Interest on tenant's capital @ 6%		27 277	
Remaining balance (divisible balance – interest on tenant's capital)		164 723	
Profit @ % of remaining balance (£p.a.)	45%	74 125	
Total tenant's share (profit + interest on capital) (£p.a.)			101 402

12.4.1.1.3 Replacement cost method (or contractor's test)

The contractor's test is used to value properties that do not usually let in the open market such as schools, universities, petrol chemical works, hospitals, lighthouses, clinics, town halls and fire stations. The test is derived from cost information rather than rents or profits.

In the case of new buildings, the method is straightforward to employ because building costs close to the antecedent valuation date can be examined. But when valuing older buildings, an adjustment needs to be made to reflect probable depreciation in value because of deterioration and obsolescence. In such cases, the method starts to become unreliable as there is no ready market information to help ascertain what reductions should apply.

The contractor's test involves estimating the current cost of replacing the hereditament with a functionally equivalent building at the antecedent valuation date (including any rateable plant and machinery) and then deducting an allowance for age and obsolescence. The value of the site, cleared but with all services available for the existing use, will then be added to the replacement building cost. An adjustment may then be made, usually by applying a percentage reduction, to allow for general difficulties with the hereditament such as a confined site or poor access. The capital value thus produced will then be brought to an annual equivalent by applying a de-capitalisation rate appropriate to the Rating List in question.

These rates may be set by government and usually vary depending on type of property. The final stage is to 'stand back and look' at the valuation, to take account of any items or matters not already considered such as the economic health of an industry, business or organisation.

Consider a large county hospital built in the 1970's and valued using the contractor's test:

Cost of new buildings totalling $15\,487\,m^2$ @ £1150 per m^2 (£)	17810050
Less 23% depreciation allowance for age and obsolescence (£)	(4096312)
Plus value of land totalling 5 ha for existing use @ £250000/ha (£)	1250000
Capital Value (£)	14963739
De-capitalise (multiply) capital value at rate of say 3.67%	× 0.0367
	549169
'Stand back and look': reflect buildings in poor run-down area, say RV (£ p.a.)	500000

Wherever possible, it is useful to use more than one method to assess rateable value: the contractor's test for an 'awkward' building with rental comparison applied to those buildings within the hereditament that could be let at market rents. A seaside pier might be valued using the profits method for the fairground element, contractor's test on the non-profit-making elements and rental comparison for the kiosks.

12.4.1.2   Council tax

The Council Tax is the tax on occupiers of residential dwellings in England. The tax base is assessed by reference to the capital value of residential dwellings. It is, therefore, a tax based on the value of land and improvements. Government valuers allocate each dwelling to one of eight value bands (A–H) according to their estimated value as at 1991 prices, shown in Table 12.2. This is because the tax was introduced in 1993 and there has not been a revaluation in England since then (there has been a revaluation in Wales). Dwellings constructed since 1991 are allocated to a value band based on an estimate of what they would have been worth in 1991.

The valuation task, therefore, involves placing each dwelling in a value band. By adopting a banding approach, the valuation process is simplified and this makes the tax easier to administer.

Table 12.2   Council Tax bands in England.

Band	Applies to properties with values in 1991 that would have been	Council Tax (as a % of B and D)
A	Less than £40000	67%
B	£40001–£52000	78%
C	£52001–£ 68000	89%
D	£68001–£88000	100%
E	£88001–£120000	122%
F	£120001–£160000	144%
G	£160001–£320000	167%
H	£320001+	200%

The Council Tax List is kept up to date by receiving information from local authorities when dwellings are altered, sending questionnaires to residents, visiting dwellings and accessing transaction data. Information collected includes age, type of dwelling, size, situation of property and features of the locality. A dwelling can be moved from one band to another if improvements identified at time of sale justify such a shift in value. The improvements can be in relation to an improvement in the locality but only upwards, whereas physical or material changes to the dwelling itself can lead to a drop in band.

Local authorities, which are responsible for administering the Council Tax, set the Band D tax; then the tax for the other bands is fixed in the proportions shown in Table 12.2. There are various exemptions and discounts (25% if one resident, certain exemptions for empty dwellings and for certain types of residents such as students).

12.4.2 Transfer and wealth taxes

Valuations for CGT, Corporation Tax and IHT are based on statutory definitions of MV. Although the definitions are to be found in different statutes, they are broadly consistent with the internationally recognised definition of MV but subject to the following assumptions that have been laid down by case law over the years (RICS 2019: UK VPGA 15):

- The sale is hypothetical.
- The vendor and purchaser are hypothetical, prudent and willing parties to the transaction.
- All preliminary arrangements necessary for the sale to take place have been carried out prior to the date of valuation and there is adequate publicity or advertisement before the sale takes place so that it is brought to the attention of all likely purchasers.
- The property is offered for sale on the open market by whichever method of sale will achieve the best price and the vendor would divide the property into whatever natural lots would achieve the best overall price (prudent lotting).
- The valuation should reflect the bid of any *special purchaser* in the market (provided the purchaser is willing and able to purchase).

12.4.2.1 Capital gains tax and corporation tax

CGT was introduced in 1965 as a means of taxing capital gains made by individuals and trustees on the disposal of assets, after having set off any losses incurred within a tax year. Companies are subject to the same tax regime under the name of Corporation Tax. For CGT purposes, 'assets' includes all forms of property unless exempt (and an individual's main residence is exempt). 'Disposal' (which includes disposal of part of an asset) can be the sale, gift, receipt of compensation for damage to an asset, insurance compensation or payment for surrender of rights. Tax is only paid on disposal of an asset if a 'chargeable gain' was made in the preceding financial year. This is calculated by taking the sale proceeds and deducting the original acquisition cost and associated fees, any enhancement expenditure and disposal fees. The first part of the chargeable gain is exempt for individuals (half for trustees), but the exact amount varies from year to year. Enhancement expenditure is permitted as an allowable deduction so long as it is reflected in the state or nature of the property at the time of disposal, thus excluding improvements that have worn out by the time the property is disposed of. Disposal costs are also allowable and these, like acquisition costs, include professional and legal fees and any other costs reasonably incurred in marketing the property, including the cost of a valuation and any apportionment for CGT purposes.

In most cases, where the disposal is by way of an open market sale, the disposal proceeds are the amount received from selling the asset but sometimes the MV may need to be estimated if the sale was not made at arm's length or was a gift. The Taxation of Chargeable Gains Act, 1992 defines MV as the price for which those assets could be sold on the open market with no reduction for the fact that this may involve an assumption that several assets are to be sold at the same time. Stock-in-trade is not regarded as capital for CGT purposes, so property companies' developments are not subject to CGT (Hayward 2009). Certain disposals are exempt including transfers between husband and wife and gifts to charity. Also, certain organisations are exempt from CGT including charities, local authorities, friendly societies, scientific research associations, pension funds and non-resident owners (Johnson et al. 2000).

If a disposal is made where only *part of the property* was used as a business asset during the relevant period of ownership, then the chargeable gain must be apportioned between the gain on the business asset and the gain on the non-business asset. The grant of a lease is regarded as a part disposal of a property asset and is liable to CGT.

12.4.2.2   Inheritance tax

The Inheritance Tax Act 1984 requires IHT to be paid at a rate of 40% on the transfer value (net of costs and CGT) of a person's estate held at death, on certain lifetime gifts and some transfers in and out of trusts. The value of these transfers is calculated by reference to the reduction in value of the remaining estate and in most cases, this is the same as the value of the transferred estate. The valuation date is the date of transfer and, regarding transfers on death, the exact valuation moment is that immediately before death, although, where a property is sold shortly after death, this is normally taken as good evidence of MV at death. The reduction in value is estimated by valuing the transferor's estate before and after the transfer. For example, John owns two small prime shop units in Oxford Street; together they are worth £1 million, but individually each is worth £300 000. John leaves one of the units to his daughter. The gift for the purposes of IHT is the loss to John's estate, in other words £1 million less £300 000.

No reduction is made to the valuation due to the sale of the estate 'flooding the market' but if a higher price is achievable by selling in smaller lots then this can be assumed (Johnson et al. 2000). IHT is payable in cases where the transfer value is over the threshold of £325 000 (at the time of writing). It is often avoided due to family trusts and taper relief for lifetime gifts (exempt after seven years). Agricultural land and woodlands are exempt. Certain (generally low-value) gifts are exempt, including those made between husband and wife, those not exceeding £3000 in any tax year, maintenance payments, wedding gifts, small gifts to many people and gifts out of income tax. Of more relevance to property are gifts to UK-based charities, registered housing associations, qualifying parliamentary political parties, national museums, universities, the National Trust and certain other bodies.

If an outright gift is made to someone during the estate owner's lifetime, it is a 'potentially exempt transfer' and will only become chargeable to IHT if the transferor dies within seven years of making the gift. An outright gift is one in which the transferor does not retain any benefit or value. A gift with reservation of benefit is one that is not fully given away so that either the person getting the gift does so with conditions or restrictions attached, or the person making the gift retains some benefit. Where this happens to gifts made on or after 18 March 1986, the assets are included in the estate but there is no seven-year limit as there is for outright gifts. To complicate matters even more a gift may begin as a gift with reservation but sometime later the reservation may cease. For example, if an estate owner gives a shop to a son or daughter but continues to run the business there rent free, that would be a gift with reservation. If after two years the transferor starts to pay a market rent, the reservation ceases. The gift becomes outright at that point and the seven-year period runs from the date the reservation ceased. Conversely, a gift may start as an outright gift and then become a gift with reservation.

12.4.2.3   Stamp duty land tax

Stamp Duty Land Tax is a transaction tax paid by purchasers of freehold and leasehold interests[6] in land and property. The tax does not require any valuation input since it is calculated as a proportion of price paid. Transaction taxes like this are widely regarded as inefficient because they discourage mobility and penalise the improvement of land. As a result of these inefficiencies, the tax is subject to government intervention in the form of exemptions and differential rates. Tax revenue will fluctuate according to market conditions.

12.4.3   Betterment taxation in England

Following several attempts to levy betterment taxes at a local level, a national betterment tax was introduced as part of the 1947 Town and Country Planning Act. This was a tax on the entire uplift in value that would be realised when planning permission was granted for development, hence it was known as a development charge. The tax failed because landowners opted to hold rather than sell their land, and so the charge was abolished in 1953. Then, in 1967, a 'betterment levy' was introduced at a rate of 40% on 'projects of material development' (sales, lettings, development activity, the receipt of compensation for planning and compulsory purchase decisions). This was more like CGT than a development tax as it was not payable on grant of planning permission. Nevertheless, it was not successful and was abolished in 1971. The 1974 Finance Act introduced another higher rate CGT called a 'development gains tax', but this was replaced in 1976 when the government introduced a Development Land Tax at 80% of the land-value uplift. The rate was reduced to 60% in 1979 before being abolished altogether in 1985. Several decades later, a tax on the 'windfall gain' enjoyed by landowners when planning permission is granted for residential development (known as Planning Gain Supplement) was passed into law in 2007 but repealed two years later before implementation following lobbying from landowners, developers and local government.

These betterment taxes all failed for a mixture of political and practical reasons: political uncertainty resulting from changes in government was undoubtedly a significant factor; the taxes were often introduced following a boom but implemented *after* or *as* markets collapsed; and the complexity and potential for avoidance also took its toll. No betterment tax seems capable of successfully capturing the full unearned land value from all landowners because the tax is event-based, relying on the action of landowners to apply for planning permission for development. Landowners can opt to wait, and in a rising and perhaps uncertain market, this option has value, particularly if the betterment tax policy is perceived as a transient one. Landowners who do not seek planning consent earn implicit land-value gain regardless of whether it is realised or not.

England has now given up with these national attempts to tax betterment. Instead, there are land-value capture (LVC) instruments, which are administered at the discretion of individual local authorities:

- *Planning obligations or 'Section 106 (S106) agreements':* These are negotiated legal contracts between the local planning authority and the applicant (landowner or developer usually) to undertake works. Over time, these works have evolved from site-specific mitigation measures into pooled financial contributions, including the provision of affordable housing. Although yielding substantial local benefits funded directly from land-value uplift, these agreements have been criticised for creating uncertainty and involving costly, lengthy, opaque negotiations. They tend to favour bigger developers who can resource the negotiation process more effectively.
- *Community infrastructure levies (CIL):* These are non-negotiable tariff-style levies by local authorities to finance off-site infrastructure projects. The levy is based on a rate per unit area of additional floor space, and the rate usually varies between land uses and locations. CIL provides up-front, relatively predictable revenue. Although CIL has been levied on a wider spectrum of development sizes and types than S106, in many parts of the country, it is not viable to charge a CIL, so these places lose out on infrastructure investment. CIL seems to be less appropriate for large development schemes constructed over long periods of time, particularly phased schemes. Developers seem unhappy with the disconnection between revenue collection on the one hand and delivery of infrastructure on the other.
- *Special levies:* For example, the London Mayor's CrossRail Levy, and local authority administered Business Improvement Districts. These are charged on a similar basis to CIL.

These LVC instruments do not specify a fixed percentage of land-value uplift to be allocated to the landowner. Instead, they state that the amount of land value to be captured by the government must not compromise the financial viability of the proposed development. A development is considered financially viable if it generates sufficient risk-adjusted returns (profit) to the landowner and developer.

In order to test for financial viability, appraisals are undertaken at two stages in the planning process: during area-wide plan-making and during site-specific decision-taking. Area-wide viability appraisals are undertaken by local planning authorities to set targets for planning obligations and CIL over a local plan period. Site-specific appraisals are undertaken by applicants seeking to challenge these area-wide targets on a scheme-by-scheme basis. These viability appraisals assess whether an area-wide policy or a site-specific scheme can deliver planning obligations and CIL as well as provide returns to the landowner and the developer.

A range of viability appraisal models has evolved but they are all based on the residual method of valuation:

Land value = Development value – Development cost
(including developer's profit and cost of planning obligations)

If the land value is sufficient to persuade the landowner to sell, this indicates viability. There are two key issues when using the residual method to undertake financial viability appraisals: how to estimate a suitable landowner's return (known as *benchmark land value*) and how to handle changes in input cost and value assumptions over the development period.

Using financial viability appraisals to set the level of land-value capture can be likened to setting, at a local level, property tax rates: any tax alters the behaviour of market participants: set too low, policy aspirations might not be met; too high and landowners may reduce supply, particularly in relation to marginal sites as they become unviable. Faced with a new tax, political uncertainty can encourage landowners to wait for a change of government (and policy), but if the tax is accepted by all political parties, then uncertainty disappears and there is no benefit to withholding land.

In England, some landowners make very big profits on some land sales, but little is known about landowner returns, so it can be difficult to judge how increasing land-value capture will affect landowner behaviour if the potential for reaping these kinds of rewards is jeopardised. A landowner's return is likely to be influenced by:

- EUV, any sunk costs, hope values;
- Market supply and demand;
- Nature of landowner: Financial and tax position, farmer (life changing), speculator, etc.;
- Whether the site is brown or green-field, urban or rural and
- Because different elements of a scheme may have different risk profiles, different margins could apply.

After years of arguments between local government, landowners, developers and their advisors, central government has directed[7] that a landowner's return (known as *benchmark land value*) must be based on EUV plus a premium.

Changes in costs and values can be handled in several ways, such as:

- Input the original or 'day one' estimates,
- Forecast inflation in costs and growth in values over the course of the development and
- Include review mechanisms so that additional land-value capture can be triggered if returns exceed certain criteria.

Forecasts and review mechanisms are now permissible under government viability appraisal guidelines. There is no prescribed format for review mechanisms; they can be negotiated on a case-by-case basis. Key considerations are likely to be:

- How many reviews there should be and how they should be triggered
- The nature of the appraisal process, including data requirement and modelling approach

Key points

- Property tax is unpopular, but most people recognise the need and rationale for taxing it in some way. Property-related taxes may account for around 5% of total tax revenue in a developed country.
- Most of the land and property value is in residential not business use, but residents vote, businesses do not.
- Occupation taxes are widespread, and most countries have some form of transaction tax. IHTs and CGTs are common in developed countries, but they are more difficult and contentious. Land-value capture is becoming a very popular revenue-raising tool, especially for local government.
- Property tax is a pseudo land tax when revalued regularly.
- Land-value uplift is an economic response to investment in land and to the grant of rights to improve or develop land. Whether and how much of that uplift is captured by the state is a policy decision.
- As far as CGT is concerned, for most properties, the gain on disposal will usually be restricted to the gain since 31 March 1982 for properties acquired before that date or the date of acquisition for properties acquired afterwards. Generally, a valuation is required to estimate the MV of the asset on the disposal date if the disposal was not at arm's length. A valuation might also be required to estimate the MV on 31 March 1982 for rebasing the gain and calculating the indexation allowance. For part disposals, property valuations may be required of the part disposed of and the part retained.
- Market valuations are required for IHT purposes when there is no evidence of an open market sale of the transferred estate. This might be because the transfer was by way of a gift or some other means that does not fit the description of an 'arm's length' transaction.

Notes

1. Demand for land determines its value. If a tax takes all that value, anyone holding it without using it would have to pay nearly what it would be worth to anyone else who wanted to use it.
2. Possible approaches: (i) take the value of the whole and deduct the value of the buildings, (ii) allocate a percentage of the sale price of property to the land, (iii) extrapolate what little evidence might be available, (iv) use the residual method to estimate land value.
3. Strong planning is needed to avoid taxing hope value on farmland near urban areas.
4. But if you tax 100% of the uplift, then you take away all incentive to develop; you only leave the option value.
5. The focus as far as rating is concerned is on rental value (as a basis for the assessment of rateable value) rather than capital value.
6. Where the lease term is seven years or more.
7. Guidance on 'Viability' published by the Department for Levelling Up, Housing and Communities. Originally published in 2014, revised in 2018, and again in 2019.

References

Evans, A. (1983). The determination of the price of land. *Urban Studies* 20: 119–129.

George, H. (1879). *Progress and Poverty: An Inquiry into the Cause of Industrial Depressions, and of Increase of Want with Increase of Wealth. The Remedy*. New York: Appleton and Co.

Hayward, R. (2009). *Valuation: Principles into Practice*, 6e. London: Estates Gazette.

Johnson, T., Davies, K., and Shapiro, E. (2000). *Modern Methods of Valuation*, 9e. London, UK: Estates Gazette.

Marshall, H. and Williamson, H. (1996). *The Law and Valuation of Leisure Property*, 2e. London: Estates Gazette.

RICS (2019). *RICS Valuation – Global Standards 2017: UK Supplement*. London, UK: Royal Institution of Chartered Surveyors.

Questions

[1] Your client occupies 47 West Street, a shop unit. On the ground floor is a sales area with a width of 6 metres and a depth of 18 metres. At the rear of this area, up a flight of five steps, is a further sales area with a width of 6 metres and a depth of 12 metres. This rear sales area has a frontage of and a door to East Street, which is also a shopping area. On the first floor is storage accommodation with an area of 60 square metre net internal area. Recently, 45 West Street, a shop in the same parade, was let for £53 000 per annum on a full repairing and insuring lease for a term of 10 years with a rent review to MV after 5 years. This shop has a width of 5 metres and a depth of 16 metres on the ground floor. On the first floor, there are 30 square metre net internal area of storage accommodation. The zone A rate for West

Street has been agreed with the Valuation Officer at £1000 per square metre. The assessment of a shop in East Street adjoining the part of your client's property fronting that street has been agreed at £27 000 rateable value. This shop has ground floor sales only; the width is 6 metres and the depth is 12 metres. Assuming 6-metre zones, advise your client of the likely rateable value of 47 West Street.

[2]   Your client has recently taken occupation of Unit A, a modern single-storey warehouse with a gross internal floor area of 2000 square metres on an industrial estate. Six years ago, your client agreed a rent of £75 000 per annum on a full repairing and insuring lease for a term of 10 years with a rent review to MV after 5 years. Your client also has the right to determine the lease at the date of the first rent review. Unit B on the same estate has a gross internal floor area of 500 square metres. In April last year, the rateable value of this warehouse was agreed at £40 900. Included in the assessment was rateable plant and machinery with an agreed estimated capital value for rating purposes of £15 000, with a cap rate of 6%. In April last year, Unit C, another warehouse on the same estate, was let for £70 000 per annum on a full repairing and insuring lease for a term of five years with no rent reviews. The warehouse has a gross internal floor area of 1000 square metres. Advise your client of the likely rating assessment of Unit A for the current Rating List.

[3]   In March of this year, your client took occupation under a lease of the third floor of a modern four-storey office building. All the floors are serviced by a lift and each floor has a net internal area of 1000 square metres. To let the floor, the landlord has waived the service charge. At around the time of the antecedent valuation date, the fourth floor of the same building was let at a rent of £40 000 per annum under a lease for a term of 10 years with a rent review to MV at the end of the first 5 years. The tenant was responsible for internal repairs and the landlord recovered the cost of external repairs and insurance from the tenant under a service charge fixed at 10% of the rent. The lease of this floor was assigned two years ago for a premium of £25 000 of which £5000 was in respect of fixtures and fittings. Assuming a yield of 8%, estimate the rateable value of the third floor.

Answers

[1]

Analysis of comparable evidence:
45 West Street:

Ground floor:

Zone A:	5 m × 6 m = 30 m²
Zone B:	(5 m × 6 m)/2 = 15 m²
Zone C:	(5 m × 4 m)/4 = 5 m²

Area ITZA = 50 m²
Area ITZA 50 m² @ £1000/m² = £50 000

Assume storage area has a value of A/10, i.e. £100/m² to give
store value = £3000
Rent of shop in terms of RV = £53 000 p.a.

East Street Comparable:

Ground floor:
Zone A: 6 m × 6 m = 36 m²
Zone B: 6 m × 6 m = 18 m²
Area ITZA: = 54 m²
So £27 000 p.a./54 m² = £500/m² ITZA

Valuation of 47 West Street:

West Street frontage:
Zone A: 6 m × 6 m = 36 m²
Zone B: (6 m × 6 m)/2 = 18 m²
Zone C: (6 m × 6 m)/4 = 9 m²
Area ITZA: = 63 m²
So, 60 m² × £1000/m² = £63 000 p.a.

East Street frontage:
Zone A: 6 m × 6 m = 36 m²
Zone B: (6 m × 6 m)/2 = 18 m²
Area ITZA   = 54 m²
So, 54 m² × £500/m² = £27 000 p.a.

Storage on first floor is twice size of comparable, so 2 × £3000	6000
Total so far: £63 000 + £27 000 + £6000	96 000
End adjustment: difference in levels −5%	(4800)
RV estimate (£p.a.)	91 200

[2]

Analysis of comparable evidence

Unit B rateable value (£ p.a.)	40 900
Interest on plant and machinery £15 000 @ 6% (£ p.a.)	(900)
	40 000
Area:	500 m²

So RV/m² = £40 000 p.a./500 m² = £80/m²

Note: This was agreed at the time of the last rating list. Unit B is not straightforward comparable – this small warehouse has £15 000 worth of P + M. It may be an expensively built property to house the P + M. Therefore, use Unit C.

Unit C rent = £70 000 / 1000 m² = £70 / m², so current rent lower than rating list level of rent!

Valuation of Unit A

$2000\,m^2 @ £70 / m^2 = £140000$. In view of size difference, deduct 5%.
So the rateable value is £133000 p.a.

[3]   The lift provides easy access to all floors so they are all assigned the same value, although there is variation on ground floor. The service charge was only recently waived and so will not affect the rental level as at the antecedent valuation date.

Rent from comparable evidence = £40000 p.a.

Analysis of premium: £20000/YP three years @ 8%: 20000/2.5771 = £7760 p.a.

Estimate of rateable value: £40000 + £7760 = £47760 p.a.

Chapter 13
Valuations for Expropriation

13.1 Introduction

Expropriation[1] refers to the action by a government or state-sanctioned entity of taking property from its owner for public use or benefit. Property might be expropriated for many reasons: to make provision for infrastructure and other development projects, to plan new areas of urban settlement or to reallocate property for restitution or consolidation purposes for example. Regardless of the reason, the use of expropriation powers can have a substantial impact on the livelihoods of those affected, and therefore fair compensation is regarded as a just response.

When assessing fair compensation, the aim is to place the affected party in a position after expropriation that is no better or worse than before, and the valuer's role is central to this aim. For example, in Nigeria the law stipulates that compensation for expropriated buildings and installations must be assessed using the depreciated replacement cost (DRC) method. Egbenta and Udoudoh (2018) argue that this does not reflect market value and is inadequate to put claimants in position they were in before acquisition.

Similarly, focusing on the Dutch compulsory purchase process, Holtslag-Broekhof et al. (2018) explore the variation between the compensation offered by the expropriator (the acquiring authority) and by the courts. Dutch compensation heads of claim are comparable to the United Kingdom – market value of the acquired property, any diminution in value of retained property (assessed via a before-and-after valuation) and any loss of income resulting from compulsory acquisition. Examining 94 compensation claims that had ended in legal proceedings, it was found that the final offer of compensation in court was on average 56.7% higher than the last compensation offer from the expropriator. Previous research has also shown that settled compensation amounts are higher than offers.

Property Valuation, Third Edition. Peter Wyatt.
© 2023 John Wiley & Sons Ltd. Published 2023 by John Wiley & Sons Ltd.
Companion website: www.wiley.com/go/wyatt/propertyvaluation3e

The reasons suggested for this difference are related to different systems of valuation and different interpretations of the legislation.

Usually, when a property is expropriated, compensation is payable in respect of rights taken or extinguished, the effect on any retained rights and for losses to livelihoods of affected parties. Compensation may also be paid to landowners where no tenure rights have been acquired but there has been a reduction in value because of nearby public works, such as noise from a new road. Valuations are required to quantify all these items, and valuers may act for acquiring authorities, landowners, and other affected parties.

13.2   Valuation for expropriation

13.2.1   Valuing property rights that are to be taken or extinguished

Expropriation legislation should state what tenure rights can be expropriated and which are entitled to compensation. These will include perpetual rights (e.g. freehold interests) and terminable rights (e.g. leasehold interests). For the latter, it will be necessary to determine the earliest termination date: compensation would then be assessed based on any profit rent that the tenant may enjoy during the period until termination plus the value of any improvements that have been made at the occupier's expense.

In the case of land and property where more than one party holds tenure rights, some means of allocating compensation will be needed. For example, a landowner may lease land to a tenant who then sub-leases to another party. Each interest can be valued individually and compensation estimated accordingly. However, there may be additional value attributable to the land or property if one or more of these interests were to be merged. Some means of allocating this *synergistic value* will be necessary and should reflect the value of both the land and improvements to the land (the way in which synergistic value can be estimated is discussed in Chapter 14). An investor landlord whose tenure rights have been expropriated should be able to claim for the costs of reinvestment in another property.

There should be compensation for occupation agreements that are less formal than a lease, such as licences to occupy and rights to use. Compensation should cover the cost of relocation, disturbance and any loss of goodwill. These are sometimes referred to as a disturbance *payment* as opposed to disturbance *compensation* because the affected party has not had an interest compulsorily acquired but has been dispossessed.

Legislation generally does not compensate holders of informal rights for the value of expropriated land, but there may be compensation in respect of improvements made to the land. The extent of this compensation is likely to depend on the specific circumstances of the informal occupation – duration, degree of permanence, extent of acceptance for example. If occupiers are to be relocated, then compensation should also cover all relevant costs.

In addition to infrastructure and public service provision, a country's land policy may permit economic development as a legitimate ground for expropriation. In such cases, the affected party should be entitled to a share of the development value (sometimes referred to as hope value) of the land. If development land value is to form the basis for compensation, then the existing-use value (which presumably would be lower anyway) is disregarded. Compensation for development land value is a controversial issue. Some argue that if land could be expropriated at existing-use value, then public authorities could afford to fund infrastructure investment. Others oppose this view.

Special assumptions attached to the market value basis should prevent holders of tenure rights benefiting from value that could only be attributed to the actions of the expropriating authority, since it is not a component of value that the owner could have realised in the market. This is known as the "no-scheme world" assumption. It can be difficult to obtain evidence of market values if the expropriation order has been around for a while because the impending project may have influenced values in the area. There should also be regard to the period for which occupied land would have likely remained available for its existing use and to the availability of other suitable land.

13.2.2 *Valuing retained property rights*

There may be situations where only a portion of a person's property rights is to be expropriated but the retained land becomes less valuable as a result; a new road that severs a land parcel in two for example. Compensation should be paid for the reduction in value of the retained land as well as for land taken for the road construction itself.

The value of retained land may fall because of the *construction* of the works, where rights of access, light and support are temporarily taken for example, or *operation* of works, perhaps due to noise, vibration, smell, fumes, smoke, artificial lighting or the discharge of any substance.

The value of tenure rights might be affected by easements or wayleaves that permit certain authorities the right to lay and maintain cables and pipes on, under or over land. Compensation in these circumstances is usually in the form of an annual rental payment plus, in some cases, compensation to cover land lost for crop production and extra costs associated with the presence of equipment.

13.2.3 *Valuing compensation for disturbance*

The owner of expropriated tenure rights has the expense of finding new accommodation and moving. More importantly, livelihoods, social networks, family connections and a sense of belonging are all likely to be affected. As noted above, compensation in respect of expropriated land is usually based on a definition of market value that assumes the seller is 'willing', but this is clearly not the case and loss is suffered as a result of being dispossessed and having to relocate. To address this, compensation for *disturbance* should be paid in respect of:

- Relocation costs if all land is expropriated, reorganisation costs if only part of land is expropriated or total extinguishment costs if a business operation is to permanently close.

- Loss incurred in relation to work-in-progress. In the case of agricultural land, this would be the market value of trees and perennial crops and the value of the harvest for annual crops. Similarly for industrial processes, this would be the value of any non-replaceable stock and unfinished production materials.
- Loss of goodwill; this relates to financial value over and above market value that a person or entity may obtain as a result of owning or occupying the specific property. This usually manifests itself in the form of customer loyalty.
- Other losses or damages suffered as a result of being forced to move.

13.2.4   Valuing customary and informal land for expropriation purposes

Compensation processes must include informal and customary interests and rights and should include allowances for improving sub-standard living conditions (Roberts 2018). When valuing customary and informal land for expropriation purposes, all holders of affected property rights should be identified. The valuer should understand the relevant customs and practices and how these might influence or even facilitate markets in land and property.

If property rights are communally held, it is essential that value be attributed to the correct individuals or groups in the correct proportions. Sometimes, apportionment of financial compensation among members of a community may not be appropriate. Instead, it may be possible to offer to relocate a community to a suitable alternative site. Although disruptive, such an approach can help to retain community linkages.

If the land can be leased from the community rather than expropriated outright, this can also help the community retain some link to their land. Compensation would be based on a temporary loss of rights and benefits. A solution of this type might be more appropriate for shorter, fixed-term projects – expropriation of mining rights for example. An alternative solution might be a profit-sharing arrangement. If a part of a community's land is to be expropriated, in addition to compensation for land taken, it may be possible to agree a programme of mitigation that could include additional community facilities such as education, healthcare, infrastructure and amenities on the retained land.

Informally occupied land can be especially challenging to value. Dwellings may have no addresses, so identification is problematic. Establishing the size of land holdings requires physical measurement, but some places may be difficult to access. If land holdings are occupied by extended families, it can be difficult to establish the legitimate recipient of compensation and disputes can result. Due to the lack of market price information, engagement and consultation are essential to reach a consensus on what is to be compensated and how much the compensation should be. Compensation need not be monetary, especially in informal markets.

Whatever approach or mix of approaches to estimating compensation is used, negotiations should proceed sensitively and with due regard to the non-market interests of the holders of customary tenure rights. Consultation with affected communities should take place from an early stage and involve men and women. Affected parties should be made aware of the compensation guidelines, preliminary compensation figures and their right to object if they feel that the amounts offered are not fair. Disruption to livelihoods should be minimised and affected parties should be placed in the same position as before, or better. This requires compensation for loss of use rights, such as cropping and grazing, and for non-transferable or permanent improvements, including buildings, wells, boreholes and fruit trees. Compensation for cropping land may be limited to the cost of preparing virgin land, including de-bushing, clearing, stumping and surface levelling. Regarding grazing land, compensation may be limited to legally fenced-off grazing land. Transferable improvements should be replaced on a 'new-for-old' basis whenever possible. If rights holders are to be relocated, they should be offered a plot within the new scheme or alternative land of similar size in the vicinity.

Compensation should be paid before the acquiring authority takes possession and should include an allowance for disturbance. It is also important to ensure that payments retain their value in real terms by linking them to inflation. Procedures should be put in place to allow affected parties the opportunity to appeal to an independent body against compensation arrangements they have been offered.

There is a need for fair and transparent policies and procedures for expropriation of land; expropriating authorities must ensure that livelihoods of affected persons are not negatively affected by expropriation.

13.2.5 Expropriation and non-market value

Market value does not compensate for livelihood replacement. Also, communities place non-monetary value on land. How can these social, cultural and environmental values be compensated? Rao et al. (2017) argue that the well-being contribution of land extends beyond its market value and compensation for expropriation should cover these losses. Rao (2019) identifies various land 'functionings' (ways that land contribute to well-being) under five headings:

[1]   Being able to secure necessities of life
[2]   Being financially secure
[3]   Being able to protect oneself from discrimination, exploitation, violence and assault as a human right
[4]   Being able to establish social associations and harness personal, familial and societal interests through these associations
[5]   Being able to maintain and enhance self-respect and identity

Although difficult to quantify, compensation for non-market value might take the form of a discretionary payment that would be dependent upon the length of time a claimant has occupied the land and inconvenience likely to be suffered when dispossessed. Issues related to cultural norms, values and beliefs, particularly in relation to customary land, should be taken into account.

Religious sites such as sacred trees, shrines and mountains have spiritual value to resident populations, but they are not traded so there is no concept of value-in-exchange. An alternative basis of compensation is therefore required. This might be an equivalent building in the case of a meeting place for religious worship. Alternatively, the compensation might cover the expense of identifying and acquiring a new site plus the cost of constructing a suitable building and appropriate disturbance compensation. In the case of burial grounds and other sacred places, financial compensation is unlikely to be appropriate. Instead, avoidance and/or mitigation measures should be considered.

The following general guidelines can be stated:

- Independent and impartial valuers should be appointed to value expropriated tenure rights and any reduction in the value of land affected by expropriation.
- The valuation process should be participatory and minimise conflict and stress on affected parties.
- If an affected party wishes to appoint a valuer to value affected tenure rights, the cost should be borne by the expropriating authority.
- The process should provide for resolution of disputes over value and valuation process.
- If the acquiring authority is empowered by a specific law, the valuation date should be set by law, not by the acquiring authority.
- Valuers should ensure that compensation is fairly and expediently agreed, and:
  ○ Provide, if possible, an early indication of the amount of compensation likely to be awarded.
  ○ Base estimates of market value on openly agreed transaction prices where possible. (Again noting the comment in Chapter 1 regarding the context-specific nature of value, compensation should not be based on values set by the State for other purposes such as land and property taxation as these may not fully reflect the 'loss' of tenure rights.)
  ○ Take non-market value into consideration.

Roberts (2018) recommends that, where possible, licenced professional valuers should conduct valuations, engaging with the community including women and vulnerable groups in a free prior informed consent (FPIC) framework. By doing so, traditional leaders and community members can explain how they use and value land, thus ensuring that any informal and non-market values are incorporated into valuations. Questions that may be posed to community members when determining the value of resources might include: is the community willing to lease the land and, if so, for what duration? Land-based resources to be compensated include current and future mature and growing crops, improvements and natural resources (such as firewood, fruits, fungi) as well as communal rights such as commoners' rights.

13.3 Valuations for compulsory purchase and planning compensation in England

13.3.1 Legal background

The government and organisations responsible for the utility networks in England have the legal power to compulsorily acquire property for specific purposes. This might be to build a new road, a wind farm or a nuclear power station. Specific compulsory purchase powers may be contained within an Act of

Parliament for a particular function, such as a new high-speed railway. General Acts of Parliament may confer compulsory purchase powers on specific bodies. Examples include the Transport and Works Act 1992, or the Harbours Act 1964, or the Planning Act 2008 for Nationally Significant Infrastructure Projects. Public bodies with statutory compulsory purchase powers include local authorities, statutory undertakers (such as the Highways Agency, utilities, transport infrastructure providers), some executive agencies (such as Homes England) and health service bodies. Also, landowners can initiate a compulsory purchase process via Purchase Notice or Blight Notice (see below).

The legal basis of the right to claim compensation in these respects can be found in the *Land Compensation Act 1961* (LCA61), the *Compulsory Purchase Act 1965* (CPA65) and the *Land Compensation Act 1973* (LCA73), as amended by the *Planning and Compensation Act 1991* (PCA91), the *Planning and Compulsory Purchase Act 2004*, the Localism Act 2011, the Housing and Planning Act 2016 and the Neighbourhood Planning Act 2017 (NPA17). A substantial body of case law provides legal interpretation of these statutes. Together, this legislation and case law is referred to as the *Compensation Code*.

The guiding principle of the legislation in respect of property owners who have been affected by compulsory purchase is to ensure, financially at least, that they are restored to the position before acquisition took place. Known as the *equivalence principle*, a landowner should be paid neither more nor less than their loss. Consequently, the owner of an interest being compulsorily acquired is entitled to compensation equivalent to the value of the land being acquired. Where some land is retained and its value drops, the owner is entitled to be compensated for this diminution whether it is caused by severance of the two parts of land or by injurious affection to the retained land. An owner will also be entitled to losses that are a consequence of being compelled to vacate the land, known as *disturbance*. Denyer-Green (2019) provides a detailed discourse of the statutory framework and case law that has built up around compulsory purchase and compensation; here we discuss the topic from the valuer's perspective.

Valuers are often appointed to estimate the value of compulsorily acquired property and to estimate any diminution in value of land resulting from either construction activity or use of the finished development (smell from sewage works for example). The legislation referred to above includes a statutory definition of market value so, when valuing for compulsory purchase and compensation, valuers need to depart from the Red Book definition of market value and follow statutory regulations instead. That said, the statutory definition of market value is essentially the same as the internationally recognised definition.

The following sections consider these various situations in which compensation will be payable, known as 'heads of claim'. In many cases, owners of property interests will be entitled to more than one 'head of claim'.

13.3.2   Compensation for land[2] taken (compulsorily acquired)

All interests in land may be acquired including freehold, leasehold and equitable interests (such as a mortgagee), land rights such as "compulsory works orders" (open-cast mining), and wayleaves (electricity and water). There are limited rights to compensation for occupation agreements that are less formal than a lease, such as tenancies at will, tenancies on sufferance and licences. These rights include compensation for relocation costs and any loss of goodwill. Regard is had to amount of time the land occupied would have been likely to have remained available for the purposes of the business and to the availability of other suitable land. If a tenant is holding over under statutory security of tenure provisions, disturbance compensation under those provisions can be chosen as the basis for compensation if it is more than the compulsory purchase order (CPO) disturbance payment.

The acquiring authority sets out the nature and extent of the property interest to be acquired in a 'notice to treat', which may be served on owners of all property interests except holders of periodic tenancies of one year or less. There are other methods of obtaining possession too: by agreement or via a General Vesting Declaration (this is like a notice to treat but title in the property interest is conveyed to the acquiring authority as well as the right to enter and take possession).

Where a tenant has a contractual or statutory right to renew a lease, that right will form part of the value of his leasehold interest (Johnson et al. 2000) but the leasehold interest should be valued based on the earliest termination date (Denyer-Green 2019).

The relevant valuation date is either the date of entry and taking possession if the acquiring authority has served a notice to treat and notice of entry, or the vesting date if the acquiring authority has executed a general vesting declaration. When compensation is assessed by the Lands Tribunal, the valuation date is the last day of the hearing if possession has not already been taken by then. The date of the notice to treat fixes the nature of interest to be acquired, but the extent and condition are fixed at the date of entry. Items claimed under the disturbance head of claim may pre-date the notice to treat and post-date entry.

The preferred basis for the assessment of compensation is Market Value. In other words, the fact that the acquisition is compulsory must be disregarded. The additional compensation heads of claim (disturbance, loss payments) are there to reflect the fact that the acquisition is compulsory (not a willing seller). The market value can reflect the existing use, or it can be development value (discussed further below). Either way, the use must be lawful. Market value can have regard to a bid from a special purchaser but must disregard any effect on value attributable to the acquiring authority's scheme (the 'no-scheme' principle). Sections 6A to 6E of the LCA61 (inserted by S32 NPA17) set out how land should be valued using the 'no-scheme' principle:

- S6A: any increases or decreases in value caused by the scheme or prospect of the scheme must be disregarded.
- S6B: increases in value of claimant's other land (adjacent or contiguous to land taken) must be deducted from compensation amount.
- S6C: if the claimant received compensation for injurious affection on retained land, and that land is subsequently acquired for the purposes of the scheme, compensation is reduced by the injurious affection amount.
- S6D: defines the scheme for the purposes of the 'no-scheme' world. The default is the scheme underlying the acquisition. There are special provisions for new

towns and development corporations.[3] In these areas, the scheme is the development of any land for the purposes for which the area is or was designated. Also, where land is acquired for regeneration or redevelopment which is facilitated or made possible by a 'relevant transport project', the 'scheme' includes the relevant transport project.

- S6E: sets out qualifying conditions and safeguards

New transport projects often raise land values in the vicinity of stations or hubs, which can facilitate regeneration and redevelopment schemes. Where land is acquired for regeneration or redevelopment, which is facilitated or made possible by a relevant transport project, the effect of Section 6D is that the scheme to be disregarded includes the relevant transport project – subject to the qualifying conditions and safeguards in section 6E. The intention of this special provision is to ensure that an acquiring authority should not pay for land it is acquiring at values that are inflated by its own or others' public investment in the relevant transport project. Where it applies, the land in question will be valued as if the transport project as well as the regeneration scheme had been cancelled on the relevant valuation date.

Valuing under the no-scheme principle is difficult because market prices are likely to be influenced by the prospect of the scheme, particularly if it has been known about for a while. The impending scheme may have influenced values in the area over some period. Blight notices can help here (see below).

In assessing market value, special suitability or adaptability of the land for any purpose must be disregarded if it is one that could be applied only in pursuance of statutory powers, or for which there is no market apart from the requirements of an authority possessing compulsory purchase powers. This element of value is not part of market value because it is not an element the owner could have realised in the open market. A scheme should be identified in narrower rather than broader terms.

If it is not possible to estimate a market value, perhaps because there is no market for the property being acquired, compensation is based on the estimated cost of equivalent reinstatement or resettlement, unless development value is higher. This approach is common in developing economies where markets in many types of property are either small or non-existent. In such circumstances, it can be difficult to assure comparability of replacement land in terms of size, quality, access and other value-significant attributes. Nevertheless, it is common to offer compensation that is part money and part land/premises, but it should be appreciated that the process can be expensive, time consuming and challenging to get right.

Equivalent reinstatement is difficult to justify if the interest is a short lease and is uncommon in relation to business premises. There are four general tests: land must be used for a purpose that would continue, there is no market for that use, there is a *bone fide* intention to reinstate, and if the reinstatement cost is disproportionate then it may not be allowed.

Compensation for land taken can be a significant proportion of the total claim value in rural areas, the rest being severance and injurious affection. In urban areas, business loss can be a big component.

Johnson et al. (2000) suggest that the methods employed to estimate the value of property that is compulsorily acquired are no different from those adopted in other market valuations, just subject to the statutory rules. In most cases the valuation is likely to be on an existing-use basis using the comparison or investment method. Care must be exercised when selecting comparable evidence because transactions would have taken place in the 'scheme' world. If the valuer feels that the scheme has influenced the evidence obtained from these comparables, then they may need to be adjusted to give a value in the 'no-scheme' world. After commencement of a compulsory purchase order, land values can be affected either negatively (blight) or positively (betterment).

13.3.3 Identifying the planning position

Development value may be considered alongside existing-use value but, in many compulsory purchase cases, an impending acquisition will mean no planning permission for development will be forthcoming (Denyer-Green 2019). Therefore, it is necessary to make certain planning assumptions so that an accurate assessment of development value can be made. The legal extent of these assumptions is set out Ss 14-17 of the LCA61 (as amended, most recently by S232 of the Localism Act 2011):

- Existing planning permission on the relevant land or other land. Planning permission may also be assumed for the acquiring authority's scheme but disregarding any purpose that could only be possible in pursuance of statutory powers. The extent of the scheme that can be considered is that which is included in the CPO. A wider scheme can be argued in some circumstances or with the agreement of the Lands Tribunal but only if the CPO (or documents published with it) identifies a wider scheme.
- Planning permission for development specified in a Certificate of Appropriate Alternative Development (a hypothetical planning permission) as at the valuation date. A Certificate of Appropriate Alternative Development can be obtained from the relevant local planning authority to indicate any planning permissions that could have been obtained. The definition of alternative appropriate development is development for which, had the scheme been cancelled when the CPO was made, consent could 'reasonably have been expected to be granted' if considered at the valuation date or later. The assumptions regarding cancellation of the scheme are the same as set out in S6A LCA61 (op cit). It is necessary to identify *development* rather than just *use* of the land.
- Prospective planning permission on relevant or other land on or after the development date in the absence of the scheme and which is not included in the Certificate of Appropriate Alternative Development. The phrase 'relevant land or other land' means that planning permission on land not being acquired can be considered, and 'on or after the development date' means that the valuer must account for hope value. In other words, the prospect of development value having regard to the forward planning context to the extent that it is reflected in market transactions.

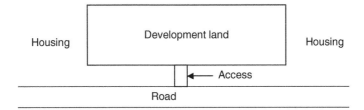

Figure 13.1   Development value.

Although planning permission for the scheme can be assumed, the effect of the scheme itself must be disregarded when assessing development value. This is set out as four 'cancellation' assumptions in S14 of the LCA61 (as amended):

- The scheme underlying the compulsory acquisition was cancelled on the launch date.
- No action has been taken by the acquiring authority for the purposes of the scheme.
- There is no prospect of the same scheme, or any other project to meet the same or substantially the same need, being carried out in the exercise of a statutory function or by the exercise of compulsory purchase powers.
- If the scheme was for highway construction, then no highway will be constructed to meet the same or substantially the same need.

If additional development is permitted within 10 years of acquisition, the owner is entitled to the difference between the compensation paid and the amount that would have been paid assuming the permission was in force at the time (Denyer-Green 2019).

Development value can include synergistic value or ransom value provided they would have existed in the 'no-scheme world'. For example, if the parcel of land labelled 'access' in Figure 13.1 is being compulsorily acquired to provide access to the development land, the owner gets a proportion of the value of the development land. This principle was laid down in the landmark case of *Stokes v Cambridge Corporation (1961)* in which the proportion was one third. If the development land is being acquired too but can only be developed if satisfactory access can be provided, the market value will be the full development value less the estimated cost of acquiring the necessary additional land (Denyer-Green 2019).

13.3.4   Compensation for severance and injurious affection

Severance is where retained land loses value because it has been severed from the acquired land and injurious affection is where retained land loses value due to proposed construction on and use of acquired land for the scheme. The latter is payable where either part of an owner's land is acquired, or when no land is taken at all, although the compensation right in this latter circumstance is very limited. These two circumstances are now considered in turn.

13.3.4.1   Compensation where part of an owner's land is acquired

Where only part of an owner's property is taken, the CPA65 allows compensation for severance of and injurious affection to the part retained. Severance is where retained land loses value because it has been severed from the acquired land. Compensation for severance is based on the reduction in value of the retained land, which need not be contiguous but must be in the same ownership and functionally related. While a drop in value due to severance is easy to explain, injurious affection to the retained land is slightly harder to envisage. Essentially it is injury or damage caused by construction works, including disturbance for having to vacate premises. But it also covers any diminution in value caused by subsequent use of the works. It is difficult to separately quantify diminutions in value resulting from severance and injurious affection. Therefore, to estimate these figures a valuer would value the land as it was before the CPO and then value the same land on completion of the works. The difference between the before and after valuations represents the drop in value. If the value of the land taken is then deducted from difference between the before and after valuations, this gives the compensation for severance and injurious affection. For example, the market value of a property before the acquiring authority's scheme is £250 000 and afterwards it is £200 000, then compensation is therefore £50 000. If the market value of the land taken is £30 000, then the loss in value of retained land due to severance and injurious affection is £20 000.

Now consider a more detailed case. A local authority wishes to redesign access to an industrial estate in preparation for its expansion. To enable this, it has served a CPO on the industrial unit at the entrance to the estate giving notice of the planned acquisition of part of its land. Once the redesigned access is complete in three years' time, the unit will benefit from improved access arrangements plus additional storage land. The tenant of the unit has 8 years remaining on a 15-year full repairing and insuring (FRI) lease with five-year upward-only rent reviews. The current rent is £100 000 per annum, the (no-scheme) market rent for the whole unit is estimated to be £120 000 per annum, and for the retained part after severance it is £80 000 per annum. Injurious affection caused by carrying out of works will reduce the market rent of the retained land to £70 000 per annum but it is estimated that its market rent will rise to £90 000 per annum once the works are complete. The local authority has stated that it will pay for the new access and storage land. Assuming a freehold yield of 8% and a leasehold yield of 10%, compensation for the landlord and tenant is assessed as follows:

Landlord's interest

'Before' valuation:

Term rent received (£ p.a.)	100 000	
YP 3 years @ 8%	2.5771	
		257 710
Reversion to market rent (£ p.a.)	120 000	
YP perpetuity @ 8%	12.5	
PV £1 3 years @ 8%	0.7938	
		1 190 700
'Before' capital value (£)		1 448 410

'After' valuation:

Term rent (100 000 × 80 000/120 000) [a] (£ p.a.)	66 667	
YP 3 years @ 8%	2.5771	
		171 808
Reversion to market rent (£ p.a.)	90 000	
YP perpetuity @ 8%	12.5	
PV £1 3 years @ 8%	0.7938	
		893 025
'After' capital value (£)		1 064 833

[a] This calculation determines the *current* rent for the retained part using the evidence of *market* rents for the retained part and the whole.

Therefore, the drop in value resulting from part of the land being acquired and from injurious affection is the difference between the before and after valuations, £1 448 410 – £1 064 833 = £383 577. The following calculation determines the value of land taken only:

Term rent lost (100 000 – 66 667) (£ p.a.)	33 333	
YP 3 years @ 8%	2.5771	
		85 902
Reversion to market rent lost (120 000 – 80 000) (£ p.a.)	40 000	
YP perpetuity @ 8%	12.5	
PV £1 3 years @ 8%	0.7938	
		396 900
Capital value of land taken (£)		482 802

Therefore, compensation for severance and injurious affection (betterment in this case) is £383 577 – £482 802 = –£99 225. In other words, the value of the land taken (£482 802) is reduced by the capital value of the enhancement to the unit resulting from the works, i.e. an increase in market rent from £80 000 to £90 000 per annum on reversion, when capitalised into perpetuity at 8% deferred three years, produces a betterment (or improvement in capital value) of £99 225.

Tenant's interest

'Before' valuation:

Market rent of whole unit (£ p.a.)	120 000	
Less contract rent (£ p.a.)	(100 000)	
Profit rent (£ p.a.)	20 000	
YP 3 years @ 10%	2.4869	
Valuation (£)		49 738

'After' valuation:

Market rent of retained part (£ p.a.)	70 000
Less contract rent for retained part [a] (£ p.a.)	(66 667)
Profit rent (£ p.a.)	3333
YP 3 years @ 10%	2.4869
Valuation (£)	8290

[a] Calculated as above

Therefore, the difference in value is £49 738 – £8290 = £41 448. This represents the value of the land taken plus the diminution in value resulting from severance and injurious affection. Separating these amounts can be undertaken as follows:

Value of land taken:

Profit rent (20 000 – [20 000 x 80 000/120 000]) (£ p.a.)	6667
YP 3 years @ 10%	2.4869
Valuation (£)	16 580

Therefore, compensation for severance and injurious affection is £41 448 – £16 580 = £24 868.

In cases like the one above, where part of a property subject to a lease is taken, the rent needs to be apportioned between the part taken and the part left, and this was done in the ratio of rental value of the part retained to the rental value of the whole. In cases where only a small part of a property is taken, a nominal apportionment of, say, £1 per annum on land taken may be agreed. The tenant then continues to pay full rent under the lease for the remainder of the term but receives full compensation for loss of rental value from the acquiring body while the landlord is compensated for injury to his reversion (Johnson et al. 2000).

The CPA65 provides the owner with an option to request the acquiring authority purchases the whole property, and the success of such a request depends on whether there has been a material detriment to the retained part. LCA73 requires whole proposed works to be considered (including those off-site) when assessing detriment.

13.3.4.2   Compensation where no land is taken

Property owners can also claim compensation where none of their land is taken. There are two ways that this can be done: under Section 10 of the CPA65 compensation can be claimed for *execution* (construction) of works and under Part 1 of the LCA73 compensation can be claimed for the (subsequent) *use* of public works. Generally, compensation is assessed as a percentage of the existing-use value of the property before the scheme/effect.

Section 10 of the CPA65 provides for compensation where rights of access, light and support are taken. To successfully claim compensation for injurious affection caused by execution of works, four rules must be satisfied. These are known as the 'McCarthy Rules' because they resulted from a House of Lords decision in the case of *Metropolitan Board of Works v McCarthy (1874)*:

- The works must be authorised by statute.
- If the works were not authorised by statute, the injury caused would be actionable at law (as a nuisance).
- The injury arises from a physical interference with some right which is attached to the land and which has a market value. In cases where the interference is temporary a decrease in rental value is sufficient to sustain a claim even where the capital value, after conclusion of the works, is unaffected (Denyer-Green 2019).
- The injury must be caused by execution of works, not subsequent use.

Section C

The basis of compensation is the diminution in value of the interest. The usual measure of compensation is the reduction in the existing-use value of the affected land attributable to the injury that gave rise to the claim (Marshall and Williamson 1996). The valuation date is the date of the loss.

Part 1 of the LCA73 provides a code for compensation for use of public works such as roads, airports, and so on. Owners qualifying property interests, whose rights have been affected, have a right to claim compensation (referred to as making a 'Part 1 Claim'). This is for the reduction in the existing-use value of their interest caused by certain physical factors, namely noise, vibration, smell, fumes, smoke, artificial lighting or the discharge of any substance. The first date of claim is one year after the use of public works first commenced, and the last day of claim is six years from the first claim day (note, phasing of work). The claimant must own the freehold or leasehold interest in a property, the latter having at least three years remaining, and a rateable value over a specified level. The basis of compensation is the diminution in the existing-use value of the interest and, in most cases, the practical approach to the valuation is to estimate a 'no-scheme world' value of the affected property and then make a judgement as to the percentage depreciation that can be attributed to the physical factors (Denyer-Green 2019). Compensation can be reduced if the compensating authority mitigates the effects.

13.3.5 Compensation for disturbance and other losses

The owner of a compulsorily acquired business property can claim either the costs of relocation (including removal costs, loss of stock, new stationery, loss of goodwill) or the cost of winding up the business, known as 'total extinguishment'. In most cases the business occupier will only be granted relocation costs, but a sole trader aged 60 years or over in a property with a rateable value over a specified level has a statutory right to opt for total extinguishment.

Disturbance compensation is only payable if compensation is based on existing-use value (as opposed to development value) and is usually payable in respect of any item that is not too remote and is a natural and reasonable consequence of the acquisition of the owner's interest. The amount of disturbance compensation is normally calculated by valuing existing fixtures from the perspective of an incoming tenant in the same line of business plus, if the business is to be extinguished, the loss on forced sale (the difference between value to an incoming tenant and the price achieved on sale) (Johnson et al. 2000). Typical relocation costs that can be claimed for include:

- Removal;
- Legal, surveyor's and architect's fees and Stamp Duty relating to acquisition of new premises;
- Special adaptations to replacement premises;
- Loss of profits during move;
- Diminution of goodwill following move (reflected in gross profits);
- Depreciation in value of stock;
- Notification of new address to customers and new stocks of stationery due to change of address.

Typical extinguishment costs would be value of business goodwill, loss on forced sale of stock, vehicles and plant and machinery, redundancy costs and administration costs of winding up the business.

S35 of the NPA17 inserts a new S47 into the LCA73, bringing the assessment of compensation for disturbance for minor and unprotected tenancies into line with that for licensees and protected tenancies (a tenancy with the protection of Part II of the Landlord and Tenant Act 1954).

Regard should be had to the likelihood of either continuation or renewal of the tenancy, the total period for which the tenancy might reasonably have been expected to continue, and the likely terms and conditions on which any continuation or renewal would be granted. For protected tenancies, the right of a tenant to apply for a new tenancy is also to be considered.

Investors have limited rights to compensation introduced by the PCA91 for the costs of re-investment in another UK property for up to one year from the date of entry.

For residential occupiers, *home loss payments* are payable if the dwelling was occupied as a main residence for one year before the expropriation. For business occupiers, a *basic loss payment* is payable to all owners and an *occupier's loss payment* is payable to owners who have been in occupation for a year or more.

The LCA73 authorises disturbance *payments* to claimants in cases where disturbance *compensation* is not payable because the claimant has not had an interest compulsorily acquired but has been dispossessed. This situation would arise if the acquiring authority compulsorily acquired a freehold interest subject to a short lease. The authority is unlikely to renew the lease, so a disturbance payment is made to cover reasonable removal expenses and, where relevant, loss sustained by the tenant for the business having to quit the land (Johnson et al. 2000).

In the case of short tenancies, there is no requirement to serve a notice to treat but compensation arrangements are similar to those that apply to other interests. For land taken, compensation is payable in respect of market value of the leasehold interest and should reflect any renewal rights. If only part of the land is to be acquired, there is a right to compensation for the diminution in the value of retained land even if it is held under a separate lease, provided it is adjoining or adjacent. For disturbance, only losses relating to the period between date of entry and expiry of term are recoverable. If the tenant has statutory security of tenure, disturbance compensation can be claimed under Part II of the Landlord and Tenant Act 1954 if it is considered to be higher than a disturbance payment under CPO legislation.

There are limited rights to compensation for occupation agreements that are less formal than a lease, such as tenancies at will, tenancies on sufferance and licences. Compensation is for relocation costs and any loss of goodwill. Regard is had to the amount of time the land occupied would have been likely to have remained available for the purposes of the business and to the availability of other suitable land.

13.4 Planning compensation in England

Compensation may also be paid to property owners when certain planning decisions are made and these adversely affect property value. The objective of the valuer in such cases is to estimate the reduction in value, usually by adopting a before-and-after valuation approach.

13.4.1 Revocation, modification and discontinuance orders

The Town and Country Planning Act, 1990 (TCPA90) provides for compensation if a planning permission that was previously granted is revoked, modified or discontinued by a local planning authority. The order must be made before building or other work is completed or before a change of use has taken effect (Johnson et al. 2000). Compensation covers abortive expenditure and for loss or damage directly attributable to the order, including a drop in property value, calculated in accordance with Section 5 of the LCA61 (i.e. a before-and-after valuation to reveal the difference between the market value of land with the benefit of the planning permission and with the permission revoked or modified [Johnson et al. 2000]). The Planning (Listed Buildings and Conservation Areas) Act 1990 provides for compensation on the same basis as the TCPA90 but in respect of loss caused by the refusal, revocation, modification or the grant of conditional listed building consent or by the issue of a Building Preservation Notice.

13.4.2 Purchase notices

Under the TCPA90, where planning permission is refused or granted subject to conditions or where a local planning authority serves a revocation, modification or discontinuance order or refuses, modifies or grants a conditional listed building consent, this may entitle the owner to serve a *Purchase Notice* as an alternative to a compensation claim as described above (Johnson et al. 2000). The property owner must serve the notice on the local authority within one year of the planning decision with proof that the property is incapable of reasonable beneficial use and requiring it to purchase the property. Once the purchase notice is confirmed, the acquiring authority is deemed to have served a notice to treat, and normal compulsory purchase rules apply (Marshall and Williamson 1996).

13.4.3 Blight compensation

Similarly, planning proposals that could eventually involve compulsory acquisition may well depreciate the value of affected property or even render it valueless. As a result, under certain circumstances, the owner-occupier can require the acquiring authority to purchase the property by serving a *Blight Notice*. An owner-occupier must be a freeholder or lessee with three or more years unexpired lease term who has occupied for the last 6 months or 6 months in the previous 12 months and the property has been unoccupied since vacated. Investor-owners are not entitled to serve blight notices. The owner must be able to show that reasonable efforts to sell the property were unsuccessful except at a price substantially lower than might reasonably be expected in a market without the threat of compulsory acquisition. If the acquiring authority accepts the blight notice, then a notice to treat is deemed to have been served and the valuation principles and assessment of compensation are the same as those that apply to the compulsory acquisition of land. Alternatively, the acquiring authority may reject the notice or propose to acquire only part of any land.

Key points

- Valuation for expropriation purposes is a complex area that is influenced by a large body of statutes and case law. Valuers working in the private sector on behalf of property owners and valuers representing the government and other acquiring authorities may be requested to provide opinions of market value or, with sufficient knowledge and experience, to negotiate compensation claims on behalf of either party.
- Although the law is complex, two fundamental points are worth reiterating. First, market value is central to the assessment of compensation for land taken and diminution in market value is central to the assessment of compensation for severance and injurious affection. Second, market valuations must be undertaken in the *no-scheme world*, a concept that lends itself more to theoretical understanding than practical application!
- Care should be taken to ensure that compensation adequately reflects loss of livelihood, particularly in the case of small farms, subsistence agriculture and other marginal economic activities. These are difficult to replace due to a lack of suitable alternative land. Insufficient payment will not sustain the family for long, forcing the affected party to seek an alternative livelihood, often in the unskilled labour market.

Notes

1. Also known as compulsory purchase (UK) or eminent domain (US).

2. Although compulsory purchase legislation refers to *land* being acquired or the value of *land* being affected by compulsory acquisition and public works, the legislation and therefore valuation rules apply to property interests in general.

3. Delineated urban areas designated for regeneration and are allocated specific and usually time-limited incentives for developers, investors and occupiers.

References

Denyer-Green, B. (2019). *Compulsory Purchase and Compensation*, 11e. UK: Routledge.

Egbenta, R. and Udoudoh, F. (2018). Compensation for land and building compulsorily acquired in Nigeria: a critique of the valuation technique. *Prop. Manag.* 36 (4): 446–460.

Holtslag-Broekhof, S., Beunen, R., Van Marwijk, R., and Wiskerke, J. (2018). Exploring the valuation of compulsory purchase compensation. *J. Eur. Real Estate Res.* 11 (2): 187–201.

Johnson, T., Davies, K., and Shapiro, E. (2000). *Modern Methods of Valuation*, 9e. London, UK: Estates Gazette.

Marshall, H. and Williamson, H. (1996). *The Law and Valuation of Leisure Property*, 2e. London: Estates Gazette.

Rao, J. (2019). A 'capability approach' to understanding loses arising out of the compulsory acquisition of land in India. *Land Use Policy* 82: 70–84.

Rao, J., Tiwari, P., and Hutchison, N. (2017). Capability approach to compulsory purchase compensation: evidence of the functionings of land identified by affected landowners in Scotland. *J. Prop. Res.* 34 (4): 305–324.

Roberts, B. (2018). *Land valuation and compensation primer*. In: *Part of the Responsible Investment in Property and Land (RIPL) Guidebook*. Landesa Rural Development Institute.

Questions

[1] A tenant of a shop (ground and upper floors) is to be compulsorily acquired. The tenant has lived in the upper part and run a bakery on the ground floor for the past five years. The rent is £70 000 per annum for the whole property on an internal repairing (IR) lease with 10 years unexpired. The market rent is £100 000 per annum, of which £60 000 per annum can be attributable to the shop part. The rateable value of the shop is £40 000. The net profit for the last financial year was £180 000 after deducting rent of £70 000, mortgage interest of £10 000, repairs of £5000 and rates of £20 000, all relating to the whole building. The previous two years unadjusted net profits have been £160 000 and £170 000, but remuneration to the tenant (who works full-time for the business) and her husband (who works half-time) has not been deducted. The tenant is 62 years old and does not wish to buy another business. Prepare a claim for compensation.

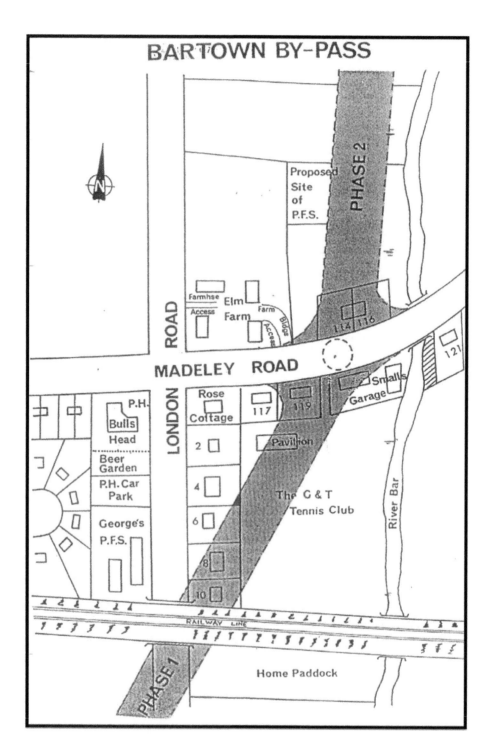

[2]   A local authority is carrying out a town centre retail development for which land is being compulsorily acquired. The scheme will take approximately two years to complete. Xenon Bikes is the tenant of a shop held on a 20-year lease,

with 5-year rent reviews, which began 8 years ago. The two-storey shop has a single-storey rear extension and rear yard with access to a service road from which goods are delivered. The rear extension and rear yard are due to be acquired. When complete, the new scheme will allow for service access via a covered service way from nearby unloading bays. The rent passing for the shop is £50 000 per annum and the current market rent in the absence of the scheme is £60 000 per annum, of which £55 000 per annum can be attributed to the retained part of the shop. When the scheme is complete, the rent for the retained part is estimated to be £48 000 per annum. Estimate the amount of compensation payable to the tenant, Xenon Bikes, for the land taken, severance and injurious affection. Assume a leasehold yield of 10%.

Answers

[1]

Land taken (Rule 2, Section 5, LCA61):

Market Rent (£ p.a.)	100 000	
Plus landlord's expenses (estimates);		
▪ External repairs (£ p.a.)	8000	
▪ Insurance (£ p.a.)	2000	
IRI rental value (£ p.a.)	110 000	
Less rent paid (£ p.a.)	(70 000)	
Profit rent (£ p.a.)	40 000	
YP 10 years @ 8%	6.7101	
Valuation (£)		268 404

Disturbance (Rule 6, Section 5, LCA61):

The claimant is over 60 years old so a claim for total extinguishment under S46 of the LCA73 stands. The average of the last three years' earnings is taken as the best evidence of profitability.

Net profit (£ p.a.)			170 000
Mortgage interest (£ p.a.)			10 000
Repairs for upper part, say (£)			1000
Less (hypothetical) part-time assistant (£ p.a.)			(40 000)
Less profit rent in respect of shop part, say (£ p.a.)			(30 000)
Less interest on capital (£ p.a.):			
▪ Fittings		(15 000)	
▪ Stock		(5000)	
▪ Cash		(3000)	
Total capital (£)		(23 000)	
Amortised at 8%		0.08	
			(1840)
Adjusted net profit (£ p.a.)			109 160
Capitalised in perpetuity at a target rate return of 20%			5.0000
Valuation of business (£)			545 800

Additional items (£) [a]:
- Sale of fittings to acquiring authority
- Notification to suppliers
- Loss on stationery
- Disconnection of services
- Removal costs
- Finding new living accommodation
- Home loss payment

Sale of fittings to acquiring authority	10 000
Notification to suppliers	1000
Loss on stationery	1000
Disconnection of services	500
Removal costs	3000
Finding new living accommodation	4000
Home loss payment	5000
Compensation estimate based on total extinguishment (£)	570 300

[a] Business is a bakery so there is no forced sale of stock

[2]

Compensation for land taken, severance and injurious affection

'Before' valuation		
MR of whole property (£ p.a.)	60 000	
Less rent paid (£ p.a.)	(50 000)	
Profit rent (£ p.a.)	10 000	
YP 2 years @ 10%	1.7355	
		17 355
'After' valuation		
MR of retained part (£ p.a.)	48 000	
Less rent paid (£ p.a.) [a]	(45 833)	
Profit rent (£ p.a.)	2167	
YP 2 years @ 10%	1.7355	
		3761
Valuation of land taken, severance and injurious affection (£)		13 594

[a] Ratio of MR for whole (£60 000) to MR of retained part (£55 000) can be used to estimate the rent passing in relation to the retained part: (£55 000/£60 000) × £50 000 = £45 833.

To split the compensation for the land taken from the compensation for severance and injurious affection:

MR of land taken (no scheme): £60 000 – £55 000 (£ p.a.)	5000	
Less rent passing in respect of land taken: £50 000 – £45 833 (£ p.a.)	(4167)	
Profit rent in respect of land taken (£ p.a.)	833	
YP 2 years @ 10%	1.7355	
Value of land taken (£)		1446

So, compensation in respect of severance and injurious affection: £13 594 – £1446 = £12 148

Chapter 14
Valuation Variance, Risk and Optionality

14.1 Introduction

This final chapter discusses ways that valuations, and valuation reports, might be improved by considering how the valuation figure might vary. It begins with a review of research into the twin issues of valuation accuracy and valuation variance. Valuation accuracy refers to the difference between a valuation of a property and its subsequent sale price, whereas valuation variance is the difference between one valuer's valuation and another's. Both of these concepts are ways of measuring the performance of valuers. Indeed, they have become central to valuation negligence claims. Often, the legal profession refers to a 'margin of error' within which valuations would be expected to fall and outside of which might indicate cause for concern. A key question for the legal profession is margin of error around what? The eventual sale price (valuation accuracy) or between valuations (valuation variance).

The chapter then considers ways of analysing risk associated with a valuation. From the outset, it is important to distinguish valuation uncertainty from risk. The purpose for which a valuation is commissioned may involve a lot of risk, a development scheme for example. Valuation uncertainty will be often, but not always, closely related to the level of risk. However, valuation uncertainty could be very low where several very good direct comparable transactions are available as comparables, even where the actual risk and uncertainty attached to a development or investment is very high.

Finally, the chapter discusses the concepts of optionality, flexibility and uncertainty. Conventional static valuation methods and techniques struggle to deal with these concepts. Therefore, this final section summarises research on how they might be identified and handled within a valuation context.

Property Valuation, Third Edition. Peter Wyatt.
© 2023 John Wiley & Sons Ltd. Published 2023 by John Wiley & Sons Ltd.
Companion website: www.wiley.com/go/wyatt/propertyvaluation3e

14.2   Valuation accuracy and valuation variance

Valuation accuracy is the difference between a valuation of a property and its subsequent sale price. The MSCI real estate valuation and sale price study has been running for over 20 years. It analyses performance of valuers in the UK property market by tracking the difference between investment valuations and sale prices. The weighted average absolute differences between sale prices and valuations (which have been adjusted for growth and capital expenditure) was +9.1% in 2017, compared to +8.7% in 2016 (RICS 2019). The weighting is by capital value and the positive sign means that valuations were below sale prices. There were differences at the sector level, with valuations of London offices and industrial properties in the south-east being less accurate than the average. In a wider study of valuation accuracy around the world, Walvekar and Kakka (2020) found the weighted average absolute difference in 2019 to be +9.5% globally. This compares to 8.1% in the United Kingdom for 2019. This study also reported the 10-year weighted difference, which was 9.3% globally and 9.4% in the United Kingdom.

Valuation variance is the difference between valuers' valuations. Although the profession has sought to enforce rigorous standards and guidance, valuations of the same property conducted by different valuers will vary. Brown (1985) examined valuation variance by taking a sample of 26 properties, which had been valued by two different firms of valuers over a four-year period. It was found that the valuations from one firm were a good proxy for the valuations of the other and that there was no significant bias between the two firms' valuations. Hutchison et al. (1996) undertook research into variance in property valuation, involving a survey of major national and local firms. The average overall variation was found to be 9.53% from the mean valuation of each property. They also found evidence to suggest that valuation variation may be a function of the type of company that employs the valuer and, specifically, whether it is a national or local firm. The study revealed that national practices produced a lower level of variation (8.63%) compared with local firms (11.86%) perhaps due to the level of organisational support, especially in terms of availability of transactional information.

The main task of a valuer is to assemble known facts and make assumptions about the unknown variables to estimate value. The valuer's job is to minimise uncertainty as much as possible by being careful, exercising due diligence, checking information and calculations, and justifying assumptions. Most valuations result in a spot estimate of value, but this cannot be regarded as an absolute; there is likely to be a degree of uncertainty surrounding it. This uncertainty may be due to:

- Incomplete input information (and assumptions turn out to be unrealistic). The more facts that are known, the less the uncertainty. For example, a valuation of an investment property that has a tenant paying a rent will be more certain than a valuation of an empty property where the rent would be an estimate.
- Incorrect input details, such as wrong areas. This can be exacerbated in opaque markets where information is difficult to obtain and verify.

- Incorrect method or application of valuation method or methods. This might be due to incorrect input details or mistakes or poor judgement in the valuation itself.
- Bias. Valuers act on client instructions, make judgements and respond to different pressures when preparing a valuation, and these processes can provide opportunities for valuers to respond differently.
- Property type. Perhaps the location or the physical characteristics of the property are unusual or the property is of a type for which there is little or no comparable evidence. Some types of property are more heterogeneous than others and are harder to value such as trade-related properties and development land.
- Market volatility and market inefficiency. A valuation is not a permanent part of the property. Analysis of market data only suggests what happened in the past and it is for the valuer to interpret these data to assess current market value.
- Market inactivity. In some property markets, transactions occur infrequently and perhaps on a confidential basis. This makes it difficult to interpret market movements.

Kinnard et al. (1997) found that valuers conducting valuations for lending purposes experienced significant pressure from certain types of client, especially mortgage brokers and bankers. Gallimore and Wolverton (1997) found evidence of bias in valuations resulting from knowledge of the asking price or pending sale price. Gallimore (1994) found evidence of confirmation bias where valuers make an initial valuation, 'anchor' to this estimate of value and then find evidence to support it. The initial opinion of value or asking price was found to significantly influence the valuation outcome. In a survey of 100 lenders, finance brokers, valuers and investors Bretten and Wyatt (2001) found that most factors believed to cause variance related to the individual 'behavioural characteristics' of the valuer. Erikson et al. (2019) suggested that the primary source of bias in valuing residential properties for lending purposes in United States is in the weights valuers assign to selected comparable evidence after adjusting for observable attributes. According to their empirical research, this unequal weighting resulted in an additional 23% of appraisal values being at least equal to the contract price. They also found that valuers were also more likely to bias valuations for the properties associated with loan officers and real-estate brokers they worked with more frequently.

Variance can enter the valuation process at any stage from the issuing of instruction letters and negotiation of fees through to external pressure being exerted on the valuer when finalising the valuation figure. Following the Carsberg Report (RICS 2002) the RICS Red Book now contains stricter guidelines to reduce the likelihood of external pressure, and the adoption of quality assurance systems in the workplace can help maintain acceptable standards. For example, terms of engagement must include a statement of the firm's policy on the rotation of valuers responsible and a statement of the quality-control procedures in place. If a property has been acquired within the year preceding the valuation and the valuer or firm has received an introductory fee or negotiated the purchase for the client, the valuer/firm shall not value the property unless another firm has provided a valuation in the intervening period.

14.3   Analysing risk

In the context of a valuation, risk can be defined as ambiguity regarding the inputs. Some valuation inputs are key to the final valuation figure, the estimate of rental value and yield in an income capitalisation valuation, for example. It is therefore important to estimate these as precisely as possible. Single 'spot' estimates might give the impression of decisive estimation but that may be an illusion that the valuer would rather explore more explicitly. This is where risk-analysis techniques have a role to play.

Often, risk analysis focuses on the downside, what happens if things turn out worse than expected, what is the likelihood of making a loss? However, it is good practice to set the context of a spot estimate with a two-sided analysis of risk.

For many types of property, risk can be categorised as either systematic or unsystematic. Systematic risk is more general and would be expected to affect all properties, for example inflationary pressure, economic downturns, interest rate fluctuations, and so on. Unsystematic (or specific) risk affects specific properties and might be caused by business, financial or liquidity risks. Sources of specific risk include:

- Tenant, e.g. non-payment of rent or other contractual obligations,
- Sector of the market,
- Location,
- Physical structure and
- Legislation, e.g. landlord and tenant legislation, fiscal policy, planning policy

The illiquidity and 'lumpiness' of property accentuate these risks. Development property has additional risks (RTPI 2018):

- Land risk: The site may have unforeseen problems such as contamination or archaeological remains
- Planning risk: Planning permission may not be granted for the requested scheme, time taken to obtain permission may be longer than anticipated and planning obligations and conditions may be more onerous than expected.
- Development risk: Costs may be higher than expected or there may be delays.
- Sales risk: The market may decline during development.

Static valuation models presume that the future or, more accurately, valuers' expectations of the future, can be predicted with a high level of confidence. Yields, market rents, the exercising of break options and the lengths of void periods thereafter are all input as single estimates. There is a need to consider ways to model uncertainty in valuations, more so now than ever before because of the greater diversity of lease arrangements encountered in the market.

Perhaps the greatest level of uncertainty is encountered with development valuations. This is because they are based on projections of lots of cost and value inputs. So the following explanation of various modelling techniques will use development valuation as an example. But the techniques can be applied to other types of property and other valuation methods too.

14.3.1 Sensitivity analysis

Where uncertain market conditions or other variable factors could have a material impact on the valuation, it may be prudent to provide a sensitivity analysis to illustrate the effect that changes to these variables could have on the reported valuation. Sensitivity analysis investigates the impact of uncertainty on key input variables by examining the degree of change in the valuation caused by a change in one or more of the key input variables. Usually, a change of 5 or 10% either side of the expected values of the key variables is tested to measure the effect on value. A more sophisticated analysis may apply more realistic variations to the key variables, for example, more upside variation in rent in a rising market. Or different positive and negative percentage changes may be applied depending on the variable, for example plus or minus 10% for rental value and plus or minus 2% for rental growth.

Sensitivity analysis does not consider the likelihood of outcomes and the input variables are usually altered one at a time. Although simplistic, the process does require the valuer to think about the realistic limits on shifts in the input variables and produces a range of valuations within which the actual price would be expected to fall.

Sensitivity analysis should be accompanied by a narrative describing the cause and nature of uncertainty. Analysis does not mean forecasting worse-case scenario. It should address the impact of reasonable and likely alternative assumptions. There is a need to consider interdependence or correlation between significant inputs, otherwise the degree of uncertainty may be over-estimated.

Univariate sensitivity analysis seeks to quantify the effect of changes in the values of certain input variables on the output variable, one variable at a time. Bivariate sensitivity analysis extends univariate analysis by examining the impact of changes to two variables at the same time. For example, Table 14.1 shows a sensitivity analysis of the effect on the land value when key input variables are altered.

Bivariate analysis does not take account of any possible correlation between the input variables; instead, they are assumed to move independently. But logic

Table 14.1 Sensitivity analysis – impact on land value.

Univariate sensitivity analysis					
(a) Rent			**(b) Yield**		
Original value	£200	£711 492	Original value	7.00%	£711 492
−5%	£190	£563 922	+5%	7.35%	£569 051
−10%	£180	£416 352	+10%	7.70%	£439 560

Bivariate sensitivity analysis: rent and yield			
		Yield	
		7.00%	7.35%	7.70%
	£200	£711 492	£569 051	£439 560
Rent	£190	£563 922	£428 603	£305 586
	£180	£416 352	£288 155	£171 613

Table 14.2 Break-even analysis.

Input variable	Original value	Break-even value	Change
Rent	£200	£190	a drop of 5.00%
Yield	7.00%	7.25%	a rise of 3.57%
Building cost	£969/m²	£1105/m²	a rise of 14.04%
Finance rate	10.00% p.a.	14% p.a.	a rise of 40%
Void period	0.25 years	1.00	a rise of 300%

suggests that as rents rise, yields should fall and vice versa. Some of the output is repeated from the univariate sensitivity analysis but bivariate analysis provides more information about what happens when changes coincide, such as an increase in yield and a drop in rent.

A variation of sensitivity analysis is break-even analysis. This is the recalculation of a valuation in which the output is set to zero by altering key inputs. Table 14.2 illustrates a simple example from a development valuation where the developer's profit is set to zero by altering each of the listed inputs one at a time.

Sensitivity analysis models uncertainty in a very simplistic way but it does encourage the valuer to think about how assumptions and point estimates of key input variables might vary.

14.3.2 Scenario modelling

Scenario modelling examines the value impact of changes in several inputs at the same time. The valuer constructs several scenarios that reflect different possible futures, perhaps corresponding to optimistic, realistic and pessimistic circumstances, and then examines the impact on value of each scenario. The difference between sensitivity analysis and scenario testing is that the latter examines the impact on value of simultaneous changes to several variables and therefore begins to give a more realistic representation of how the key variables might respond to economic changes. It creates specific pictures (scenarios) of the future as a means of reflecting uncertainty.

Extending the example in Table 14.2, scenarios for combinations of values of rent, yield and building costs can be created. Numerous scenarios using different combinations of values can be constructed, but it is perhaps better to think carefully about practical combinations of values rather than try and input every permutation. Table 14.3 reports land value under three scenarios: realistic, best and worst case.

Scenario modelling allows the valuer to 'bookend' the valuation, but it still does not give any idea of the likelihood that any of these discrete outcomes might occur. To do that, we need to assign some measure of probability or likelihood to each scenario. For example, assume 40% probability for realistic outcome, 30% for the worst and 30% for the best outcomes. The mean or expected outcome is calculated as follows:

$$\text{Expected land value} = (0.4 \times 776913) + (0.3 \times 1049494) + (0.3 \times 428923) = £754290$$

Table 14.3   Scenario modelling.

Scenario	Realistic	Best	Worst
Input variables:			
Rent (£/m²)	150	152	148
Yield (%)	8.00	7.80	8.25
Building costs (£/m²)	800	790	820
Output variable:			
Land value (£)	776 913	1 049 494	428 923

Table 14.4   Risk and discrete probability modelling.

Property 1			Property 2		
Valuation (£)	Probability	Weighted valuation (£)	Valuation (£)	Probability	Weighted valuation (£)
2 800 000	2%	56 000	(80 000)	5%	(4000)
3 000 000	18%	540 000	2 000 000	20%	400 000
3 125 000	60%	1 875 000	3 500 000	50%	1 750 000
3 200 000	15%	480 000	3 700 000	20%	740 000
3 300 000	5%	165 000	4 600 000	5%	230 000
Weighted average valuation (£)		3 116 000	Weighted average valuation (£)		3 116 000

Neither the distribution of valuations nor the probabilities themselves need be symmetrical about the middle or realistic valuation.

A drawback of this type of analysis is a lack of market evidence on which to base selection of probabilities, but the process does focus the mind on the likelihood of achieving predicted returns. For example, a prime shop property and an old factory may yield the same return but how likely is the latter to be achieved relative to the former? In other words, how volatile or uncertain is the return? Discrete probability modelling does not properly reflect the uncertainty or risk that might be associated with the expected cash flows – it calculates an expected value rather than a measure of variation or uncertainty. To illustrate what this means, consider two property investments, Property 1 and Property 2, in Table 14.4.

The weighted average valuations are identical and, at first glance, the most probable outcome for Property 2 is £3 500 000 compared to £3 125 000 for Property 1, but closer inspection reveals that the range (volatility) of valuations for Property 1 is £500 000 and for Property 2 it is £4 680 000 and with a 5% probability of making a loss. Property 1 is likely to be more attractive to a risk-averse investor.

If a valuer can reasonably foresee different values arising under different circumstances, another approach would be to provide alternative valuations based on special assumptions reflecting those different circumstances, but only if they are realistic, relevant and valid in connection with the circumstances of the valuation (RICS 2011: VS 2.2).

This is a slight improvement on scenario modelling, but input variables are *individually* probabilistic so probabilities for specific combinations may be unrealistic. It is also important to consider the entire distribution of future possibilities and recognise that input values are uncertain, not discretely but continuously.

14.3.3 Simulation

It is unrealistic to assume a small number of discrete possible valuation outcomes. There is likely to be a range of outcomes and this would be best represented by a continuous probability distribution. If the probability distributions for predicted valuation outcomes for Properties 1 and 2 in Table 14.4 are assumed to be 'normally distributed' around their mean values, Property 1 would have a narrower, more peaked distribution indicating lower volatility whereas Property 2 would have a flatter, wider distribution indicating higher volatility. Standard deviation measures this volatility; the smaller the standard deviation of a distribution, the less volatile it is.

For example, assume 50 valuers have been asked to value Properties 1 and 2 and the mean valuation for Property 1 was £3 200 000 with a standard deviation of £500 000 and for Property 2 the mean valuation was £3 500 000 but with a much higher standard deviation of £1 000 000. The 'coefficient of variation' measures standard deviation relative to the mean and is a useful measure of volatility because allows comparison of valuations whose mean values are not equal. It is calculated by dividing the standard deviation by the mean. The coefficient of variation for Property 1 is 15.63% and for Property 2 it is 28.57%. Property 1 is less volatile by both standard deviation and coefficient of variation measures.

The example above shows how valuations from many valuers can be represented by a continuous probability distribution. However, usually, it is not the valuation output that is the focus of risk analysis, it is the inputs into the valuation itself. A technique known as simulation refers to the modelling of probability distributions for input variables. Simulation enables valuers to assign probabilities to input variables in the valuation and run simulations of likely combinations of values of these inputs to produce a probability distribution and associated confidence range for the output valuation. Statistics that summarise the uncertainty surrounding the valuation output can then be calculated. Usually, these would be the mean valuation and a measure of dispersion, such as the standard deviation.

Simulation involves a series of steps:

[1] Build a valuation model and identify key variables
      The valuation would be undertaken using the best estimates of the input variables. These input variables can be classified as either deterministic, which can be predicted with a high degree of certainty, or stochastic, which cannot be predicted with a high degree of certainty. Generally, the stochastic variables that have a significant impact on the valuation are the ones on which simulation is likely to be run. Deterministic variables might include the rent review period, purchase costs and management costs. Key stochastic variables will include the yield, the market rent, lease-break and lease-renewal options, including void periods and associated costs.

[2] Ascribe a probability distribution for each key stochastic input variable
Each stochastic variable needs to be represented as a probability distribution rather than a point estimate. A probability distribution is a way of presenting the quantified risk for the variable. Ideally the estimation of probability distributions would be based on empirical evidence, but often data are not available in a sufficient quantity to allow this. A pragmatic alternative is to gather opinions of possible values of each variable, along with their probability of occurrence, from experts. These expert opinions could then be used to select an appropriate probability function, of which there are many. The probability functions that are typically chosen are the continuous 'normal' distribution (in which case a mean and standard deviation would need to be specified) and the closed 'triangular' distribution (in which case the mode, minimum and maximum values would need to be specified). A useful characteristic of the triangular distribution is that, unlike the normal distribution, symmetry does not have to be assumed; the maximum and minimum values do not have to be equally spaced each side of the mode. In this way, the triangular distribution might offer a more realistic representation than the normal distribution if more upside or downside risk is expected.

The input variables may also be independent or dependent. An independent variable is unaffected by any other variable in the model whereas a dependent variable is determined in full or in part by one or more other variables in the model. Different degrees of interdependence can significantly affect the simulation result. It is therefore necessary to specify the extent to which the input variables are correlated.

[3] Run simulation
Having selected the key variables and their probability distributions, a simulation run can begin. A run refers to the process whereby a distribution of valuation outcomes is generated by recalculating the valuation many times, each time using different randomly sampled combinations of values from within the parameters of the probability distributions of the key stochastic variables. In this way, the process selects input values in accordance with their probabilities (values are more likely to be selected from areas of the distribution that have higher probabilities of occurrence) and correlations (more likely combinations of values will be selected). Havard (2002) illustrates how this process works in the case of two variables, rental growth rate and exit yield, to which discrete probabilities have been assigned, as shown in Table 14.5.

The simulation program randomly selects from the cumulative probability distribution for each variable. If we assume 22 was randomly selected for rental growth and 67 for the exit yield, this would equate to 3% rental growth rate and an exit yield of 9.25%. These sample values are then input into a run of the valuation model.

In other words, because some values of key variables will have a greater probability of being achieved than others, the sample-selection procedure ensures that they appear more frequently. This simulation process determines the range and probability of the valuation outcome.

Table 14.5 Stochastic variable value selection.

Annual rental growth rate			Exit yield		
%	Probability	Cumulative probability	%	Probability	Cumulative probability
0	2	1–2	7.75	1	1
1	5	3–7	8.00	4	2–5
2	7	8–15	8.25	7	6–12
3	10	16–25	8.50	10	13–22
4	15	26–40	8.75	15	23–37
5	21	41–61	9.00	21	38–58
6	15	62–76	9.25	15	59–73
7	10	77–86	9.50	10	74–83
8	7	87–93	9.75	7	84–90
9	5	94–98	10.00	5	91–95
10	2	99–100	10.25	5	96–100

[4] Output

When setting up the simulation program, the output variable in the valuation model would have been specified and, invariably, this will be the valuation figure. The simulation results will provide information about the distribution of the valuation, including its central tendency (mean, median, mode), spread (range, standard deviation) and measures of symmetry (skewness) and peakedness (kurtosis). Regression analysis can also be undertaken to rank the input variables in terms of their impact on the output valuation.

For example, consider a freehold property investment that is let on a lease that has a break clause in two years' time. The rent passing is £200 000 per annum. Rather than a point estimate, the market rent is modelled as a stochastic variable that has a normal distribution with a mean value of £210 000 per annum and a standard deviation of £10 000. In addition, the maximum rent is considered to be £225 000 and a minimum of £190 000 per annum. It is not known whether the tenant will exercise the break option but if it is exercised, then there is likely to be a void period. Again, this is modelled as a stochastic variable but this time as a triangular distribution with a mode of one year, a maximum of two years and a minimum of six months.

It is also possible, and advisable, to consider correlations between these stochastic variables. This ensures that when combinations of input values are selected, they do so in accordance with their correlation. For example, if the two variables were market rent and yield, a negative correlation would be expected between these two variables, because a market upturn might be expected to stimulate demand in both the occupier market, thus raising rents, and the investor market, thus lowering yields. In the valuations below, correlations are not considered necessary between the two relatively unrelated variables of market rent and break option exercise. Therefore, the inputs for this valuation are as follows:

Inputs

Yield	5.00%
Rent passing (£)	200 000
Period to break/lease end (years)	2.00
Void costs (% market rent)	60%
Discount rate for void costs	6.00%
Purchase costs (% purchase price)	6.50%

Stochastic inputs

Market rent (£)	223 077
Market rent (£) – mean	210 000
Market rent (£) – standard deviation	10 000
Market rent – min	190 000
Market rent – max	225 000
Void period (years)	1.53
Void period – most likely	1.00
Void period – min	0.5
Void period – max	2.00

The market rent of £223 077 and the void period of 1.53 years are selected at random from within the parameters set by their probability distributions. Another run of the valuation would generate different values for these two inputs. So, using these inputs, the two valuations, one assuming the break option is not exercised and the other assuming that it is, would be as follows:

No void or break exercised

Rent passing (£ p.a.)	200 000	
YP for 2 years @ 5%	1.8594	
		371 882
Market rent – new lease (£ p.a.)	223 077	
YP perpetuity @ 5%	20.0000	
PV £1 for 2 years @ 5%	0.9070	
		4 046 748
Valuation gross of purchaser's costs (£)		4 418 630
Valuation net of purchase costs @ 6.50% (£)		4 148 948

Void or break exercised

Rent passing (£ p.a.)	200 000	
YP for 2 years @ 5%	1.8594	
		371 882
Market rent – new lease (£ p.a.)	223 077	
YP perpetuity @ 5%	20.0000	
PV £1 for 3.53 years @ 5%	0.8638	
		3 854 045
Void costs (£)	(126 000)	
PV £1 from mid-way through void period: 2.5 years @ 6%	0.8644	
		(108 920)
Valuation gross of purchaser's costs (£)		4 225 927
Valuation net of purchase costs @ 6.50% (£)		3 968 007

To handle the uncertainty as to whether the break option would be exercised or not, market evidence is examined to see how many tenants on average do exercise a break. Assume the result is 30%, so this percentage is used to weight the two valuations above and this produces a weighted average valuation of £4 094 666, i.e. $(0.7 \times £4\,148\,948) + (0.3 \times £3\,968\,007)$. However, this valuation does not reflect the uncertainty surrounding the market rent and void period. To do this, the valuations must be run many times.

One thousand iterations of the valuations were run and the summary statistics for the output weighted average valuation are shown in Table 14.6. The optimistic skew of the exit-yield distribution has increased the mean valuation of both properties approximately £15 000 above the original point estimates. In both cases, the standard deviation around the mean was just under £100 000. Figure 14.1 and the skewness value in Table 14.6 reveal that both output distributions are positively skewed, the property let under standard lease terms slightly

Table 14.6   Summary statistics.

	Valuation outputs
Mean (£)	3 859 924
Standard deviation (£)	137 375
Skewness	−0.1168
Kurtosis	2.2891
Maximum (£)	4 125 976
Minimum (£)	3 539 808

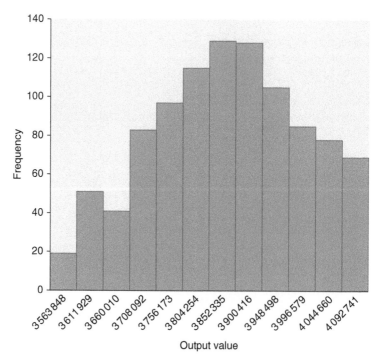

Figure 14.1   Valuation probability distribution.

more so. This is because the exit yield, which is itself positively skewed, explains more of the variation in value of the standard let investment.

It should be noted that specifying the forms of probability distributions for the inputs and their correlations is a challenge. The problem with this sort of analysis is being unable to confidently predict distributions and correlations of input variables. Statistical confidence requires sample sizes that are significantly larger than the typical pool of comparable evidence available when valuing a property. More research is needed to confidently base the choice of probability distributions and selection of co-relationships between variables on empirical evidence.

14.4 Flexibility and options

So far in this chapter, the discussion has centred on the modelling of a range of possible scenarios or simulations of inputs into a valuation and examining how this might affect the valuation output. The modelling assumes that these scenarios and simulation runs, once set in train, cannot be altered. But the future is not just a selected set of inputs; it is also a collection of decisions that can be made along the way. For example, the valuation of a parcel of development land may suggest that immediate development is feasible but the landowner may choose to hold it for a period of time before developing it. As Titman (1985) noted: 'The fact that investors choose to keep valuable land vacant or underutilized for prolonged periods of time suggests that the land is more valuable as a potential site for development in the future than it is for an actual site for constructing any particular building at the present time' (p. 505). It is important to understand how land is valued for *immediate* development and as a *potential* development site. Regarding the latter, uncertainty about the optimal development in the future is an important determinant of the value of the vacant land, because the landowner can opt to wait. Waiting removes downside risk. Property has *option* value.

The more uncertain the value resulting from the option, the higher the option value (Cunningham 2006; Bulan et al. 2009). This is because a development option can be used to limit downside risk and depends on price volatility for upside potential; the more volatility, the greater the upside potential. Geltner and De Neufville (2018) provide an example to illustrate this: a property is valued at £100 today and has two equally possible future scenarios, an increase in value to £110 or a drop in value to £90. A call option to buy this property would yield a profit of £10 because the option would not be exercised in the downside scenario, limiting the loss to £0. If the two alternative scenarios were £150 and £50, i.e. there was more volatility, then the call option would yield a profit of £50.

The longer an option lasts the higher the option value. Unlike financial options, real estate options are often perpetual and more flexible, and so have a higher value, all else equal. Furthermore, properties are expensive, and this increases the value of call options and decreases the value of put options.

Conventional valuation approaches fail to explicitly reveal option value (Ott 2002). Indeed, when there is greater uncertainty, this usually means valuers adopt higher yields and target rates of return, and this reduces value. This is counter to the intuition from option value theory, which suggests higher option value in times of greater uncertainty.

Realistic scenarios spanning a range of likely combinations of futures can be constructed to investigate the effect on value of exercising options. These scenarios can supplement a single valuation and, in doing so, transform a market valuation into an investment valuation, where the exercise of options that might trigger certain scenarios is contingent upon the investor's decision.

Real estate development is probably the sector that is most amenable to optionality. Developers can opt to develop now, delay or even abandon projects. Even when a project is underway, developers can phase construction or alter the product before completion, perhaps by switching the end use or by choosing to expand later on, based on favourable outcomes at early stages, or stop at a later date based on unfavourable outcomes at early stages. Development is different to other options because it takes time to realise the exercise price, adding to uncertainty and therefore increasing the value of the option.

For example, a residential property that is considered to have development potential has just been let at a rent of £1000 per annum, reviewable each year. The investor can redevelop the site in any of the next five years and receive a redevelopment value of £100 000, but when is the optimum time? If the investor's target rate of return is 4%, the rent is expected to grow at 4% per annum and the expected growth in redevelopment value is also 4% per annum, then it makes no difference. This is shown below.

Redevelopment at the end of year...	1	2	3	4	5
Rent (£ p.a.)	1040	1082	1125	1170	1217
Projected redevelopment value (£)	104 000	108 160	112 486	116 986	121 665
Total (£)	105 040	109 242	113 611	118 156	122 882
Valuation (£)	101 000	101 000	101 000	101 000	101 000

The present value of the investment is the same regardless of year the investor chooses to redevelop. This is because expectations of rental growth and redevelopment value growth (4% per annum) exactly reconcile with the investor's target return of 4%. But if expectations change then so do the valuations. Take two scenarios, the first is where rental growth is 3% per annum.

Redevelopment at the end of year...	1	2	3	4	5
Rent (£ p.a.)	1030	1061	1093	1126	1159
Projected redevelopment value (£)	104 000	108 160	112 486	116 986	121 665
Total (£)	105 030	109 221	113 579	118 111	122 825
Valuation (£)	100 990	100 981	100 971	100 962	100 953

Here it makes sense to redevelop as soon as possible because the year one valuation is the highest. Conversely, if the growth in redevelopment value is 5% per annum, then it makes sense to delay the redevelopment option because the highest valuation is in year five.

Redevelopment at the end of year. . .	1	2	3	4	5
Rent (£ p.a.)	1040	1082	1125	1170	1217
Projected redevelopment value (£)	105 000	110 250	115 763	121 551	127 628
Total (£)	106 040	111 332	116 887	122 720	128 845
Valuation (£)	101 962	102 932	103 912	104 902	105 901

In terms of value, the base-case scenario (expectations are exactly met) valuation is £101 000. The downside scenario (expectations are not met) valuation is £100 990, assuming the investor opts to redevelop at year one. The upside scenario (expectations are exceeded) valuation is £105 901, assuming the investor opts to redevelop at year five. If these scenarios are considered equally likely to occur, the weighted average valuation is:

$$(0.5 \times £100 990) + (0.5 \times £105 901) = £103 446$$

The optionality that the investor has adds £103 446 – £101 000 = £2446 to the valuation. This is because the downside risk can be minimised by exercising the redevelopment option as soon as possible and the upside potential can be maximised by delaying the redevelopment option for as long as possible. Conventional valuations undervalue property investments and developments that have optionality.

Optionality occurs everywhere and real estate decisions are no different. Tenants can opt to exercise a break clause, renew a lease or vacate a property. Landlords can opt to increase the rent at a rent review, to sell a property interest or refurbish a property at the end of a lease.

Many *real* options (i.e. options that relate to real estate) are irreversible, such as developing a parcel of land. As seen from the previous example of the five-year investment, delaying the exercise of an option can be valuable because, once exercised, it is irreversible.

14.5   Uncertainty

In addition to any quantitative analysis of risk and flexibility, the limitations of valuation approaches may also require qualitative reflection on the valuation outcome. Thorne (2021) argues that those relying on a valuation need alerting to any issue that could affect the reliability of the figure. Such reflection relates more to uncertainty (unknown outcomes) than it does to risk and flexibility (measurable outcomes). The definition of valuation uncertainty in the International Valuation Standards is '[t]he possibility that the estimated value may differ from the price that could be obtained in a transfer of the subject asset or liability taking place on the valuation date on the same terms and in the same market' (IVSC 2013).

The single estimate valuation could be accompanied by a qualitative comment in cases where uncertainty is thought to materially affect the valuation. The comment would indicate the cause of the uncertainty and the degree to which it is reflected in the reported valuation. The valuer might also comment on the robustness of the valuation, perhaps noting the availability and relevance of comparable market evidence, so that the client can judge the degree of confidence that the valuer has in the reported figure. It is important for valuers to communicate valuation uncertainty to clients as it may affect how that valuation is used in a decision, such as a lending assessment.

Thorne (2021) goes on to argue that it is a matter of judgement as to when a valuation should be accompanied by a valuation uncertainty caveat, i.e. when the uncertainty is 'material'. Useful indicators might be:

- whether the uncertainty could be expected to influence decisions and expose to significant loss,
- whether the valuation is for internal or external use, and
- extent to which a portfolio's value is affected.

The caveat should take the form of an explanatory narrative, explaining the source of the uncertainty, the effect on the market, the valuation, steps taken to mitigate and maybe a view on how long uncertainty may last for.

For development property, valuation uncertainty represents not only the impact of variation within the inputs but also the options inherent in the process that are not necessarily picked up within the valuation approaches. This reinforces the need to compare valuation outcomes with market transactions wherever possible and to fully explore alternative scenarios and other potential outcomes.

Uncertainty surrounding estimates of current levels of costs and revenues and future cost and price inflation introduces scope for justifiable variations in estimation of the key inputs into a development appraisal. This will, in turn, produce intrinsic uncertainty in the output. Rarely will development appraisals by different appraisers produce identical findings. Development appraisals are prone to uncertainty because there is uncertainty in assumptions about current levels of the inputs and about how these variables will change over the uncertain development period. As noted in Byrne et al. (2011), there are two key types of uncertainty: defensible disagreement between modellers about model composition and inputs, and unanticipated changes affecting revenues and costs.

Key points

- 100% valuation accuracy is an unattainable goal. Valuation variance has been identified in empirical studies of valuation practice. The courts accept that a degree of variance is inevitable through the adoption of the *margin of error* principle.
- Sensitivity analysis, scenario modelling and simulation are all recognised methods of analysing risk.

- Simulation is a logical extension of sensitivity analysis, scenario testing and discrete probability modelling that adds a quantitative measure of risk to a single point estimate of value. It does this by assigning probability distributions to key input variables. The drawback with this type of analysis is a lack of evidence on which to base these distributions and any correlations between them. Nevertheless, the discipline of building a 'risk-aware' simulation model can lead to a deeper understanding of influences on property value.
- Option pricing is not a feature of current mainstream valuation, and the binomial example presented here does not adequately represent real-world uncertainty and optionality; it simply demonstrates that option value exists. To be more realistic, an option pricing model must include many scenarios of the future.

References

Bretten, J. and Wyatt, P. (2001). Variance in commercial property valuations for lending purposes: an empirical study. *J. Prop. Investment Finance* 19 (3): 267–282.

Brown, G. (1985). Property investment and performance measurement: a reply. *J. Valuation* 4: 33–44.

Bulan, L., Mayer, C. And Somerville, C. (2009) Irreversible investment, real options, and competition: evidence from real estate development, J. Urban Econ., 65(3), 237–251.

Byrne, P., McAllister, P. and Wyatt, P. (2011) Precisely wrong or roughly right? An evaluation of development viability appraisal modelling. Journal of Financial Management of Property and Construction, 16, 3, 249–271.

Cunningham, C. (2006). House price uncertainty, timing of development, and vacant land prices: evidence for real options in Seattle. *J. Urban Econ.* 59 (1): 1–31.

Erikson, M., Fout, H., Palim, M., and Rosenblatt, E. (2019). The influence of contract prices and relationships on appraisal bias. *J. Urban Econ.* 111: 132–143.

Gallimore, P. (1994). Aspects of information processing in valuation judgement and choice. *J. Prop. Research* 11 (2): 97–110.

Gallimore, P. and Wolverton, M. (1997). Price-knowledge-induced bias: A cross-cultural comparison. *J. Prop Valuation Investment* 15 (3): 261–273.

Geltner, D. and De Neufville, R. (2018). *Flexibility and Real Estate Valuation Under Uncertainty: A Practical Guide for Developers*. Wiley Blackwell.

Havard, T. (2002). *Investment Property Valuation Today*. London: Estates Gazette.

Hutchison, H., MacGregor, B., Nanthakumaran, N. et al. (1996). *Variations in the Capital Valuations of UK Commercial Property*. Royal Institution of Chartered Surveyors, London: Research Report.

IVSC (2013). *Valuation Uncertainty*, Information Paper 4. London: International Valuation Standards Council.

Kinnard, W., Lenk, M., and Worzala, E. (1997). Client pressure in the commercial appraisal industry: how prevalent is it? *J. Prop. Valuation Investment* 15 (3): 233–244.

Ott, S. (2002). *Real options and real estate: a review and valuation illustration. In Real Estate Valuation Theory* (ed. K. Wang and M.L. Wolverton), 411–423. Research Issues in Real Estate book Series.

RICS (2002). *The Carsberg Report on Property Valuations*. London: Royal Institution of Chartered Surveyors.

RICS (2011). *RICS Valuation – Professional Standards 2012 (the 'Red Book')*. London: Royal Institution of Chartered Surveyors.

RICS (2019). *Valuation and Sale Price*. London: Royal Institution of Chartered Surveyors.

RTPI (2018) Planning risk and development: how greater planning certainty would affect residential development. Royal Town Planning Institute. RTPI Research Paper. April 2018.

Thorne, C. (2021). Valuation uncertainty – when and why this is important. *J. Prop. Investment Finance* 39 (5): 500–508.

Walvekar, G. and Kakka, V. (2020). *Private Real Estate: Valuation and Sale Price Comparison 2019*. MSCI.

Questions

[1]   A development site has planning consent for 2000 square metres (gross internal area) of office space with an efficiency ratio of 85%. Research indicates a build cost of £969 per square metre. An estimated £120 000 should cover external works (highways, landscaping, car parking). Professional fees are estimated at 13% of building and external works. Miscellaneous costs are assumed to be in the order of £80 000 and a figure of 3% of all these costs is assumed for contingencies. Comparable evidence suggests a rent of £200 per square metre and an investment yield of 7%. Disposal costs are 5.75% of NDV and site acquisition costs are 5.75% of the acquisition price. Site preparation costs are estimated to be £25 000. Fees for a full planning application are currently £2.50 per square metre of gross internal area. Building regulation fees are £20 000. The lender's legal fees, loan arrangement fee and developer's legal fees total £95 238. The letting agent's fee is 10% of the first year's rent, the letting legal fee is 5% of the first year's rent and the marketing fee is estimated to be £10 000. A lead-in period of six months is considered appropriate, and finance has been secured at 10% per annum. Comparable evidence from developments schemes at Bristol Business Park indicates a building period of 15 months and a void period of 3 months can be assumed. Assume a developer's profit requirement of 20% of all costs.

a)   Calculate the value the site.

b)   Calculate the amount of developer's profit.

c)   Calculate the percentage return on NDV and development yield (rent as a percentage of total costs).

d)   Calculate the payback period, rent cover, interest cover and rent:debt ratio. Also, calculate the rent that would be required if the land was purchased for £1 m and all other revenue and cost inputs remained the same.

e)   Conduct the following risk analyses:

i)   Sensitivity analysis: Model the effect on developer's profit of 5 and 10% shifts in rent and 5 and 10% shifts in yield.

ii)   Scenario modelling: Construct a pessimistic scenario (5% upward shift in yield, 5% downward shift in rent and six-month void period) and an optimistic scenario (5% upward shift in rent, a yield of 6.75% and no void) and report effect on developer's profit.

[2]   The tables below provide an outline of revenues and costs for a mixed-use development, together with the internal rate of return (IRR) and other performance metrics.

Section C

Inputs

	£/m²	NIA (m²)		
Revenues				
Residential – market dwellings (price, area)	2000	5000		£9 756 098
Residential – affordable dwellings (price, area)	1000	1000		£975 610
Commercial space (rent, area, yield)	200	4000	5.00%	£15 609 756
Other revenue				£3 000 000
Costs				
Site acquisition price, including acquisition costs		£5 281 642	6.80%	£5 640 793
Site preparation, infrastructure, utilities				£1 000 000
Residential – market dwellings (£/m², gross: net)	1000	80%		£6 250 000
Residential – affordable dwellings (£/m², gross: net)	1000	80%		£1 250 000
Commercial space	800	85%		£3 764 706
Abnormal costs				£200 000
Professional fees (% total construction costs)			10.00%	£1 146 471
Contingency (% total construction costs)			3.00%	£343 941
Planning fees				£20 000
Building control, NHBC, etc.				£50 000
S106 planning obligations				£500 000
CIL				£500 000
Other fees (e.g. legal, loan, valuation)				£200 000
Marketing				£300 000
Other assumptions				
Site acquisition costs (% acquisition price)	6.80%			
Sale transaction costs (% sale price)	2.50%			
Letting transaction costs (% annual rent)	15.00%			

Cash Flow		0	1	2	3	4	5	6	7	8
Revenue										
Market dwellings	9 756 098	—	—	—	—	—	—	—	4 878 049	4 878 049
Affordable dwellings	975 610	—	—	—	—	—	—	—	487 805	487 805
Commercial space	15 609 756	—	—	—	—	—	—	—	7 804 878	7 804 878
Other revenue	3 000 000	—	—	—	—	—	—	—	—	3 000 000
Total revenue (development value)	29 341 463	—	—	—	—	—	—	—	13 170 732	16 170 732
Costs										
Site acquisition	5 640 793	(5 640 793)	—	—	—	—	—	—	—	—
Site preparation	1 000 000	(1 000 000)	—	—	—	—	—	—	—	—
Market dwellings	6 250 000	—	(312 500)	(312 500)	(625 000)	(1 250 000)	(1 875 000)	(1 250 000)	(625 000)	—
Affordable dwellings	1 250 000	—	(62 500)	(62 500)	(125 000)	(250 000)	(375 000)	(250 000)	(125 000)	—
Commercial space	3 764 706	—	(188 235)	(188 235)	(376 471)	(752 941)	(1 129 412)	(752 941)	(376 471)	—
Abnormal costs	200 000	—	(10 000)	(10 000)	(20 000)	(40 000)	(60 000)	(40 000)	(20 000)	—
Professional fees	1 146 471	—	(57 324)	(57 324)	(114 647)	(229 294)	(343 941)	(229 294)	(114 647)	—
Contingency	343 941	—	(17 197)	(17 197)	(34 394)	(68 788)	(103 182)	(68 788)	(34 394)	—
Planning fees	20 000	—	(1 000)	(1 000)	(2 000)	(4 000)	(6 000)	(4 000)	(2 000)	—
Building fees	50 000	—	(2 500)	(2 500)	(5 000)	(10 000)	(15 000)	(10 000)	(5 000)	—
Planning obligations	500 000	—	(25 000)	(25 000)	(50 000)	(100 000)	(150 000)	(100 000)	(50 000)	—
CIL	500 000	—	(25 000)	(25 000)	(50 000)	(100 000)	(150 000)	(100 000)	(50 000)	—
Other fees	200 000	—	(10 000)	(10 000)	(20 000)	(40 000)	(60 000)	(40 000)	(20 000)	—
Marketing	300 000	—	(15 000)	(15 000)	(30 000)	(60 000)	(90 000)	(60 000)	(30 000)	—
Total costs	21 165 911	(6 640 793)	(726 256)	(726 256)	(1 452 512)	(2 905 024)	(4 357 535)	(2 905 024)	(1 452 512)	—
Net cash flow		(6 640 793)	(726 256)	(726 256)	(1 452 512)	(2 905 024)	(4 357 535)	(2 905 024)	11 718 220	16 170 732

IRR	33.05%
Equity invested (£)	19713399
Profit (£)	8175552
Equity multiple	1.41
Profit on cost	39%
Profit on value	28%

a) Use a spreadsheet to replicate the cash flow and calculation of the performance metrics.
b) Set up random *uniform* distributions for the following inputs, run 1000 simulations and report the resulting mean IRR:
- Residential market dwelling prices between £1900 and £2200 per square metre
- Commercial rent between £190 and £215 per square metre
- Commercial yield between 4.5 and 5.5%
c) As (b) but this time use random *normal* distributions as follows:
- Residential market dwelling prices have a mean of £2000/m² and a standard deviation of £50/m².
- Commercial rent has a mean of £200/m² and a standard deviation of £10/m².
- Commercial yield has a mean of 5% and a standard deviation 0.5%.
d) Comment on the resulting IRRs from (a), (b) and (c).
e) If you have access to a simulation add-on for Excel, set up the inputs as follows:

Simulation inputs	Mean	SD	Min	Max
Residential – market dwellings (£/m²)	2000	50	1900	2200
Commercial rent (£/m²)	200	10	190	215
Commercial yield	5.00%	0.50%	4.50%	5.50%

Correlation matrix	Market dwelling	Commercial rent	Commercial yield
Market dwelling	1		
Commercial rent	0.75	1	
Commercial yield	–0.6	–0.4	1

Run 1000 simulations and report the mean IRR and variance, together with the maximum and minimum IRR values. Also, show the resulting frequency distribution of IRRs.

Answers

[1]

a) Residual valuation to calculate site value

Development value

Net internal area (NIA) (m²)	1700		
Estimated rental value (ERV) (£/m²)	200		
		340 000	
Net initial yield	7.00%	14.2857	
Gross development value (GDV) before sale costs (£)			4 857 143
Net development value (NDV) after sale costs (£)			4 761 905

Development costs

Site preparation (£)		(25 000)	
Building costs (£/m² GIA)	969	(1 938 000)	
External costs (£)		(120 000)	
Professional fees (% building costs and external works)	13.00%	(267 540)	
Miscellaneous costs (£)		(80 000)	
Contingency allowance (% construction costs)	3.00%	(72 166)	
Planning fees (£)		(5000)	
Building regulation fees (£)		(20 000)	
Planning obligations (£)		0	
Other fees, e.g. legal, loan, valuation (£)		(95 238)	
Finance on costs and fees for HALF building period @	10.00%	(160 993)	
Finance on costs and finance for void period @	10.00%	(67 131)	
Letting agent's fee (% ERV)	10.00%	(34 000)	
Letting legal fee (% ERV)	5.00%	(17 000)	
Marketing (£)		(10 000)	
Developer's profit on total development costs (%):	20.00%	(582 414)	
Total development costs (TDC) (£)			(3 494 482)
NDV − TDC (£)			1 267 423
Land costs (£)			
Developer's profit on land costs (%)	20.00%	(211 237)	1 056 185
Finance on land costs over total development period	10.00%	2.00	0.8264
Residual land value before purchase costs (£)			872 881
Residual land value after purchase costs (£)			819 606

b)   Residual valuation to calculate developer's profit

Development value

Net internal area (NIA) (m²)	1700		
Estimated rental value (ERV) (£/m²)	200		
		340 000	
Net initial yield	7.00%	14.2857	
Gross development value (GDV) before sale costs (£)			4 857 143
Net development value (NDV) after sale costs (£)			4 761 905

Development costs

Land price (£)		(819 606)	
Land purchase costs (% land price)	6.50%	(53 274)	
Finance on land costs for total development period @	10.00%	(183 305)	
Site preparation (£)		(25 000)	
Building costs (£/m² GIA)	969	(1 938 000)	
External costs (£)		(120 000)	
Professional fees (% building costs and external works)	13.00%	(267 540)	
Miscellaneous costs (£)		(80 000)	
Contingency allowance (% construction costs)	3.00%	(72 166)	
Planning fees (£)		(5000)	
Building regulation fees (£)		(20 000)	
Planning obligations (£)		0	
Other fees, e.g. legal, loan, valuation (£)		(95 238)	
Finance on building costs and fees for HALF building period @	10.00%	(160 993)	
Finance on building costs, fees and interest to date for void period:	10.00%	(67 131)	
Letting agent's fee (% ERV)	10.00%	(34 000)	
Letting legal fee (% ERV)	5.00%	(17 000)	
Marketing (£)		(10 000)	
Total development costs (TDC) (£)			(3 968 254)
Developer's profit on completion (£)			793 651

c)   Return measures
- Return on NDV: 16.67%
- Income yield: 8.57%

d)   Risk measures
- Payback period (costs/ERV), i.e. inverse of income yield: 11.67 years
- Rent cover (profit/rent): 2.33 years
- Interest cover (profit/annual mortgage payment on costs): 2.33 years*
- Rent-to-debt ratio (rent/annual mortgage payment): 1.04
- Rent needs to be £212/m² if the land price is £1m (found using goal seek).

*Assuming mortgage term is 25 years at an interest rate of 7%, then the multiplier will be 0.0858 and the annual mortgage payment on the costs will be £340 518.

e) Risk analyses
   i) Sensitivity analysis

		Rent				
		£180	**£190**	**£200**	**£210**	**£220**
Yield	7.70%	(67 050)	146 850	360 751	574 651	788 551
	7.35%	118 479	342 686	566 894	791 101	1 015 308
	7.00%	322 561	558 106	793 651	1 029 196	1 264 741
	6.65%	548 124	796 201	1 044 278	1 292 354	1 540 431
	6.30%	798 751	1 060 751	1 322 752	1 584 752	1 846 752

ii) Scenario model

	Current values	**Pessimistic profit**	**Optimistic profit**
Changing cells			
ERV	£200	£190	£210
Yield	7.00%	7.35%	6.75%
Void	0.25	0.50	0.00
Result cells:			
Developer's profit	793 651	248 468	1 306 381

[2]
a) As above
b) By way of example, here is one set of values:

	£/m²	**NIA (m²)**	**Yield**	**Value**
Residential – market dwellings	1919	5000		£9 360 976
Commercial space	208	4000	5.10%	£15 915 830

And a sample from the 1000 runs:

	IRR	
Trial	32.71%	Ranked
1	33.91%	26.94%
2	36.18%	27.09%
3	33.88%	27.42%
4	35.94%	27.44%
5	36.99%	27.54%
6	38.49%	27.56%
7	30.86%	27.69%
8	36.32%	27.77%
9	37.70%	27.86%
10	29.63%	27.96%
. . .	. . .	. . .

990	32.08%	42.06%
991	34.48%	42.07%
992	33.91%	42.09%
993	33.84%	42.09%
994	31.39%	42.11%
995	27.56%	42.14%
996	30.05%	42.14%
997	34.28%	42.19%
998	35.94%	42.38%
999	38.21%	42.40%
1000	29.24%	43.04%
Mean IRR	34.88%	

c)

	IRR	
Trial	38.75%	Ranked
1	32.21%	12.08%
2	38.94%	13.23%
3	41.89%	15.17%
4	34.64%	16.55%
5	43.53%	18.30%
6	33.19%	18.38%
7	25.53%	18.42%
8	27.86%	18.49%
9	32.08%	18.60%
10	20.70%	18.97%
. . .	. . .	. . .
990	38.33%	51.55%
991	28.02%	51.75%
992	40.52%	52.05%
993	27.93%	52.89%
994	26.38%	52.90%
995	37.09%	53.33%
996	40.44%	55.03%
997	31.63%	55.49%
998	42.44%	56.34%
999	27.12%	56.61%
1000	34.82%	57.61%
Mean IRR	33.85%	

d) In (c), because the tails of the distributions have not been truncated, the model produces much lower and higher IRRs at the extremes compared to the uniform distribution model in (b), but the bell shape of the normal distribution results in a mean that is closer to the point estimate IRR in (a).

Section C

e)

Mean IRR = 33.68%
Variance = 0.256%
Maximum IRR = 46.39%
Minimum IRR = 23.05%
Frequency distribution of IRRs:

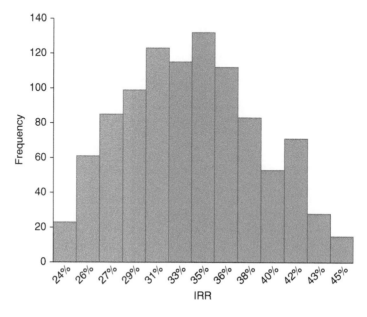

Appendix
Land Uses and Valuation Methods

A.1 Agriculture and fisheries

Various agricultural land uses are possible in addition to crops and livestock. Abattoirs and slaughterhouses might be valued using a comparison or replacement cost method depending on the degree of specialisation. Comparison might be based on the value of similar industrial buildings for example. Game farms include facilities such as runs, pens, hatcheries and ancillary buildings, and these might be valued using either a comparison or cost approach. Grain silos are likely to be valued using a cost approach. Livestock markets can generate revenue from several sources so a comparison approach or profits method might be appropriate. The number and size of boxes, size of the yard, ancillary facilities and access would influence the value of a stud farm, and a valuation might be undertaken using a comparison or cost approach or profits method. Stables include boxes, yard, tack rooms, feed stores, barns and other stores. A comparison method is likely to be possible. Finally, there might be farm diversification into leisure activities, and these would most likely be valued using the profits method.

A.2 Forests and woodland

Trees and perennial crops (non-orchard trees with long growing periods) may be valued by estimating the value of one year's produce and capitalising that figure at a rate of return that will be influenced by the nature and lifecycle of the tree or crop. Larger areas can be divided by crop type, age, species, composition or condition and valued separately. Two valuation methods predominate. First, sales comparison; this involves comparing prices achieved for similar forests with similar-growing stock per hectare. Care is needed due to paucity of sales and

Property Valuation, Third Edition. Peter Wyatt.
© 2023 John Wiley & Sons Ltd. Published 2023 by John Wiley & Sons Ltd.
Companion website: www.wiley.com/go/wyatt/propertyvaluation3e

heterogeneity of forests. This approach is more useful for small amenity woods. Second, income capitalisation; here, land and standing timber would be valued separately. Land can be valued as 'planting land' using comparable evidence, and the income approach can be used to value standing timber. The method might depend on the age of the trees at the valuation date: 'Present Market Value' should be used to value mature or nearly mature woods (the volume of standing timber is assessed and then multiplied by the standing timber price), whereas an 'Expectation Value' should be used to value crops that are not yet mature (the volume yield from thinnings and final crop is forecast, multiplied by the appropriate unit values based on standing timber prices at relevant points in time, and then then discounted to present value at an appropriate discount rate).

A.3   Natural resource extraction – water, minerals and other materials

The value of onshore oil and gas fields usually comprises a royalty payment applied to the quantity of oil and gas produced, a land value for the site and the value of site improvements (buildings, plant and machinery). The site value can be estimated using a comparison method and the site improvements using the replacement cost method.

A.4   Recreation and leisure

Outdoor amenity and open space are most likely valued using a profits method. There may be multiple sources of income, including licences and concessions, all of which must be accounted for when using the profits method to estimate a net profit prior to capitalization. For some uses, rental and sales evidence may exist so a comparison approach may be possible and preferable to the profits method. Some leisure attractions, such as museums, art galleries, libraries and historic properties, rarely make a profit and are usually valued using the replacement cost method. Sports facilities and grounds can be valued using a comparison method if there is sufficient rental or sales evidence; otherwise they are likely to be valued using the profits method. If facilities are non-profit making – either they are owned/operated by charitable organisations or the state (usually local government) – then the replacement cost method would be appropriate. Hostels, holiday centres and outdoor activity centres would be valued using the profits method.

A.5   Utilities and infrastructure

The way in which utilities are valued will depend to a large extent on how they are owned and operated. In some countries, the water suppliers, power generators and the power distribution networks may be privately owned and can be valued using the profits method. In other countries they may be state owned and so, alongside infrastructure that includes a large amount of plant such as sewage treatment works, oil refineries and pipelines, they may be valued using the replacement cost method.

With regard to water supply, there may be a statutory requirement for water to be supplied to consumers within defined geographical areas. There may also be private water supply on a discretionary basis, perhaps to farmers or large industrial facilities. This is likely to be from wells, boreholes, springs, watercourses or lakes. Valuation assumptions are likely to address title, licencing, infrastructure and quality, quantity and surety of course and supply. Special assumptions may also be appropriate in respect of charging, future regulation, environmental impacts, branding and reputation and alternative uses. The profits method can be used to value these non-statutory undertakings or the replacement cost method as a last resort.

The sales comparison method could be used to value refuse disposal facilities, as could the income approach, capitalising actual/notional royalties over the life of the resource/void space. Plant and equipment would usually be valued using the replacement cost method. This method is also appropriate for valuing post and telecommunication facilities, cemeteries and burial grounds. Crematoria, on the other hand, are capable of being operated at a profit and so the profits method can be used.

Transport infrastructure should be valued using the replacement cost method. Car parks, vehicle (other than cars) storage, toll roads and bridges including ferries, moorings, marinas, boat yards and anchorages are more likely to be valued using a comparison method or the profits method if they are capable of making a profit.

A.6   Residential

Dwellings can usually be valued using a sales or rental comparison method because there is sufficient transaction evidence available and adequate homogeneity in the stock. Hotels, self-catered holiday accommodation, serviced apartments and student accommodation can be valued using rental comparison. Residential property that has been purpose-built for renting is usually valued using the income approach, having regard to parties' obligations, length of tenancies, any rent reviews and breaks (RICS 2018). The way that affordable housing is valued depends on its tenure. Social rent, affordable rent and intermediate rent dwellings are valued by capitalising the net rental income (RICS 2016). The net social/affordable/intermediate rent may be derived by reducing an estimate of net market rent by a suitable proportion. Equity share dwellings are valued by adding the initial equity sale to future staircasing receipts. Shared ownership dwellings are valued by adding the capital value of the proportion sold[1] to the capitalised value of the estimated rental income on the unsold part.[2] For example, assume a dwelling is estimated to have a market rent of £300 per week (£15 600 per annum) and a market value of £300 000. If this dwelling were to be let at a social rent, this would be, say £145 per week gross (£7540 per annum) based on average incomes. To arrive at a net rent, deduct say 10% for management and maintenance, 2.5% for voids and bad debts and 5% for a repair fund. This produces a net rent of £6221 per annum. If this is capitalised at a yield of 6%, the capital value is £103 675 (35% of market value). If this dwelling were to be let at an affordable rent, this would be say 80% of market rent or £240 per week (£12 480 per annum) gross. The net rent would be £10 296 per annum after 5% for management and maintenance, 2.5% for voids and debts and 5% for repairs. Capitalised at a yield of 6% produces a capital value of £71 600 (57% of market value). Finally, if the dwelling were to be valued on a shared ownership basis, assume 35% (£105 000) was sold on day one, and the remainder was let to the occupier at a rent that represents a return of 2.75% on the remaining equity of £195 000, i.e. £5363 per annum gross. With 5% management costs and 2% for voids and debts, this produces a net rent of £4988 per annum. Capitalised at a 6% yield, this equates to £83 133, which, when added to the £105 000 of initial capital, produces a value of £188 133 (63% of market value).

A.7   Community services

Community services are not run as profitable entities and therefore the replacement cost method is appropriate. For medical and health care services such as hospices, hospitals and health care facilities, the replacement cost method is appropriate, although if the facilities are privately run, then a profits method may be used. Similarly, the replacement cost method is appropriate for valuing nurseries, schools, colleges, universities and other education facilities. Once again, if they are privately run, the profits method may be appropriate.

A.8   Land and buildings with (Re)development potential

The residual method is used to value development potential in land and property.

Notes

1. Estimate the market capital value for the subject dwelling and multiply this by the percentage of equity sold.
2. Estimate the rent from the retained equity (this is typically obtained by amortising the sum at a very low yield, say 2.75%, in perpetuity) and capitalise that rent at a typical investment yield.

References

RICS (2016). *Valuation of Land for Affordable Housing*. Guidance Notes, 2e. Royal Institution of Chartered Surveyors April, 2016.

RICS (2018). *Valuing Residential Property Purpose Built for Renting*. Guidance Note, 1e. Royal Institution of Chartered Surveyors July 2018.

Glossary

Ad valorem tax a tax where the assessed amount is based on the value of a transaction or of property.

Asset an item of property owned by a person or company, regarded as having value.

Asset stripping the practice of taking over a company in financial difficulties and selling each of its assets separately at a profit without regard for the company's future.

Assessed value usually a statutorily defined basis of value for tax purposes, sometimes referred to as *cadastral* value.

Betterment tax levied on any increase in value attributable to public infrastructure investment or the granting of land-use rights.

Break-up value the value of individual land and property assets that are sold following the closure of a business or enterprise.

Capital gains tax accrues to assets that have appreciated in value over time and is payable on sale or transfer.

Capital value capitalised *Rental Value* (see below).

Conflict of interest include a valuer acting for both buyer and seller of a property in the same transaction, valuing on behalf of a lender while providing advice to the borrower, or valuing a property recently valued for another client. Should such a conflict arise, the valuer must decide whether to accept the valuation instruction. If the instruction is accepted the valuer should inform the client about the possibility and nature of the conflict, recommend that independent advice is sought and agree how the conflict is to be managed.

Contingent valuation a method of estimating value based on an assumption that users of the product can accurately reveal their perception of value by stating willingness to pay a certain amount for the product.

Cost the expense of producing something (a building on a piece of land for example); it is a production-related concept and an important component of many valuations.

Customary tenure generally understood to refer to the local rules, institutions and practices governing land, fisheries and forests that have, over time and use, gained social legitimacy and become embedded in the fabric of a society. Although customary rules are not often written down, they may enjoy widespread social sanction and may be generally adhered to by members of a local population. Customary tenure systems are extremely diverse, reflecting different ecosystems, economies, cultures and social relations.

Deliberative and inclusionary process (DIP) a set of methodological approaches aimed at creating better-informed decisions that are owned by and have the broad consent of all relevant actors and stakeholders. The process includes participatory appraisal, focus groups, Delphi approach, consensus conferences and citizen's juries. DIPs seek to build a process of defining and redefining interests that stakeholders introduce as the collective experience of participation evolves. As participants become more empowered, i.e. more respected and more self-confident, they may become more

Property Valuation, Third Edition. Peter Wyatt.
© 2023 John Wiley & Sons Ltd. Published 2023 by John Wiley & Sons Ltd.
Companion website: www.wiley.com/go/wyatt/propertyvaluation3e

ready to adjust, to listen, to learn and to accommodate a greater consensus.

Development value the value of a site assuming it is developed or redeveloped to its **highest and best use**.

Discount rate reflects time preference (the preference for current rather than future consumption) and perceived risk.

Externality side effect from one activity, which has consequences for another activity, but is not reflected in market prices – these effects may be beneficial (positive externalities) or detrimental (negative externalities) for which there is no payment through the markets – e.g. cost to society of pollution for which an entity does not pay.

Fair value '. . .the estimated price for the transfer of an asset or liability between identified knowledgeable, and willing parties that reflects the respective interests of those parties'. (IVSC)

the value of the land with no improvements. Improved land value is the value of land plus improvements to the land. Sometimes, the distinction between unimproved and improved land can be difficult to discern. For example, unimproved land may include certain improvements, such as clearing and drainage, which have merged into the land and so are valued with it.

Highest and best use the use that maximises potential while being possible, permissible and financially feasible.

Inheritance tax charged on the value of property owned at death.

Income approach a valuation approach that is used to value properties held as investments. Such properties generate a rental income and capital value is estimated as a multiple of this income.

Informal tenure may be described as tenure rights that are neither derived from statute nor any customary tenure regime. People living in informal settlements often do not purport to claim legal ownership of the land from either customary tenure systems or from statutes but rather rely on their investment in the land for the time being.

International Property Measurement Standards a standardised and globally applicable method for measuring property (www.ipms.org).

Investment method an application of the income approach applied to investment properties that are widely held as investments such as shops, office and industrial properties. More specialised properties are valued using the **profits method** (see below).

Investment value '. . .the value of an asset to the owner or prospective owner for individual investment or operational objectives'. (IVSC)

Market a forum in which buyers and sellers interact. An 'open' market is one that has few barriers to entry, information and trading. A market is often defined in terms of its geographical extent, the type of commodity being traded (farmland, dwellings or office space for example) or the characteristics of the buyers and sellers (such as investors, occupiers or developers) but can be defined in very specific terms – commercial mining operations in the Pacific islands for example. Markets vary in terms of access to information and the costs associated with buying and selling. Poor access to information and high transaction costs can constrain market activity. Markets in tenure rights that relate to land, fisheries and forests are different to markets in other tradable commodities, mainly due to their diversity, geographical distribution, degree of state intervention and opacity of trading activity. This has a detrimental impact on the level of access to market information, particularly transaction prices.

Market rent '. . . the estimated amount for which a property, or space within a property, should lease on the date of valuation between a willing lessor and a willing lessee on appropriate lease terms, in an arm's-length transaction, after proper marketing wherein the parties had each acted knowledgeably, prudently and without compulsion'. (IVSC)

Market value synonymous with the concept of *exchange* value, this is '. . . the estimated amount for which a property should exchange on the date of valuation between a willing buyer and a willing seller in an arm's length transaction after property marketing wherein the parties had each acted knowledgeably, prudently and without compulsion'. (IVSC)

Mass valuation also known as **mass appraisal**, this refers to the valuation of large numbers of land and property units for taxation purposes, usually conducted using algorithms and statistical models.

Non-market value an intrinsic value that can be attributed to the social and environmental benefits that a property may offer.

Occupation tax a recurrent tax usually assessed with reference to the value of land or the value of land and improvements, payable by occupiers or owners of land and property.

Ownership tax see occupation tax.

Participatory valuation a method of estimating value of non-market assets that involves key stakeholders discussing and deciding on quanta of value.

Premium a capital sum in lieu of rent payments, usually paid at the start of a lease.

Price an exchange-related concept that refers to the amount requested, offered or paid for something. It is an exchange-related concept and, once paid, a fact. Price relies on the existence of a *market* in which commodities are exchanged. In a single transaction there might be an asking price advertised by the seller, a bid price offered by the potential buyer and finally, usually after some period of negotiation, an agreed exchange or sale price at which the property is transacted.

Profit rent the difference between the actual rent paid by a tenant and the rent that the tenant could receive if the tenure rights were sublet at market rent. A profit rent can be notional if the tenant chooses not to sublet, i.e. the financial benefit of paying a rent below market levels is internalised.

Profits method an application of the income approach that is used to value specialised trade-related properties. They are valued by considering rent as surplus payable out of net profit.

Property a word that has many meanings depending on context. It is used in this book principally to refer to physical land and buildings owned or occupied by an individual or an entity.

Real estate land, infrastructure, buildings and other improvements, minerals and other subterranean natural resources.

Receipts and expenditure method see profits method.

Rental value an estimate of annual payment for the holding of tenure rights.

Reversionary value the value that may be realisable by the holder of tenure rights when subsidiary rights in the same real estate asset (a leasehold for example) end.

Risk-free discount rate rate of return that a person can expect on a completely riskless asset. Most people normally use a government bond rate as a risk-free rate, because it is perceived to have very little risk, if any. Consequently, such a rate can be thought of as representing the time value of money or pure time preference.

Roll the list of taxable units of land and property held and maintained by a state.

Sales tax see transfer tax.

Social discount rate (SDR) a rate used to capitalise the annual value of non-market assets in order to try and reflect long-term value to present and future generations. It reflects society's relative valuation of today's wellbeing versus wellbeing in the future. It is a discount rate used in computing the value of funds spent on social projects. Choosing an appropriate SDR is crucial for cost–benefit analysis and thus has important implications for resource allocations. Used in estimating, inter alia, the value of enforcing environmental protection. A higher SDR implies greater risks to the assumption that the benefits will materialise.

Solatium a form of compensation for emotional rather than physical or financial harm.

Special value '. . .an amount that reflects particular attributes of an asset that are only of value to a special purchaser'. (IVSC)

Synergistic value '. . .an additional element of value created by the combination of two or more assets or interests where the combined value is more than the sum of the separate values'. (IVSC)

Transfer tax is assessed as a percentage of reported price on transfer of ownership.

Valuation the process of forming an opinion of value.

Value an estimate of price that can be based on various definitions or **bases**.

Index

Property Valuation, Third Edition. Peter Wyatt.
© 2023 John Wiley & Sons Ltd. Published 2023 by John Wiley & Sons Ltd.
Companion website: www.wiley.com/go/wyatt/propertyvaluation3e